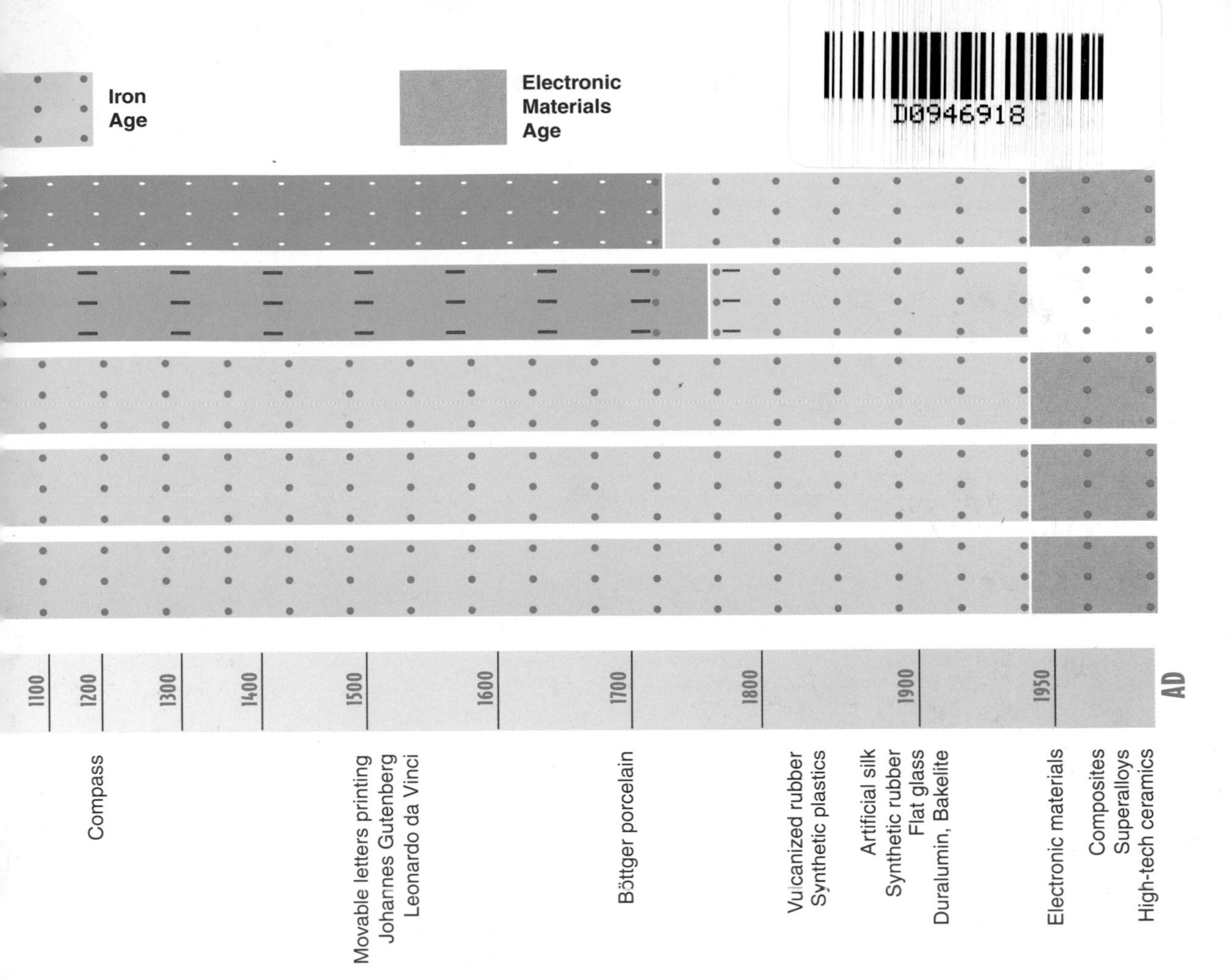

Materials ◆ Civilizations ◆ Ages

Byzantine
Medieval
Islamic and Ottoman Empires
Renaissance
Western Exploration and Colonization
Sung | Mongols Yuan | Ming | Quing

Understanding Materials Science

Second Edition

Springer
New York
Berlin
Heidelberg
Hong Kong
London
Milan
Paris
Tokyo

Rolf E. Hummel
University of Florida

Understanding Materials Science

History • Properties • Applications

Second Edition

Springer

Rolf E. Hummel
College of Engineering
University of Florida
Gainesville, FL 32601
USA

Cover illustration: Rubens N Vulcano Cat. 1676. © Museo Nacional del Prado-Madrid. Reproduced with permission.

Library of Congress Cataloging-in-Publication Data
Hummel, Rolf E., 1934–
Understanding materials science / Rolf E. Hummel.—2nd ed.
p. cm.
Includes bibliographical references and index.
ISBN 0-387-20939-5 (alk. paper)
1. Materials science. I. Title.
TA401.6.A1H86 2004
620.1′1—dc22 2004041693

ISBN 0-387-20939-5 Printed on acid-free paper.

Printed in the United States of America. (MP/HAM)

9 8 7 6 5 4 3 2 1 SPIN 10960202

Springer-Verlag is a part of *Springer Science Business Media*

springeronline.com

Understanding the history of materials means understanding the history of mankind and civilization.

Preface to Second Edition

My concept for this book, namely to show the connection between the technical and the cultural, economic, ecological, and societal aspects of materials science seems to have been realized judging from the enthusiastic reception of readers coming not only from the hard sciences, but also from the humanities and medicine. Indeed, the first printing sold out less than two years after publication, which made reprinting necessary and gave me the opportunity to make some changes and additions. The comments of many students who took a course based on this book and by colleagues from all over the world confirmed my hunch that many individuals would be interested not only in the physical and technological aspects of materials but also in how materials and the goods which were made of them shaped the development of mankind. By relating history and technology in this text many students were disabused of the idea that "most everything was invented in the past hundred years" (and in the USA).

When contemplating a second edition, I withstood the temptation to substantially expand the book by including major new topics and a more detailed treatment of several subjects. If substantial in-depth treatment of a given topic is wanted, reference should be made to one of the many encyclopedic materials science books that have two or three times more pages. One should realize, however, that the contents of these larger books cannot be covered in a customary three-credit semester course. This necessitates that each teacher select those topics which he/she deems to be most important, that is, with which he/she is most familiar. I have chosen to restrict the presented material and the depth of its treatment to that which is realistically digestible in an introductory course in materials science without overburdening the students with too much detail. Clearly, this text is not meant to be encyclopedic, but instead aims to whet the appetite of its readers and to inspire them to further explore the properties and applications of metals, alloys, ceramics, plastics, and electronic materials by means of easily understandable explanations and entertaining historical facts. It is also intended to raise the readers' awareness of their obligations to society as practicing engineers and scientists.

What has been changed compared to the first edition? Naturally, there is always room for improvement. Accordingly, a large number of additions, corrections, and clarifications have been made on almost each page. Furthermore, the treatment of "high-tech ceramics" has been substantially expanded (mostly at the suggestions of my colleagues) by including topics such as silicon nitride ceramics, transformation-toughened zirconia, alumina, ultra-hard ceramics, and bioceramics. A separate section on composite materials has been added, including fiber-reinforced composites, particular composites, and laminar composites. A section on advanced fabrics seemed to be of interest to the readers. Most of all, however, Chapter 18 (Economic and Environmental Considerations) has been rewritten and expanded in many places by updating the statistical information on prices of materials, production figures, world reserves, consumption (particularly oil), recycling (particularly plastics, paper, household batteries, electronic scrap, automobiles), the possible use of "biodiesel" (rape plant oil), waste prevention, lead-free solder, energy savings through recycling, efficient design, and stability of materials. The iron and steel production statistics were updated in Chapter 7, and new figures on gold production and consumption were included in Chapter 17. Finally, considerations on new trends such as "nanomaterials by severe plastic deformation," a rendition of Moore's law, and more philosophical remarks on the expected ethical behavior of engineers have been incorporated into Chapter 19.

A few readers have suggested that I should provide the complete solutions for the homework problems. I am against this. The exercises should be challenges (some more, some less). Giving the detailed solution of the problems (rather than just the numerical results) would tempt many students not to work the problems and in turn would deprive them of an important learning experience and the satisfaction of having succeeded through perseverance. I can assure the readers, however, from my own experience that all problems are solvable.

Those readers who like interactive communication and animated visualizations by using the computer are directed to the post scriptum of the Preface to the First Edition that follows.

My thanks go to many of my students who, through their kind words of praise and their challenging questions, helped me to clarify many points. My colleagues, such as Professor Emeritus Gerold (MPI Stuttgart/Germany), Professor Emeritus Petzow (MPI Stuttgart/Germany), Professor Emeritus Hench (Imperial College London), and Professors Ebrahimi, Sigmund, and Mecholsky (University of Florida) helped with valuable suggestions which are much appreciated.

Rolf E. Hummel
Gainesville, FL
June 2004

Preface to the First Edition

It is a challenging endeavor to trace the properties and the development of materials in the light of the history of civilization. Materials such as metals, alloys, ceramics, glass, fibers, and so on have been used by mankind for millennia. Actually, materials have shaped entire civilizations. They have been considered of such importance that historians and other scholars have named certain ancient periods after the material which was predominantly utilized at that respective time. Examples are the Stone Age, the Bronze Age, and the Iron Age. As time progressed the materials became increasingly sophisticated. Their properties were successively altered by man to suit ever-changing needs. We cannot but regard with utmost respect the accomplishments of men and women who lived millennia ago and who were capable of smelting, shaping, and improving the properties of materials.

Typical courses on world history expose students mainly to the description of major wars, the time span important rulers have reigned, and to the formation, expansion and downfall of world empires. Very little is generally said about the people who lived and toiled in ancient times and about the evolution of civilizations. This book traces the utilization, properties, and production techniques of materials from the Stone Age via the Bronze Age and the Iron Age up to modern times. It explains the physical properties of common materials as well as those of "exotic materials" such as superalloys, high-tech ceramics, optical materials, electronic materials, and plastics. Likewise, natural and artificial fibers and the technique of porcelain- and glass-making are covered. Moreover, this book provides a thorough introduction into the science and engineering of materials, covering all essential features that one would expect to find in a horizontally integrated introductory text for materials science. Specifically, the book presents the mechanical, electrical, magnetic, optical, and thermal properties of all materials including textiles, fibers, paper, cement, and wood in a balanced and easily understandable way. This book is not an encyclopedia of materials science. Indeed, it is limited in its depth so that the content can be conveniently taught in a one-semester (15-week), three-credit-hour course. Nevertheless, the topics are considered to be essential for introducing engineers and other interested readers to the fascinating field of materials science.

Plenty of applied problems are given at the end of the technical chapters. The solutions for them are listed in the Appendix. The presentation follows an unusual sequence, starting with a description of the properties of the first materials utilized by man, such as stone, fiber, and copper. Subsequently, the differences between these materials are explained by considering their atomistic structure, the binding forces between the atoms, and their crystallography. A description of the Bronze Age is followed by the treatment of alloys and various strengthening mechanisms which are achieved when multiple constituents are blended to compounds. The properties of iron and steel are explained only after an extensive history of iron and steel making has been presented. In Part II, the electronic properties of materials are covered from a historical, as well as from a scientific, point of view. Eventually, in Part III the historic development and the properties of ceramics, glass, fibers and plastics as we understand them today are presented. The book concludes with a chapter on economics, world resources, recycling practices, and ecology of materials utilization. Finally, an outlook speculating on what materials might be utilized 50 years from now is given. Color reproductions of relevant art work and artifacts are included in an insert to show the reader how materials science is interwoven with the development of civilization.

This book is mainly written for engineering, physics, and materials science students who seek an easily understandable and enjoyable introduction to the properties of materials and the laws of physics and chemistry which govern them. These students (and their professors) will find the mixture of history, societal issues, and science quite appealing for a better understanding of the context in which materials were developed. I hope, however, that this book also finds its way into the hands of the general readership which is interested in the history of mankind and civilization as it relates to the use and development of materials. I trust that these readers will not stop at the end of the historical chapters, but instead will continue in their reading. They will discover that the technical sections are equally fascinating since they provide an understanding of the present-day appliances and technical devices which they use on a daily basis. In other words, I hope that a sizeable readership also comes from the humanities. Last, but not least, future archaeo-metallurgists should find the presentation quite appealing and stimulating.

A book of this broad spectrum needs, understandably, the advice of many specialists who are knowledgeable in their respective fields. It is my sincere desire to thank all individuals who in one way or another advised me after I wrote the first draft of the manuscript. One individual above all stands out particularly: Dr. Volkmar Gerold, Professor Emeritus of the University of Stuttgart and the Max-Planck-Institut for Materials Research who read the manuscript more than once and saw to it that each definition and each fact can stand up to the most rigid scrutiny. My sincere thanks go to him for the countless hours he spent on this project.

Other colleagues (most of them from the University of Florida) have read and advised me on specific chapters. Among them, Dr. R.T. DeHoff (diffusion and general metallurgy), Dr. A. Brennan (polymers), and Dr. E.D. Verink (corrosion) are particularly thanked. Further, Drs. C. Batich and E. Douglas (polymers), Drs. D. Clark, J. Mecholsky, and D. Whitney (ceramics), Dr. C. Beatty (recycling), Dr. J.D. Livingston (MIT; magnetism), Dr. C. Pastore (Philadelphia College of Textiles and Science; fibers), Mr. E. Cohen (Orlando, FL) and Mr. R.G. Barlowe (U.S. Department of Agriculture, World Agricultural Outlook Board) need to be gratefully mentioned. Ms. Tita Ramirez cheerfully typed the manuscript with great skill and diligence. Finally, Dr. M. Ludwig carried on my research work at those times when my mind was completely absorbed by the present writings. To all of them my heartfelt thanks.

Rolf E. Hummel
Gainesville, FL
April 1998

P.S. For those readers who want to deepen their understanding in selected technical topics covered in this book and who have a propensity to the modern trend of playing with the computer, I recommend considering a computer software entitled *"Materials Science: A Multimedia Approach"* by J.C. Russ (PWS Publishing Comp. Boston/Ma; http://www.pws.com). This CD-ROM provides animated visualizations of physical principles and interactive sample problems.

Contents

Preface to the Second Edition vii

Preface to the First Edition xi

PART I: MECHANICAL PROPERTIES OF MATERIALS

1 The First Materials (Stone Age and Copper–Stone Age) 3

2 Fundamental Mechanical Properties of Materials 12

3 Mechanisms 24

3.1 The Atomic Structure of Condensed Matter 24
3.2 Binding Forces Between Atoms 26
3.3 Arrangement of Atoms (Crystallography) 31
3.4 Plasticity and Strength of Materials 47
3.5 Summary 61
3.6 Concluding Remarks 62

4 The Bronze Age 66

5 Alloys and Compounds 74

5.1 Solid Solution Strengthening 74
5.2 Phase Diagrams 75
5.3 Precipitation Hardening (Age Hardening) 89
5.4 Dispersion Strengthening 95
5.5 Grain Size Strengthening 96
5.6 Control of Strength by Casting 97
5.7 Concluding Remarks 99

6 Atoms in Motion **102**

6.1 Lattice Defects and Diffusion 102
6.2 Diffusion in Ceramics and Polymers 114
6.3 Practical Consequences 115
6.4 Closing Remarks 122

7 The Iron Age **125**

8 Iron and Steel **141**

8.1 Phases and Microconstituents 141
8.2 Hardening Mechanisms 142
8.3 Heat Treatments 144
8.4 Alloyed Steels 149
8.5 Cast Irons 150
8.6 Closing Remarks 153

9 Degradation of Materials (Corrosion) **155**

9.1 Corrosion Mechanisms 155
9.2 Electrochemical Corrosion 159
9.3 Practical Consequences 165
9.4 Degradation of Polymers and Glass 167

PART II: ELECTRONIC PROPERTIES OF MATERIALS

10 The Age of Electronic Materials **173**

11 Electrical Properties of Materials **185**

11.1 Conductivity and Resistivity of Metals 187
11.2 Conduction in Alloys 193
11.3 Superconductivity 194
11.4 Semiconductors 197
11.5 Conduction in Polymers 206
11.6 Ionic Conductors 210
11.7 Thermoelectric Phenomena 211
11.8 Dielectric Properties 214
11.9 Ferroelectricity and Piezoelectricity 218

12 Magnetic Properties of Materials **223**

12.1 Fundamentals 223
12.2 Magnetic Phenomena and Their Interpretation 227
12.3 Applications 235

13 Optical Properties of Materials 245

13.1 Interaction of Light with Matter 245
13.2 The Optical Constants 247
13.3 Absorption of Light 252
13.4 Emission of Light 259
13.5 Optical Storage Devices 267

14 Thermal Properties of Materials 271

14.1 Fundamentals 271
14.2 Interpretation of the Heat Capacity by Various Models 275
14.3 Thermal Conduction 278
14.4 Thermal Expansion 282

PART III: MATERIALS AND THE WORLD

15 No Ceramics Age? 287

15.1 Ceramics and Civilization 287
15.2 Types of Pottery 290
15.3 Shaping and Decoration of Pottery 294
15.4 The Science Behind Pottery 296
15.5 History of Glass-Making 302
15.6 Scientific Aspects of Glass-Making 307
15.7 Cement, Concrete, and Plaster 315
15.8 High-Tech Ceramics 317

16 From Natural Fibers to Man-Made Plastics 326

16.1 History and Classifications 326
16.2 Production and Properties of Natural Fibers 333
16.3 Tales About Plastics 345
16.4 Properties of Synthetic Polymers 354
16.5 Composite Materials 361
16.6 Advanced Fabrics 363

17 Gold 366

18 Economic and Environmental Considerations 373

18.1 Price 373
18.2 Production Volumes 374
18.3 World Reserves 376
18.4 Recycling and Domestic Waste 383
18.5 Substitution of Rare and Hazardous Materials 400

18.6 Efficient Design and Energetic Considerations 401
18.7 Safety 403
18.8 Stability of Materials (in Light of the World Trade Center collapse) 404
18.9 Closing Remarks 404

19 What Does the Future Hold? 407

Appendices

I. Summary of Quantum Number Characteristics 415
II. Tables of Physical Constants 417
III. Periodic Table of the Elements 421
IV. Solutions to Selected Problems 422

Index 427

PART I
MECHANICAL PROPERTIES OF MATERIALS

1

The First Materials (Stone Age and Copper–Stone Age)

Materials have accompanied mankind virtually from the very beginning of its existence. Among the first materials utilized by man were certainly stone and wood, but bone, fibers, feathers, shells, animal skin, and clay also served specific purposes.

Materials were predominantly used for tools, weapons, utensils, shelter, and for self-expression, that is, for creating decorations or jewelry. The increased usage and development of ever more sophisticated materials were paralleled by a rise of the consciousness of mankind. In other words, it seems to be that advanced civilizations generally invented and used more elaborate materials. This observation is probably still true in present days.

Materials have been considered of such importance that historians and other scholars have named certain ancient periods after the material which was predominantly utilized at that respective time. Examples are the *Stone Age*, the *Copper–Stone Age (Chalcolithic*[1] *Period)*, the *Bronze Age*, and the *Iron Age*. The Stone Age, which is defined to have begun about 2.5 million years ago, is divided into the *Paleolithic* (Old Stone Age), the *Mesolithic* (Middle Stone Age), and the *Neolithic* (New Stone Age) phases. We will consider on the following pages mostly the Neolithic and Chalcolithic periods. Surprisingly, these classifications do not include a *Ceramics Age*, even though pottery played an important role during extended time periods (see Chapter 15).

---

[1]*Chalcos* (Greek) = copper; *lithos* (Greek) = stone.

The names of some metals have entered certain linguistic usages. For example, the Greeks distinguished the *Golden Age* (during which supposedly peace and happiness prevailed) from the *Silver Age*. Rather than being descriptive of the materials that were used, these distinctions had more metaphorical meanings. Specifically, gold has always been held in high esteem in the eyes of mankind. Medals for outstanding performances (sport events, etc.) are conferred in gold, silver, or bronze. Specific wedding anniversaries are classified using gold, silver, and iron.

Until very recently, the mastery of materials has been achieved mainly by empirical means or, at its best, by a form of alchemy. Only in the nineteenth and twentieth centuries did systematic research lead to an interdisciplinary field of study that was eventually named *materials science*. This will be explained and demonstrated in detail in later chapters.

Materials often have to be cut, shaped, or smoothed before they reach their final form and designation. For this, a tool that is harder than the work piece has to be set in action. As an example, flint stone having a sharp edge was used by early man for cutting and shaping other materials such as wood.

The simplest and most common method of making stone tools from bulk rocks was by *percussion flaking*. Specifically, a lump stone was struck with another stone to detach small pieces from it. If these flakes happened to have sharp edges, they could be used as cutting tools. In early times, the tools were hand-held. Later (probably 5,000–10,000 years ago), stone flakes were attached to wooden handles using fibers or vegetable resin. This provided for better leverage, thereby amplifying their impact. Other flakes may have been used as spear or arrow tips, etc. (see Plate 1.1).

Recent excavations in the Gona Valley of Ethiopia yielded about 3,000 tools consisting of hammer stones and knives probably used to sharpen sticks or to cut meat. They are said to be two and a half million years old (!) and have likely been split from volcanic rocks. No remains of the toolmakers were found.

Fishing hooks were made from shell and bone. Ground mineral pigments were used for body painting. Grass fibers (e.g., flax, hemp, etc.) or animal hair (wool) served as clothing and for holding loose objects together. Jade, greenstone, and amber were utilized for adornments. This list could be continued.

Stones, particularly flint and obsidian (a dark gray natural glass that precipitated from volcanic emissions, see Plate 1.2) were available to Neolithic man in sufficient quantities at certain locations. Because of their abundance and their sharp edges, stones filled the needs as tools and weapons. Thus, it is not immediately evident why mankind gradually switched from a stone-using society to the metals age.

FIGURE 1.1. Copper pendant found in a cave in northeast Iraq; about 9500 B.C. The shape was obtained by hammering native copper or by carving copper ore. (Reprinted by permission from C.S. Smith, *Metallurgy as a Human Experience* (1977), ASM International, Materials Park, OH, Figure 2.)

This transition, incidentally, did not occur at the same time in all places of the world. The introduction of metals stretched over nearly 5,000 years, if it occurred at all, and seems to have begun independently at various locations. For example, metals were used quite early in Anatolia, the bridge between Asia and Europe (part of today's Turkey),[1] where a highly developed civilization existed which cultivated seed-bearing grasses (wheat and barley) and domesticated such animals as cattle, sheep, and goats. The transition from a nomadic to a settled society left time for activities other than concerns for everyday gathering of food. Thus, man's interest in his environment, for example, in native copper, gold, silver, mercury, or lead, is understandable.

Neolithic man must have found out that metals in their *native* state (that is, not combined with other elements, as in ores) can be deformed and hardened by hammering or can be softened by heating. Pieces of native metals were probably quite valuable because they were rare. Still, these pure metals were generally too soft to replace, to a large extent, tools and weapons made of stone. Thus, pure metals, particularly *copper, silver,* and *gold,* were mostly used for ceremonial purposes and to create ornaments or decorations. As an example, one of the very earliest copper artifacts, a 2.3-cm long, oval-shaped pendant is shown in Figure 1.1. It was found in a cave in northeast Iraq (Shanidar). It is believed that it has been created around 9500 B.C. by hammering native copper or possibly by carving copper ore. Utensils made of metal must have lent some prestige to their owner. Copper, in particular, played an outstanding role because of its appearance and its relative abundance (especially after man learned how to smelt it). In short, the stone and

[1]See the map on the rear endpaper for locations cited in the following discussion.

copper ages coexisted for a long time. This led to the above-mentioned name, Chalcolithic, or Copper–Stone Age.

The exact time when Neolithic man begun to use copper will probably never be exactly known, but it is believed that this was about 8000 B.C. Copper weapons and utensils were found in Egyptian graves dating about 5000 B.C. The epics of Shu Ching mention the use of copper in China at 2500 B.C. Native copper for ornaments is believed to have been used in the Lake Superior area in Michigan (USA) starting A.D. 100–200 where rich deposits of native copper are present. (Other scholars date Native American copper use as early as 4000 B.C.)

Eventually, native copper and other metals must have been nearly exhausted. Thus, Neolithic man turned his attention to new sources for metals, namely, those that were locked up in minerals. A widely used copper ore is malachite (Plate 1.3). It is plentiful in certain regions of the earth such as in Anatolia, or on the Sinai peninsula. Other regions, such as Cyprus, contain chalcopyrite (a copper-iron sulfide). Now, the smelting of copper from copper ore, that is, the separation of copper from oxygen, sulphur, and carbon, was (and is), by no means, a trivial task. It requires intense heat, that is, temperatures above the melting point of pure copper (1084°C) and a "reducing atmosphere"; in other words, an environment that is devoid of oxygen and rich in carbon monoxide. The latter is obtained by burning wood or charcoal. When all conditions are just right, the oxygen is removed from the copper ore and combines with carbon monoxide to yield gaseous carbon dioxide, which is allowed to escape. Finally, a *fluxing agent*, for example, iron ore, assists in

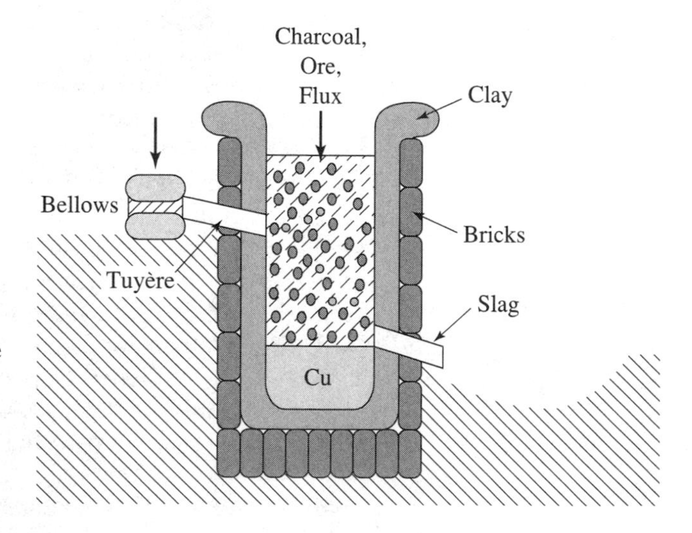

FIGURE 1.2. Schematic representation of an ancient copper smelting furnace which was charged with a mixture of charcoal, copper ore, and flux (e.g., iron ore). The oxygen was provided by forcing air into the furnace by means of foot-operated bellows.

the reduction process. It also aids eventually in the separation of the molten copper from the slag once the melt has cooled down. Specifically, iron ore combines with the unwanted sand particles that just happen to be contained in the ore.

The immense heat was accomplished by burning charcoal combined with blowing air into the furnace either by mechanically activated bellows and/or through blow tubes (called tuyères) (Figure 1.2), or by placing the furnace near the top of a mountain where the updraft winds were utilized. It is still a mystery today how Neolithic man could have found this chain of procedures without a certain degree of intuition or possibly the help of initiates.

Archaeo-metallurgists have recently ruled out the hypothesis that copper could have been accidentally formed in campfires whose enclosures may have consisted of copper-ore–containing rocks. The temperatures in campfires (600–700°C) are known to be too low for smelting copper and the reducing atmosphere does not persist for a long enough time. (However, lead, which has a lower melting temperature, can be smelted this way from its ore.)

It is believed today that the "technology" of copper smelting was probably borrowed from the art of making pottery, which was developed nine or ten thousand years ago or perhaps even earlier at certain locations. Indeed, the oldest known artifact made of baked clay is a fertility figurine called the "Venus of Vestonice," which was found in the Czech Republic and supposedly dates back to about 23,000 B.C. (see Figure 15.1). In general, however, copper smelting and pottery seem to appear at comparable times in history. Specifically, Neolithic man had observed that mud bricks harden when dried in the sun and soften when again exposed to rain. A deliberate attempt to accelerate the drying process by exposing the mud bricks to the heat of a fire probably led to the observation that an irreversible hardening process had occurred. A chemical transformation near 500°C causes a permanent consistency of clay which makes it water-resistant. It can be reasonably assumed that this observation eventually led to the systematic development of the art of pottery and the design of kilns instead of drying clay over or under an open fire. Neolithic man must have observed that stacking pots on top of wood fuel and covering this pile with fragments of pottery and earth would increase the temperature. Eventually, kilns with permanent walls were developed, parts of which still exist today, dating back to the beginning of the sixth millennium B.C. We shall return to this subject in Chapter 15.

Neolithic people have decorated some pottery utilizing probably the same ground-up metal ores (mixed with a lead oxide binding agent) that were used customarily for tribal body painting. Various metal oxides produce different colors. Pigments of copper oxide, for example, yield a blue color, chromium oxide gives green, antimony salts yield yellow, and iron yields pink hues after a second firing of these "glazes." Could it have been that the overfiring of glazed pots accidentally produced small droplets of metals, that is, caused some smelting of metal ores in the glazed areas?

Another question remains to be answered. Was copper smelting conceived of independently in different parts of the world, or was this technology transferred from neighboring regions through trading contacts? Possibly both happened. Among the first civilizations to utilize copper smelting were probably the inhabitants of Anatolia (Catal Hüyük) and of the Sinai peninsula (Timna Valley), both blessed with rich and abundant copper ores on or near the surface. On the other hand, copper (and gold) objects have been found in graves at Varna on the Black Sea dating back to about 4300 B.C. (Plate 1.4).

Naturally, raw copper needed to be transported to other places where goods were produced from it. For standardization, copper ingots were cast in a peculiar form that resembled the shape of an ox hide, as shown in Figure 1.3. A vivid depiction of ancient copper smelting and casting has been found on a mural in the tomb of an Egyptian nobleman; see Figure 1.4.

Seemingly independent from this development, Europeans had turned, out of necessity, to underground copper mining even before 4000 B.C. (for example, at Rudna Glava in Yugoslavia). One mine in Bulgaria was found to have shafts about 10 meters deep. The copper mines on the Balkans are the earliest so far discovered in the world. Other indigenous copper workings were discovered in southern Spain (Iberia) and northern Italy. The dislodging of rocks in mines was accomplished by burning wood at the end of a tunnel and then quenching the hot rock with water. This caused the rock to crack so that small pieces could be loosened with a pick. Underground mining must have been a large-

FIGURE 1.3. Copper ingots were traded in the Mediterranean region in an *ox-hide* shape having a length of about 30 cm. Specimens have been found in shipwrecks off the south coast of Turkey and in palace storerooms in Crete. (See also Figure 4.1.) Incidentally, raw gold was traded in the form of large rings.

FIGURE 1.4. A portion of a mural from the tomb of the Vizier Rekh-Mi-Re at Thebes depicting metal melting and casting during the second millenium B.C. in Egypt. Note the foot-operated bellows, the heaps of charcoal, and the "green" wood sticks with which the hot containers were held. (Reprinted by permission from B. Scheel, *Egyptian Metalworking and Tools*, Shire Publications, Aylesbury, U.K.)

scale operation that involved workers who supplied the fuel, others who were involved in transportation, and naturally the actual miners.

Subsurface ores are often more complex in composition than those found on the face of the earth. In particular, they contain sulfur that needs to be removed before smelting. For this, a separate heating process, which we call today "roasting," needed to be applied.

Among the earliest metalworkers in Europe were people in whose graves characteristic bell-shaped clay cups have been found, and who therefore are called the *Bell Beaker Folk*. They were superb potters and coppersmiths. They traveled across the continent from Poland to the west and north to Scandinavia and the British Isles, offering their services as makers of knives, spear heads, hammers, axes, and as tinkerers. They spread the knowledge of metalworking across Europe during the second and third millennia B.C. However, it is not quite clear where the Bell Beaker Folk came from, but it is assumed that they originated in Spain.

In this context, it is interesting to know about a Stone-Age man (named by the press, "Ötzi"), whose well-preserved, mummified body was found in 1991 in a glacier of the Austrian–Italian Alps (Tyrol) at an altitude of 3200 meters. Carbon-14 dating (taken on his bones and soft tissue) places his age at approximately 3300 B.C., which is in the European Chalcolithic period. Among his possessions was an axe with a wooden handle of yew that had a small blade of copper (not bronze) whose size is 9.5 cm in length and 3.5 cm in breadth (Plate 1.5). Further, he possessed a small knife with a stone blade attached to a wooden handle and a bow made of yew with several flint-tipped arrows. Several items appear to be remarkable. First, tools made of copper were apparently in use during the Chalcolithic period even though copper is relatively soft and thus could not have been used for cutting down trees. Second, copper had an apparently wider use than previously assumed and was therefore not in possession of the privileged people only. Third, the find, unlike those experienced in burial sites, shows equipment that Chalcolithic man considered to be vital for his endeavor in the high Alpine mountains. Copper and stone were certainly parts of these necessities.

Surprisingly enough, copper smelting technologies, quite similarly to those in Timna, began in northern Peru not before the year 800, that is, about 5000 years later. And the Aboriginal people in Australia and Tasmania, the North American Indians, and the South Pacific Islanders never engaged in copper smelting (or any other major metal technologies) until the Europeans arrived, despite the rich mineral resources that slumbered on and in their mountains. One may conclude, therefore, that various cultures had different interests and needs that should not be compared nor their values judged.

It can be reasonably assumed from the above considerations that Chalcolithic man intuitively understood some of the basic mechanical properties of materials. Stone (and many other glassy and ceramic materials) is *hard* and *brittle*. Copper (and many other metals as well as wet clay) is *ductile*[1]; that is, these materials can be permanently deformed (to a certain limit) without breaking. Copper is soft in its native or freshly molten state, but eventually hardens when *plastically*[2] (i.e., permanently) deformed. Finally, wood is to a large degree an *elastic* material; its original shape is restored when a moderate pressure that was applied to it has been removed. Many metals can likewise be elastically deformed, until, upon exceeding a critical load, they undergo permanent deformation. In short, some of the fundamental mechanical properties of materials such as hardness, ductility, elastic or plastic deformability, brittleness, and strength must have been known for a long time. Early man has utilized these different properties of materials to best suit a particular purpose. Nothing has changed in this respect during the past ten thousand years; see Plate 1.6. Therefore, it seems to be quite appropriate to explain these fundamental properties of materials, as we understand them today, in the chapters to come.

---

[1]*Ducere* (Latin) = to shape, to draw out, to lead.
[2]*Plasticos* (Greek) = to shape, to form.

Suggestions for Further Study

R.W. Cahn, *The Coming of Materials Science*, Pergamon/Amsterdam (2001).

B. Cunliffe (Editor), *The Oxford Illustrated Prehistory of Europe*, Oxford University Press, New York (1994).

R.J. Harrison, *The Beaker Folk—Copper Age Archaeology in Western Europe*, Thames and Hudson, London (1980).

D. Lessem, *The Iceman*, Crown, New York (1994).

R.F. Mehl, *Brief History of the Science of Metals*, AIME (1984).

J.G. Parr, *Man, Metals, and Modern Magic*, Greenwood Press, Westport, CT (1958).

R. Raymond, *Out of the Fiery Furnace—The Impact of Metal on the History of Mankind*, The Pennsylvania State University Press, University Park, PA (1984).

B. Scheel, *Egyptian Metalworking and Tools*, Shire Publications, Aylesbury, UK (1989).

C.S. Smith, *Metallurgy as a Human Experience*, ASM International (formerly American Society of Metals), Materials Park, OH (1977).

K. Spindler, *The Man in the Ice*, Harmony, New York (1994).

A.J. Wilson, *The Living Rock*, Woodhead Publications, Cambridge, UK (1994).

2

Fundamental Mechanical Properties of Materials

The goal of the following pages is to characterize materials in terms of some of the fundamental mechanical properties that were introduced in Chapter 1.

A qualitative distinction between ductile, brittle, and elastic materials can be achieved in a relatively simple experiment using the *bend test*, as shown in Figure 2.1. A long and comparatively thin piece of the material to be tested is placed near its ends on two supports and loaded at the center. It is intuitively obvious that an elastic material such as wood can be bent to a much higher degree before breakage occurs than can a brittle material such as stone or glass. Moreover, elastic materials return upon elastic deformation to their original configuration once the stress has been removed. On the other hand, ductile materials undergo a permanent change in shape above a certain threshold load. But even ductile materials eventually break once a large enough force has been applied.

To quantitatively evaluate these properties, a more sophisticated device is routinely used by virtually all industrial and scientific labs. In the *tensile tester*, a rod-shaped or flat piece of the material under investigation is held between a fixed and a movable arm as shown in Figure 2.2. A force upon the test piece is exerted by slowly driving the movable cross-head away from the fixed arm. This causes a **stress**, σ, on the sample, which is defined to be the force, F, per unit area, A_0, that is,

$$\sigma = \frac{F}{A_0}. \tag{2.1}$$

Since the cross section changes during the tensile test, the *ini-*

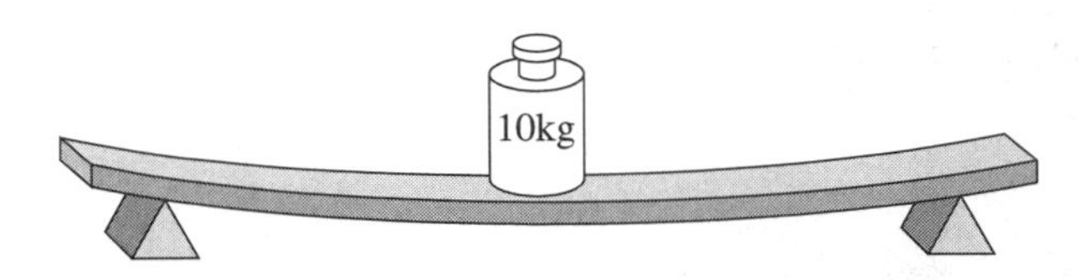

FIGURE 2.1. Schematic representation of a bend test. Note that the convex surface is under tension and the concave surface is under compression. Both stresses are essentially parallel to the surface. The bend test is particularly used for brittle materials.

tial unit area, A_0, is mostly used; see below. If the force is applied parallel to the axis of a rod-shaped material, as in the tensile tester (that is, perpendicular to the faces A_0), then σ is called a **tensile stress**. If the stress is applied parallel to the faces (as in Figure 2.3), it is termed **shear stress**, τ.

Many materials respond to stress by changing their dimensions. Under tensile stress, the rod becomes longer in the direction of the applied force (and eventually narrower perpendicular to that axis). The change in longitudinal dimension in response to stress is called **strain**, ϵ, that is:

$$\epsilon = \frac{l - l_0}{l_0} = \frac{\Delta l}{l_0}, \tag{2.2}$$

where l_0 is the initial length of the rod and l is its final length.

The absolute value of the ratio between the lateral strain (shrinkage) and the longitudinal strain (elongation) is called the **Poisson ratio**, ν. Its maximum value is 0.5 (no net volume change). In reality, the Poisson ratio for metals and alloys is generally between 0.27 and 0.35; in plastics (e.g., nylon) it may be as large as 0.4; and for rubbers it is even 0.49, which is near the maximum possible value.

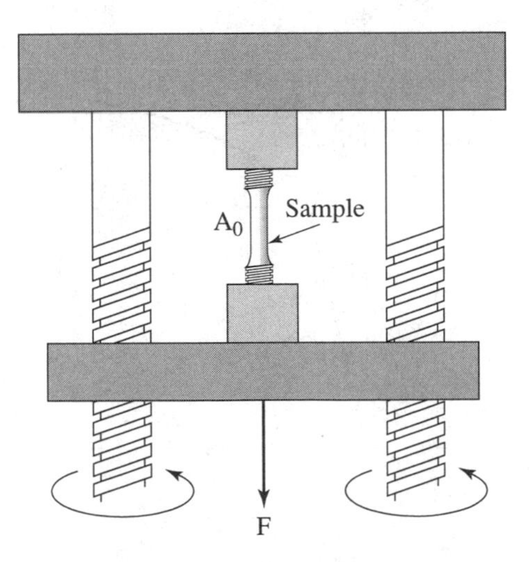

FIGURE 2.2. Schematic representation of a tensile test equipment. The lower cross-bar is made to move downward and thus extends a force, F, on the test piece whose cross-sectional area is A_0. The specimen to be tested is either threaded into the specimen holders or held by a vice grip.

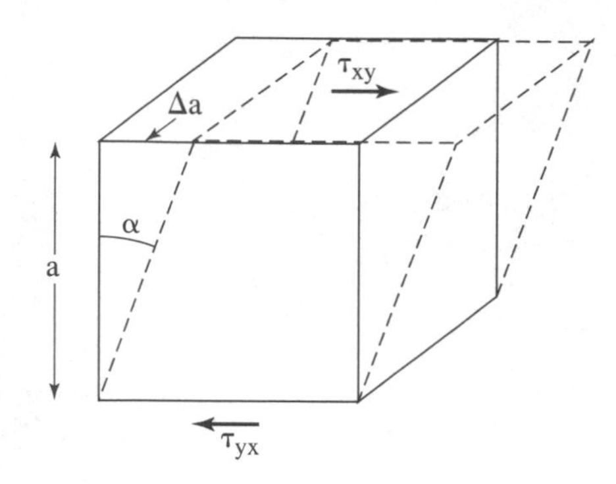

FIGURE 2.3. Distortion of a cube caused by shear stresses τ_{xy} and τ_{yx}.

The force is measured in newtons ($1\ N = 1\ kg\ m\ s^{-2}$) and the stress is given in $N\ m^{-2}$ or pascal (Pa). (Engineers in the United States occasionally use the *pounds per square inch* (psi) instead, where $1\ psi = 6.895 \times 10^3$ Pa and 1 pound = 4.448 N. See Appendix II.) The strain is unitless, as can be seen from Eq. (2.2) and is usually given in percent of the original length.

The result of a tensile test is commonly displayed in a *stress–strain diagram* as schematically depicted in Figure 2.4. Several important characteristics are immediately evident. During the initial stress period, the elongation of the material responds to σ in a linear fashion; the rod reverts back to its original length upon relief of the load. This region is called the *elastic range*. Once the stress exceeds, however, a critical value, called the **yield strength**, σ_y, some of the deformation of the material becomes permanent. In other words, the yield point separates the elastic region from the plastic range of materials.

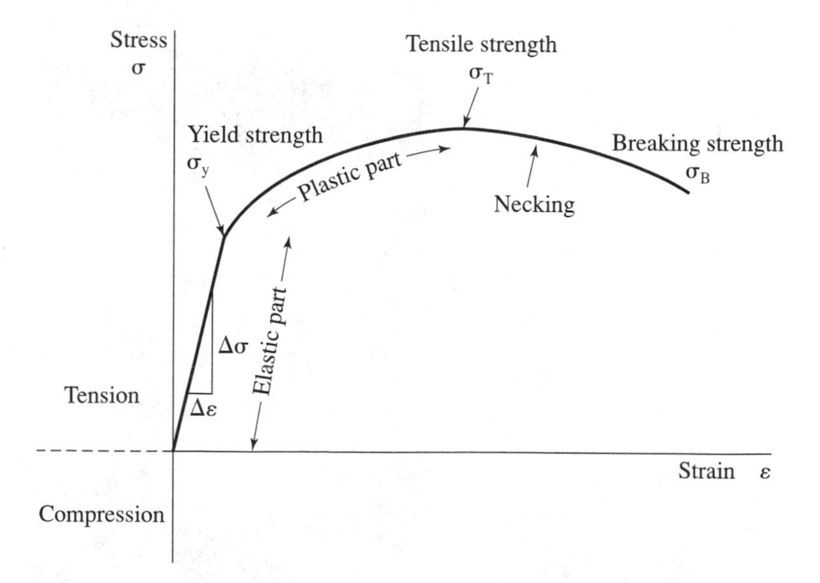

FIGURE 2.4. Schematic representation of a stress–strain diagram for a ductile material. For actual values of σ_y and σ_T, see Table 2.1 and Figure 2.5.

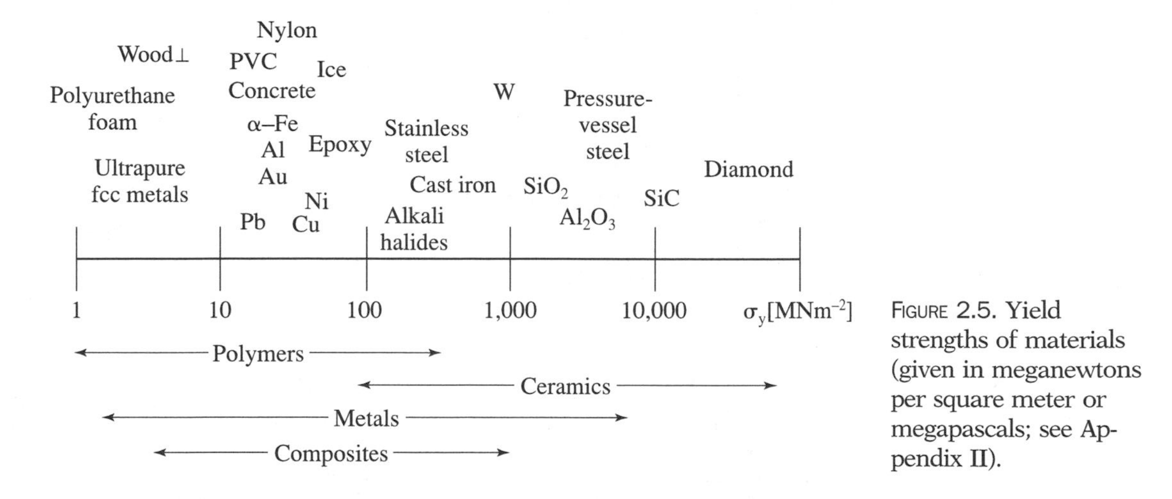

FIGURE 2.5. Yield strengths of materials (given in meganewtons per square meter or megapascals; see Appendix II).

This is always important if one wants to know how large an applied stress needs to be in order for plastic deformation of a workpiece to occur. On the other hand, the yield strength provides the limit for how much a structural component can be stressed before unwanted permanent deformation takes place. As an example, a screwdriver has to have a high yield strength; otherwise, it will deform upon application of a large twisting force. Characteristic values for the yield strength of different materials are given in Table 2.1 and Figure 2.5.

The highest force (or stress) that a material can sustain is called the **tensile strength**,[1] σ_T (Figure 2.4). At this point, a localized decrease in the cross-sectional area starts to occur. The material is said to undergo *necking*, as shown in Figure 2.6. Because the cross section is now reduced, a smaller force is needed to continue deformation until eventually the **breaking strength**, σ_B, is reached (Figure 2.4).

The slope in the elastic part of the stress–strain diagram (Figure 2.4) is defined to be the **modulus of elasticity**, E, (or **Young's modulus**):

$$\frac{\Delta\sigma}{\Delta\epsilon} = E. \tag{2.3}$$

Equation (2.3) is generally referred to as **Hooke's Law**. For shear stress, τ [see above and Figure 2.3], Hooke's law is appropriately written as:

$$\frac{\Delta\tau}{\Delta\gamma} = G, \tag{2.4}$$

[1]Sometimes called *ultimate tensile strength* or *ultimate tensile stress*, σ_{UTS}.

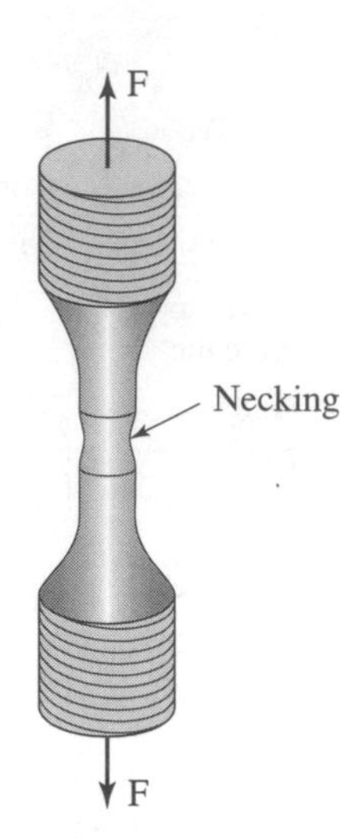

FIGURE 2.6. Necking of a test sample that was stressed in a tensile machine.

where γ is the **shear strain** $\Delta a/a = \tan \alpha \approx \alpha$ and G is the **shear modulus**.

The modulus of elasticity is a parameter that reveals how "stiff" a material is, that is, it expresses the resistance of a material to elastic bending or elastic elongation. Specifically, a material having a large modulus and, therefore, a large slope in the stress–strain diagram deforms very little upon application of even a high stress. This material is said to have a high **stiffness**. (For average values, see Table 2.1.) This is always important if one requires close tolerances, such as for bearings, to prevent friction.

Stress–strain diagrams vary appreciably for different materials and conditions. As an example, *brittle materials*, such as glass, stone, or ceramics have no separate yield strength, tensile strength, or breaking strength. In other words, they possess essentially no plastic (ductile) region and, thus, break already before the yield strength is reached [Figure 2.7(a)]. Brittle materials (e.g., glass) are said to have a very low **fracture toughness.** As a consequence, tools (hammers, screwdrivers, etc.) should not be manufactured from brittle materials because they may break or cause injuries.

Ductile materials (e.g., many metals) on the other hand, withstand a large amount of permanent deformation (strain) before they break, as seen in Figure 2.7(a). (**Ductility** is measured by the amount of permanent elongation or reduction in area, given in percent, that a material has withstood at the moment of fracture.)

Many materials essentially display no well-defined yield strength in the stress–strain diagram; that is, the transition between the elastic and plastic regions cannot be readily determined [Figure 2.7(b)]. One therefore defines an *offset yield strength* at which a certain amount of permanent deformation (for example,

TABLE 2.1. Some mechanical properties of materials

Material	Modulus of elasticity, E [GPa]	Yield strength, σ_y [MPa]	Tensile strength, σ_T [MPa]
Diamond	1,000	50,000	same
SiC	450	10,000	same
W	406	1000	1510
Cast irons	170–190	230–1030	400–1200
Low carbon steel, hot rolled	196	180–260	325–485
Carbon steels, water-quenched and tempered	~200	260–1300	500–1800
Fe	196	50	200
Cu	124	60	400
Si	107	—	—
10% Sn bronze	100	190	—
SiO_2 (silica glass)	94	7200	about the same
Au	82	40	220
Al	69	40	200
Soda glass	69	3600	about the same
Concrete	50	25*	—
Wood ∥ to grain	9–16	—	33–50*; 73–121$^+$
Pb	14	11	14
Spider drag line	2.8–4.7	—	870–1420
Nylon	3	49–87	60–100
Wood ⊥ to grain	0.6–1	5*	3–10*; 2–8$^+$
Rubbers	0.01–0.1	—	30
PVC	0.003–0.01	45	—

*compression; $^+$tensile.

Note: The data listed here are average values. See Chapter 3 for the directionality of certain properties called *anisotropy*; see also Figure 2.4.) For glasses, see also Table 15.1.

0.2%) has occurred and which can be tolerated for a given application. A line parallel to the initial segment in the stress–strain curve is constructed at the distance $\epsilon = 0.2\%$. The intersect of this line with the stress–strain curve yields $\sigma_{0.2}$ [Figure 2.7(b)].

Some materials, such as rubber, deform elastically to a large extent, but cease to be linearly elastic after a strain of about 1%. Other materials (such as iron or low carbon steel) display a sharp yield point, as depicted in Figure 2.7(c). Specifically, as the stress is caused to increase to the *upper yield point*, no significant plastic deformation is encountered. From now on, however, the material will yield, concomitantly with a drop in the *flow stress*, (i.e., the stress at which a metal will flow) resulting in a *lower yield point* and plastic deformation at virtually constant stress [Figure 2.7(c)]. The lower yield point is relatively well defined but fluc-

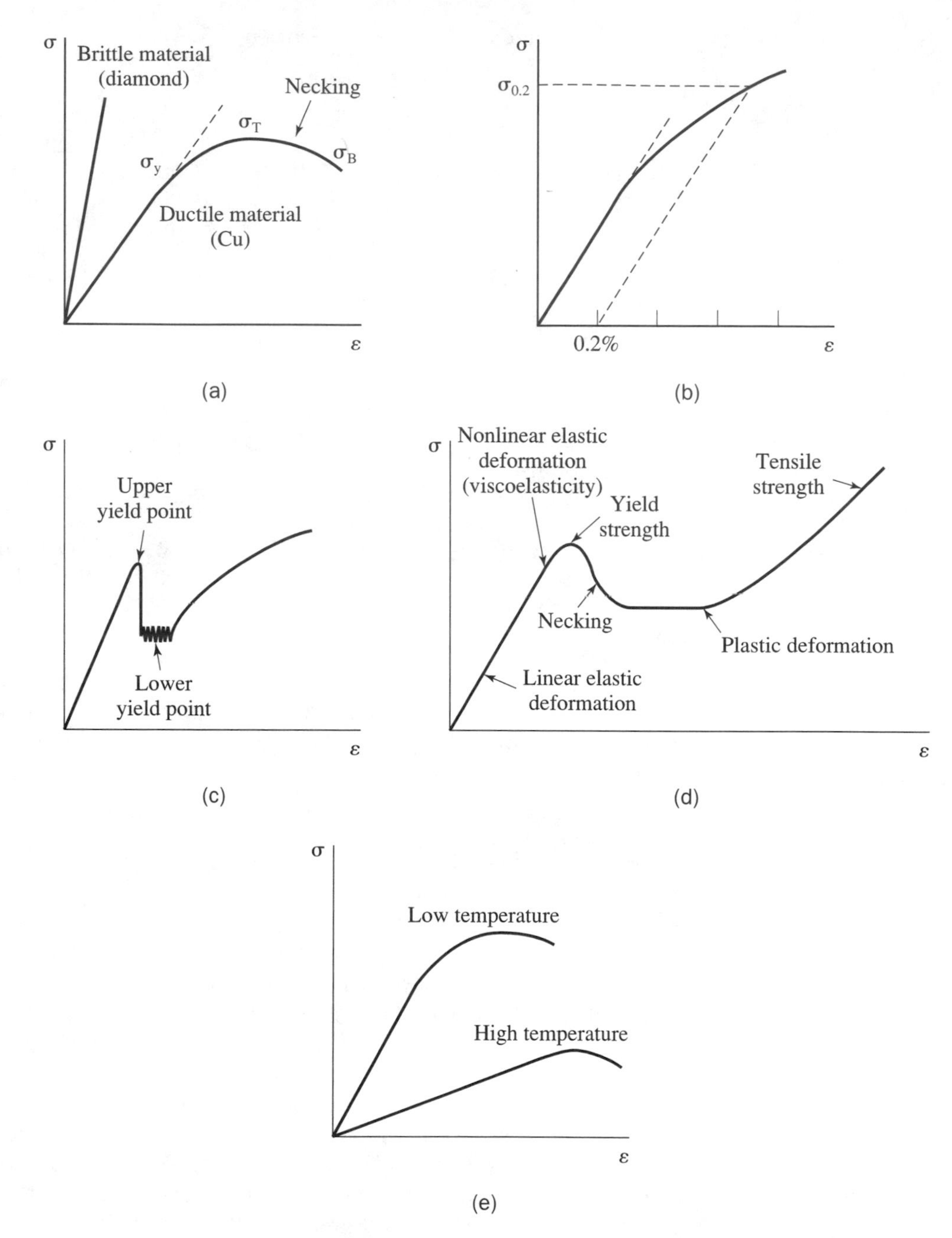

FIGURE 2.7. Schematic representations of stress–strain diagrams for various materials and conditions: (a) brittle (diamond, ceramics, thermoset polymers) versus ductile (metals, alloys) materials; (b) definition of the offset yield strength; (c) upper and lower yield points observed, for example, in iron and low carbon steels; (d) thermoplastic polymer; and (e) variation with temperature.

tuates about a fixed stress level. Thus, the yield strength in these cases is defined as the average stress that is associated with the lower yield point. Upon further stressing, the material eventually hardens, which requires the familiar increase in load if additional deformation is desired. The deformation at the lower yield point starts at locations of stress concentrations and manifests itself as discrete bands of deformed material, called *Lüders bands*, which may cause visible striations on the surface. The deformation occurs at the front of these spreading bands until the end of the lower yield point is reached.

A few polymeric materials, such as nylon, initially display a linear and, subsequently, a nonlinear (viscoelastic) region in the stress–strain diagram [Figure 2.7(d)]. Moreover, beyond the yield strength, a bathtub-shaped curve is obtained, as depicted in Figure 2.7(d).

Stress–strain curves may vary for different temperatures [Figure 2.7(e)]. For example, the yield strength, as well as the tensile strength, and to a lesser degree also the elastic modulus, are often smaller at elevated temperatures. In other words, a metal can be deformed permanently at high temperatures with less effort than at room temperature. This property is exploited by industrial rolling mills or by a blacksmith when he shapes red-hot metal items on his anvil. The process is called *hot working*.

On the other hand, if metals, alloys, or some polymeric materials are *cold worked*, that is, plastically deformed at ambient temperatures, eventually they become less ductile and thus harder and even brittle. This is depicted in Figure 2.8(a), in which a material is assumed to have been stressed beyond the yield strength. Upon releasing the stress, the material has been permanently deformed to a certain degree. Restressing the same material [Figure 2.8(b)] leads to a higher σ_T and to less ductility. The plastic deformation steps can be repeated several times until eventually $\sigma_y = \sigma_T = \sigma_B$. At this point the workpiece is brittle, similar to a ceramic. Any further attempt of deformation would lead to immediate breakage. The material is now **work hardened** (or *strain hardened*) to its limit. A coppersmith utilizes cold working (hammering) for shaping utensils from copper sheet metal. The strain hardened workpiece can gain renewed ductility, however, by heating it above the *recrystallization temperature* (which is approximately 0.4 times the absolute melting temperature). For copper, the recrystallization temperature is about 200°C.

The **degree of strengthening** acquired through cold working is given by the *strain hardening rate*, which is proportional to the slope of the plastic region in a true stress–true strain curve. This needs some further explanation. The **engineering stress** and the

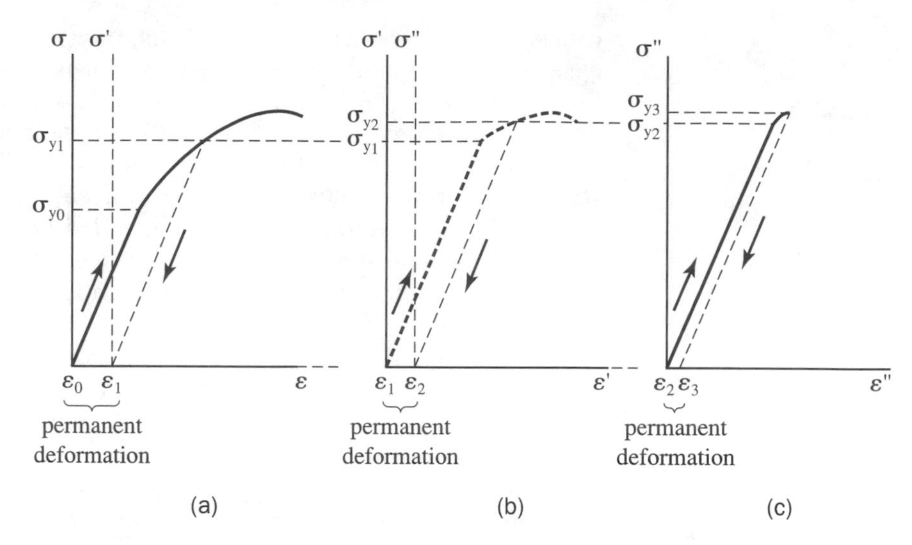

FIGURE 2.8. Increase of yield strength (and reduction of ductility) by repeated plastic deformation. (a) Sample is moderately stressed until some plastic deformation has occurred, and then it is unloaded, which yields permanent deformation. (b) The sample is subsequently additionally permanently deformed. Note that the coordinate system has shifted after unloading from ϵ_0 to ϵ_1. (c) Limit of plastic deformation is reached after renewed stressing.

engineering strain, as defined in Equations (2.1) and (2.2), are essentially sufficient for most practical purposes. However, as mentioned above, the cross-sectional area of a tensile test specimen decreases continuously, particularly during necking. The latter causes a decrease of σ beyond the tensile strength. A **true stress** and **true strain** diagram takes the varying areas into consideration (Figure 2.9(a)). One defines the true stress as:

$$\sigma_t = \frac{F}{A_i}, \tag{2.5}$$

where A_i is now the instantaneous cross-sectional area that varies during deformation. The true strain is then:[1]

$$\epsilon_t = \int_{l_0}^{l_i} \frac{dl}{l} = \ln\left(\frac{l_i}{l_0}\right) = \ln\left(\frac{A_0}{A_i}\right). \tag{2.6}$$

[1]See Problems 8 and 9.

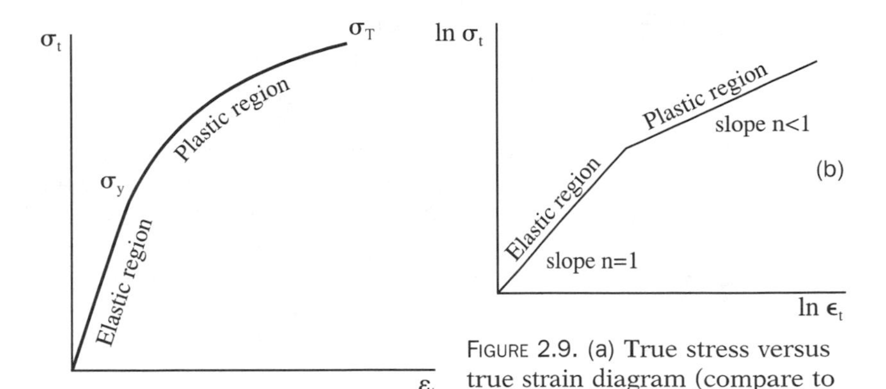

FIGURE 2.9. (a) True stress versus true strain diagram (compare to Figure 2.4). (b) $\ln \sigma_t$ versus $\ln \epsilon_t$ diagram.

In many cases, and before necking begins, one can approximate the true stress–true strain curve by the following empirical equation:

$$\sigma_t = K(\epsilon_t)^n, \tag{2.7}$$

where n is the strain hardening exponent (having values of less than unity) and K is another materials constant (called the *strength coefficient*) which usually amounts to several hundred MPa. Taking the (natural) logarithm of Eq. (2.7) yields:

$$\ln \sigma_t = n \ln \epsilon_t + \ln K, \tag{2.8}$$

which reveals that the **strain hardening exponent** (or **strain hardening rate**), n, is the slope in the plastic portion of an $\ln \sigma_t$ versus $\ln \epsilon_t$ diagram, see Figure 2.9(b).

The tensile test and the resulting stress–strain diagrams have been shown above to provide a comprehensive insight into many of the mechanical properties of materials. For specialized applications, however, a handful of further tests are commonly used. Some of them will be reviewed briefly below.

The **hardness test** is nondestructive and fast. A small steel sphere (commonly 10 mm in diameter) is momentarily pressed into the surface of a test piece. The diameter of the indentation is then measured under the microscope, from which the *Brinell hardness number* (BHN) is calculated by taking the applied force and the size of the steel sphere into consideration. The BHN is directly proportional to the tensile strength. (The *Rockwell* hardness tester uses instead a diamond cone and measures the depth of the indentation under a known load whereas the *Vickers* and *Knoop* microhardness techniques utilize diamond pyramids as indenters.)

Materials, even when stressed below the yield strength, still may eventually break if a large number of tension and compression cycles are applied. The **fatigue test** measures the number of bending cycles that need to be applied for a specific load until failure occurs. Fatigue plays a potentially devastating role in airplane and automobile parts.

When subjected to a sudden blow, some materials break at a lower stress than that measured using a tensile machine. The **impact tester** investigates the *toughness* of materials by striking them at the center while fixing both ends. **Toughness** is defined as the energy (not the force) required to break a material. A heavy pendulum usually is utilized for the blow. The absorbed energy during the breakage is calculated from the difference in pendulum height before and after impact.

The **creep test** measures the continuous and progressive plastic deformation of materials at high temperatures while a constant stress or a constant load below the room temperature yield strength is applied. The temperature at which creep commences varies widely among materials but is generally above 0.3 times the absolute melting temperature. Lead creeps already at room temperature. We will return to creep in Chapter 6.

Leonardo da Vinci (1452–1519) invented already a wire testing device in which sand is poured into a bucket (acting as tensile load) until the wire breaks.

In conclusion, the mechanical properties of materials include ductility, yield strength, elasticity, tensile strength, hardness, toughness under shock, brittleness, fatigue behavior, stiffness, and creep. The question certainly may be raised whether or not it is possible to explain some or all of these diverse properties by one or a few fundamental concepts. We shall attempt to tackle this question in the next chapter.

Problems

2.1. What was the original length of a wire that has been strained by 30% and whose final length is 1 m?

2.2. The initial diameter of a wire is 2 cm and needs to be reduced to 1 cm. Calculate the amount of cold work (reduction in area in percent) which is necessary.

2.3. Calculate the initial diameter of a wire that has been longitudinally strained by 30% and whose final diameter is 0.1 cm. Assume no volume change.

2.4. What force is needed to plastically deform a wire of 2 cm diameter whose yield strength is 40 MPa?

2.5. Calculate the ductility of a wire (that is, its percent area reduction at fracture during tensile stressing) whose initial diameter was 1 cm and whose diameter at fracture is 0.8 cm.

2.6. Calculate the true stress at fracture for a metal rod whose engineering fracture strength is 450 MPa and whose diameter at fracture was reduced by plastic deformation from 1 to 0.8 cm.

2.7. Calculate the strain hardening exponent for a material whose true stress and true strain values are 450 MPa and 15%, respectively. Take $K = 700$ MPa.

2.8. In Eq. (2.6), the relation

$$\epsilon_t = \int_{l_0}^{l_i} \frac{dl}{l} = \ln\left(\frac{l_i}{l_0}\right) = \ln\left(\frac{A_0}{A_i}\right)$$

is given. Show in mathematical terms for what condition (pertaining to a possible change in volume) this relation is true.

2.9. Show that the true and engineering stress and strain are related by

$$\sigma_t = \sigma(1 + \epsilon)$$

and

$$\epsilon_t = \ln (1 + \epsilon)$$

for the case when no volume change occurs during deformation, that is, before the onset of necking.

2.10. Compare engineering strain with true strain and engineering stress with true stress for a material whose initial diameter was 2 cm and whose final diameter at fracture is 1.9 cm. The initial length before plastic deformation was 10 cm. The applied force was 3×10^4 N. Assume no volume change during plastic deformation.

2.11. A metal plate needs to be reduced to a thickness of 4 cm by involving a rolling mill. After rolling, the elastic properties of the material cause the plate to regain some thickness. Calculate the needed separation between the two rollers when the yield strength of the material after plastic deformation is 60 MPa and the modulus of elasticity is 124 GPa.

2.12. A cylindrical rod of metal whose initial diameter and length are 20 mm and 1.5 m, respectively, is subjected to a tensile load of 8×10^4 N. What is the final length of the rod? Is the load stressing the rod beyond its elastic range when the yield strength is 300 MPa and the elastic modulus is 180 GPa?

2.13. Calculate the Poisson ratio of a cylindrical rod that was subjected to a tensile load of 3500 N and whose initial diameter was 8 mm. The modulus of elasticity is 65 GPa, and the change in diameter is 2.5 μm. Assume that the deformation is entirely elastic.

2.14. Calculate the Poisson ratio for the case where no volume change takes place.

Suggestions for Further Study

See the end of Chapter 3. Further, most textbooks of materials science cover mechanical properties.

3

Mechanisms

3.1 • The Atomic Structure of Condensed Matter

This chapter strives to explain the fundamental mechanical properties of materials by relating them to the atomistic structure of solids and by discussing the interactions which atoms have with each other.

Atoms[1] that make up condensed matter are often arranged in a three-dimensional periodic array, that is, in an ordered manner called a *crystal*. The periodic arrangement may differ from case to case, leading to different *crystal structures*. This fact has been known since the mid-nineteenth century through the work of A. Bravais (1811–1863), a French physicist at the École Polytechnique. His concepts were confirmed (1912), particularly after X-ray diffraction techniques were invented, by Laue and Bragg, and routinely utilized by Debye–Scherrer, Guinier, and others. (Periodic arrays of atoms act as three-dimensional gratings which cause X-rays to undergo interference and thus produce highly symmetric diffraction patterns from which the periodic arrangement of atoms can be inferred.) Since about a decade ago, the arrangement of atoms also can be made visible utilizing high-resolution electron microscopy (or other analytical instruments). Specifically, a very thin foil of the material under investigation is irradiated by electrons, yielding a two-dimensional projection pattern of the three-dimensional atomic arrangement, as seen in the lower part of Figure 3.1.

---

[1]*Atomos* (Greek) = indivisible. This term is based on a philosophical concept promulgated by Democritus (about 460–370 B.C.) postulating that matter is composed of small particles that cannot be further divided.

FIGURE 3.1. High-resolution electron micrograph of silicon in which a boundary between a crystalline (lower part) and an amorphous region (upper part) can be observed. (Courtesy of S.W. Feng and A.A. Morrone, University of Florida.)

A perfect crystal in which all atoms are equidistantly spaced from each other is, however, seldom found. Instead we know from indirect observations that some atoms in an otherwise ordered structure are missing. These defects are called *vacancies*. Their number increases with increasing temperature and may reach a concentration of about 1 per 10,000 atoms close to the melting point. In other cases, some extra atoms are squeezed in between the regularly arranged atoms. They are termed *self-interstitials* (or sometimes *interstitialcies*) if they are of the same species as the host crystal. These defects are very rare because of the large distortion which they cause in the surrounding lattice. They may be introduced by intense plastic deformation or by irradiation effects, for example, in nuclear reactors. On the other hand, if relatively small atoms are crowded between regular lattice sites which are of a different species (impurity elements such as carbon, hydrogen, nitrogen, etc.), the term *interstitial* is used instead. Moreover, parts of planes of atoms may be absent, the remaining part ending at a line. These line imperfections are referred to as *dislocations*. (Dislocations are involved when materials are plastically deformed, as we will discuss in detail later on.)

In still other solids, one detects an almost random arrangement of atoms, as observed in the upper part of Figure 3.1. This random distribution constitutes an *amorphous* structure.

The mechanical properties of solids depend, to a large extent, on the arrangement of atoms, as just briefly described. Thus, we need to study in this chapter the microstructure and the crystallography of solids. The crystal structures, however, depend on the type of interatomic forces that hold the atoms in their position. These interatomic bonds, which result from electrostatic interactions, may be extremely strong (such as for ionic or covalent bonds; see below). Materials whose atoms are held together in this way are mostly hard and brittle. In contrast, metallic and particularly van der Waals bonds are comparatively weak and are thus responsible, among others, for the ductility of materials. In brief, there is not *one* single physical mechanism that determines the mechanical properties of materials, but instead, a rather complex interplay between interatomic forces, atomic arrangements, and defects that causes the multiplicity of observations described in Chapter 2. In other words, the interactions between atoms play a major role in the explanation of mechanical properties. We shall endeavor to describe the mechanisms leading to plasticity or brittleness in the sections to come.

In contrast to this, the electrical, optical, magnetic, and some thermal properties can be explained essentially by employing the *electron theory* of condensed matter. This will be the major theme of Part II of the present book, in which the electronic properties of materials will be discussed.

3.2 • Binding Forces Between Atoms

Ionic Bond

Atoms consist of a positively charged nucleus and of negatively charged electrons which, expressed in simplified terms, orbit around this nucleus. Each orbit (or "shell") can accommodate only a maximum number of electrons, which is determined by quantum mechanics; see Appendix I. In brief, the most inner "*K*-shell" can accommodate only two electrons, called *s*-electrons. The next higher "*L*-shell" can accommodate a total of eight electrons, that is, two *s*-electrons and six *p*-electrons. The following "*M*-shell" can host two *s*-electrons, six *p*-electrons, and ten *d*-electrons; and so on.

Filled outermost $s+p$ shells constitute a particularly stable (nonreactive) configuration, as demonstrated by the noble (inert) gases in Group VIII of the Periodic Table (see Appendix). Chemical compounds strive to reach this noble gas configuration for maximal stability. If, for example, an element of Group I of the Periodic Table, such as sodium, is reacted with an element of Group VII, such as chlorine, to form sodium chloride (NaCl), the sodium gives up its only electron in the *M*-shell, which is transferred to the chlorine atom to fill the *p*-orbit in its *M*-shell; see Figure 3.2. The sodium atom that gave up one electron is now positively charged (and is therefore called a *sodium ion*),

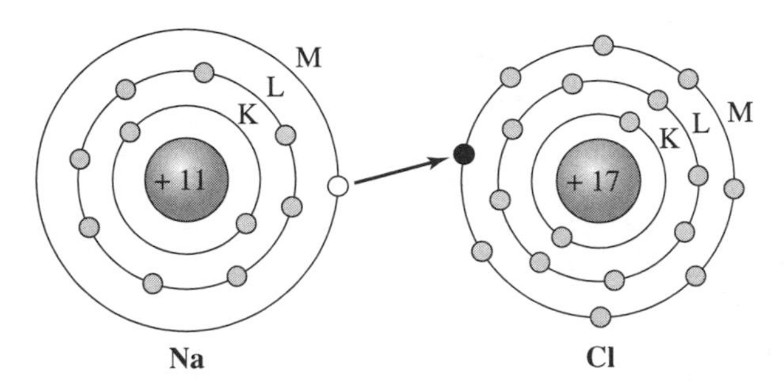

FIGURE 3.2. Schematic representation of the formation of ionically bound NaCl by transferring one electron from Na to Cl to yield Na^+ and Cl^- ions.

whereas a negatively charged chlorine ion is formed by gaining one electron. The net effect is that the two oppositely charged ions attract each other electrostatically and thus produce the ionic bond. Ionic bonds are extremely strong (see Table 3.1). Any mechanical force that tries to disturb this bond would upset the electrical balance. It is partly for this reason that ionically bound materials are, in general, strong and mostly brittle. (Still, in particular cases for which the atomic structure is quite simple, plastic deformation may be observed, such as in NaCl single crystals, which can be bent by hand under running water.)

The question arises why the oppositely charged ions do not approach each other to the extent that fusion of the two nuclei would occur. To answer this, one needs to realize that the electrons which surround the nuclei exert strong repulsive forces upon each other that become exceedingly stronger the closer the two ions approach. Specifically, the orbits of the electrons of both ions start to mutually overlap. It is this interplay between the electrostatic attraction of the ions and the electrostatic repulsion (caused by the overlapping electron charge distributions) that brings about an equilibrium distance, d_0, between the ions, as shown in Figure 3.3.

Characteristic examples of materials which are held together in part or entirely by ionic bonds are the alkali halides, many oxides, and the constituents of concrete. The strength and the tendency for ionic bonding depend on the difference in *"electroneg-*

TABLE 3.1 Bonding energies for various atomic bonding mechanisms

Bonding mechanism	Bonding energy [kJ·mol^{-1}]
Ionic	340–800
Covalent	270–610
Metallic	20–240
Van der Waals	<40

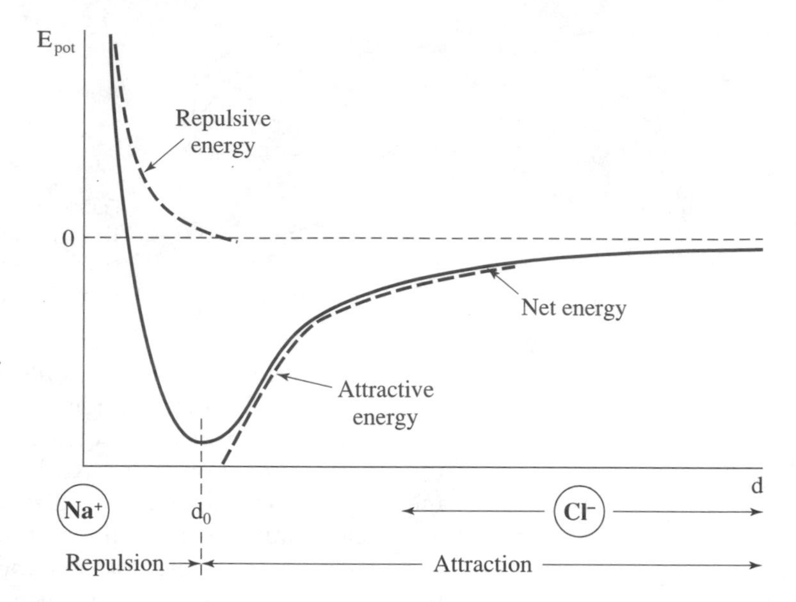

FIGURE 3.3. Schematic representation of the potential energy, E_{pot}, of an Na^+ and a Cl^- ion as a function of the internuclear distance, d. The equilibrium distance, d_0, between the two ions is the position of smallest potential energy.

ativity" between the elements involved, that is, the likelihood of an atom to accept one or more extra electrons. Chlorine, for example, is strongly electronegative because its outer shell is almost completely filled with electrons (Figure 3.2). Sodium, on the other hand, is said to be weakly electronegative (actually it is electropositive: Group I of the Periodic Table!) and, therefore, readily gives up its valence[1] electron.

Covalent Bond

Covalently bound solids such as diamond or silicon are typically from Group IV of the Periodic Table. Consequently, each atom has four valence electrons. Since all atoms are identical, no electrons are transferred to form ions. Instead, in order to achieve the noble gas configuration, double electron bonds are formed by *electron sharing* [see Figure 3.4(a)]. In other words, each Group IV atom, when in the solid state, is "surrounded" by eight electrons, which are depicted in Figure 3.4(a) as dots.

The two-dimensional representation shown in Figure 3.4(a) is certainly convenient but does not fully describe the characteristics of such solids. Indeed, in three dimensions, silicon atoms, for example, are arranged in the form of a tetrahedron[2] around a center atom having 109°28′ angles between the bond axes, as depicted in Figure 3.4(b). Because of this directionality and be-

[1]*Valentia* (Latin) = capacity, strength.
[2]*Tetraedros* (Greek) = four-faced.

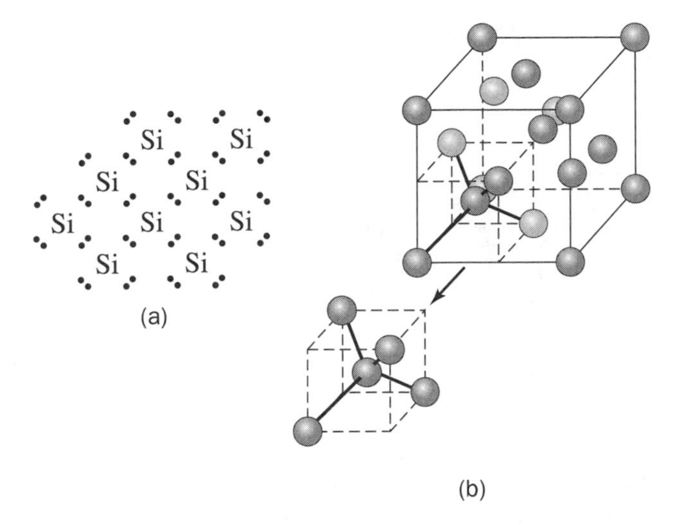

FIGURE 3.4. (a) Two-dimensional and (b) three-dimensional representations of the covalent bond as for silicon or carbon (diamond cubic structure). The charge distribution between the individual atoms is not uniform but cone-shaped. The angle between the bond axes, called the *valence angle*, is 109°28′. (See also Figure 16.4(a).)

cause of the filled electron shells, covalently bound materials are hard and brittle. Typical representatives are diamond, silicon, germanium, silicate ceramics, glasses, stone, and pottery constituents.

In many materials a mixture of covalent and ionic bonds exists. As an example, in GaAs, an average of 46% of the bonds occur through electron transfer between Ga and As, whereas the remainder is by electron sharing. A range of binding energies is given in Table 3.1.

Metallic Bond

The outermost (that is, the valence) electrons for most metals are only loosely bound to their nuclei because of their relative remoteness from their positively charged cores. All valence electrons of a given metal combine to form a "sea" of electrons that move freely between the atom cores. The positively charged cores are held together by these negatively charged electrons. In other words, the free electrons act as the bond (or, as it is often said, as a "glue") between the positively charged ions; see Figure 3.5. Metallic bonds are nondirectional. As a consequence, the bonds do not break when a metal is deformed. This is one of the reasons for the high ductility of metals.

Examples for materials having metallic bonds are most metals such as Cu, Al, Au, Ag, etc. Transition metals (Fe, Ni, etc.) form mixed bonds that are comprised of covalent bonds (involving their 3d-electrons; see Appendix I) and metallic bonds. This is one of the reasons why they are less ductile than Cu, Ag, and Au. A range of binding energies is listed in Table 3.1.

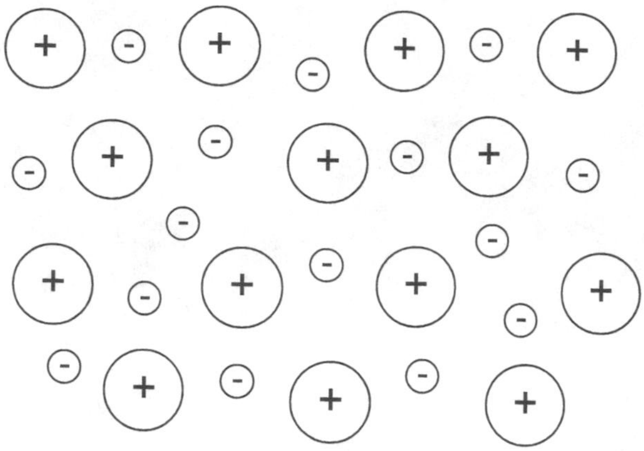

FIGURE 3.5. Schematic representation of metallic bonding. The valence electrons become disassociated with "their" atomic core and form an electron "sea" that acts as the binding medium between the positively charged ions.

Van der Waals Bond

Compared to the three above-mentioned bonding mechanisms, van der Waals bonds[1] are quite weak and are therefore called *secondary bonds*. They involve the mutual attraction of dipoles. This needs some explanation. An atom can be represented by a positively charged core and a surrounding negatively charged electron cloud [Figure 3.6(a)]. Statistically, it is conceivable that the nucleus and its electron cloud are momentarily displaced with respect to each other. This configuration constitutes an electric dipole, as schematically depicted in Figure 3.6(b). A neighboring atom senses this electric dipole and responds to it with a similar charge redistribution. The two adjacent dipoles then attract each other.

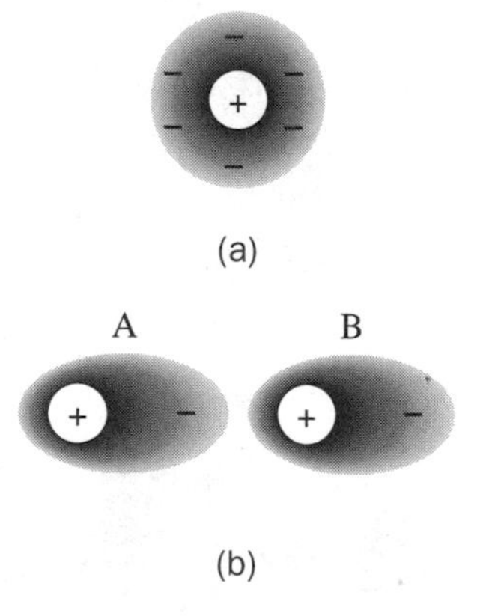

FIGURE 3.6. (a) An atom is represented by a positively charged core and a surrounding negatively charged electron cloud. (b) The electron cloud of atom 'A' is thought to be displaced, thus forming a dipole. This induces a similar dipole in a second atom, 'B'. Both dipoles are then mutually attracted, as proposed by van der Waals.

[1]Johannes Diederik van der Waals (1837–1923), Dutch physicist, received in 1910 the Nobel Prize in physics for his research on the mathematical equation describing the gaseous and liquid states of matter. He postulated in 1873 weak intermolecular forces that were subsequently named after him.

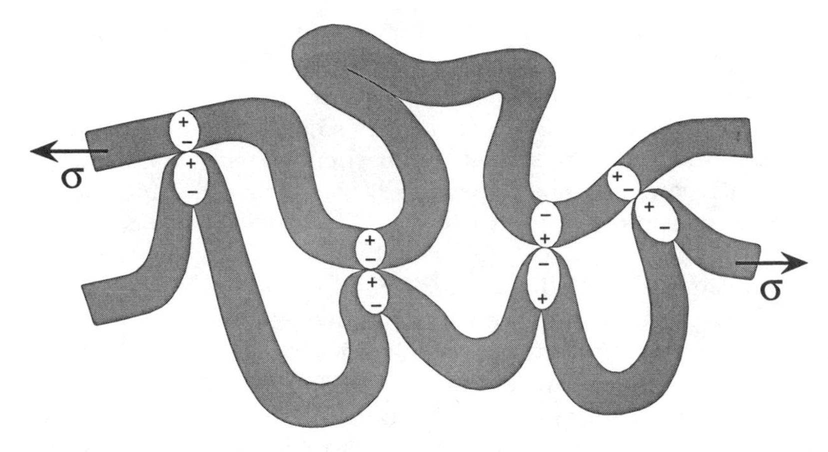

FIGURE 3.7. Two polymer chains are mutually attracted by van der Waals forces. An applied external stress can easily slide the chains past each other.

Many polymeric chains which consist of covalently bonded atoms contain areas that are permanently polarized. The covalent bonding within the chains is quite strong. In contrast to this, the individual chains are mutually attracted by weak van der Waals forces (Figure 3.7). As a consequence, many polymers can be deformed permanently since the chains slide effortlessly past each other when a force is applied. (We will return to this topic in Section 16.4.)

One more example may be given. Ice crystals consist of strongly bonded H_2O molecules that are electrostatically attracted to each other by weak van der Waals forces. At the melting point of ice, or under pressure, the van der Waals bonds break and water is formed.

Mixed Bonding

As already mentioned above, many materials possess atomic bonding involving more than one type. This is, for example, true in compound semiconductors (e.g., GaAs), which are bonded by a mixture of covalent and ionic bonds, or in some transition metals, such as iron or nickel, which form metallic and covalent bonds.

3.3 • Arrangement of Atoms (Crystallography)

The strength and ductility of materials depend not only on the binding forces between the atoms, as discussed in Section 3.2, but also on the arrangements of the atoms in relationship to each other. This needs some extensive explanations.

The atoms in crystalline materials are positioned in a periodic, that is, repetitive, pattern which forms a three-dimensional grid called a *lattice*. The smallest unit of such a lattice that still possesses the characteristic symmetry of the entire lattice is called a *conventional unit cell*. (Occasionally smaller or larger unit cells are used to better demonstrate the particular symmetry of a unit.) The entire lattice can be generated by translating the unit cell into three-dimensional space.

Bravais Lattice

Bravais[1] has identified 14 fundamental unit cells, often referred to as *Bravais lattices* or translation lattices, as depicted in Figure 3.8. They vary in the lengths of their sides (called *lattice constants*, a, b, and c) and the angles between the axes (α, β, γ). The characteristic lengths and angles of a unit cell are termed *lattice parameters*. The arrangement of atoms into a regular, repeatable lattice is called a *crystal structure*.

The most important crystal structures for metals are the face-centered cubic (**FCC**) structure, which is typically found in the case of soft (ductile) materials, the body-centered cubic (**BCC**) structure, which is common for strong materials, and the hexagonal close-packed (**HCP**) structure, which often is found in brittle materials. It should be emphasized at this point that the HCP structure is *not* identical with the simple hexagonal structure shown in Figure 3.8 and is *not* one of the 14 Bravais lattices since HCP has three extra atoms inside the hexagon. The unit cell for HCP is the shaded portion of the conventional cell shown in Figure 3.9. It contains another "base" atom within the cell in contrast to the hexagonal cell shown in Figure 3.8.

The lattice points shown as filled circles in Figure 3.8 are not necessarily occupied by only *one* atom. Indeed, in some materials, several atoms may be associated with a given lattice point; this is particularly true in the case of ceramics, polymers, and chemical compounds. Each lattice point is equivalent. For example, the center atom in a BCC structure may serve as the corner of another cube.

We now need to define a few parameters that are linked to the mechanical properties of solids.

c/a Ratio

The separation between the basal planes, c_0, divided by the length of the lattice parameter, a_0, in HCP metals (Figure 3.9), is theoretically $\sqrt{8/3} = 1.633$, assuming that the atoms are completely spherical in shape. (See Problem 3.6.) Deviations from this ideal ratio result from mixed bondings and from nonspherical atom shapes. The *c/a* ratio influences the hardness and ductility of materials; see Section 3.4.

[1]See Section 3.1.

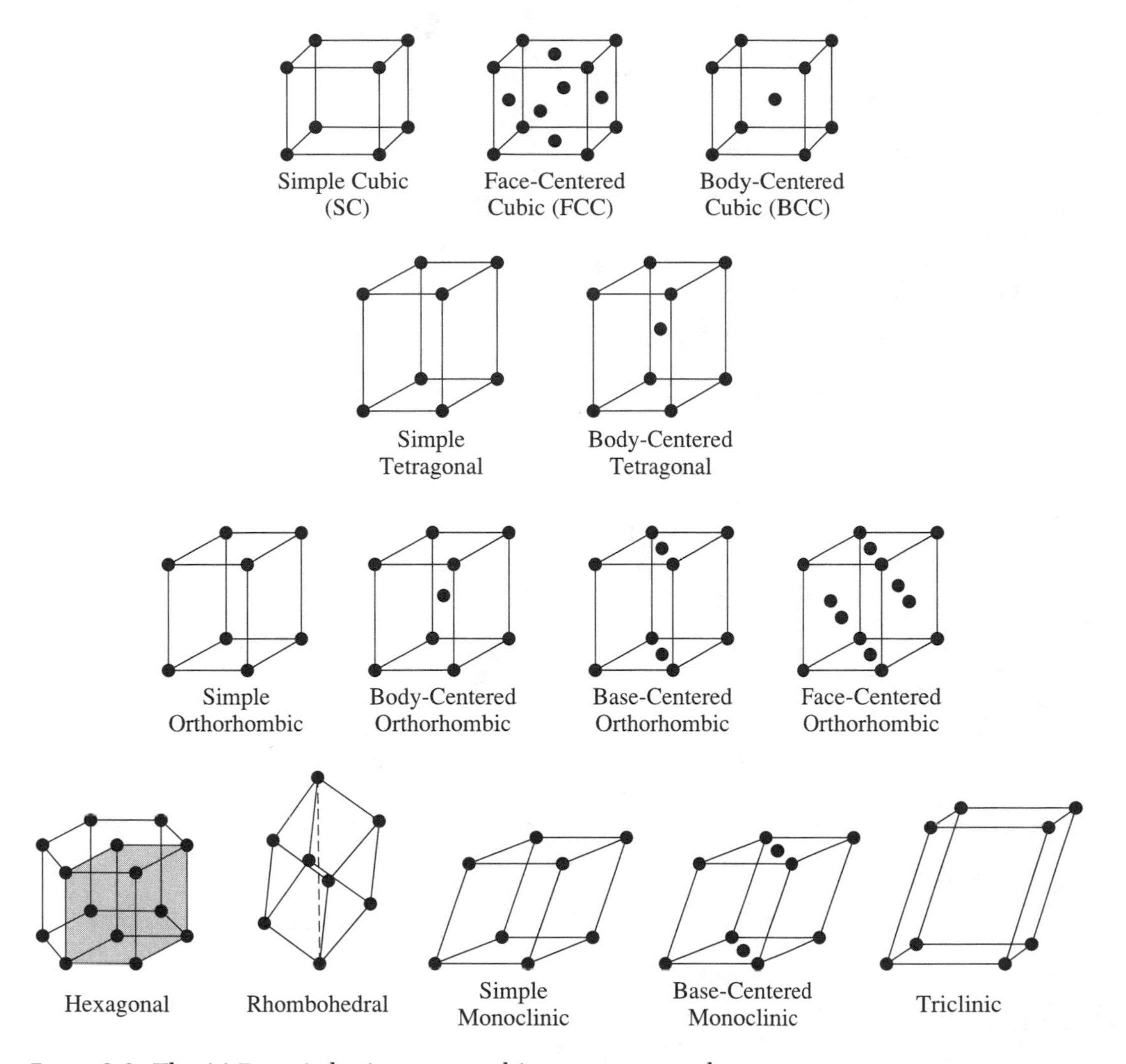

FIGURE 3.8. The 14 Bravais lattices grouped into seven crystal systems:

First row: $a = b = c$, $\alpha = \beta = \gamma = 90°$ (cubic);
Second row: $a = b \neq c$, $\alpha = \beta = \gamma = 90°$ (tetragonal);
Third row: $a \neq b \neq c$, $\alpha = \beta = \gamma = 90°$ (orthorhombic);
Fourth row: at least one angle is $\neq 90°$. *Specifically:*
Hexagonal: $\alpha = \beta = 90°$, $\gamma = 120°$ $a = b \neq c$ (the unit cell is the shaded part of the structure);
Rhombohedral: $a = b = c$, $\alpha = \beta = \gamma \neq 90° \neq 60° \neq 109.5°$;
Monoclinic: $\alpha = \gamma = 90°$, $\beta \neq 90°$, $a \neq b \neq c$;
Triclinic: $\alpha \neq \beta \neq \gamma \neq 90°$, $a \neq b \neq c$.

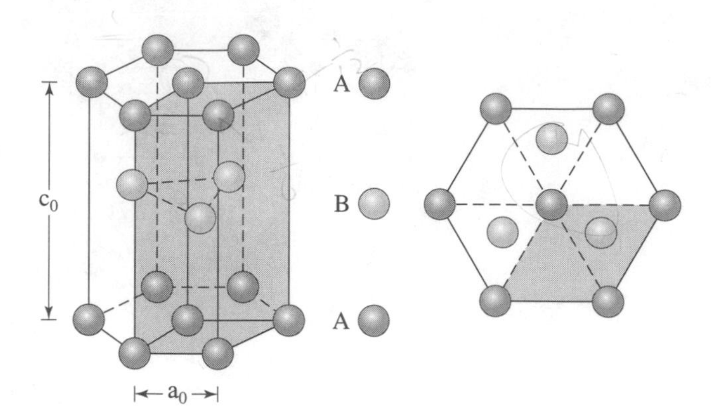

FIGURE 3.9. Hexagonal close-packed (HCP) crystal structure. The sixfold symmetry of the lattice is evident. One unit cell is shaded for clarity. There are three crystallographically equivalent possibilities for this unit cell. The stacking sequence (ABA), which will be explained later, is also depicted.

Coordination Number

The **coordination number** is the number of nearest neighbors to a given atom. For example, the center atom in a BCC structure [Figures 3.8 and 3.10(a)] has eight nearest atoms. Its coordination number is therefore 8. The coordination numbers for some crystal structures are listed in Table 3.2.

Atoms per Unit Cell

The **number of atoms per unit cell** is counted by taking into consideration that corner atoms in cubic crystals are shared by eight unit cells and face atoms are shared by two unit cells. They count therefore only 1/8 and 1/2, respectively. As an example, the number of atoms associated with a BCC structure (assuming only one atom per lattice point) is $8 \times 1/8 = 1$ corner atom and one (not shared) center atom, yielding a total of two atoms per unit cell. In contrast to this, an FCC unit cell has four atoms ($8 \times 1/8 + 6 \times 1/2$). The FCC unit cell is therefore more densely packed with atoms than the BCC unit cell; see also Table 3.2.

Packing Factor

The **packing factor**, P, is that portion of space within a unit cell which is filled with spherical atoms that touch each other, i.e.:

$$P = \frac{A_{uc} * V_A}{V_{uc}}, \tag{3.1}$$

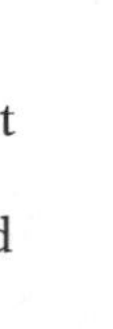

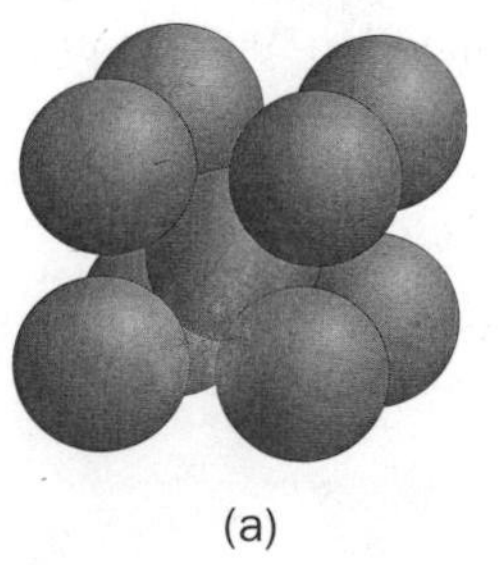

(a)

(b)

FIGURE 3.10. Geometric arrangement of atoms considered to be hard spheres for a BCC structure (a) and an FCC structure (b).

TABLE 3.2. Some parameters and properties of different crystal structures

Crystal structure	Coordination number	Atoms per unit cell	Packing factor	Mechanical properties
HCP	12*	2	0.74*	Brittle
FCC	12	4	0.74	Ductile
BCC	8	2	0.68	Hard
Simple cubic (SC)	6	1	0.52	No representative materials

*Assuming a *c/a* ratio of 1.633.

where A_{uc} is the number of atoms per unit cell (see above), $V_A = \frac{4}{3}\pi r^3$ is the volume of an atom (assuming hard spheres of radius r), and V_{uc} is the volume of the unit cell. The packing factor in FCC and HCP structures is 0.74, whereas a BCC crystal is less densely packed, having a packing factor of 0.68. (Note that $P = 0.74$ for HCP structures is only true if the c/a ratio equals the ideal value of 1.633.)

Linear and Planar Packing Fractions

The **linear packing fraction** is the portion of a line through the centers of atoms in a specific *direction* that is filled by atoms. For example, the linear packing fraction in the direction of the face diagonal of an FCC unit cell in which the corner atoms and the face atom touch each other [Figure 3.10(b)] is 100%. Similarly, a **planar packing fraction** can be defined as that portion of a given *plane* that is filled by atoms.

Density

The **density**, δ, of a material can be calculated by using:

$$\delta = \frac{A_{uc}{*}M_a}{V_{uc}{*}N_0} = \frac{N_a{*}M_a}{N_o}, \tag{3.2}$$

where M_a is the atomic mass of the atom (see Appendix III), N_a is the number of atoms per volume, and N_0 is a constant called the Avogadro number (see Appendix II). Experimental densities of materials are given in Appendix III.

Stacking Sequence

A **stacking sequence** describes the pattern in which close-packed atomic planes are piled up in three dimensions. To explain this, let us assume for simplicity that the atoms are represented by hard spheres and are arranged on a plane in such a way to take up the least amount of space, as shown in Figure 3.11(a). The atoms are said to be situated in a close-packed plane which we designate here as plane A. We now attempt to construct a regular and repetitive array of atoms in three dimensions. First we

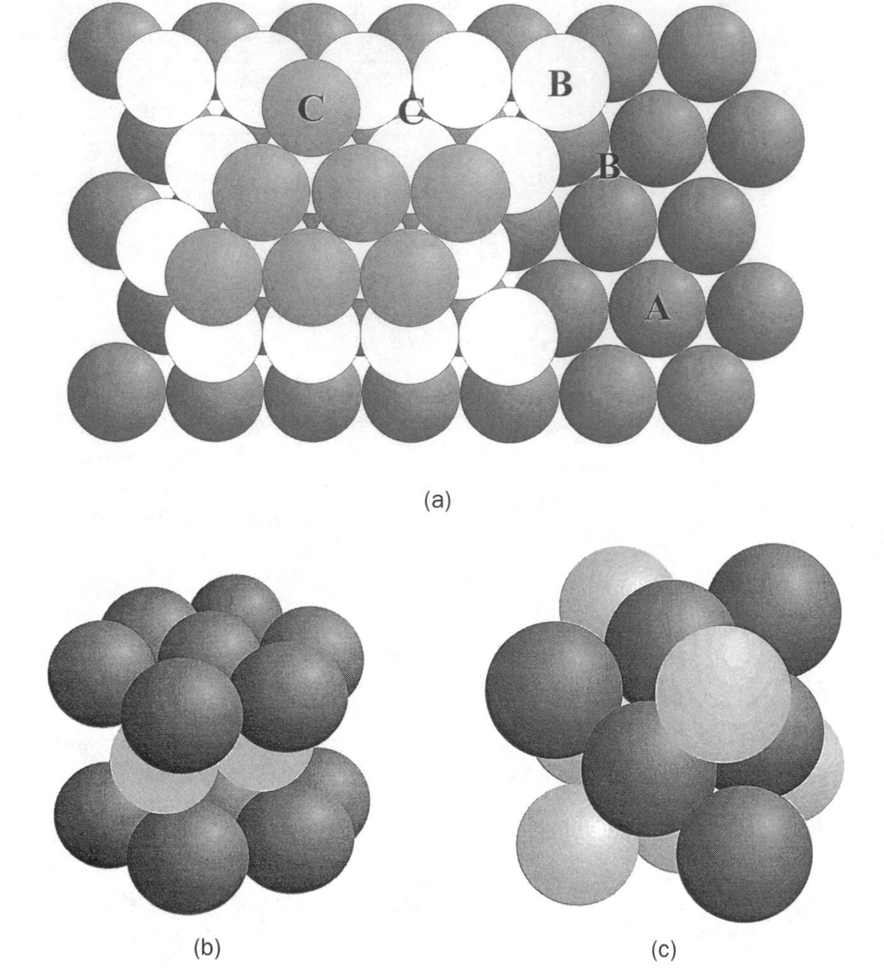

FIGURE 3.11. (a) A two-dimensional representation of a three-dimensional stacking sequence for close-packed lattice planes. (b) Close-packed planes for an HCP crystal structure. (c) Close-packed planes in an FCC structure.

fill the depressions marked B, which are formed by each set of three atoms in plane A. This constitutes the second plane of atoms. When filling the third level with atoms, one has two choices. One may arrange the third level identical to the first one. The stacking sequence is then said to be ABAB, which represents the HCP crystal structure depicted in Figures 3.9 and 3.11(b). In other words, atoms in the first and third levels are exactly on top of each other. As an alternative, the hitherto unoccupied spaces, C, may be occupied on the third level. Only the fourth close-packed atomic plane is finally identical to the first one, which leads to a stacking sequence called ABC ABC. The diagonal planes in an FCC crystal structure have this stacking sequence [Figure 3.11(c)].

In conclusion, HCP and FCC structures both have the highest possible packing efficiency (Table 3.2), but differ in their stacking sequence. Both crystal structures are quite common among metals. The preference of one structure over the other is rooted in the tendency of atomic systems to assume the lowest possible energy level.

Stacking Faults and Twinning

In some instances, the regular stacking sequence may be interrupted. For example, a stacking sequence ABC AB ABC may occur instead of ABC ABC ABC. This configuration, where one layer is missing, is called an *intrinsic stacking fault*. On the other hand, the sequence ABC B ABC, where one B layer has been *added*, is termed an *extrinsic stacking fault*. Another defect may involve the stacking sequence ABC ABC AB$\underline{C}$ BAC BA$\underline{C}$ ABC ABC (Figure 3.12). Here, the order is inverted about the layers marked $\underline{C}$ due to a *twin* which is incorporated in the FCC crystal. Such configurations may be formed during the heating of a previously deformed FCC material (*annealing twins*) or by plastic deformation (*deformation twins*). We shall discuss the implications of stacking faults and twinning at a later point.

Polymorphic and Allotropic Materials

Some materials have different crystal structures at lower temperatures than at higher temperatures. (The inherent energy of a crystal structure is temperature-dependent.) Materials that undergo a transformation from one crystal structure to another are called *polymorphic* or, when referring to elements, *allotropic*. A well-known example is the allotropic transformation of room

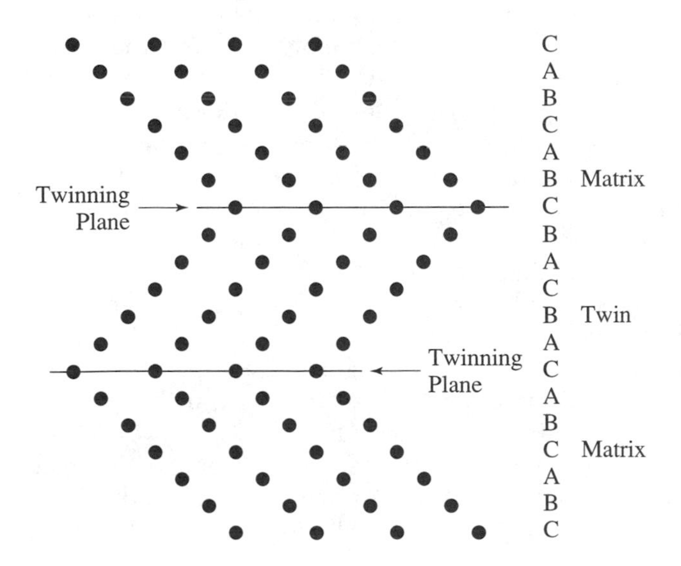

FIGURE 3.12. Stacking faults in an FCC metal to yield twinning. The matrix and the twin match at the interface, called the twinning plane. The figure shows a $(1\bar{1}0)$ plane normal to the (111) twin plane, as we shall explain below.

temperature, ductile, tetragonal, white, β-tin to its low-temperature, brittle, diamond cubic, gray, α modification. The transformation occurs slowly at 286.2 K and causes a 27% expansion and thus a disintegration into powder (tin plague). As a consequence of this transformation, old organ pipes made of tin (or tin alloys) may disintegrate in severe winter temperatures. A similar breakdown has been reported for the tin-based coat buttons of Napoleon's army in a harsh Russian winter. Another example is the temperature-dependent FCC to BCC allotropic transformation of pure iron at 912°C. The prevailing crystal structure also may depend on the external pressure. As an example, graphite is stable at ambient conditions whereas diamond is the high-pressure modification of carbon.

Rotation Axis and Symmetry

A plane in a cubic crystal is said to have *fourfold symmetry* because, when rotated about an axis normal to its faces, four rotations of 90° can be achieved at which the cell remains invariant. The cubic unit cell has three of those fourfold axes in addition to four threefold axes that pass through two opposite corners. Further, there are six twofold axes that pass through two opposite edge centers. In general, an axis is termed n-fold when the angle of rotation is $2\pi/n$.

Miller Indices

Miller indices are sets of numbers which identify particular *planes* or sets of parallel planes in unit cells. In order to determine them for a cubic lattice, one sketches a Cartesian coordinate system having its axes parallel to the edges of a unit cell. Several sets of coordinates (separated by commas) are noted in Figure 3.13(a). To arrive at the Miller indices (hkl) for a plane, one has to locate the intercepts of this plane with the coordinate axes and then form the reciprocal of the intercepts. If one or more of these numbers is a fraction, *clear the fraction* to obtain three integers. If a plane is parallel to a coordinate axis, the intercept is considered to be at infinity and its reciprocal value is consequently zero. In those cases where a plane passes through the origin of a coordinate system, the origin has to be moved. [See the primed axes in Figure 3.13(c).] Negative numbers are denoted by a bar on top of the number. Selected examples are given in Figures 3.13(a)–(c).

The faces of a cube designated as (100), (010), (001), $(\bar{1}00)$, $(0\bar{1}0)$, and $(00\bar{1})$ planes are crystallographically equivalent and can therefore be collectively described by {100}, utilizing curly brackets. The face-diagonal planes (110), (101), (011), $(1\bar{1}0)$, $(\bar{1}01)$, $(0\bar{1}1)$ (and the equivalent planes having negative Miller indices) are summarized as {110}, where $\bar{1}$ signifies generally a shift of the coordinate axis and consequently an intercept at -1; see

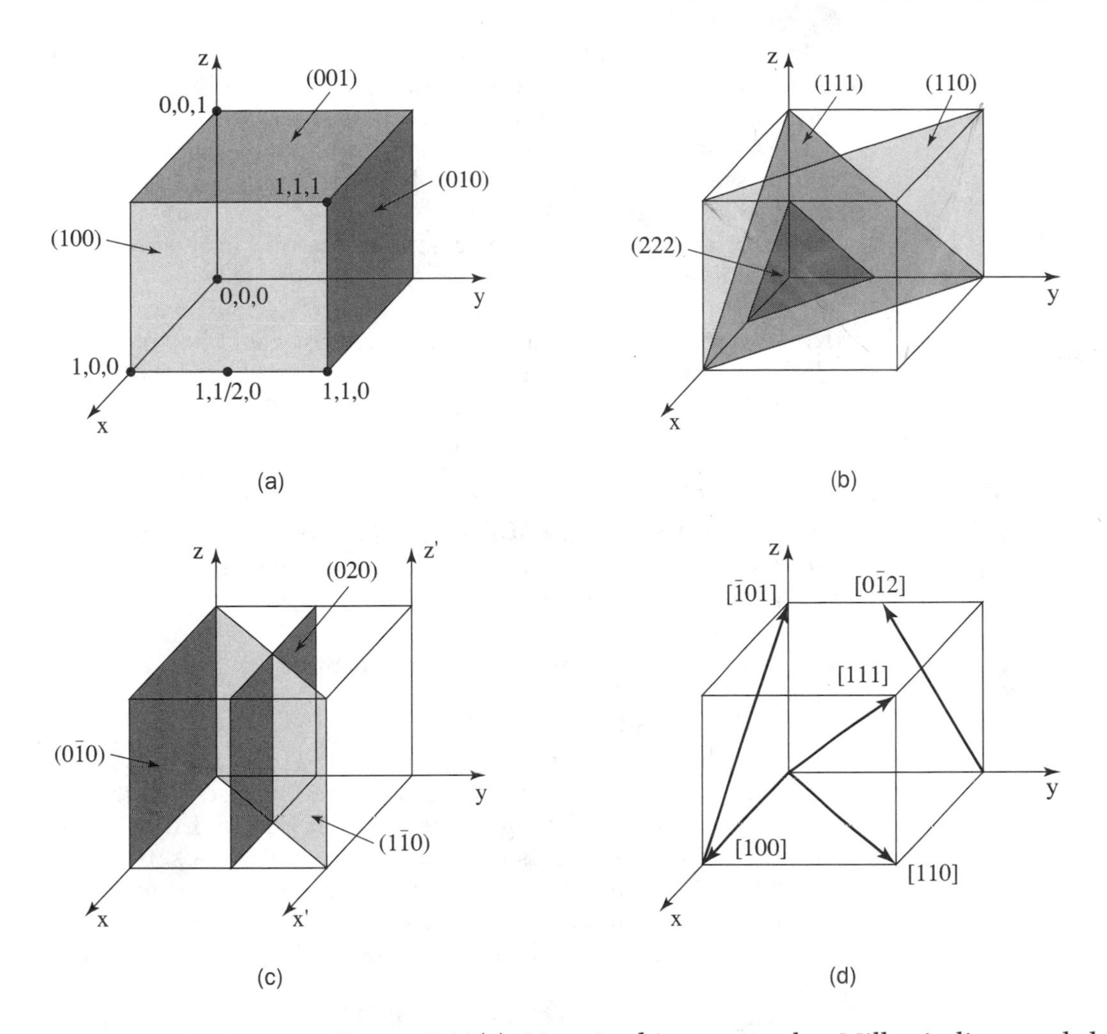

FIGURE 3.13. (a–c) Examples for Miller indices for planes. [Note the primed coordinate system in (c).] (d) Indices for directions for a cubic unit cell.

Figure 3.13(c). Note in this context that Miller indices and their negatives describe the same plane, that is, $(110) \equiv (\overline{1}\overline{1}0)$; however, the vectors normal to these planes point in opposite directions. Further, multiples of Miller indices do not describe the same set of planes, even though the planes in question are parallel to each other. This can be seen in Figure 3.13(b), where (111) describes a set of planes which has twice the interplanar spacing than the set designated by (222). See in this context Eq. (3.3) below.

It should be pointed out that some planes may not necessarily be occupied by any atoms in certain crystal structures. For example, in the cubic primitive structure, only every second (020) plane contains atoms. In contrast, in the FCC lattice, *every* plane of the (020) set is occupied.

In the BCC lattice, the (111) planes pass only through the corner atoms but not through the center of the body-centered ones. In order to accommodate the body-centered as well as the corner atoms, the (222) set of planes needs to be taken into account; see Figure 3.13(b). Note, however, that the (222) plane sketched in Figure 3.13(b) passes through the body-centered atoms of adjacent cubes, for example, the one immediately below the depicted one.

Interplanar Spacing

The distance between parallel and closest neighboring lattice planes which have identical Miller indices is given for cubic materials by:

$$d_{hkl} = \frac{a_0}{\sqrt{h^2 + k^2 + l^2}}, \tag{3.3}$$

where a_0 is the lattice constant (see above) and *hkl* are the Miller indices of the planes in question. More complex expressions exist for other crystal systems.

Directions

Directions in unit cells are identified by subtracting the coordinates of the tail from the coordinates of the tip of a distance vector (i.e., "head minus tail"). The set of numbers thus gained is then reduced to the smallest integers and inserted into square brackets. As above, negative numbers are denoted by a bar on top of the number. Examples are given in Figure 3.13(d). A set of equivalent directions is summarized in angle brackets; for example, the ⟨100⟩ directions are all parallel to the coordinate axes. The indices for directions are, strictly speaking, not "Miller" indices. Despite this fact, the term "Miller indices" is often used for directions as well as for planes.

It needs to be emphasized that the [100] direction points opposite to the [$\bar{1}$00] direction. In other words, mutually negative directional indices are not identical. Further, two directional indices, where one is a multiple of the other, are identical. Finally, directions and planes in cubic systems which have the same indices are mutually perpendicular to each other, as can be inferred from Figure 3.13.

Hexagonal Planes and Directions

The designation of **planes** in the **hexagonal** system is equivalent. However, because of the specific symmetry in hexagonal structures, one usually utilizes a coordinate system that contains four axes, a_1, a_2, a_3, and c, and consequently four-digit *Miller–Bravais indices* (*hkil*), as depicted in Figure 3.14(a). (The index i is defined by the relation $h + k + i = 0$.) The Miller–Bravais indices for the top (or bottom) planes, called *basal planes*, are there-

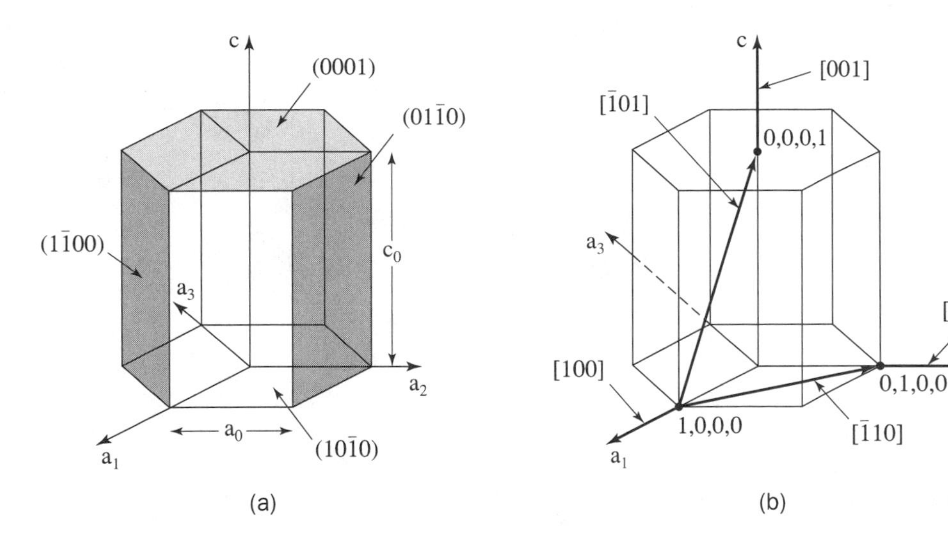

FIGURE 3.14. (a) Four-digit Miller–Bravais indices for *planes* in a hexagonal crystal structure. (b) Selected examples for *directions* in the hexagonal crystal system using three-digit indices. The coordinate intercepts are given with commas separating them.

fore (0001), whereas the indices for the six vertical, external surfaces (called *prism planes*) are of the type $\{1\bar{1}00\}$.

The **directions** in **hexagonal** systems are more conveniently expressed by three-digit indices (whereas the a_3-axis is considered to be redundant and is therefore omitted). The procedure for arriving at these indices is the same as for cubic systems. Selected examples are given in Figure 3.14(b). (For completeness, an example for the four-digit notation is given. The [100] direction may be listed as 1/3 $[2\bar{1}\bar{1}0]$, which commands: move from the origin 2 units into the a_1 and -1 units each into the a_2 and a_3 directions and divide the result by 3 if the same length of the vector is desired.)

Polycrystallinity

So far we tacitly assumed that the lattice is continuous throughout an entire piece of material, that is, we considered only what is called a *single crystal*. This arrangement of atoms is, however, rarely found in practical cases unless the growth environment is carefully controlled, as, for example, when producing silicon single crystals used for microelectronic devices or blades for turbine engines made of superalloys. Instead, solids generally consist of a number of individual single crystallites, called *grains*. They range in size between millimeters and micrometers. (Recently, even nanocrystals have become of interest.) Their orientations are somewhat rotated with respect to the neighboring

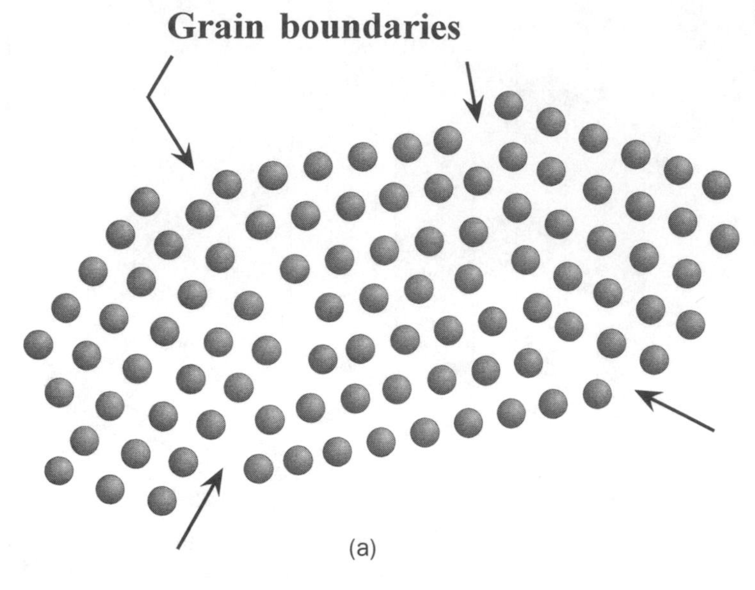

(a)

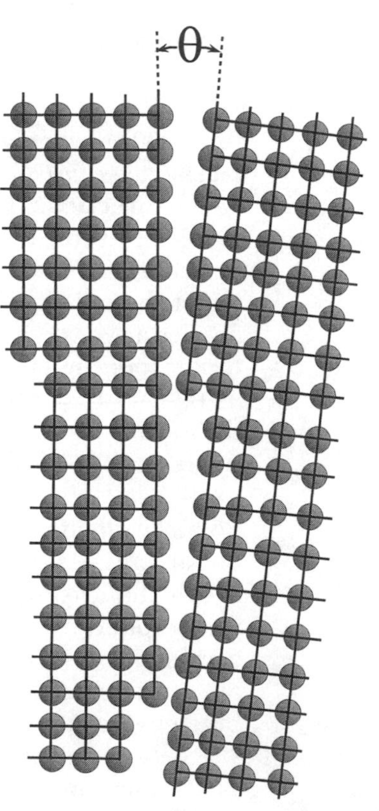

(b)

FIGURE 3.15. Schematic representation of polycrystalline materials consisting of (a) three single crystals which meet on mutual large-angle grain boundaries, and (b) a small-angle grain boundary.

grain; see Figure 3.15. These materials are then called *polycrystals*. The individual grains are separated by *grain boundaries* in which some atomic mismatch occurs. In essence, the grain boundaries are less densely packed than the grains themselves, which results in important properties, as we shall discuss later. One distinguishes between *large-angle grain boundaries* [Figure 3.15(a)] and *small-angle grain boundaries*. For the latter, the mismatch is relatively small, that is, on the order of only a few degrees [θ in Figure 3.15(b)]. Other types of grain boundaries will be discussed in Section 3.4.

Anisotropy

Many physical properties of materials, such as the modulus of elasticity, or some optical, electrical, or magnetic constants may depend on the crystallographic direction in which they were measured. For example, the elastic modulus for copper varies between 66.7×10^9 Pa in the [100] direction and 191×10^9 Pa in the [111] direction, i.e., by a factor of about 3 (with an intermediate value of 130×10^9 Pa for the [110] direction). This dependence of a physical property on the crystallographic direction is called *anisotropy*. It is caused by the variation in distance between the atoms (or ions) in different directions. Materials for which the physical properties are identical in each direction are termed *isotropic*. For example, the electrical conductivity for materials having a cubic lattice structure is completely isotropic.

Atomic Structure of Polymers

Polymers consist of groups of atoms which often are arranged in a chainlike manner, as already discussed in Section 3.2 and Figure 3.7. The atoms which partake in such a chain (or macromolecule) are regularly arranged along the chains. Several atoms combine and thus form a particular building block, called a *monomer*. Thousands of monomers join to make a *polymer*. If one out of four hydrogen atoms in polyethylene is replaced by a chlorine atom, polyvinylchloride (PVC) is formed upon polymerization [Figure 3.16(b)]. In polystyrene, one hydrogen atom in polyethylene is replaced by a benzene ring. More complicated macromolecules may contain side chains which are attached to the main "backbone." They are appropriately named *branched polymers*.

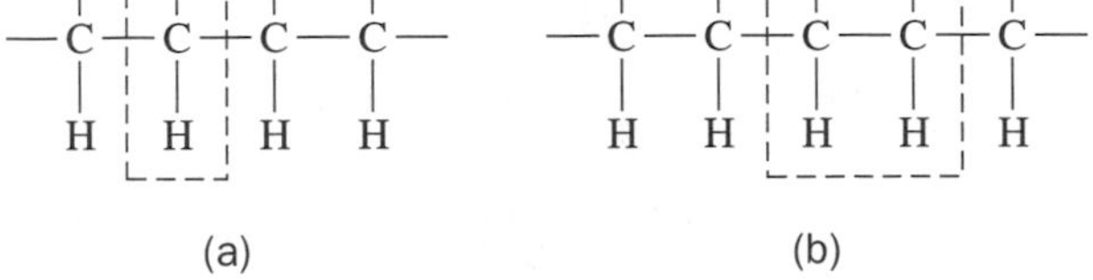

(a) (b)

FIGURE 3.16. (a) Polyethylene, (b) polyvinylchloride. (The dashed enclosures mark the repeat unit.)

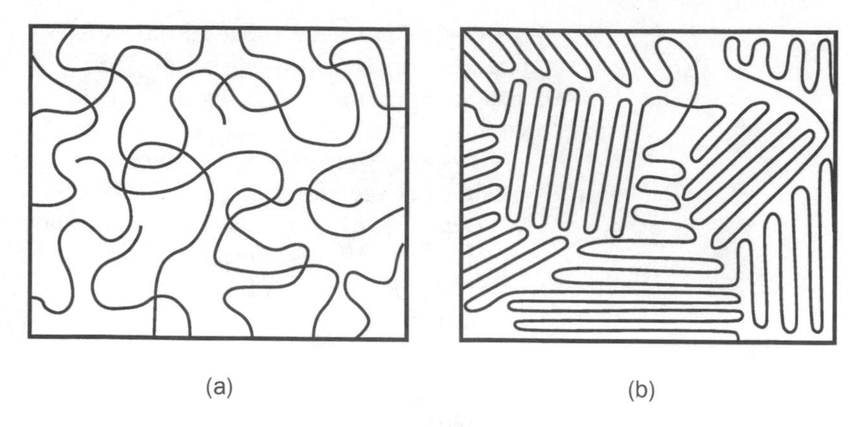

FIGURE 3.17. Simplified representation of (a) an amorphous polymer, and (b) a crystalline polymer (folded chain model) as found in polyethylene and many other thermoplastic materials. Each loop is about 100 carbon atoms long and extends in three dimensions forming plates or lamellae. See Section 16.4.

Macromolecules whose backbones consist largely of carbon atoms, as in Figure 3.16, are called *organic* polymers. Frequently, these carbon atoms are substituted by other elements, such as silicon, oxygen, nitrogen, sulfur, etc. As mentioned in Section 3.2, the binding forces which hold the individual atoms in polymers together are usually covalent, but ionic bonding likewise plays a significant role in polymeric structures. Thus, the repeat units which make up the chains are arranged in an ordered manner. In contrast to this, a large number of polymeric chains often are randomly arranged with respect to each other, that is, they are not laid out in three-dimensionally ordered patterns. In this case, the material is appropriately considered to be noncrystalline; i.e., amorphous; see Figure 3.17(a). There are, however, polymers in which the individual chains are folded forward and backward, thus leading to some symmetry and crystallinity [Figure 3.17(b)]. As an example, polyethylene [Figure 3.16(a)], which consists of *repeat units* of one carbon atom and two hydrogen atoms, has an orthorhombic crystal structure. In still other polymers, crystalline and amorphous regions are mixed. We shall return to this topic in Chapter 16.

Crystal Structures of Ceramics

Ceramics are defined to be inorganic nonmetals. Examples are oxides (SiO_2, MgO, BeO), halides (NaCl, CsCl), sulfides (ZnS, etc.), nitrides, carbides, and other compounds. They are mainly ionically/covalently bound (see Section 3.2) and, therefore, have

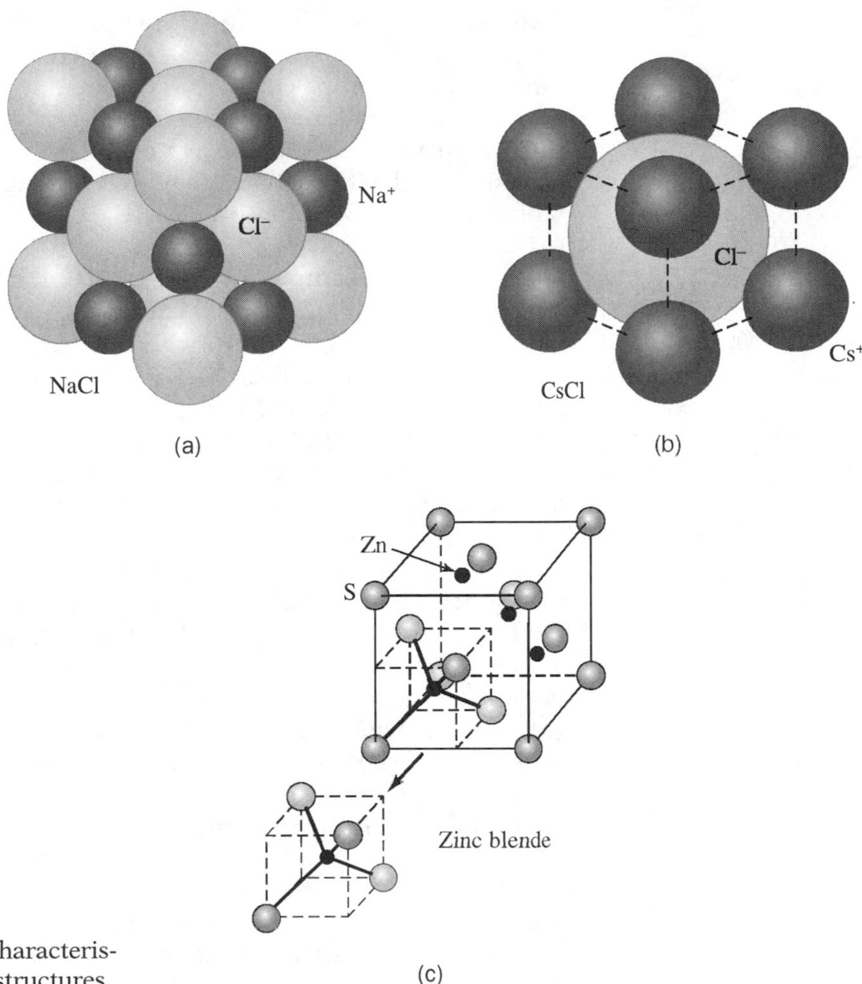

FIGURE 3.18. Some characteristic ceramic crystal structures.

crystal structures that assure electrical neutrality concomitant with the highest possible packing efficiency. Since the radii of the involved ions often differ considerably, a multitude of crystal structures having characteristic names are listed in handbooks of crystallography. Among them are the notorious *Sodium chloride structure*, a cubic structure where the Na^+ and Cl^- ions alternate their positions. Specifically, the Cl^- ions may occupy the FCC positions, as shown in Figure 3.8(a), whereas the Na^+ ions occupy the centers of the cube edges and the center of the cubic cell, thus forming a second FCC lattice which is shifted by the coordinates 1/2, 1/2, 1/2 relative to the FCC lattice of the Cl^- ions. An appreciable number of compounds, such as NiO, MnO, FeO, MgO, BaO, and CaO, crystallize in this structure.

In another crystal structure, the *Cesium chloride structure*, the large Cl^- ions reside in the center of a cube and the smaller Cs^+ ions occupy the cube corners, as depicted in Figure 3.18(b). (Expressed in a different way, one can describe the CsCl structure as cubic primitive, wherein two ions, that is, a Cs and a Cl ion, are associated with each lattice point.) Among ceramics that crystallize in the CsCl structure are KCl, CsBr, CsI, and, naturally, CsCl.

As a further example, the *zinc blende structure* consists of a unit cell in which all face and corner positions of a cube are occupied by, say, sulfur atoms, whereas zinc atoms are positioned on interior tetrahedral sites. In other words, each Zn ion is bonded to four S atoms, as depicted in Figure 3.18(c). (For better visualization, the atoms are not shown in contact to each other. However, some ions touch each other along 1/4 of the body diagonal.) Examples of compounds that crystallize in the zinc blende structure are ZnS, ZnTe, GaAs, SiC, and CdTe.

The *diamond cubic* (DC) structure is quite similar to the just-mentioned zinc blende structure, except that in this case all of the lattice sites are occupied, for example, by carbon or silicon atoms, which are covalently bound to their four closest neighbors; see Figure 3.4. (Alternatively, one can state that a *pair of atoms* is associated with a given FCC lattice site.) Representatives of materials having the DC structure are Si, Ge, and, of course, carbon in its diamond modification. Other ceramic crystal structures will be presented in later parts of this book when they are needed to explain certain properties of materials. They may be rather complex in some cases.

Crystalline silica (SiO_2, quartz, or β-cristobalite) is partly ionic and partly covalently bound in tetrahedral form (see Figure 15.2). In amorphous noncrystalline silica (called *glass, fused silica*, or *vitreous silica*), the silicon and oxygen atoms are mainly linked by covalent bonds. Additions of sodium and limestone which yield the commercially important soda-lime-silica glass (windows, bottles) loosen up the strong bonds and make the material pliable at lower temperatures (700° compared to above 1000°C for glassy silica). We shall return to glasses in Chapter 15.

Closing Remarks

Up to this point, we have studied and classified the mutual relationships which atoms may assume with respect to each other. However, we still have not answered the key question, namely, why certain materials that have a specific crystal structure lend themselves readily to plastic deformation while others that have a different crystal structure are hard and brittle. We shall address this topic in the next section.

3.4 • Plasticity and Strength of Materials

The Role of Dislocations and Their Movement

We learned in Section 3.1 that the strength and the ductility of materials depend on the forces which hold the individual atoms together. This is, however, as we said before, only one part of the story. Indeed, one can estimate the strength of a material taking solely the binding forces between the atoms into consideration, and compare this finding with experimental results. It is then observed that for ductile, pure materials the theoretical force which appears to be necessary to rupture a piece of metal by breaking the binding forces between atoms [Figure 3.19(a)] would be between three and five orders of magnitude larger than those actually found in experiments. Moreover, the discrepancy between calculation and experiment varies for different crystal structures. Specifically, BCC materials are, as a rule, much stronger than FCC materials, whereas considerations based purely on atomic binding forces would predict approximately the same strengths.

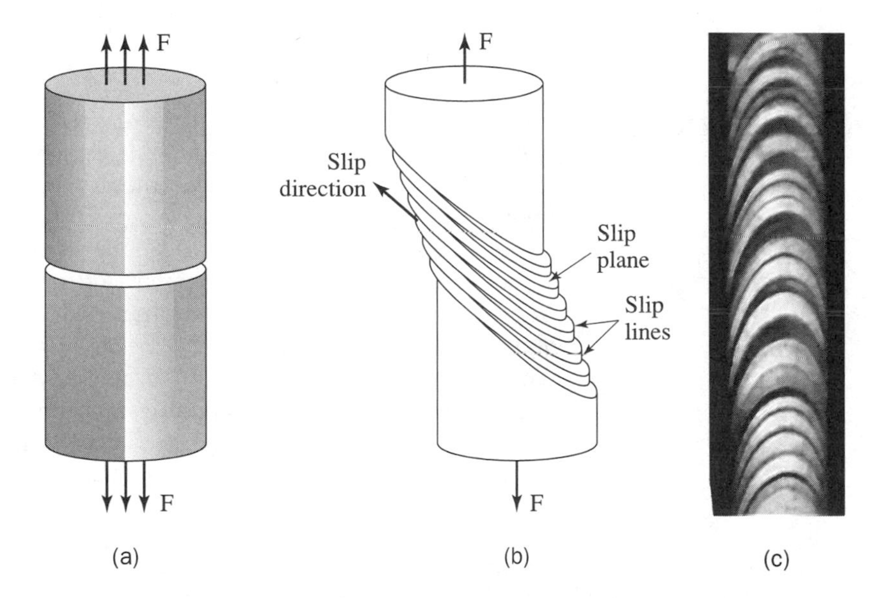

FIGURE 3.19. Schematic representation of two mechanisms by which a single crystal is assumed to be stretched when applying a force, *F*: (a) breaking interatomic bonds, (b) considering slip, and (c) photomicrograph of slip bands in a single crystal of zinc stressed at 300°C. Adapted from *Z. Phys.* **61,** 767 (1930).

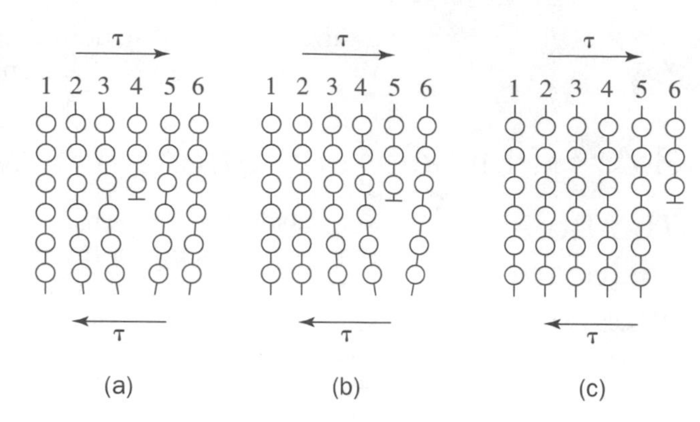

FIGURE 3.20. Simplified two-dimensional representation of an edge dislocation and of dislocation movement under the influence of a shear stress, τ, in a single crystalline, cubic-primitive lattice. A {100} plane is depicted. The shear stress is shown to be applied parallel to the slip plane and normal to the dislocation line (see Figure 3.21).

There exists, of course, an interrelationship between these binding forces and the plastic deformation of materials, but this interrelationship is rather complex. To better understand this, one has to know that crystalline materials deform predominantly by a process called **slip**, which occurs along specific lattice planes, called *slip planes*, as depicted for a single crystal rod in Figures 3.19(b) and (c). The experimentally observed shear stress which needs to be applied in order that slip commences (called the *critical resolved shear stress*, τ_0, see below) is more than three orders of magnitude smaller than that which one would expect from calculations.

Dislocations

The source for this surprising discrepancy was ascribed in 1934 (by Orowan, Polanyi, and Taylor) to certain imperfections in crystals, called **dislocations** and their preferred movement (slip) along the slip planes. This concept initially met with considerable skepticism. It was not until the late 1950s that this notion was finally accepted, in particular once dislocations were seen in the transmission electron microscope. (Precursors of the dislocation theory were developed by Masing, Polanyi, Prandtl, and Dehlinger in the late 1920s. Mathematical models treating the elastic behavior of homogeneous, isotropic media were promulgated even earlier by Volterra and Love but remained essentially unnoticed until the late 1930s.)

Edge Dislocations

Dislocations are *line-defects* in crystals which are created, for example, during solidification in the form of *networks*. We shall discuss here mainly dislocations with respect to plastic deformation. One particular type is the **edge dislocation**. Its geometry is described by an extra half-plane which is said to be inserted into an otherwise ideal lattice as shown in Figure 3.20(a). This

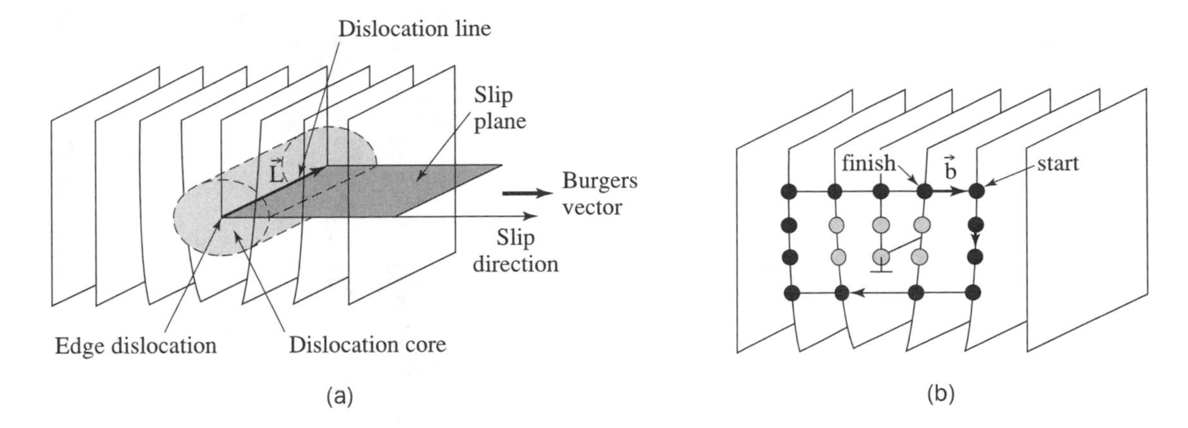

FIGURE 3.21. (a) Definition of slip direction, slip plane, Burgers vector, dislocation line, and dislocation core for an edge dislocation in a cubic primitive lattice [compare to Figure 3.19(b)]. Note that the edge dislocation slips in its slip plane, which is situated *between* two neighboring atomic planes. (b) Construction of a "right-hand finish to start Burgers vector."

causes a far-reaching elastic distortion of the lattice as well as a severe distortion in the vicinity of the end point of the inserted plane. The locus where the half-plane terminates is defined to be the *dislocation line* [Figure 3.21(a)]. It is depicted by the symbol ⊥, where the vertical line represents the extra half-plane and the horizontal line the *slip plane*; see Figure 3.21. (An edge dislocation involving an extra half-plane of atoms in the bottom portion of a crystal is designated by ⊤.)

The insertion of an extra half-plane is by no means limited to the atomic scale. Indeed, these types of deviations from regular structures can be found quite frequently in nature. Figure 3.22 presents, as an example, a corn cob which shows an extra half-row of kernels squeezed between an otherwise nearly regular pattern.

If a shear stress above the critical resolved shear stress is applied to a crystal as indicated in Figure 3.20, the distortion of the lattice in the vicinity of an edge dislocation allows the bonds between the involved atoms to break. This is shown, as an example, for plane 5 in Figures 3.20(a) and (b) whose lower part joins, after breaking, with half-plane number 4. As a consequence, plane 5 now contains the dislocation, as depicted in Figure 3.20(b). In other words, the dislocation has moved one atomic distance to the right. Repeated shifts of this kind lead to incremental movements of the dislocation and eventually to the formation of a step at the exterior surface of the material as shown in Figures 3.20(c) and 3.19(b) and (c). The material is now considered to be plastically deformed. The applied shear stress which is necessary to move a straight dislocation line in an otherwise undistorted crystal lattice is termed the *Peierls stress* (which is usually much smaller than the corresponding applied shear stress in the slip system; see below).

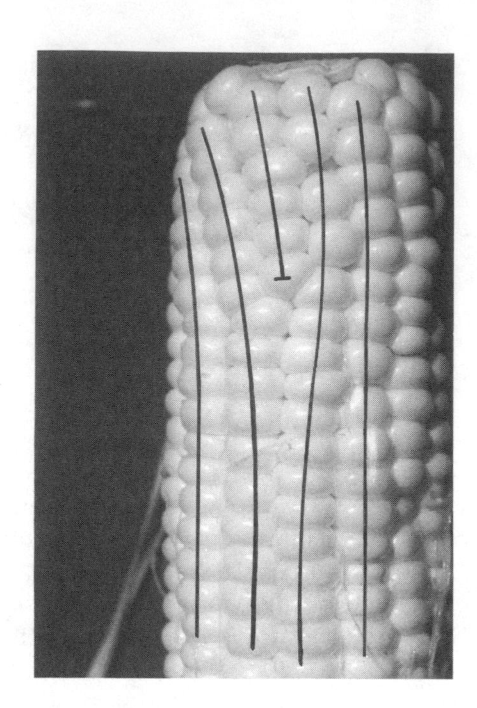

FIGURE 3.22. A corn cob featuring an "edge dislocation."

The direction in which the edge dislocation has moved is called the *slip direction*; see Figure 3.21(a). The slip direction and the dislocation line form the *slip plane*, as depicted in Figure 3.21(a). The slip plane is situated *between* two neighboring atomic planes.

In the immediate vicinity of the dislocation line the lattice is considerably distorted. This area is called the *dislocation core*; see Figure 3.21(a). It has a diameter of several atomic distances. Its size depends on the type of crystal lattice. The shape of the dislocation core influences the mobility of the dislocation line, that is, the Peierls stress.

Burgers Vector

The size and the direction of the main lattice distortion caused by a dislocation is expressed by the **Burgers vector**, **b**. It is always parallel to the slip direction, for example, as seen in Figure 3.21(a). In general, **b** is one of the shortest, periodic atomic distances in the lattice. For an edge dislocation, the Burgers vector is perpendicular to the dislocation line; see Figure 3.21(a). A dislocation is completely described when the Burgers vector and the orientation of the dislocation line (characterized by a *direction vector*, **L**) are known. To construct the Burgers vector, one follows a full clockwise circle around the dislocation line (assumed to be positive when **L** points into the paper plane, as indicated), counting the same number of atomic distances in all directions. The vector that closes the loop is the Burgers vector, specifically called in the present case to be a *right-hand, finish-to-start Burgers vector*; see Figure 3.21(b).

A Burgers vector can be described by the indices of its direction [Figure 3.13(d)] and the length of this vector. For example, in the above cubic primitive lattice, the Burgers vector is represented by ⟨100⟩. The close-packed direction in an FCC crystal is the face diagonal and the distance between the nearest atoms is one-half of this length. In other words, the Burgers vector in the direction of the face diagonal is written as **b** = 1/2 ⟨110⟩. Similarly, a Burgers vector in the direction of a space diagonal for the BCC lattice is **b** = 1/2 ⟨111⟩.

Slip in FCC Materials

The movement of dislocations caused by an external force preferentially occurs when the least amount of energy is consumed. This condition is generally fulfilled when the slip planes are close-packed, as, for example, the {111} planes in **FCC materials**. The slip directions which are energetically most preferred are the close-packed directions (⟨110⟩ in FCC) in which the distance between atoms (and thus the Burgers vector) are smallest. In this case, the mobility of the dislocations in FCC materials is quite large for a given amount of applied energy.

Slip System

The combination of a preferred slip plane {*hkl*} and a preferred slip direction ⟨*uvw*⟩ is called a **slip system** and is described by {*hkl*} ⟨*uvw*⟩. FCC materials possess four close-packed {111} planes and three close-packed ⟨110⟩ directions within each of these planes (see Section 3.3), which yields a total of 12 slip systems. They are collectively denoted by {111} ⟨110⟩. If a force is applied to an FCC metal, shear stresses in essentially all slip systems may occur. They have, however, different magnitudes. The system which is oriented in space so that it experiences the maximum shear stress becomes particularly active and plastic deformation begins in this system. With increasing stress other slip systems become additionally agile. Eventually, however, the moving dislocations in intersecting planes obstruct each other, and, thus, further slip is almost impossible. The material is now said to have been *work hardened* or *strain hardened*, as shown in Figure 2.7. Any further attempt to additionally deform the metal would then result in its failure. If more deformation is still required, the work piece first needs to be heated above 0.4 times the absolute melting temperature (called *annealing*) in order to remove the entangled dislocations (called *recovery*) or to create new grains (called *recrystallization*), which lowers the dislocation density. Subsequently, the work piece is ductile again and can be newly deformed.

In summary, the number of slip systems in FCC materials seems to be fairly optimal for the following reasons. On the one hand, there are enough possible slip systems so that at least one of them can always be activated under essentially any direction of applied stress. On the other hand, the number of slip systems is not excessively large so as to cause initially mutual blocking of adjacent dislocation movements. Since FCC metals have an optimum number of slip systems, they can be easily deformed and are therefore ductile.

HCP Materials

The situation is somewhat different in single crystalline **HCP materials**. Actually, HCP metals possess only three coplanar slip systems resulting from only one close-packed plane [i.e., the (0001) or basal plane] and three close-packed directions (the **a** directions; see Figure 3.14). Some hexagonal metals such as zinc, cadmium, or magnesium indeed slip in their basal planes, called *basal slip*. Their critical resolved shear stresses are comparable to those for FCC metals; see Figure 3.23. Thus, they are relatively ductile in

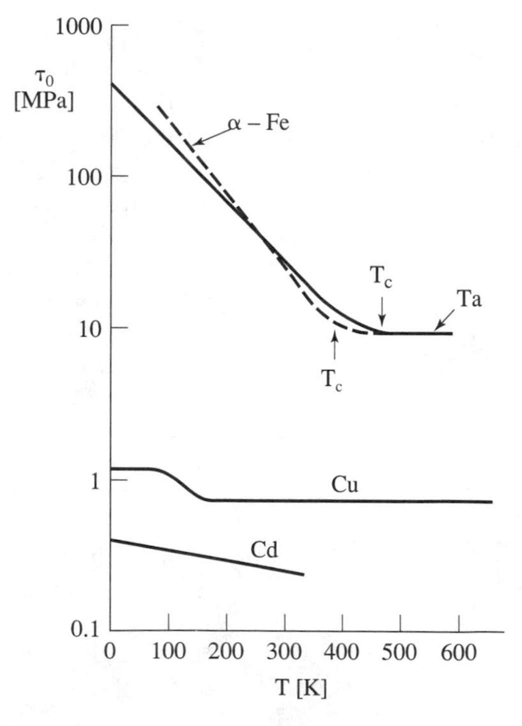

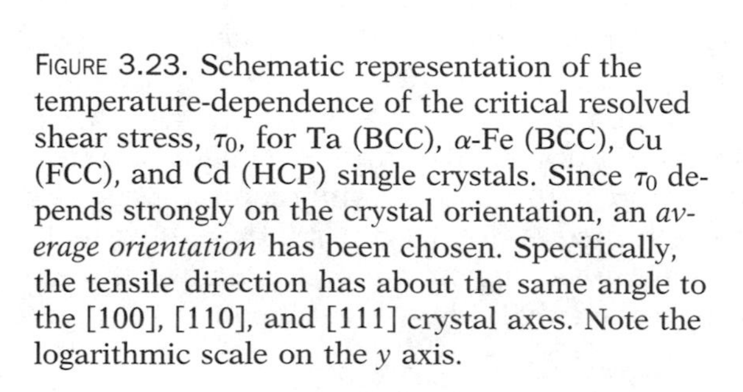
FIGURE 3.23. Schematic representation of the temperature-dependence of the critical resolved shear stress, τ_0, for Ta (BCC), α-Fe (BCC), Cu (FCC), and Cd (HCP) single crystals. Since τ_0 depends strongly on the crystal orientation, an *average orientation* has been chosen. Specifically, the tensile direction has about the same angle to the [100], [110], and [111] crystal axes. Note the logarithmic scale on the *y* axis.

TABLE 3.3. *c/a* ratios for some HCP metals

Cd	1.886
Zn	1.856
Mg	1.624
Zr	1.590
Ti	1.588
Be	1.586

their single crystalline form if proper crystal orientations for stressing are selected. There are, however, other hexagonal metals, such as titanium, zirconium, and beryllium, whose critical resolved shear stresses for basal slip at room temperature are more than two orders of magnitude higher than for Zn, Cd, or Mg. They have been found to preferentially slip instead in non–close-packed planes. Specifically, these metals slip on prism planes, i.e., $\{10\bar{1}0\}$ planes, and in the close-packed directions parallel to the **a** axes (see Figure 3.14) because these slip systems have for Ti, Zr, and Be the lowest critical resolved shear stress. The difference between these two groups of hexagonal metals is their *c/a* ratios (Section 3.3). For Zn, Mg, and Cd, they are mainly above the theoretical value of 1.633, whereas they are below 1.633 for Zr, Ti, and Be; see Table 3.3. In other words, hexagonal metals that have a small separation between basal planes display a high critical resolved shear stress for basal slip and are therefore quite hard and brittle (and vice versa).

The statements just promulgated are essentially true for single crystalline materials only. Polycrystalline Zn, Mg, or Cd instead have lower ductilities because some grains may be oriented in such a way that basal slip cannot occur and because of interactions between neighboring grains. Further, it should be noted in passing that the plasticity of HCP metals can be improved by mechanical *twinning* (Figure 3.12), which may be introduced by a properly applied shear stress.

BCC Materials

The plastic deformation of **BCC metals** differs considerably from that of FCC and HCP materials. This can be concluded from the temperature-dependence of the critical resolved shear stress, τ_0, for extremely pure single crystals. As depicted in Figure 3.23, τ_0 increases substantially toward lower temperature for BCC metals, whereas τ_0 for FCC and HCP metals rises only slightly. Interestingly enough, the critical resolved shear stress for BCC metals is fairly constant above a characteristic temperature, T_c.

The plastic deformation of *BCC metals* is often explained by pointing out that there are no close-packed planes in BCC materials. Slip has to occur, therefore, in "near close-packed" slip systems, and there are 48 of them available. (More precisely, the slip *directions* are close-packed but the slip *planes* are not.) As a consequence, not only one, but several of these slip systems may be activated simultaneously when applying a force in a chosen direction. (Examples for those slip systems are {110} ⟨111⟩ or {112} ⟨111⟩ or {321} ⟨111⟩.) This results quickly in the mutual blocking of the dislocation movements and therefore in less ductility. By the same token, the mutual interaction of dislocation motions leads to a strengthening of the material. BCC metals are therefore strong but still somewhat ductile.

This explanation was revised in the 1970s as a result of the just-mentioned detailed investigations involving the temperature-dependence of the critical resolved shear stress of extremely pure, single crystalline, BCC metals. A strong sensitivity to interstitially dissolved impurity atoms, the existence of different strengthening mechanisms at low and high temperatures, and an orientation dependency of τ_0 also have been found for BCC metals. Additionally, electron microscopy studies have been conducted which revealed that *screw dislocations* (see below), rather than edge dislocations (as in FCC and HCP metals), are predominantly involved when plastically deforming BCC metals. Even though this information has been available in the literature for about 20 years, it has not entered many general textbooks because of the complexity of the mechanisms involved. A few explanatory words may help to elucidate this issue.

Screw Dislocation

Up to this point, we only discussed the edge dislocation which we described as a straight line *perpendicular* to its Burgers vector. Generally, however, dislocation lines are curved [see Figure 3.24(b)]; that is, they may change their direction in the slip plane with respect to the Burgers vector. Thus, the case may occur where the dislocation line is parallel (or antiparallel) to its Burgers vector. The latter represents a screw dislocation [Figure 3.24(a)]. Specifically, a **screw dislocation** can be described by cutting a cylindrical crystal halfway along its length axis to its center. The opening is then displaced parallel to the axis by one atomic distance (i.e., by the Burgers vector) as depicted in Figure 3.24(a). This results in a spiral plane which winds in a helical form around an axis, that is, around the dislocation line. The screw dislocation was proposed in 1939 by J.M. Burgers. The involvement of screw dislocations is not restricted to BCC metals only.

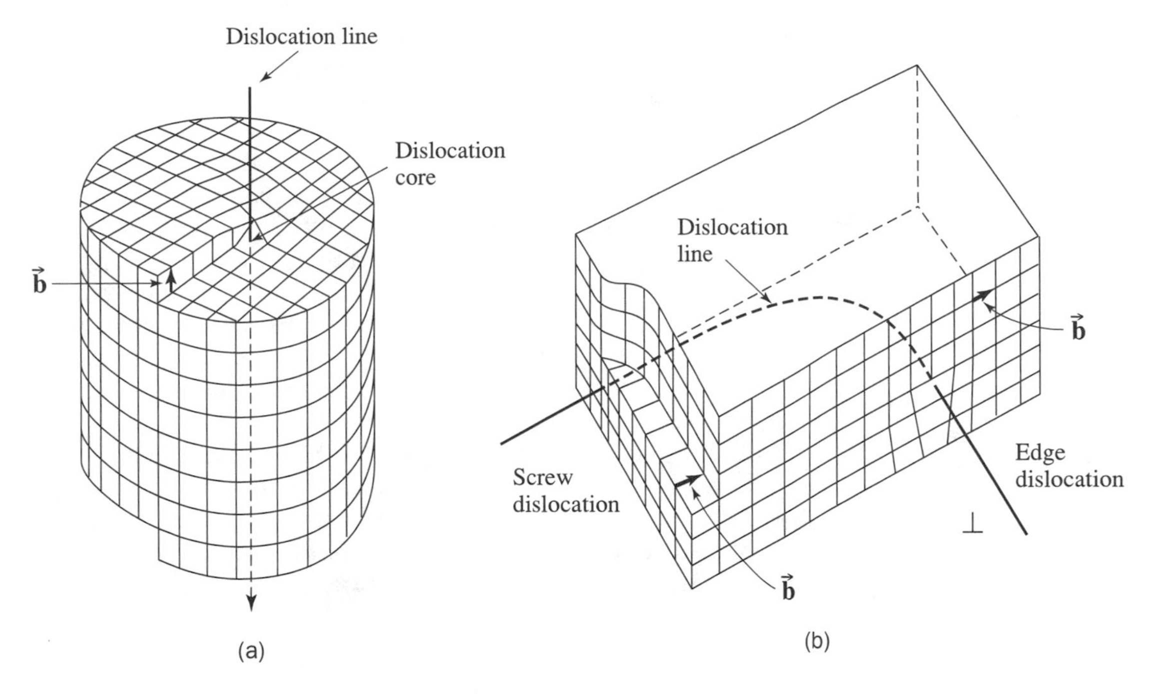

FIGURE 3.24. Schematic representations of (a) a screw dislocation and (b) mixed dislocations for which the edge and the screw dislocations are extreme cases.

Mixed Dislocations

In general, dislocations are thought to be composed of an edge part, and a screw part and are then called **mixed dislocations**. If the edge part is larger, the dislocation is termed *edge-like;* in the other case, it is called *screw-like*.

As just mentioned, plastic deformation of BCC metals below the characteristic temperature T_c is predominantly governed by the movement of screw dislocations, involving the 1/2 $\langle 111 \rangle$ Burgers vector. Screw dislocations have, however, no unique slip plane (as edge dislocations) since **b** and **L** are parallel to each other (Figure 3.24). As a result, a number of slip planes are possible. For BCC metals, the mobility of a screw dislocation is small and decreases even further at lower temperatures. Moreover, the shape of the core of a screw dislocation possesses a threefold symmetry in the $\langle 111 \rangle$ direction of its Burgers vector. This three-dimensionality prevents the dislocations from moving easily. However, when heat-induced thermal vibrations of the core atoms are involved, the dislocation movement is enhanced. In other words, the motion of screw dislocations in BCC metals is a thermally activated process that is similar to atomic diffusion (see Chapter 6). It should be added that screw dislocations may frequently change their glide planes during motion. This explains why a large number of slip systems are observed experimentally.

Refinements

Some refinements to the above-stated general rules should be mentioned at this point which are valid for all lattice types discussed so far. First, heating a solid may activate additional slip systems.

Second, additions of substitutional elements that have considerably larger or smaller atomic diameters than the host material distort the lattice. As a consequence, they may cause some obstacles for the movement of dislocations and thus some increase in yield stress (see Chapter 5).

Third, vacancies, interstitial atoms, other dislocations, boundaries between individually oriented grains, stacking faults, and other lattice defects may influence, in various degrees, the movement of dislocations, thus exercising an influence on strength and ductility. As an example, a portion of a moving dislocation line

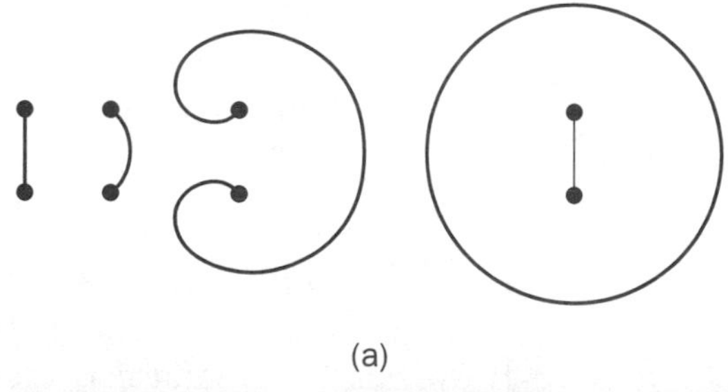

(a)

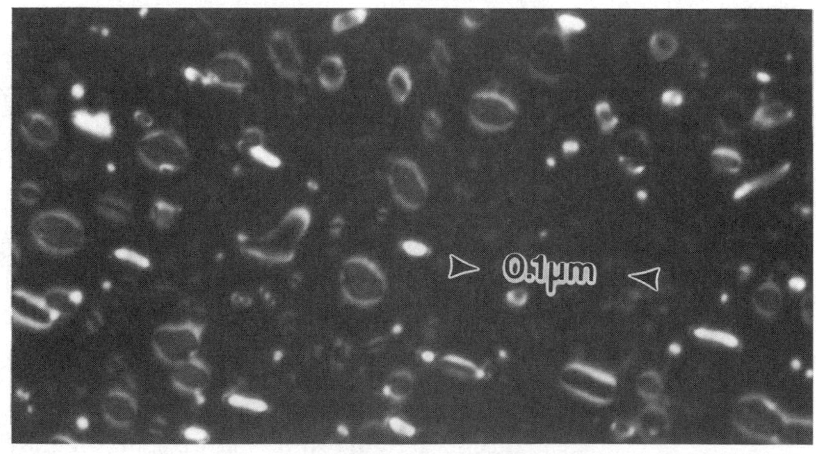

(b)

FIGURE 3.25. (a) Schematic representation of dislocation multiplication by pinning of dislocations at lattice defects, as proposed by Frank and Read. (b) Electron micrograph of dislocation loops. (Si^+ ion–implanted Si, annealed 800°C for 30 min.) Courtesy of Prof. K. Jones, University of Florida.

may be **pinned** at its ends in particular by intersecting dislocations. As a consequence, the center of the dislocation line bows out under continued stress until eventually a dislocation loop and a new pinned dislocation are formed, as depicted in Figure 3.25. A dislocation line, thus, behaves like a flexible violin string which can be bent by an applied force. The bending changes the angle between the line direction **L** and the Burgers vector **b**.

Frank–Read Source

A continuous *dislocation multiplication* mechanism as depicted in Figure 3.25(a) is called a **Frank–Read source** (after its inventors). The process may be repeated up to several hundred times until the source is blocked, i.e., until the loops cannot extend any further. Dislocation sources such as the Frank–Read source may increase the concentration of dislocations (typically 10^6 cm of dislocation line per cm^3) by six orders of magnitude, which leads to more mutual interference and thus to enhanced strength. As a rule of thumb the dislocation density, ρ, that is, the total length of all dislocation lines per unit volume (given in cm/cm^3), is related to the applied shear stress, τ, through

$$\tau = \alpha\, G\, b\, \sqrt{\rho}, \tag{3.4}$$

where α is a constant (≈ 1), G is the shear modulus [see Eq. (2.4)], and b is the magnitude of the Burgers vector. The dislocation multiplication mechanism just described explains why the experimentally observed plastic shear strain in an annealed crystal is much larger than calculated, assuming simple dislocation slip of a straight dislocation line.

Cross Slipping

Fourth, **cross-slipping** of screw dislocations may occur. This mechanism involves the shift (changeover) of a dislocation movement to an intersecting slip system once a dislocation has encountered an obstacle and thus would be blocked from further movement. Cross-slip is observed in FCC as well as in BCC materials. It is less frequent in HCP materials for which the slip planes are mostly parallel, that is, not intersecting. As an example, a screw dislocation in an FCC lattice having a Burgers vector $\mathbf{b} = 1/2\,[\bar{1}01]$ may slip either on a (111) or a $(1\bar{1}1)$ slip plane. If the primary slip system is $(111)\,[\bar{1}01]$, where most of the dislocations are active, then the $(1\bar{1}1)$ slip plane is called the *cross-slip plane* and $(1\bar{1}1)\,[\bar{1}01]$ the *cross-slip system*.

Fifth, dislocations cause in their vicinity far-reaching elastic **strain fields** whose influence gradually decreases inversely proportional to the distance from the dislocation line. These strain fields strongly interact with similar fields of neighboring dislocations, thus mutually influencing their slip behavior. In essence, the strain fields hinder the dislocation motion during plastic deformation, resulting in a need for higher applied stresses if continued deformation is wanted. In the immediate vicinity of a dislocation line, that is, within a few atomic distances of it, the just-mentioned inverse distance law is no longer valid. This area is the already mentioned *dislocation core* in which the lattice is considerably distorted, as shown in Figure 3.21. Its size is about $\leq 3b$, where b is the magnitude of the Burgers vector.

This complex behavior of dislocations makes it nearly impossible at this time to calculate, for example, the stress–strain curve by a simple physical model or to predict the plasticity of materials when certain experimental conditions, such as temperature, deformation speed, etc., are changed. Thus, empirical laws are still predominantly used when assessing plastic deformation.

Nevertheless, the experimentally observed three regions in stress–strain curves, as depicted in Figure 3.26 for appropriately oriented single crystalline FCC, HCP, as well as for BCC metals above T_c, can be qualitatively explained. In Stage I, slip occurs in a single slip system only; it is called the *easy glide* range. The strain hardening is small since essentially no interference from other slip planes takes place. The size of the easy glide region depends on the purity of a metal and the geometry of the sample. In Stage II (*linear hardening region*) multiple glide in intersecting slip systems occurs and the dislocation density grows rapidly with increasing strain. The crystal hardens continuously, and the slope in the stress–strain curve steepens and is nearly constant. The rate of strain hardening in Stage III eventually decreases due to a diminished dislocation multiplication and increased cross-slipping. It is called the *dynamic recovery* stage.

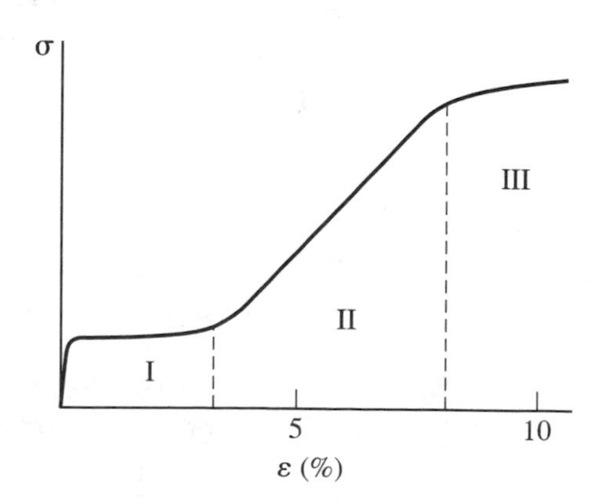

FIGURE 3.26. Schematic representation of a generalized stress–strain diagram for *single* crystals of FCC, HCP as well as for BCC metals above T_c (Figure 3.23). Compare with Figure 2.3 which depicts a stress–strain diagram of a *polycrystalline* material.

Polycrystals

So far, we have implied that the lattice is continuous throughout an entire piece of material, that is, we considered mainly *single crystals*. If a shear stress is applied to *polycrystals* (Figure 3.15), slip preferably commences in those grains in which the slip planes are tilted 45° to the load axis, as we shall see momentarily. However, slipping is somewhat impeded by the surrounding grains, which may have less favorable orientations. This results in an increase in the yield stress, σ_y, which is larger the smaller the grains are. The following relationship is often found:

$$\sigma_y = \sigma_0 + \frac{k}{\sqrt{d}} \tag{3.5}$$

(*Hall–Petch* relation), where σ_0 is the yield stress for very large grains, k is a constant, and d is the average grain size. Upon increasing the stress further, slip successively takes place in other grains that are less favorably oriented. In other words, yielding in polycrystals does not occur simultaneously in all grains at a given stress. This results in a constantly changing yield stress, σ_y, (rather than in a fixed one) as depicted in Figure 2.6(b).

One further point needs to be clarified. We said above that slip is caused by shear stress. In general, however, materials are tested in tension and not in shear for simplicity, and for better defined experimental conditions (see Chapter 2). It is thus desirable to relate the applied tensile stress to the shear stress resolved in the slip plane and in the slip direction. As a rule, the applied tensile force and the slip direction are not parallel or perpendicular to each other, but instead form an angle θ as depicted in Figure 3.27. Thus, only part of the applied tensile force becomes effective for shear. This partial force is called the *resolved shear force*:

$$F_r = F \cos \theta. \tag{3.6}$$

On the other hand, the area of the slip plane is

$$A = \frac{A_0}{\cos\phi} \tag{3.7}$$

(see Figure 3.27), where A_0 is the cross section of the single crystal rod and ϕ is the angle between the applied force and the normal to the slip plane. Knowing from Eq. (2.1) that

$$\sigma = \frac{F}{A_0}, \tag{3.8}$$

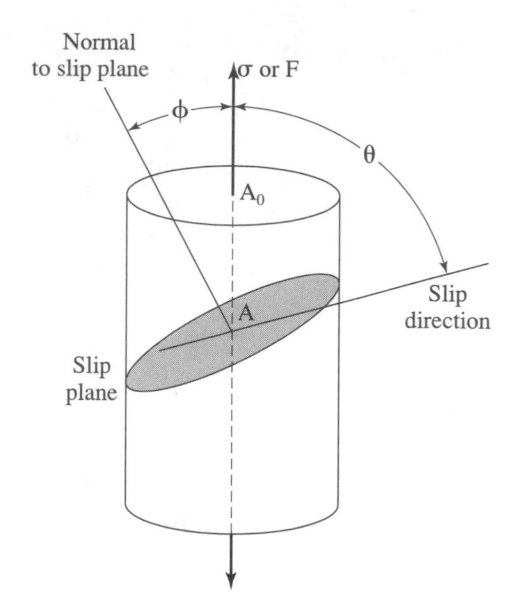

FIGURE 3.27. A slip plane normal is shown under an angle ϕ to an applied force (or stress). *Note:* $\phi + \theta$ is usually *not* equal to 90°. The two angles are not necessarily in the same plane.

and defining:

$$\tau_r = \frac{F_r}{A}, \tag{3.9}$$

and combining Eqs. (3.6)–(3.9) yields the shear stress, τ_r, resolved on the slip plane and in the slip direction, that is, the resolved shear stress:

$$\tau_r = \sigma \cos\phi \cos\theta. \tag{3.10}$$

Equation (3.10) teaches us that the largest resolved shear stress is obtained if both angles ϕ and θ are 45°, which results in $\tau_r = \sigma/2$. This means that the slip plane normal, the slip direction, and the load axis are all situated in the same plane. On the other hand, τ_r is zero for $\phi = 90°$ or $\theta = 90°$ (see Figure 3.27). The term $\cos\phi \cos\theta$ in Eq. (3.10) is sometimes called the *Schmid factor*.

Schmid's Law

The *critical resolved shear stress*, τ_0, which was introduced at the beginning of this section, is that stress which eventually causes an appreciable movement of dislocations and thus a measurable deformation of the crystal. It is a constant for a given slip system in a given crystal and varies over several orders of magnitude, as seen in Figure 3.23. τ_0 depends on the purity of a material, the crystal orientation (in the case of single crystals), and for BCC metals also heavily on temperature, as explained above in detail (Figure 3.23). Inserting σ_0 into Eq. (3.10) and rearranging it yields:

$$\sigma_0 = \frac{\tau_0}{\cos\phi \cos\theta}, \tag{3.11}$$

where σ_0 is the critical normal stress that is needed to cause plastic deformation (*Schmid's law*). For polycrystals the following approximate relationship between σ_0 for very large grains [Eq. (3.5)] and critical resolved shear stress is used:

$$\sigma_0 = 3\tau_0. \tag{3.12}$$

3.5 • Summary

Metals

It should be evident from the above description that the movement of dislocations is the key mechanism for understanding the interplay between strength, ductility, and atomic structure of materials. In other words, the ductility of pure metals depends on the number of slip systems and on whether or not these slip systems become active and eventually intersect. Much work has been expended by leading scientists to develop the above-mentioned concepts. Still, the complexity of the behavior of dislocations makes it nearly impossible at this time to predict the mechanical propertics of metals from a unified theory.

Ceramics

While dislocations are of prime importance for materials having metallic bonds, that is, for metals and alloys, one has to keep in mind that dislocations are also observed in ionic and covalent bound materials. Moreover, dislocations can even move to a small extent in ceramics. However, neither the slip directions nor the slip planes are close-packed for ionic crystals. As an example, the slip system in NiO [Figure 3.18(a)] is $\{110\}\ \langle 110 \rangle$.

The Burgers vector also needs some special consideration. Specifically, in the case where several species of atoms exist, the Burgers vector is the shortest distance between two *identical* atoms, rather than the distance between *different* species of atoms. In NiO, as in other compounds with an NaCl structure, this shortest distance is $1/2\ \langle 110 \rangle$ [see Figure 3.18(a)]. In essence, breaking the electrostatic attraction forces of oppositely charged ions or overcoming the electrostatic repulsion of identical ions makes dislocation movements in ionic crystals quite difficult. As a consequence, ceramics are hard and brittle.

Polymers

Dislocations also are found in the crystalline regions of polymers. Nevertheless, deformation in polymers largely involves the sliding of chains past each other or the uncoiling of entangled chains as outlined already in Section 3.3. The understanding of the plastic deformation of polymers will be deepened in Section 16.4.

Electronic Materials

Dislocations in silicon and other electronic materials rarely move (except at very high temperatures) because of the directionality of covalent bonds. Silicon is therefore brittle. However, dislocations play an important role in silicon wafers because of their ability to trap unwanted impurity elements. Silicon wafers are widely used by the electronics industry; see Section 11.4.

Clays

Finally, it should be briefly explained why clays or graphite are so extremely ductile without experiencing work hardening. As will be outlined in detail in Chapter 15, these materials are characterized by a layered structure whereby the individual layers (or platelets) are held together by weak van der Waals forces. As a consequence, they effortlessly slide past each other upon the application of a relatively small force. We will return to this subject in Chapter 15.

3.6 • Concluding Remarks

The question that was posed at the beginning of our discussion concerning the strength and brittleness of stone and the softness and ductility of copper can be answered now in just a few words.

Copper is characterized by moderately strong metallic bonds between its atoms. The negatively charged free electrons act as binding forces ("glue") between the positively charged atomic nuclei, which allows that the atoms can be shifted with reasonable ease under stress. Materials having metallic bonds therefore should be relatively soft and pliable. The second aspect which determines the mechanical properties of copper stems from the fact that copper is a face-centered cubic material that, because of its twelve slip systems, initially allows an almost unobstructed movement of dislocations under stress. As a consequence, copper can be deformed easily until eventually the dislocations mutually interlock, thus preventing their further movement. This causes hardening through deformation, called work hardening or strain hardening.

Stone, on the other hand—a ceramic material consisting of silicates and metal oxides—is characterized by a mixture of ionic and covalent atomic bonds, both of which are very strong. Further, dislocation movement is almost impossible because of the directionality of the covalent bonds or the tendency toward maintaining charge neutrality in the case of ionic bonds. All of this contributes to stone being hard and brittle.

Finally, the **elastic properties** which some materials may exhibit are caused by the stretching of metallic, ionic, or covalent bonds. In other words, during elastic deformation the atoms are slightly and temporarily displaced from their equilibrium positions. Since the metallic bond is relatively weak, metals and alloys have a higher elasticity compared to covalently or ionically bond materials, such as glass, ceramics, etc. When an applied stress that has induced elastic elongation is removed, the deformation is reversed and the recovery to the original shape occurs almost instantly (except in viscoelastic polymers, as we shall learn in Section 16.4).

Historically, copper was eventually replaced to a certain extent by bronze. We shall focus our attention on this material and on alloys in general in the chapters to come.

Problems

3.1. The set of {110} planes in an FCC unit cell passes through all corners of the cubic lattice and some face-centered atoms. The {110} set, however, misses the remaining face atoms. Write the Miller indices for a set of planes which is parallel to (110) but passes through all atoms of the FCC lattice.

3.2. List the Miller–Bravais indices of the six prism planes in a hexagonal lattice (show how you arrived at your result). What is the relationship to mutually opposite planes?

3.3. Show that the rhombohedral unit cell becomes an FCC or a BCC structure for specific angles α. Calculate the values of these angles. Hint: Draw the lattice vectors of the *primitive* (noncubic) unit cells of the FCC and BCC structures. The conventional (FCC, BCC) unit cells are *not* the smallest possible (primitive) unit cells.

3.4. Calculate the number of atoms for an HCP unit cell.

3.5. State the coordinates of the center atom in an HCP crystal structure (a_1, a_2, c) and show how one arrives at these values.

3.6. From the information given in Problem 3.5 above, show that the c/a ratio for the HCP structure is generally $\sqrt{8/3}$ if the atoms are assumed to be spherical. Compare this result with the experimental c/a ratios for Zn, Mg, and Zr.

3.7. State the slip plane and the slip direction in a hypothetical simple cubic lattice.

3.8. Show that a ⟨111⟩ direction lies in the (110) plane by sketching the appropriate direction and plane. Which one of the two (direction or plane) are close-packed assuming a BCC crystal structure? Discuss the implications for slip.

3.9. Determine (a) the coordination number, (b) the number of each ion per unit cell, and (c) the lattice constant (expressed in ionic radii) for the CsCl and the NaCl crystal structures.

3.10. Determine (a) the coordination number, (b) the number of each ion per unit cell, and (c) the lattice constant (expressed in ionic radii) for the zinc blende structure.

3.11. Calculate the packing factor for diamond cubic silicon.

3.12. Calculate the linear packing fraction of the [100] direction in (a) an FCC and (b) a BCC material assuming one atom per lattice point.

3.13. Calculate the planar packing fraction in the (111) plane in (a) the BCC and (b) the FCC structure.

3.14. Calculate the density of copper from its atomic mass. Compare your value with the density given in the Appendix. (The lattice parameter for Cu is $a = 3.6151$ Å.)

3.15. Sketch the $[1\bar{1}0]$ direction in the hexagonal crystal system. What is the four-digit index for this direction? Hint: The transformation equations between the 3-digit $[h'\ k'\ l']$ and the 4-digit $[h\ k\ i\ l]$ indices are:

$$h = \frac{1}{3}(2h' - k')$$

$$k = \frac{1}{3}(2k' - h')$$

$$i = -(h + k)$$

$$l = l'$$

3.16. Calculate the distance between two nearest face atoms of the conventional FCC unit cell.

Suggestions for Further Study

M.F. Ashby and D.R.H. Jones, *Engineering Materials I and II*, Pergamon, Oxford (1986).

G.E. Dieter, *Mechanical Metallurgy*, 3rd Edition, McGraw-Hill, New York (1986).

R.W. Hertzberg, *Deformation and Fracture Mechanics of Engineering Materials*, Wiley, New York (1976).

J.P. Hirth and J. Lothe, *Theory of Dislocations*, Wiley-Interscience, New York (1982).

E. Hornbogen and H. Warlimont, *Metallkunde*, 2nd Edition, Springer-Verlag, Berlin (1991).

D. Hull, *Introduction to Dislocations*, 3rd Edition, Pergamon, Elmsford, NY (1984).

Metals Handbook, American Society for Metals (1979).

D. Peckner (Editor), *The Strengthening of Metals*, Reinhold, New York (1964)

W.T. Read, Jr., *Dislocations in Crystals*, McGraw-Hill, New York (1953).

R.E. Reed-Hill and R. Abbaschian, *Physical Metallurgy Principles*, 3rd Edition, PWS-Kent, Boston (1992).

B. Sěsták and A. Seeger, *Glide and Work-Hardening in BCC Metals and Alloys*, I–III (in German), *Zeitschrift für Metallkunde*, **69** (1978), pp. 195ff, 355ff, and 425ff.

B. Sěsták, "Plasticity and Crystal Structure," in *Strength of Metals and Alloys*," Proc. 5th. Int. Conf. Aachen (Germany) August 1979, P. Haasen, V. Gerold, and G. Kostorz, Editors, Pergamon, Toronto (1980).

4

The Bronze Age

Chalcolithic man was clearly aware of the many useful features of copper that made it preferable to stone or organic materials for some specialized applications. Among these properties were its elasticity and particularly plasticity, which allowed sheets or chunks of copper to be given useful shapes. Chalcolithic man also exploited the fact that copper hardens during hammering, that is, as a result of plastic deformation. Last but not least, molten copper can be cast into molds to obtain more intricate shapes. On the negative side, surface oxidation and gases trapped during melting and casting which may form porosity were probably of some concern to Chalcolithic man. More importantly, however, cast copper is quite soft and thus could hardly be used for strong weapons or tools. Eventually, the time had come for a change through innovation. A new material had to be found. This material was *bronze*; see Fig. 4.1.

It is not known whether Chalcolithic man discovered by experimentation or by chance that certain metallic additions to copper considerably improved the hardness of the cast *alloy*. (An alloy is a combination of several metals.) In other words, cast bronze has a higher hardness than pure copper without necessitating subsequent hammering. Further, it had probably not escaped the attention of Chalcolithic man that the melting temperature of certain copper alloys is remarkably reduced compared to pure copper (by about 100°C as we know today) and that molten alloys flow more easily during casting.

Naturally, some impurities that were already present in the copper ore transferred into the solidified copper. Among them were arsenic, antimony, silver, lead, iron, bismuth, and occasionally even tin. These impurities, however, were not present in sufficient quantities that one could refer to the resulting product as an alloy. Small quantities of these impurities rarely change

FIGURE 4.1. A portion of a mural from the tomb of the Vizier Rekh-Mi-Re at Thebes (Egypt) depicting workers carrying a piece of copper in the shape of an ox hide (see Figure 1.3) and baskets containing probably tin and lead for bronze production. An overseer "supervises" the porters. Second millenium B.C. (Reprinted by permission from B. Scheel, *Egyptian Metalworking and Tools*, Shire Publications, Aylesbury, UK)

the properties of copper noticeably except in the case of bismuth, which even in small amounts makes copper brittle.

The first major and deliberate addition to copper seemed to have been arsenic (at least in the Middle East). Copper–arsenic ores were widely available in this region, but alloying also was achieved by mixing arsenic-containing ores with copper ores during smelting (see Plate 4.1). Artifacts from 3000 B.C. found in the Middle East contained up to 7% arsenic as a second constituent to copper. In this context, an archaeological find needs to be mentioned which was made in 1961. In the almost inaccessible "Cave of the Treasure," close to the cave where the Dead Sea Scrolls were unearthed, 429 objects were discovered, of which all but 13 were made of a copper–arsenic alloy. They must have been brought there around 3000 B.C. by refugees and may have belonged to a temple or shrine. The cache harbored 10 crowns, 240 elaborately decorated mace heads, chisels, and axes of different sizes and shapes.

Copper–arsenic alloys, however, were used only for a limited time. Eventually, it must have been realized that the fumes which were emitted by the arsenic during smelting killed some metalsmiths. Eventually, tin was found to be the ideal addition to copper which was alloyed in an optimal proportion of 10 mass–%. This copper-tin alloy is generally referred to as *bronze*. The melting temperature of Cu–10% Sn is about 950°C (compared to 1084°C for pure Cu). The melt flows freely into molds and no problems with gas bubbles, that is, porosity, are encountered. Most importantly, however, the alloy is hard immediately after casting and subsequent cooling but can be hardened further by hammering. Finally, copper–tin is harder and less brittle than copper–arsenic. See in this context Plate 4.2.

There are several intriguing questions that demand answers. One of them is concerned with the query about whether or not bronze was "invented" in only one region of the world (namely the Middle East, as many scholars used to believe) or independently at several places. The final word on this has not been spoken yet. However, recent archaeological evidence indicates that besides the Mediterranean area (considered by many westerners to be the "cradle of civilization"), independent bronze-producing centers existed in northern Thailand (Ban Chiang) during the third or fourth millennium B.C., and additionally in the isolation of China during the Shang dynasty starting at about 1400 B.C.

The mode of transition into the bronze age in the just-mentioned areas, which were considerably separated from each other, seemed not to have been identical. For example, Indo-China (which is said to have given to mankind a number of essential food plants such as rice, bananas, coconuts, yams, taro, and sugarcane) had a remarkable bronze production. The inhabitants of this area lived in light bamboo houses, made pottery, and domesticated pigs, chickens, and cows. In this region, bronze axes, spearheads, socket tools, bronze bracelets, clay crucibles, and sandstone molds have been found dating back as far as 3000 to 2300 B.C.

The most interesting find is that the Ban Chiang people seemed to have skipped copper production and arsenical bronze altogether and jumped immediately into the tin–bronze age. The raw materials for bronze were certainly available at essentially one and the same place (in contrast to the Near East, as we shall elucidate below). Indeed, rich alluvial[1] deposits of tin as well as copper ores are found from southern China down to Thailand and Indonesia. Another interesting observation was made by archaeologists who state that the Thai people seemed to have lived in a "peaceful bronze age" since no swords, battle axes, daggers, or mace heads have been found. Instead, bronze was mostly used for decoration and adornment. Its possession did not seem to express a status symbol since many children were buried with bronze bracelets.

In contrast to this, early Chinese bronze, made during the Shang dynasty (1600–1122 B.C.), was mainly utilized for ceremonial vessels, that is, for offering of food and wine to ancestral spirits (Plate 4.3). The bronze contained from 5 to 30% tin and between 3 and 5% lead (which makes the melt flow easier).

---

[1]Soil deposited by flowing water.

Bronze pieces from the Shang period are richly decorated by relief patterns depicting animals such as elephants, water buffalos, tigers, mythical dragons, and others. The Chinese were masters of a cast technology (Plate 4.3). They utilized fired clay molds in which the patterns were carved. No subsequent metalworking such as hammering, etc., was used. An example of their skills is a large cauldron on four legs which weighs 875 kg and whose body was cast in one piece. (It was unearthed near Anyang in 1939 and then utilized by villagers for storing pig food.)

Another astounding recent find is a set of bronze bells which were discovered in a tomb for the Marqui Yi, dated 433 B.C. (Fig. 4.2). They were shaped in a manner that, when hit with a hammer at the center, produced a lower pitch than on the edge. Modern scholars believe that the two sounds reflected the Chinese concept of a universe that is governed by two opposing, yet harmonious, forces called Yin and Yang (such as day and night, heaven and earth, sun and moon). The two-tone bells are said to have demonstrated how two forces can interact in harmony. Mu-

FIGURE 4.2. Chinese two-tone bell made of bronze. Eastern Zhou dynasty, 6th century B.C. Unearthed in 1978 at Sui Xian, Hubei Province. An entire set of those bells is called "Bian Zhong". Arthur M. Sackler Gallery, Smithsonian Institution, Washington, DC. Gift of Arthur M. Sackler S1987.285.

sic, which was important to the Chinese of that time, was a means to communicate with their forefathers, who gave them life and wealth. Music from bells and drums was not a form of entertainment but a ceremonial feast for and with their ancestors. It is still a mystery how the Chinese of the fifth century B.C. were able to cast two-tone bells. The bronze bells of the Marqui Yi weigh about 2500 kg, they are nicely decorated, and must have represented a large portion, if not all, of the wealth of their owner.

It is believed today that China developed its bronze technology only slightly later than the West and independently from the outside world. It grew, as in the West, out of the ceramic tradition. Chinese potters achieved kiln temperatures as high as 1400°C which allowed them to produce their unique porcelain. In contrast to Thailand, China briefly went through an initial copper smelting period (at around 2000 B.C.). But arsenic–bronzes seem to be absent in China. Furthermore, two types of bronzes were developed, one consisting of the usual copper–tin alloy and the other of copper with lead. Also interesting is a find of a piece of brass (copper–zinc), dating back to about 2200–2000 B.C., which was probably smelted from zinc-bearing copper ore.

As outlined above, both the Chinese and the Thais possessed ample copper as well as tin raw materials. This was definitely not true for the Chalcolithic man residing in the Near East. Major bronze-producing centers in 2000 B.C. were in Mesopotamia,[1] Assyria, Anatolia, and Cyprus.[2,3] All of these centers were blessed with abundant deposits of copper ore, as described in Chapter 1. However, no tin seemed to have been found in the vicinity of these places. As a matter of fact, the major tin deposits as known today are in China, Thailand, Malaysia, England, Germany, Nigeria, Zaire, Australia, Bolivia, and Mexico. They consist of tin oxide, or cassiterite (see Plate 4.1), which is inserted into granite in the form of veins. Cassiterite[4] is broken down by water and washed into rivers where it can be panned like gold. Minor deposits were possibly in Italy, Spain, France (at the mouth of the river *Loire*) and Sardinia.

In short, reputed archaeological evidence seems to point to the fact that, during the Chalcolithic time, no major known tin sources were situated in the Near East except possibly for some

---

[1]Mesopotamia once lay between the lower Tigris and lower Euphrates rivers and is today part of Iraq. (From Greek: *mesos* = middle and potomos = river.)
[2]*Kypros* (Greek) = copper.
[3]*Aes cyprium* (Latin) = copper; *aes* (Latin) = ore.
[4]*Kassiteros* (Greek) = tin.

native tin in the Zagros mountains on the eastern edge of the Mesopotamian plain. If they existed, however, they were quickly exhausted. Instead, documents have been found which record immense tin transporting caravans. Further, coastal ships could have carried tin from the Far East into the Mediterranean area. (An ancient shipwreck bearing tin has recently been found off the coast of Israel.) All taken, it is still a mystery today how the Bronze Age could have started already around 2000 B.C. in the Mediterranean area when no tin was readily available for experimentation or the accidental discovery of bronze. The transfer of bronze technology from the Far East or another area hitherto unknown to the West therefore should be considered to be a possibility.

One of the theories postulates that tin may have come from the southern slopes of the Caucasus (now Armenia) where both malachite (copper ore) and cassiterite (tin ore) are found. These minerals may well have been accidentally smelted together, thus yielding tin–bronze.

Other scholars believe that trade connections between the Middle East and Eastern Europe (where tin occurs in Bohemia, Saxonia, etc.) existed as early as 2500 B.C. This brings us to focus our attention on the Europeans, in particular on the Uneticians (named after a village near Prague) who by 1500 B.C. had become the dominant people in Europe. Their influence extended over a large territory, that is, from the Ukraine to the Rhine valley. The Uneticians were skillful bronzesmiths who manufactured in large quantities items such as pins (to hold garments together), tools (such as axes and plowshares), weapons, and jewelry. Actually, one item, a neck ring, was produced in such large quantities that it served as a kind of a currency, that is, it was exchanged for gold, furs, amber, and glass beads. These neck rings were, incidentally, quite similar in appearance to those found in Syria. The Uneticians traded not only with the south, but also with the British Islands, Scandinavia, and Ireland, where the fruits of their work have been found in many graves. They were ingenious inventors of new applications or copied items which they liked. Indeed, the safety pin was a product of the Uneticians. Archaeologists have found at Unetician sites such items as knitting needles, the remnants of an elaborate loom, and a strainer for the production of cheese. A large find in bronze-metal artifacts (axes, chisels, spears, knives, bracelets, pins, and a chain) was uncovered in a peat swamp located at the Federsee (a lake in southern Germany) which remarkably preserved a Bronze Age settlement.

Another area where bronze technology was practiced was located near the Indus river in ancient northwest India. There, a highly developed civilization, called the Harappan people, had settled between about 7000 and 1500 B.C. (until the Aryans invaded the land). Excavations at Mehrgarh (today's Pakistan) have demonstrated that the Harappans must have been skilled bronze workers as early as 2300 B.C., applying the lost wax casting technique, annealing, and riveting. They produced human figurines, vessels, arrowheads, spearheads, knives, and axes. The sickles which were unearthed suggest that bronze articles were utilized to support agriculture. Copper ores for these activities probably came from plains near the south end of the Indus river (Mohenjodaro) and from areas northwest of the Indus valley, which is today's Afghanistan, as evidenced by large heaps of copper slag. The copper ingots of this area had the form of a semicircle. As in the Near East case, the origin of tin is, however, not quite clear at present. Some scholars believe that it came from the Deccan Plateau in central and western India. Copper smelting was not confined to the Harappan urban area. Instead, many regional Indian cultures likewise smelted copper and bronze even though not always with the same sophistication.

A written document concerning bronze can be found in Greek mythology. Homer, in his *Iliad* (which is assumed to have been created between 800 and 700 B.C.), reports of Hephaestus, the Greek god of fire, who throws copper and tin along with silver and gold into his furnace, thus creating a superior shield for Achilles.

Finally, the early inhabitants of the central highlands of Peru before and during the Inca period engaged in some bronze technology which is believed to have started around A.D. 1450 or possibly even somewhat earlier. In contrast to the European and Asian customs, however, the maximal arsenic content of the goods (pins, chisels, axes) was only 1.5% and the tin concentration was 3% or less. Many artifacts found in this area did not contain alloy constituents in amounts that would significantly alter the mechanical properties. It is therefore questionable that the alloying was done intentionally. In any event, potential tin sources would have been close by, i.e., in northern Bolivia.

In summary, elaborate bronze technologies existed in various areas of the ancient world, not only in the Middle East, as sometimes assumed.

Now that we know from the above presentation that bronze is harder than copper and that bronze has a lower melting point than copper, we certainly should be eager to know which mechanisms govern these properties. We shall explain this in the chapters to come.

Suggestions for Further Study

W.T. Chase, *Ancient Chinese Bronze Art*, China House Gallery, China Institute in America, New York (1991).

B. Cunliffe (Editor), *The Oxford Illustrated Prehistory of Europe*, Oxford University Press, Oxford (1994).

O. Dickinson, *The Aegean Bronze Age*, Cambridge University Press, Cambridge, UK (1994).

W. Fong (Editor), *The Great Bronze Age of China*, Knopf, New York (1980).

C. Higham, *The Bronze Age of Southeast Asia*, Cambridge University Press, Cambridge, UK (1996).

N.G. Langmaid, *Bronze Age Metalwork in England and Wales*, Shire Archaeology Series, Shire Publications, Aylesbury, UK (1976).

J. Mellaart, *The Chalcolithic and Early Bronze Ages in the Near East and Anatolia*, Khayats, Beirut (1966).

5

Alloys and Compounds

Pure materials have a number of inherent mechanical properties, as discussed in Chapter 3. These features, such as strength or ductility, can be altered only to a limited degree, for example, by work hardening. In contrast to this, the properties of materials can be varied significantly if one combines several elements, that is, by alloying. In this chapter, we shall unfold the multiplicity of the mechanical properties of *alloys* and *compounds* with particular emphasis on the mechanisms which are involved. Specifically, we shall discuss a number of techniques which increase the strength of materials. Among them are solid solution strengthening, precipitation hardening (age hardening), dispersion strengthening, and grain size strengthening. In order to understand these mechanisms, we need to study the fundamentals of phase diagrams.

5.1 • Solid Solution Strengthening

When certain second constituents such as tin or nickel are added to copper, the resulting alloy has a noticeably larger yield strength than pure copper, as depicted in Figure 5.1. The added atoms (called *solute* atoms) may substitute up to a certain limit regular lattice atoms (called *solvent* or *matrix*). The resulting mixture is then said to be a *substitutional solid solution*. In many cases, the size of the solvent atoms is different from the size of the solute atoms. As a consequence, the lattice around the added atoms is distorted, as shown in Figure 5.2. The movement of dislocations upon application of a shear stress is then eventually obstructed and the alloy becomes stronger but less ductile. (Only in copper–zinc alloys do both strength *and* ductility increase simultaneously.) This mechanism is called *solid solution strengthening*. Solid solution strengthening is greater (up to a certain limit) the more solute atoms are added (see Figure 5.1). Specifically, the yield strength of an alloy increases parabolically with the solute

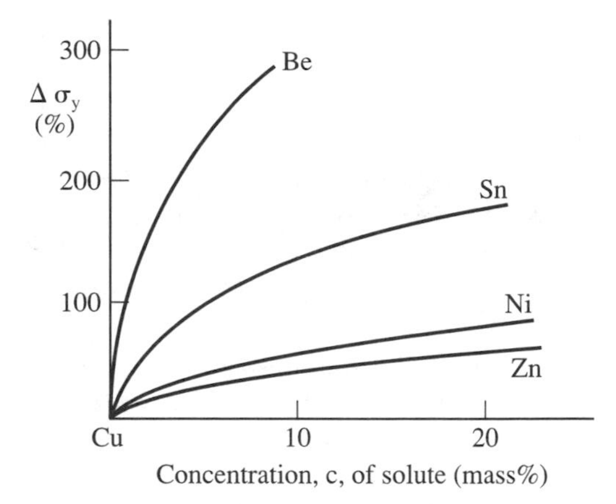

FIGURE 5.1. Change in yield strength due to adding various elements to copper. The yield strength, σ_y, increases parabolically with the solute concentration.

concentration, c, that is, as $c^{1/2}$. Further, solid solution strengthening is larger the greater the size difference between the solute and solvent atoms. Finally, the ability to lose strength at high temperatures by creep (Chapter 6) is less of a problem in solid solution strengthened alloys compared to their constituents. Many alloys, such as *bronze* or brass, receive their added strength through this mechanism.

5.2 • Phase Diagrams

At this point, we need to digress somewhat from our main theme (namely, the description of strengthening mechanisms through alloying). In order to better understand the mechanical (and other) properties of alloys or compounds, materials scientists frequently make use of diagrams in which the proportions of the involved constituents are plotted versus the temperature. The usefulness of these *phase diagrams* for the understanding of strengthening mechanisms will become evident during our discussions. (A *phase* is defined to be a substance for which the structure, composition, and properties are uniform.)

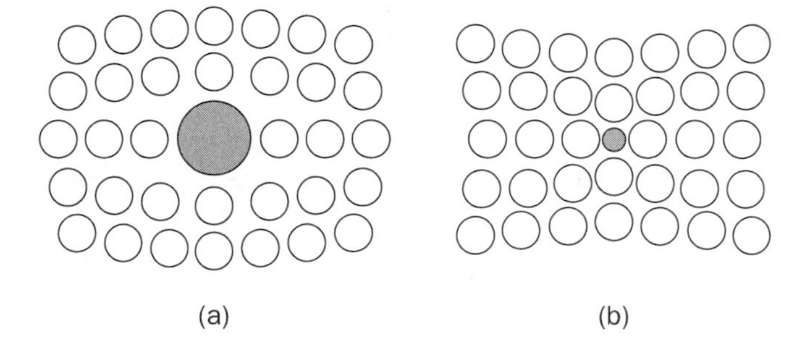

FIGURE 5.2. Distortion of a lattice by inserting (a) a larger (such as Sn) and (b) a smaller (such as Be) substitutional atom into a Cu matrix. The distortions are exaggerated.

5.2.1 Isomorphous Phase Diagram

A particular phase diagram in which only two elements (such as copper and tin) or two compounds (such as MgO and NiO) are involved is called a *binary phase diagram*. If complete solute solubility between the constituents of a compound or an alloy is encountered, the term *isomorphous binary phase diagram* is utilized. Figure 5.3 depicts the Cu–Ni isomorphous binary phase diagram as an example. Referring to Figure 5.3, we note that at temperatures above 1455°C (i.e., the melting point of nickel), any proportion of a Cu–Ni mixture is liquid. On the other hand, at temperatures below 1085°C (the melting point of copper), alloys of all concentrations should be solid. This solid solution of nickel in copper is designated as *α-phase*.

In the cigar-shaped region between the liquid and the *α*-solid-solution areas, both liquid and solid copper–nickel coexist (like coffee and solid sugar may coexist in a cup). This area is appropriately termed a *two-phase region*. The upper boundary of the two-phase region is called the *liquidus line*, whereas the lower boundary is referred to as the *solidus line*. Within the two-phase region, the freedom to change the involved parameters is quite limited. The degree of freedom, F, can be calculated from the **Gibbs phase rule**:

$$F = C - P + 1, \tag{5.1}$$

where C is the number of components and P is the number of phases.[1] In the present case, there are two components (Cu and Ni) and two phases (α and liquid), which leaves only one degree

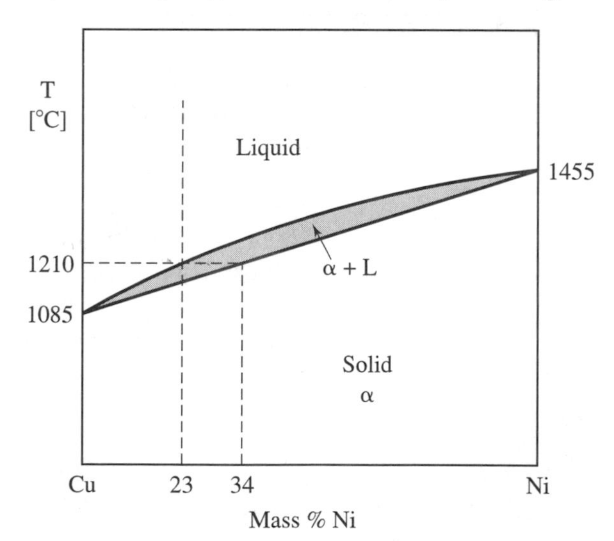

FIGURE 5.3. Copper–nickel isomorphous binary phase diagram. The composition here is given in mass percent (formerly called weight percent) in contrast to atomic percent. This section uses exclusively mass percent. For simplicity, the latter is generally designated as %.

[1]In (5.1) the pressure is assumed to be constant as it is considered in virtually all of the cases in this chapter. If the pressure is an additional variable, then the Gibbs phase rule has to be modified to read

$$F = C - P + 2. \tag{5.1a}$$

of freedom for varying any parameter. In other words, if we choose the temperature, then the composition of the α and the liquid phases is determined. As an example, if we specify the temperature to be 1210°C, then the composition of the liquid is about Cu–23% Ni (or more precisely, 23 mass percent) and the composition of the α-solid-solution is about Cu–34% Ni, as can be inferred from Figure 5.3. The horizontal line in the two-phase region which connects the just-mentioned compositions at a given temperature is called the *tie line* (see also Figure 5.4).

The question then arises, how much of each phase is present in the case at hand? For this we make use of the **lever rule**. Let us assume a starting composition of 31% Ni and again a temperature of 1210°C, as shown in Figure 5.4. A fulcrum (▲) is thought to be positioned at the intersection between the tie line (considered here to represent the lever) and the chosen composition (31% Ni). Similarly as for a mechanical lever, the amount of a given phase is proportional to the length of the lever at the *opposite* side of the fulcrum. Accordingly, by inspecting Figure 5.4 we conclude that 3/11 or 27% of this alloy at 1210°C is liquid ((34–31)/(34–23)) and 8/11 or 73% is solid ((31–23)/(34–23)).

The composition of a solid (and the corresponding liquid) changes continuously upon cooling, as demonstrated in Figure 5.5. At the start of the solidification process of, say, a 28% Ni alloy, the solid (as little as there may be) contains about 40% Ni. At this instance, the alloy is still almost entirely in the liquid state (see Tie Line Number 1 in Figure 5.5). As the temperature is lowered (by cooling the alloy, for example, in a mold), the nickel concentration of the α-solid-solution decreases steadily and the *amount* of solid increases; see Tie Line 2 as an example. Eventually, upon further cooling, the entire alloy is in the solid state and the α-solid-solution now has the intended composition of 28% Ni (Tie Line 3).

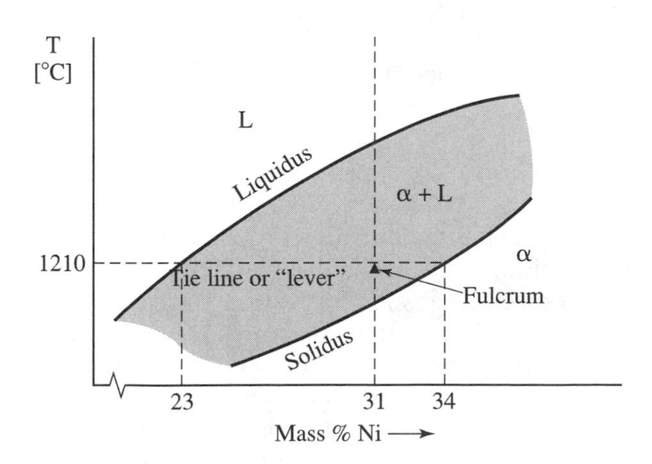

FIGURE 5.4. Part of the Cu–Ni phase diagram to demonstrate the lever rule. The section of the lever between the fulcrum and the solidus line represents the amount of liquid. (The analogue is true for the other part of the lever.)

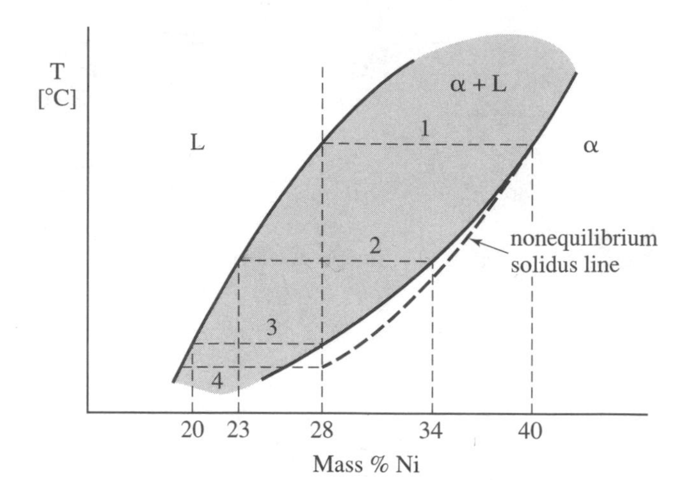

FIGURE 5.5. Equilibrium (solid lines) and nonequilibrium (dashed line) solidus curve in a part of the Cu–Ni binary phase diagram.

The just-described mechanism tacitly implies that the composition of the solid changes instantly upon cooling, as depicted in Figure 5.5. This occurs, however, only when the Cu and Ni atoms are capable of freely exchanging their lattice positions until eventually the respective equilibrium state has been reached. The mechanism by which atoms change their position is called *diffusion*. Diffusion occurs preferentially once the temperature is high and when the difference in concentration between two parts of an alloy is large. (We shall deal with diffusion in more detail in Chapter 6.) The equilibrium concentration of the alloy, as depicted in Figure 5.5, is only achieved upon extremely slow cooling. In most practical cases, however, the composition of an alloy during cooling is governed by a *nonequilibrium solidus line*, an example of which is depicted in Figure 5.5 by a dashed curve. Actually, this nonequilibrium solidus line is different for each cooling rate. Specifically, a faster cooling rate causes an increased deviation between equilibrium and nonequilibrium solidus lines.

There are practical consequences to the nonequilibrium cooling process just described. Referring back to Figure 5.5, it is evident that a Cu–28% Ni alloy, upon conventional cooling in a mold, has a composition of 40% Ni wherever the solidification has started. Next to this region, one observes layers having successively lower Ni concentrations which solidified during later cooling (comparable to the skins of an onion). This mechanism is referred to as **segregation** or **coring**. In specific cases, solidification and segregation occur in the form of tree-shaped microstructures, as depicted in Figure 5.6. This process may commence on the walls of a mold or on crystallization nuclei such as on small particles. Because of its characteristic appearance, the microstructure is referred to as **dendritic** and the respective mechanisms are called *dendritic growth* and *interdendritic segregation*. In general terms, the center of the dendrites is rich in that element which melts at high temperatures (in the present case, Ni), whereas the regions between the dendrites contain less of this element. The resulting mechanical properties have been found to be inferior to those of a homogeneous alloy. Further, the nonequilibrium alloy may have a lower melting point than that in the equilibrium state, as demonstrated by Tie Line 4 in Figure 5.5. This phenomenon is called **hot shortness** and may cause partial melting (between the dendrites) when heating the alloy slightly below the equilibrium solidus line.

It is possible to eliminate the inhomogeneities in solid solutions. One method is to heat the alloy for many hours below the solidus line. This process is called **homogenization** *heat treatment*. Other procedures involve rolling the segregated solid at

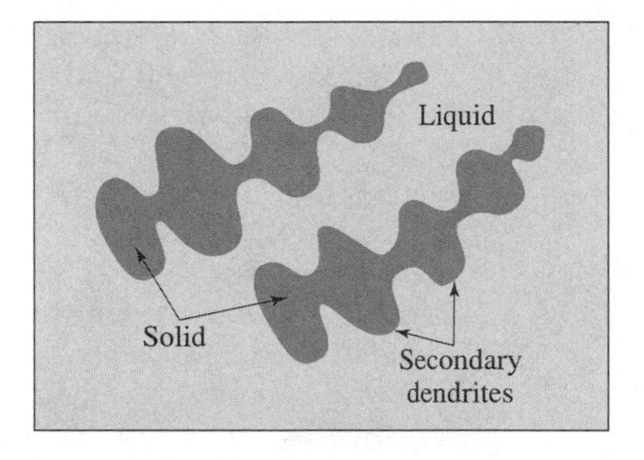

(a)

(b)

FIGURE 5.6. Microstructure of an alloy revealing dendrites: (a) schematic, (b) photomicrograph of a nickel-based superalloy.

high temperatures (called **hot working**) or alternately *rapid solidification*, which entails a quick quench of the liquid alloy to temperatures below the solidus line.

Only very few binary phase diagrams are isomorphous as just described. Indeed, unlimited solid solubility is, according to **Hume–Rothery**, only possible if the atomic radii of the constituents do not vary more than 15%, if the components have the same crystal structure and the same valence, and if the atoms have about the same electronegativity. Other restrictions may apply as well. Actually, most phase diagrams instead consist of one or more of the following six basic types known as *eutectic*, *eutectoid*, *peritectic*, *peritectoid*, *monotectic*, and *monotectoid*. They are distinguished by involving reactions between three individual phases. This will be explained on the following pages.

Complete solute solubility as discussed in this section is not restricted to selected metals only. Indeed, isomorphous phase diagrams also can be found for a few ceramic compounds, such as for NiO–MgO, or for FeO–MgO, as well as for the orthosilicates Mg_2SiO_4–Fe_2SiO_4, in which the Mg^{2+} and the Fe^{2+} ions replace one another completely in the silicate structure.

5.2.2 Eutectic Phase Diagram

Some elements dissolve only to a small extent in another element. In other words, a solubility limit may be reached at a certain solute concentration. This can be compared to a mixture of sugar and coffee: One spoonful of sugar may be dissolved readily in coffee whereas, by adding more, some of the sugar eventually remains undissolved at the bottom of the cup. Moreover, hot coffee dissolves more sugar than cold coffee; that is, the solubility limit (called **solvus line** in a phase diagram) is often temperature-dependent.

Let us inspect, for example, the copper–silver phase diagram which is depicted in Figure 5.7. When adding small amounts of copper to silver, a solid solution, called α-phase, is encountered as described before. However, the solubility of copper into silver is restricted. The highest amount of Cu that can be dissolved in Ag is only 8.8%. This occurs at 780°C. At any other temperature, the solubility of Cu in Ag is less. For example, the solubility of Cu in Ag at 400°C is only 1.2%.

A similar behavior is observed when adding silver to copper. The solubility limit at 780°C is reached, in this case, for 8% Ag in Cu. Moreover, the solubility at 200°C and lower temperatures is essentially nil. This second substitutional solid solution is arbitrarily called the β-phase.

In the region between the two solvus lines, a mixture of two solid phases exists. This two-phase area is called the *$\alpha + \beta$ region*.

To restate the facts for clarity: The α-phase is a substitutional solid solution of Cu in Ag comparable to a complete solution of

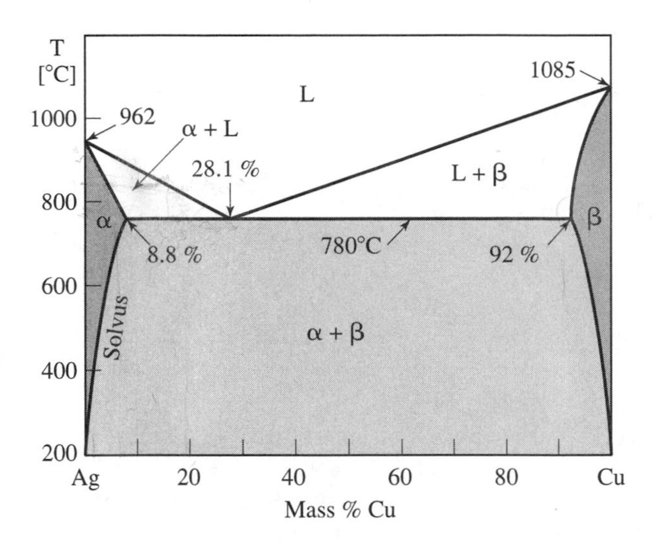

FIGURE 5.7. Binary copper–silver phase diagram containing a eutectic transformation.

sugar in coffee. Consequently, only *one* phase (sweet coffee) is present. (The analogue is true for the β-phase.) In the $\alpha + \beta$ region, *two phases* are present, comparable to a mixture of blue and red marbles. The implications of this mixture of two phases to the strength of materials will be discussed later.

We consider now a silver alloy containing 28.1% copper called the *eutectic composition* (from Greek *eutektos*, "easy melting"). We notice that this alloy solidifies at a lower temperature (called the *eutectic temperature*) than either of its constituents. (This phenomenon is exploited for many technical applications, such as for solder made of lead and tin or for glass-making.) Upon slow cooling from above to below the eutectic temperature, two solid phases (the α- and the β-phases) form simultaneously from the liquid phase according to the three-phase reaction equation:

$$L_{28.1\%\ \mathrm{Cu}} \rightarrow \alpha_{8.8\%\ \mathrm{Cu}} + \beta_{92\%\ \mathrm{Cu}}. \tag{5.2}$$

This implies that, for this specific condition, three phases (one liquid and two solid) are in equilibrium. The phase rule, $F = C - P + 1$ [Eq. (5.1)] teaches us that, for the present case, no degree of freedom is left. In other words, the composition as well as the temperature of the transformation are fixed as specified above. The *eutectic point* is said to be an **invariant point**. The alloy therefore remains at the eutectic temperature for some time until the energy difference between solid and liquid (called the **latent heat of fusion**, ΔH_f) has escaped to the environment. This results in a *cooling curve* which displays a **thermal arrest** (or plateau) quite similar to that of pure metals where, likewise, no degree of freedom remains during the coexistence of solid and liquid. A schematic cooling curve for a eutectic alloy is depicted in Figure 5.8.

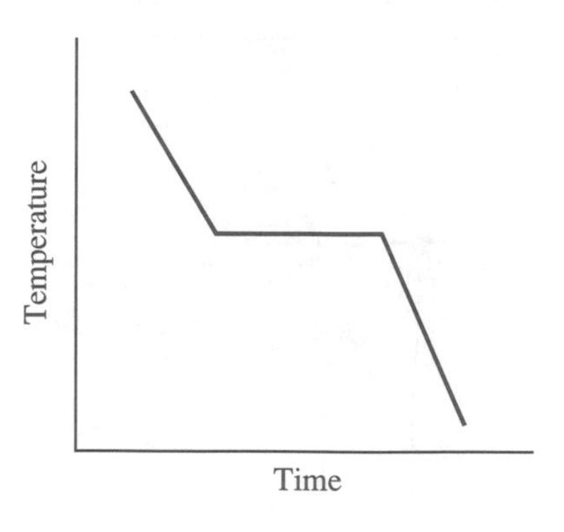

FIGURE 5.8. Schematic representation of a cooling curve for a eutectic alloy (or for a pure metal). The curve is experimentally obtained by inserting a thermometer (or a thermocouple) into the liquid alloy and reading the temperature in periodic time intervals as the alloy cools.

The microstructure, observed by inspecting a eutectic alloy in an optical microscope, reveals a characteristic *platelike* or *lamellar* appearance; see Figure 5.9. Thin α and β layers (several micrometers in thickness) alternate. They are called the *eutectic microconstituent*. (A microconstituent is a phase or a mixture of phases having characteristic features under the microscope.) This configuration allows easy interdiffusion of the silver and the copper atoms during solidification or during further cooling.

Alloys which contain less solute than the eutectic composition are called **hypoeutectic** (from Greek, "below"). Let us assume a Ag–20% Cu alloy which is slowly cooled from the liquid state. Upon crossing the liquidus line, initially two phases (α and liquid) are present, similar as in an isomorphous alloy. Thus, the same considerations apply, such as a successive change in composition during

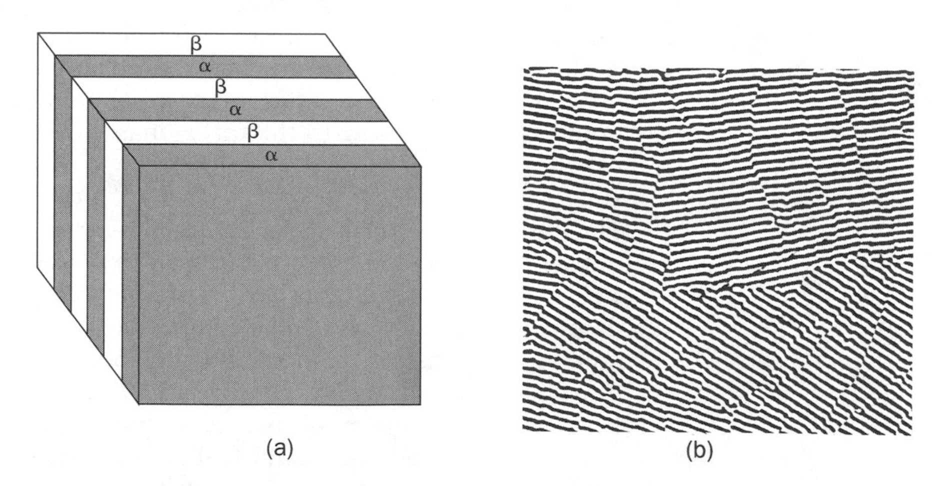

FIGURE 5.9. (a) Schematic representation of a lamellar or platelike microstructure as typically observed in eutectic alloys. (b) Photomicrograph of a eutectic alloy, 180× ($CuAl_2$–Al). Reprinted with permission from *Metals Handbook, 8th Edition*, Vol. 8 (1973), ASM International, Materials Park, OH, Figure 3104, p. 156.

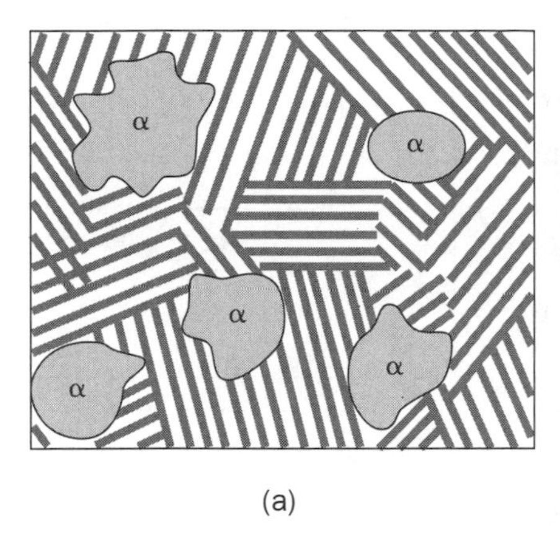

(a)

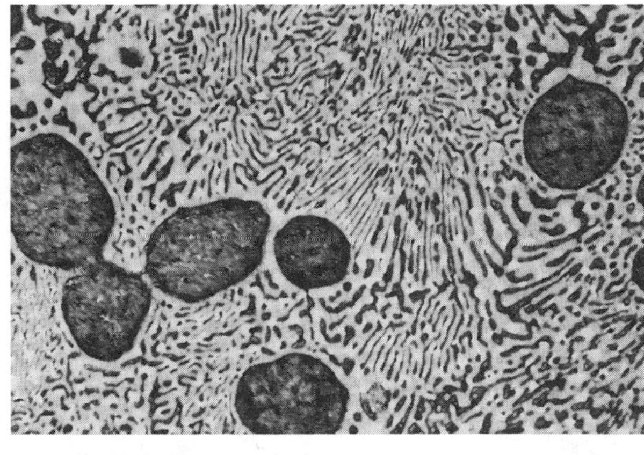

(b)

FIGURE 5.10. (a) Schematic representation of a microstructure of a hypoeutectic alloy revealing primary α particles in a lamellar mixture of α and β microconstituents. (b) Microstructure of 50/50 Pb-Sn as slowly solidified. Dark dendritic grains of lead-rich solid solution in a matrix of lamellar eutectic consisting of tin-rich solid solution (white) and lead-rich solid solution (dark) 400×, etched in 1 part acetic acid, 1 part HNO_3, and 8 parts glycerol. Reprinted with permission from Metals Handbook, 8th Ed. Vol 7, page 302, Figure 2508, ASM International, Materials Park, OH (1972).

cooling, dendritic growth, and the lever rule. When the eutectic temperature (780°C) has been reached, the remaining liquid transforms eutectically into α- and β-phases. Thus, the microstructure, as observed in an optical microscope, should reveal the initially formed α-solid-solution (called *primary* α, or *proeutectic constituent*) interspersed with lamellar eutectic. Indeed, the micrographs depicted in Figure 5.10 contain gray, oval-shaped α areas as well as alternating black (α) and white (β) plates in between. A schematic cooling curve for a Ag–20% Cu alloy which reflects all of the features just discussed is shown in Figure 5.11(a). For comparison, the cooling curve for an isomorphous alloy is depicted in Figure 5.11(b).

Silver alloys containing less than 8.8% Cu solidify similar to an isomorphous solid solution. In other words, they do not contain any eutectic lamellas. However, when cooled below the solvus line, the β-phase precipitates and a mixture of α- and β-phases is formed, as described previously in Section 5.2.1.

Hypereutectic alloys (from Greek, "above") containing, in the present example, between 28.1 and 92% Cu in silver, behave quite analogous to the hypoeutectic alloys involving a mixture of primary β-phase (appearing dark in a photomicrograph), plus plate-shaped eutectic microconstituents.

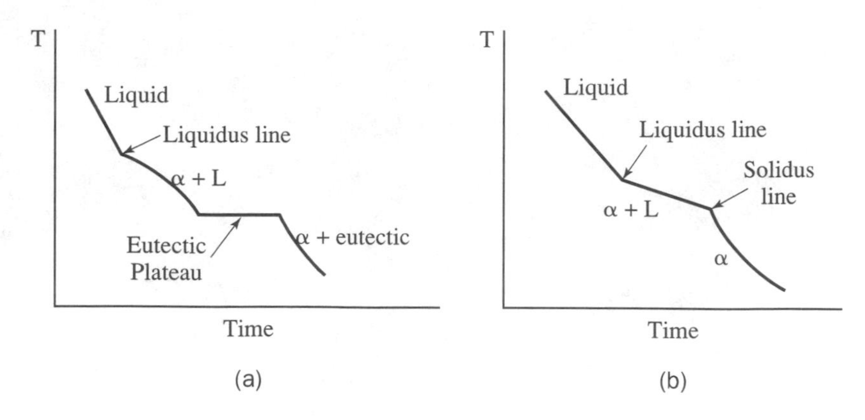

FIGURE 5.11. (a) Schematic representation of the cooling curve for a hypoeutectic alloy revealing the eutectic plateau (see Figure 5.8) and other characteristic landmarks as indicated. (b) Cooling curve for an isomorphous alloy (see Figure 5.3).

5.2.3 Eutectoid Transformation

The eutectoid reaction involves a transformation between three *solid* phases. Specifically, at the eutectoid point, a solid phase (say γ) decomposes into two other solid phases upon cooling, according to the reaction equation:

$$\gamma \rightarrow \alpha + \beta, \tag{5.3}$$

and as schematically depicted in Figure 5.12. As for the eutectic reaction described in Section 5.2.2, no degree of freedom is available at the eutectoid point, which requires that the transformation must take place at a fixed temperature and a fixed composition.

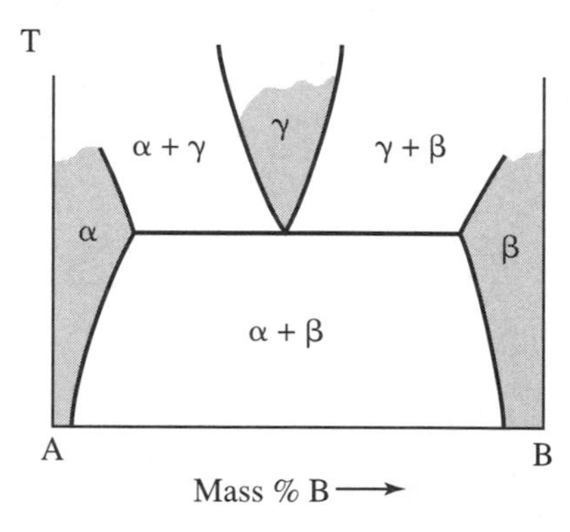

FIGURE 5.12. Schematic representation of a eutectoid three-phase reaction in a hypothetical binary phase diagram consisting of elements A and B. The high-temperature region is not shown for clarity and emphasis.

5.2.4 Peritectic and Peritectoid Transformation

A process that involves the reaction between a solid and a liquid from which eventually a new solid phase emerges is called *peritectic* (from Greek *peritekto*, "melting the environment"). The equation is, for example:

$$\alpha + L \rightarrow \beta, \tag{5.4}$$

as schematically depicted in Figure 5.13. When a peritectic A–4% B alloy is cooled from the liquid, the solidification starts by forming an α-phase containing only about 1% B. During further cool-

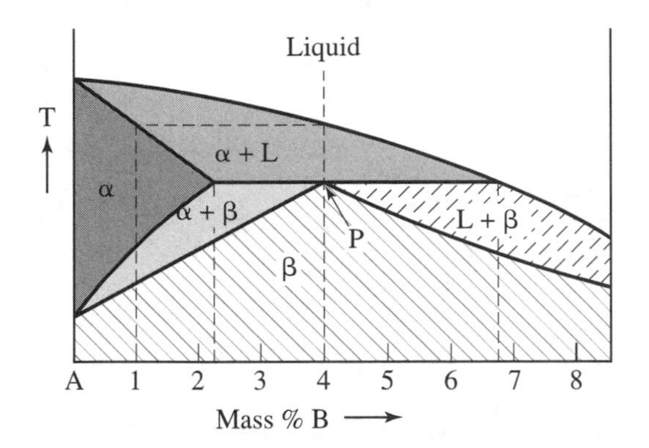

FIGURE 5.13. Part of a hypothetical phase diagram that contains a peritectic reaction.

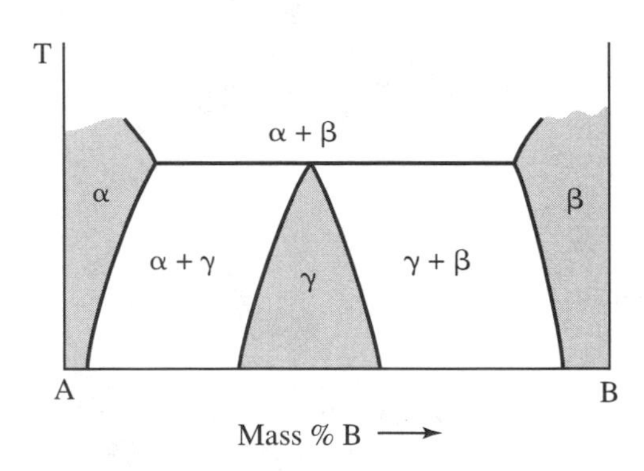

FIGURE 5.14. Schematic representation of a peritectoid reaction. The region at higher temperatures is not shown for clarity.

ing, both the solid and the liquid are enriched in B until, at the invariant peritectic point, P, α containing 2.2% B and liquid containing 6.8% B combine to form a homogeneous solid solution called β. Cooling an alloy containing between 2.2 and 6.8% B (other than the peritectic alloy) results initially in two-phase constituents ($\alpha + \beta$ or $L + \beta$), as seen in Figure 5.13.

The *peritectoid* reaction is the solid-state equivalent to the peritectic reaction, involving two solid-solution phases which combine into a new and different solid solution, as expressed, for example, by:

$$\alpha + \beta \rightarrow \gamma, \tag{5.5}$$

and as depicted in Figure 5.14.

5.2.5 Monotectic and Monotectoid Reactions

Some molten metals essentially do not mix at certain temperatures just as oil and water remain virtually separated from each other at room temperature. In other words, occasionally a *miscibility gap* in the liquid state is encountered which can be detected in a phase diagram by a dome-shaped liquidus line; see Figure 5.15. Within this dome, two liquid phases coexist. The reaction equation at the monotectic point, M, can be written as:

$$L_1 \rightarrow \alpha + L_2. \tag{5.6}$$

The *monotectoid reaction* is the solid-state equivalent to the monotectic reaction involving two solid phases (e.g., α_1 and α_2) having the same crystal structure but different compositions. They react at the monotectoid point by:

$$\alpha_2 \rightarrow \alpha_1 + \epsilon. \tag{5.7}$$

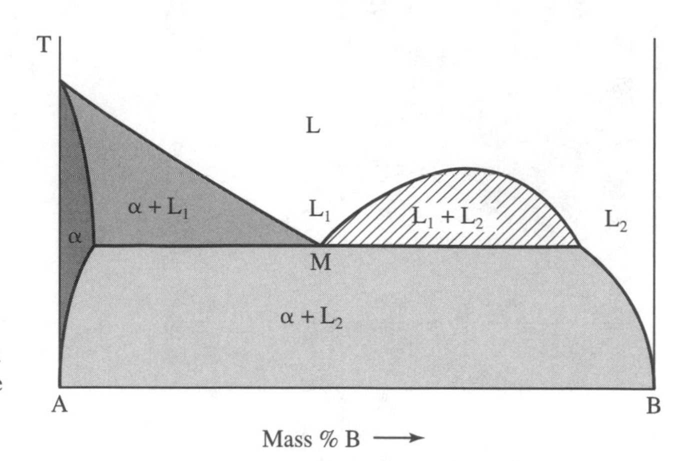

FIGURE 5.15. Schematic representation of a miscibility gap in the liquid state causing a monotectic reaction as given in Eq. (5.6).

5.2.6 Intermediate Phases (Intermetallic Compounds)

If the affinity between two constituents is particularly strong, a stoichiometric intermetallic or intermediate compound such as Mg_2Ni or $Na_2O \cdot 2SiO_2$ may be formed. This constitutes a new phase which may have a different crystal structure, a higher melting temperature, higher strength to the extent of brittleness, or better resistance to creep and corrosion. *Stoichiometric intermetallic compounds* are characterized in a phase diagram by a vertical line as seen in Figure 5.16(a). They may be formed through a peritectic or a peritectoid reaction (Section 5.2.4) or by a transformation at a congruent maximum (Figure 5.16). Actually, stoichiometric intermetallic compounds such as those found in the Mg–Pb system can be thought of as two eutectic diagrams joined back to back.

Quite often, intermediate phases possess a range of composi-

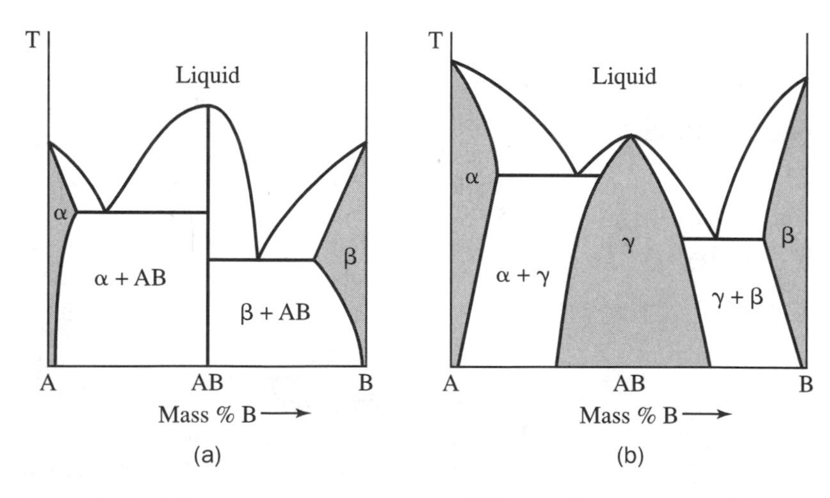

FIGURE 5.16. Schematic representation of (a) stoichiometric and (b) nonstoichiometric intermediate phases.

tions centered about a stoichiometric compound, as shown in Figure 5.16(b). These *nonstoichiometric intermediate phases* are designated by an individual Greek letter or a name rather than by a chemical formula. The copper–tin system (Figure 5.17) possesses a number of those nonstoichiometric intermediate phases. Another example is mullite ($3Al_2O_3 \cdot 2SiO_2$), which plays an important role in fine ceramics (pottery); see Figure 15.5.

5.2.7 Congruent and Incongruent Phase Transformations

Materials scientists define a **phase transformation** to be **congruent** when there is no change in composition of a given alloy during melting or solidification. As an example, the 50/50 alloy in Figure 5.16 (b) retains its composition when it is heated from the γ-phase into the liquid phase. In contrast, an **incongruent transformation** can be explained by inspecting Figure 5.14, again considering a 50/50 alloy. When this alloy is heated from the γ-phase to a temperature above the peritectoid line, two phases, having compositions equivalent to α and β are formed. In other words, a phase having the composition of the γ-phase no longer exists at elevated temperatures.

5.2.8 Copper–Tin (Bronze) Phase Diagram

At the end of this section we return to the phase diagram which is of particular interest to us in the context of our discussion on bronze. Figure 5.17 depicts the copper–tin binary phase diagram. The attentive reader will spot immediately the α-solid-solution range in which the historic Cu–10% Sn alloy is situated and, further, a temperature dependent solvus line. The reader will also certainly identify a large number of nonstoichiometric intermediate phases, some of which are formed by peritectoid and others by peritectic reactions. Of particular interest may be the (hard and brittle) ϵ-phase which is centered about the chemical formula Cu_3Sn. This ϵ-phase may precipitate upon very slow cooling of the 10% Sn alloy. However, the respective reaction is, in reality, too slow for the ϵ-phase to form, so that the 10% Sn alloy remains, in most cases, a single-phase α-solid-solution. Bell metal, which is used for its sonorous qualities when struck, contains between 14 and 25% Sn. Statuary bronze may contain, in addition to 10% Sn, small amounts of Zn, Pb, or P to enhance hardness. Technical applications of bronzes are for bearings, pump plungers, valves, and bushings. Bronze is also used for coinage (Cu with 4% Sn and 1% Zn).

5.2.9 Ternary Phase Diagrams

For many practical applications more than two constituents are mixed in an alloy. In principle, similar considerations and rules apply in binary as well as in *ternary* (three-component) systems. For example, a liquid phase could transform at the ternary eutectic point into three solid phases (e.g., α, β, and γ), or two phases could change into two different phases. Ternary phase di-

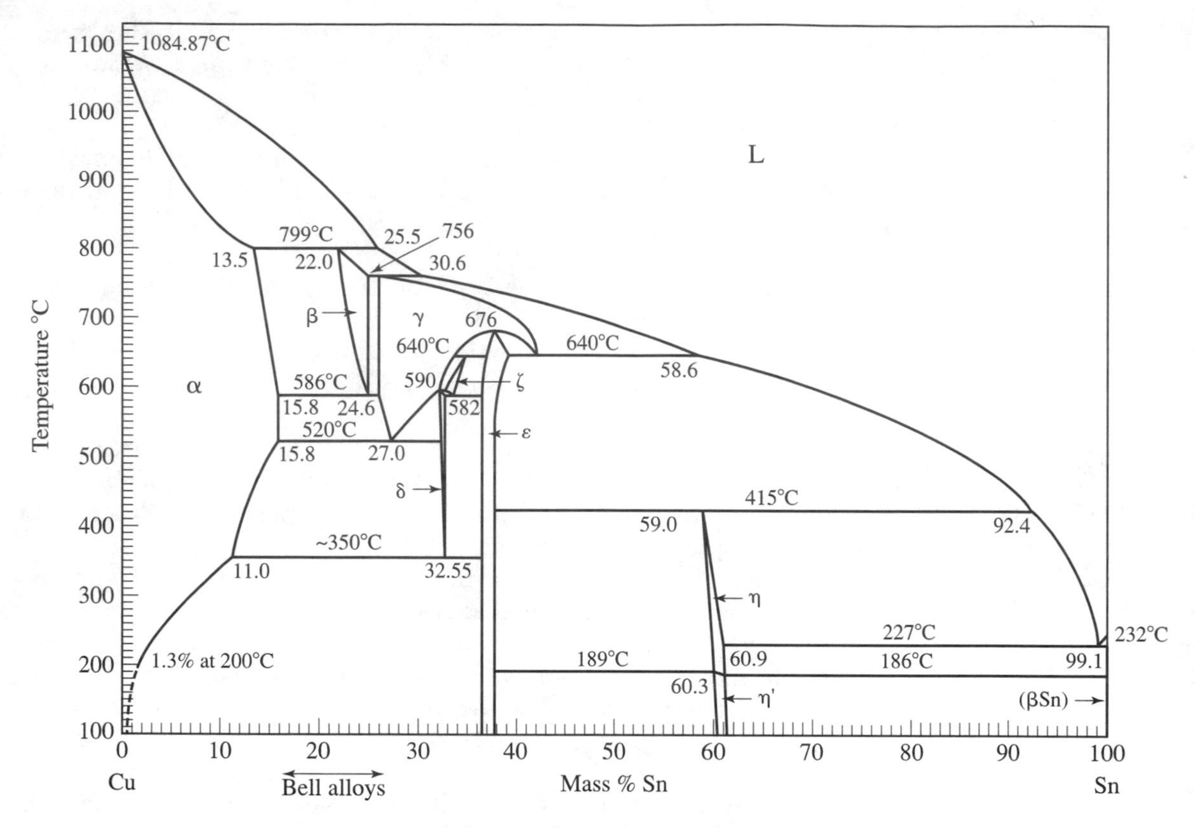

FIGURE 5.17. The copper–tin binary phase diagram. Adapted with permission from *Binary Alloy Phase Diagrams*, Vol. 1 (1986), Edited by T.B. Massalski, J.L. Murray, L.H. Bennett, and H. Baker, ASM International, Materials Park, OH.

agrams are generally plotted in the form of a triangle in which the three pure constituents occupy the corners. Since, in such a two-dimensional diagram, no axis is left for plotting the temperature, other forms for presentation need to be found. One of them is known as the *liquidus plot,* in which the freezing temperatures for each alloy are inserted in the field enclosed by the triangle [Figure 5.18(a)]. Alternately, an *isothermal plot* is often utilized [Figure 5.18(b)], which displays the phases that are encountered at a fixed temperature. (*Isos* is Greek and means "equal," and *thermos* translates into "heat").

The *isopleth plot* (*plethos*, "quantity") is a phase diagram quite similar to that of a binary system in which, however, one of the components remains fixed while the amounts of the other two constituents are allowed to vary. For example, in a ternary system of copper, tin, and lead, one could stipulate that the lead content remains 5% in all alloys. We refrain from presenting details of ternary or even *quaternary* (four-component) phase diagrams.

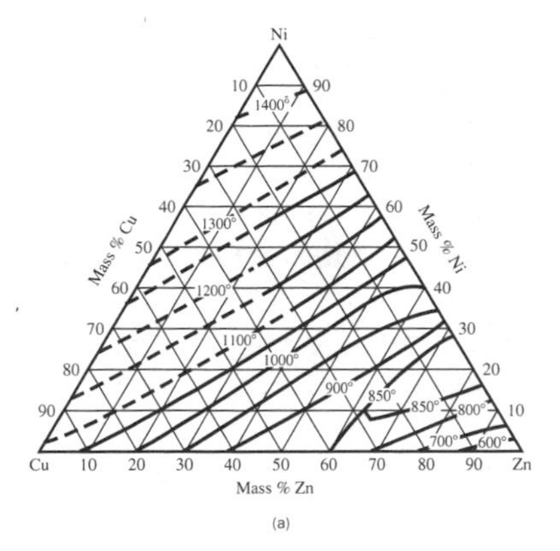

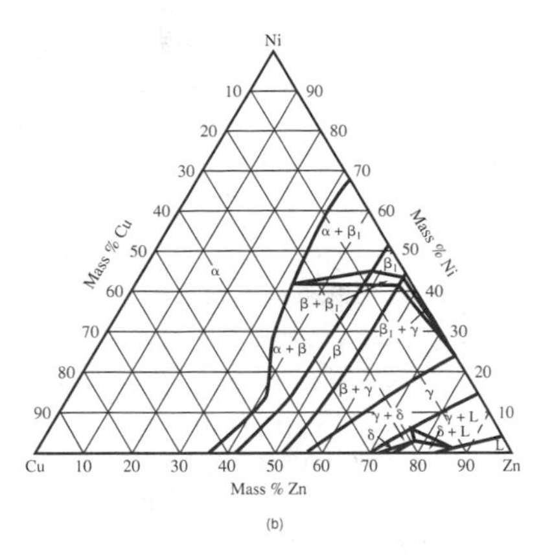

FIGURE 5.18. Examples for ternary phase diagrams (a) Cu–Ni–Zn liquidus plot, (b) isothermal plot (Cu–Ni–Zn at 650°C). Reprinted with permission from *Metals Handbook*, 8th Edition, Vol. 8 (1973), ASM International, Materials Park, OH, p. 428.

It should be mentioned, however, in closing that a particular quaternary alloy consisting of lead, tin, antimony, and some bismuth played a key role for the invention of printing of books with movable letters. This alloy that was on the one side near eutectic (low melting point) but had also a relatively high strength was the starting point for modern printing. It was invented between 1434 and 1444 by Johannes Gutenberg in Strasbourg and particularly in Mainz (Germany). The individual letters were cast by hand and eventually assembled to words and sentences. The most famous work by Gutenberg was the so-called "Gutenberg Bible" (42 lines per page) which, because of its large distribution, aided in the spread of the Bible and unified the German language.

5.3 • Precipitation Hardening (Age Hardening)

We return now to our discussion of mechanisms that are suitable to increase the strength of materials.

The strength of a few specific alloys can be considerably improved by a series of heat treatments called *age hardening* or *precipitation hardening*. The construction of major airplanes at the beginning of this century became feasible after the strength of lightweight aluminum alloys had been improved by this process. The alloy that was originally developed in 1910 by A. Wilm (a German metallurgist) essentially consisted of aluminum with about 4% copper and some magnesium and was named *Duralumin*.

We have learned in the previous section that once the solubility of one element into another exceeds a certain limit, a second phase may precipitate. When age hardenable Al alloys are *solution heat-treated* above the solvus line but below the eutectic temperature (to prevent possible nonequilibrium eutectic microconstituents from melting), a homogeneous α-solid-solution is formed (Figure 5.19). The solid-solution can be frozen-in when the alloy is rapidly cooled, that is, *quenched* from, say, 500°C to room temperature (or better, to 80°C to prevent residual stresses between the center and the surface which could cause warping or cracking). A non-equilibrium, *supersaturated solid solution* thus results. The final treatment step consists of heating the material at specific temperatures and for certain times. Examples for this annealing step, called *aging*, are shown in Figure 5.20. Because of the instability of the quenched-in structure, extremely small (<10 nm), finely and uniformly dispersed, metastable, coherent precipitates nucleate and grow at numerous places. This yields the typical condition for an increase in strength by blocking the movement of dislocations, as pointed out in Section 3.4. At prolonged aging at elevated temperatures, however, the equilibrium ($\alpha + \theta$) structure is formed which is accompanied by a decrease in hardness.

It is important to know that aging time and aging temperature are quite critical, as seen from Figure 5.20. On the one hand, it is desirable to accomplish the *artificial aging* process within a reasonable amount of time (say 30–60 min). On the other hand, annealing times that are too short leave little room for error, which could result in *overaging* and thus in a loss of strength. Further, the strength is larger and the mechanical properties are more uniformly distributed within the work piece when applying lower annealing temperatures. A reasonable strength is still achieved (and no over-aging takes place) when the precipitation is allowed to occur at room temperature, that is, for *natural aging*. However, unreasonable waiting times (up to one year or even more) may then be required.

The conditions under which age hardening takes place are (i) a decreasing solid solubility with decreasing temperature (as shown in Figure 5.19), (ii) an alloy that suppresses the precipitation of a second phase upon quenching, and (iii) the precipi-

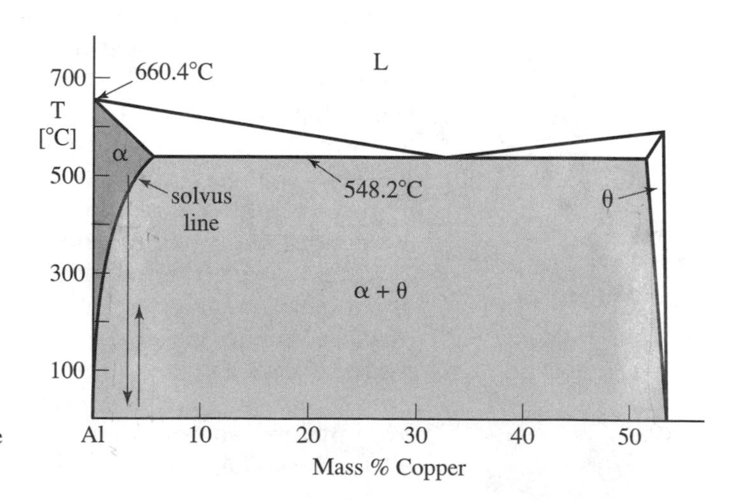

FIGURE 5.19. Part of the Al–Cu phase diagram to demonstrate age hardening.

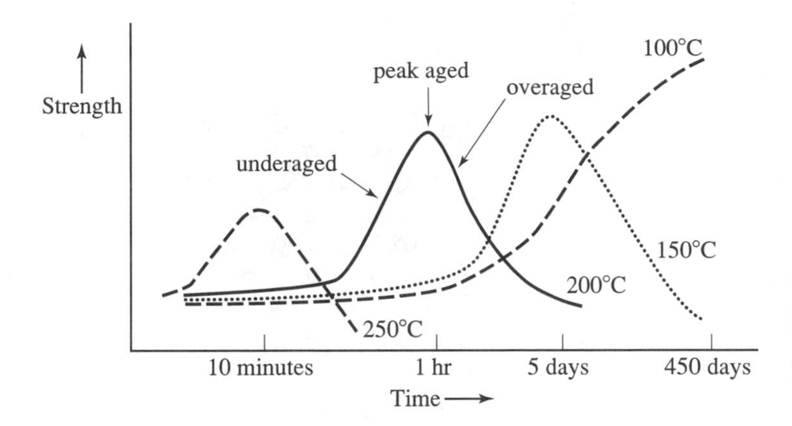

FIGURE 5.20. Schematic representation of the relationship between aging time and temperature during age hardening.

tation of a finely dispersed hard and brittle metastable phase that is coherent with the matrix (see below). Aluminum–copper alloys ideally fulfill these requirements.

Now, we know from Section 3.4 that the strength and plasticity of materials are interrelated with the slip of dislocations upon application of a shear stress to a given material. The movement of dislocations is known to be impeded when an obstacle such as the above-mentioned precipitates are encountered. This results in a stronger material compared to the matrix. The blocking of the slip of dislocations becomes most effective when the particles have a small size, are numerous, and when the lattice planes of the precipitated particles are continuous with the lattice planes of the matrix. These so-called coherent particles may have a different lattice parameter than the matrix which results in elastic lattice distortions around the particles, as shown in Figure 5.21(a). The coherency strain field surrounding the precipitates interacts with the dislocations. Specifically, the cross-sectional area for the interaction between a moving dislocation and a precipitate is larger than the particle itself. An increased strength is thus encountered. The mechanism just described is known by the term *coherent precipitation*. This is to be distinguished from *noncoherent precipitates*, for which the particles have no interrelationship with the lattice of the matrix [Figure 5.21(b)] and, therefore, do not disturb the surroundings of the particle. These particles are usually larger in size and less numerous. In this case, the slip of dislocations is only blocked if the particle lies directly within the path of the moving dislocation.

The highest strength is generally achieved when a *large amount of closely spaced*, *small*, and *round* precipitates are *coherently* dispersed throughout an alloy. These conditions provide an optimal chance for the moving dislocations to interact with many of these particles. Moreover, a soft matrix, in which the dislocations can slip more freely, renders the alloy some ductility.

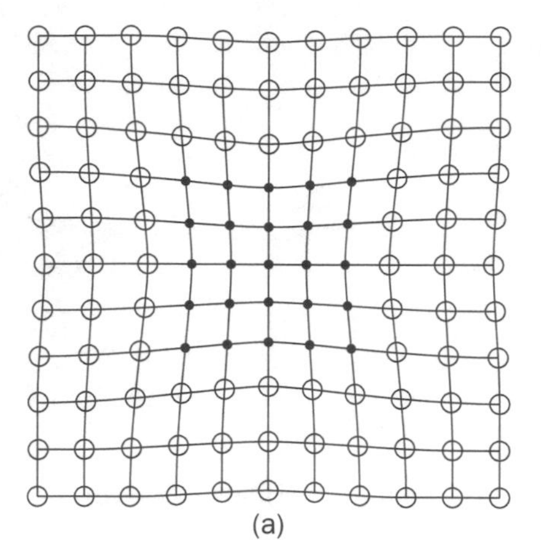

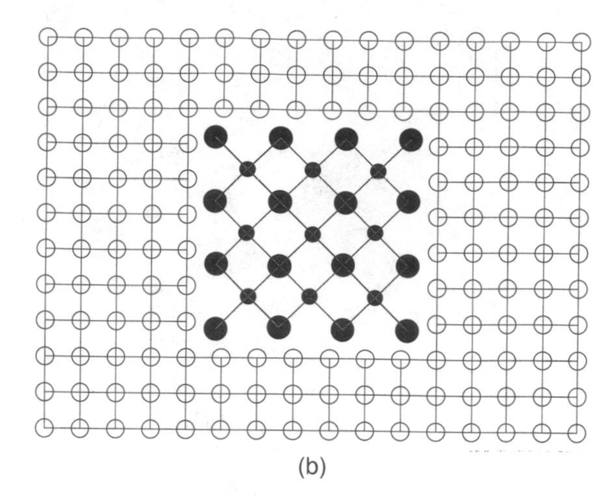

FIGURE 5.21. Schematic representation of (a) a coherent precipitate and (b) a noncoherent precipitate in a crystalline matrix.

The above statements need some refinement. For energetic reasons, coherent particles nucleate more readily than incoherent precipitates. (This is so because the specific energy of the boundary between the precipitated phase and the matrix is lower for the coherent case.) This rapid nucleation leads to many fine-dispersed precipitates having sizes between 3 and 10 nm. Moreover, the distance between coherent particles along a slip plane is then typically on the order of only 50 nm, a value that cannot be reached for the incoherent case. Indeed, theoretical considerations have shown that the spacing between particles is crucial for the hardness. Specifically, the maximal possible yield strength is inversely proportional to the interparticulate distance.

Figure 5.22(a) schematically depicts the obstruction of a moving dislocation line by an array of hard particles in a soft matrix. With increasing shear stress, certain dislocation segments are thought to be pushed between the particles and eventually bulge out. At that point, the bows touch each other and form a new dis-

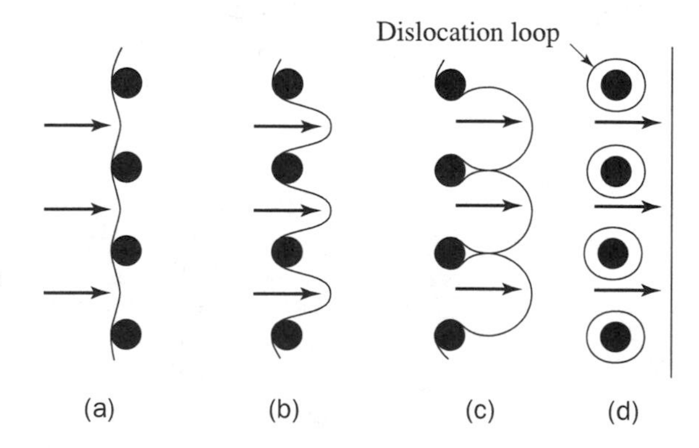

FIGURE 5.22. Schematic representation of the obstruction of a moving (edge) dislocation caused by precipitate particles (**Orowan mechanism**; compare to Figures 3.25 and 3.21 (a)).

location which continues its path on the opposite side until a new set of precipitates is encountered. As a result of this procedure, dislocation loops are left around the particles whose stress fields impede the motion of the next dislocation [Figure 5.22(d)]. This mechanism results in a yield strength that is inversely proportional to the distance between particles and therefore to the particle size. Alternately to the just-described *Orowan mechanism*, it is possible that the moving dislocations *cut* through the precipitates; see Figure 5.23. This is particularly true if the particles have a small size. In both cases, an increased stress is needed for slip to occur; that is, the strength of the material is enhanced.

So far, we have considered only spherically shaped precipitates. However, plate-shaped or needle-like particles are likewise encountered which may be aligned along specific crystallographic planes and may embrittle the alloy. They are known by the name **Widmanstätten structures**.

A word should be added about the mechanisms that govern the precipitation of second-phase particles out of a matrix. For this, nuclei initially need to be present which eventually grow with time. In the case of **homogeneous nucleation**, fluctuations in composition within the matrix may cause spontaneous formation of these nuclei throughout the crystal. Homogeneous nucleation is, however, not the process that is usually favored by nature. Instead, **heterogeneous nucleation** takes place, particularly on grain boundaries or particles. This tends to occur more rapidly, that is, before precipitation in the matrix has fully developed.

Considerable research has been expended in the past decades to study age hardening. It has been found that the aging of Al–Cu alloys proceeds via a series of individual steps. When aging commences, the copper atoms preferably nucleate on {100} planes where they form thin plates called **Guinier–Preston Zones** of type I (shortly termed GP–I zones). The shape of these zones is brought about by the large size difference between Cu and Al atoms, which causes substantial lattice strain. To minimize the strain energy, the GP zones assume a volume that is as small as possi-

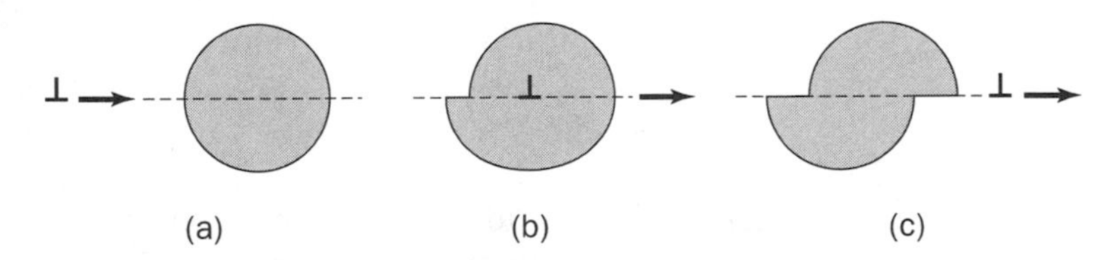

FIGURE 5.23. Schematic representation of a mechanism by which a coherent, spherical particle is cut by a dislocation.

ble. This is accomplished by two-dimensional plates that are less than 10 nm in diameter and about 0.1 nm thick. Upon further aging the plates grow into disks, called GP–II Zones or θ''-phase, which have an ordered structure and a thickness of several atomic layers. Eventually, the metastable, tetragonal, θ'-phase is formed, which is distinguished by a high degree of order and provides the highest strength in this annealing sequence. Once, upon further aging, the equilibrium θ-phase ($CuAl_2$) is finally present, the alloy already has lost some of its strength, that is, the work piece has been overaged. The θ precipitates are tetragonal and incoherent with the aluminum matrix. At this point, the size of the larger particles has increased at the expense of the smaller ones. In other words, overaging is accompanied by a decreasing number of particles. Overaging and the formation of the stable θ-phase does not occur during aging at temperatures below 100°C; see Figure 5.20.

Age hardened aluminum alloys generally allow only a moderate service temperature. That is, if the alloy is exposed to higher temperatures (welding, engine parts, etc.), overaging and thus a reduction in strength take place. Commercial age hardenable alloys include Al–Cu–Mg, Al–Zn–Mg, Al–Mg–Si, and Al–Zn–Mg–Cu. These materials are comparable in hardness to soft steel. Since they are relatively ductile before age hardening, they can be rolled, forged, or extruded into a multitude of shapes. As an example, the commercial 2014 aluminum alloy contains 4.4% Cu, 0.9% Si, 0.8% Mn, and 0.5% Mg, and has a maximum yield strength of about 413 MPa compared to 40 MPa for pure Al (see Table 2.1). Even higher strength values are obtained when employing aluminum alloys involving Mg_2Zn precipitates or Al–Li alloys, whereas the latter ones have a lower specific weight than the former. Recently developed *superalloys* (based on nickel rather than on aluminum) can be used up to nearly 1000°C before overaging occurs. As an example, in Ni–Al alloys, the precipitates have an ordered Ni_3Al structure. They are stable up to high temperatures. Most steels are hardened by precipitates of carbides.

Bronze is not age hardenable since the ϵ-phase (Figure 5.17) precipitates only very reluctantly. In essence, a Cu–10% Sn alloy remains as a single phase, even at room temperature, and derives its strength only through solid-solution strengthening and, if wanted, through work hardening.

It should be noted that the procedure that leads to precipitation hardening is not limited to metallic alloys. Indeed, it is also found in ceramic systems. As an example, quenching of specific ZrO_2-rich ZrO_2–MgO compounds from the single (cubic) phase field and subsequent aging below 1400°C cause the precipitation of MgO along grain boundaries and at intragranular voids. This improves strength and toughness. Other investigators found the formation of a metastable, intermediate, ordered compound referred to as δ-phase or $Mg_2Zr_5O_{12}$. The procedure is used commercially to increase the mechanical and shock properties of Mg-*partially stabilized zirconia*. The strengthening mechanism is, however, different than in metallic alloys, that is, it does not involve dislocations.

5.4 • Dispersion Strengthening

The strength of metals and alloys can be enhanced by still another mechanism, namely, by finely dispersing small, hard, and inert particles in an otherwise ductile matrix. These particles act again as obstacles for moving dislocations, as repeatedly explained above. They are not as effective in increasing the strength of materials compared to precipitates since their size is commonly larger than 10 nm and they are generally incoherent with the matrix. The main advantage, however, is that these foreign particles generally neither dissolve at high temperatures nor grow in size, as is known for precipitates. Thus, dispersion hardened materials maintain their strength even at high temperatures.

The particles in question may consist of oxides such as in ODS alloys (oxide dispersion strengthening). Examples are thorium oxide (ThO_2) in nickel alloys, or aluminum oxide in aluminum. Further, metallic or nonmetallic phases may serve as dispersoids. Intermetallic phases (Figure 5.16), which are often distinguished by a separate crystal structure in comparison to the matrix and which are, as a rule, hard and brittle, are also considered to be dispersoids.

In order to obtain most effective dispersion strengthening, the particles need to be hard, small, numerous (in the percentage range), and preferably round. Dispersed particles can be compared somewhat to hazelnuts contained in a chocolate matrix.

The terms "dispersion strengthening" and "precipitation hardening" are sometimes utilized interchangeably. In both cases, particles are effective in hampering dislocation motion. The difference is observed, however, in their high-temperature behavior, as explained above.

Dispersion strengthened alloys and compounds are often produced by the technique of *powder metallurgy* (mechanical alloying) whereby different kinds of powders are intimately mixed and then *sintered* together at high temperatures.

5.5 • Grain Size Strengthening

We already mentioned in Section 3.3 that a solid generally consists of a large number of individual grains whose lattice planes are somewhat rotated with respect to each other (Figure 3.15). The surfaces where these grains meet are called *grain boundaries*. The lattice planes are noncontinuous across the grain boundaries. Different work pieces may have different grain sizes that vary from several millimeters down to some micrometers. (Even nanocrystalline materials are presently under investigation, and single crystals are used for specific applications.) Grains are formed during solidification from the liquid phase. The grain boundaries represent, as the reader might have expected by now, another type of obstacle for moving dislocations and thus provide increased strength. Stressing a work piece eventually causes the dislocations to pile up behind the grain boundaries until no further slip can occur. The material then has been strengthened to its limit.

It goes almost without saying that the strengthening becomes more effective the larger is the total surface area of these grain boundaries. In other words, it is desirable for maximizing *grain size strengthening* to have a large number of small grains within a matrix. Specifically, it has been found by *Hall and Petch* that the hardness (or the yield strength) is inversely proportional to the square root of the average grain diameter (i.e., $H \approx H_0 + k\, d^{-1/2}$).

The question then arises, how to manipulate the grain size during solidification? For this, one needs to know that grains are generally formed by the previously mentioned *nucleation and growth* mechanism. We mentioned already in Section 5.3 that *heterogeneous nucleation* is energetically favored by thermodynamics. Specifically, nucleation on impurity particles or on the walls of a mold is quite common. In order to increase the number of nucleation sites (and thus the number of grains), a small amount of solid impurity particles is sometimes introduced into the liquid before solidification. They are called **inoculators**, or **grain refiners**. They consist, for example, of 0.01% solid titanium boride, which has a much higher melting point than say, an aluminum alloy. At a certain temperature near the freezing point of the host material, several atoms cluster around the inoculators, thus creating an *embryo* that already resembles the crystal structure of the solid. Upon further cooling, the embryo grows into a nucleus as increasingly more atoms are attached to the solid surface. Finally, a nucleus grows into a grain and the grain grows until it touches a neighboring grain. At this point, the grain structure is established. Another technique to obtain small grains involves fast cooling of the melt, thus undercooling the liquid and causing spontaneous nucleations on many sites.

In summary, the strength of materials can be increased by refining the grain. The grain size can be decreased by adding inoculators to the liquid metal or alloy shortly before solidification or by rapid solidification.

5.6 • Control of Strength by Casting

Solidification of liquid metals or alloys is generally allowed to occur in a mold that has an appropriate, and sometimes elaborately shaped, internal cavity. The process involved is called *casting*. The type of mold determines the strength which the material eventually will have. For example, a ceramic mold, consisting of clay or sand (particularly as used by early civilizations), allows only a slow dissipation of heat because of the poor thermal conduction of ceramics. The cooling rate is therefore small and the strength of the casting is low, as we shall explain momentarily. In contrast to this, modern molds made of metal (e.g., steel, divided into two hinged parts for reusability) provide rapid solidification through better heat dispersion and therefore yield high strength castings.

Ceramic molds are generally manufactured by utilizing the *lost wax method*. A model of the future casting is first formed out of wax which is then surrounded by clay. After the clay has dried, the wax is melted and poured out, which leaves a hollow ceramic shell. This method (also called *investment casting*) was used frequently during the Bronze Age for producing decorative art objects and is still applied today.

Another mold that can be utilized only once consists of a damp sand/clay mixture that is compacted around a wooden pattern. This wooden core is inserted in a two- or three-part box. After the pattern has been removed, the liquid metal is poured into the cavity. The raw sand can be reused (*sand casting*).

Solidification generally commences in the *chill zone*, that is, at the mold walls or at free surfaces where the liquid freezes first. This allows heterogeneous nucleation at many places and thus a substantial amount of small and randomly oriented grains (see Section 5.5). Upon further cooling, those grains in cubic structures that possess a ⟨100⟩ direction perpendicular to the mold walls preferably grow in a columnar manner toward the interior

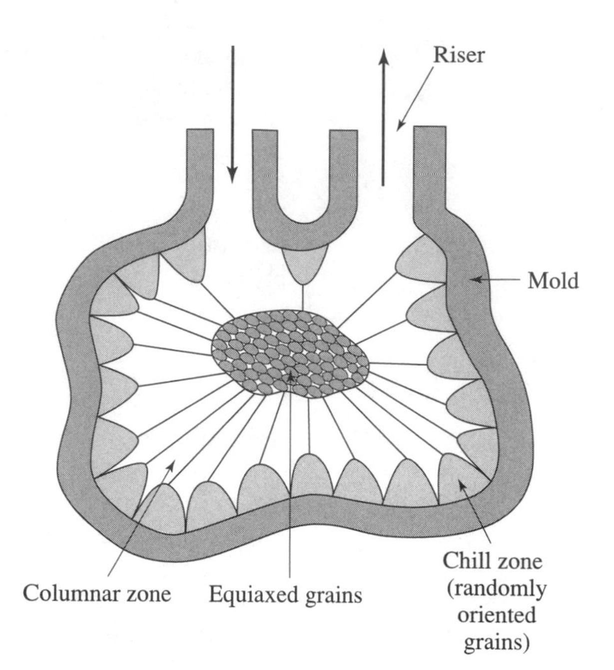

FIGURE 5.24. Schematic representation of a mold and various zones formed by cooling molten metals and alloys.

of the mold. As a consequence, a large-grained *columnar zone* is formed; see Figure 5.24. Eventually, the growth of columns in alloys (and sometimes also in pure metals) is impeded by the nucleation and subsequent growth of randomly oriented, spherical (equiaxed) grains that form during rapid solidification on inoculants or alloying constituents. The *equiaxed zone* is mostly situated in the center of the mold; see Figure 5.24. An increase of the number of small, equiaxed crystals and, concomitantly, the suppression of the growth of large, columnar (or dendritic) grains is paramount for improving the strength of alloys through grain size strengthening, as explained in Section 5.5.

A few words should be added about potential problems that could arise during casting. First, since solids are usually denser than liquids, some shrinkage occurs during solidification which may cause some cavities in the cast. To control this problem, a reservoir of liquid metal, in the form of a riser, is generally utilized which feeds liquid into the cavity, as shown in Figure 5.24. Second, shrinkage between dendrites (Figure 5.6) could cause small pores (called *interdendritic shrinkage*), which can be alleviated by rapid cooling. Third, gases dissolved by the liquid, such as hydrogen, may form bubbles and porosity during cooling. Gas pick-up can be minimized by maintaining the temperature of the melt as low as possible. Fourth, during solidification, the columnar grains push solute elements ahead of them. This causes segregation (*coring*), that is, a difference in composition between the exterior and the center of the casting. Elimination of the columnar grains, as described above, may partially alleviate this problem.

Modern industrial plants shape their materials by still another process, called *continuous casting*, which allows the subsequent hot-rolling to sheet metals, etc., without cooling to room temperature and subsequent reheating.

In summary, the strength of a cast made of alloys can be substantially improved by selecting the proper mold (metal compared to ceramic) and by utilizing suitable casting techniques. Among the latter are the introduction of inoculant particles in order to force the formation of equiaxed grains, low casting temperatures to prevent excessive gas pick-up, or rapid solidification to prevent interdendritic shrinkage. Early civilizations remarkably mastered some of these variables and thus produced relatively high strength bronze or pieces of art displaying outstanding detail.

5.7 • Concluding Remarks

We learned in this chapter that strengthening of alloys may be governed by various mechanisms which have in common that they impede the movement (slip) of dislocations through a solid. It became obvious during our discourse that second constituents provide such obstacles, particularly if the difference in atomic size between solute and solvent is large (solid-solution strengthening). Further, the precipitation of a second phase, especially when it is coherent with the matrix and when it is highly dispersed by proper heat treatments, efficiently obstructs slip, as proposed by the Orowan mechanism (precipitation hardening or age hardening). Finely dispersed, hard, and inert particles introduced, for example, by powder metallurgy provide another source for strength (dispersion strengthening). Moreover, the grain boundaries in polycrystalline materials provide additional blockers for moving dislocations and, thus, sources for strength (grain size strengthening). Materials scientists make use of all of these mechanisms, often in conjunction with work hardening (as discussed in Chapter 3), to adapt the mechanical properties of materials to specific needs and applications. One should not forget, however, that a different type of strength stems from specific binding forces (such as covalent or ionic forces), as discussed in Chapter 3. All taken, the strength of materials is governed by a rather complex interplay between several mechanisms that mainly involve the mutual interactions of atoms.

Problems

5.1. Give the amounts of the liquid and solid (α) phases (in percent) which are present when a Cu–28 mass % Ni alloy is very slowly cooled to a temperature corresponding to Tie Line #2 in Figure 5.5.

5.2. Figure 5.17 depicts the Cu–Sn phase diagram in which the single phase regions are labeled only.
   (a) Label the two-phase regions.
   (b) State the three-phase reactions that occur and give the temperatures at which they take place. Write the pertinent reaction equations.

5.3. State the maximal solubility of (a) copper in silver and (b) silver in copper. At which temperatures do these maximal solubilities occur?

5.4. What is the approximate solubility of copper in silver at room temperature?

5.5. Give the phases present and their compositions for an Al–5 mass % Cu alloy at 400°C.

5.6. Assume an Ag–10 mass % Cu alloy. To what temperature does this alloy have to be heated so that 50% of the sample is liquid?

5.7. Does a copper–nickel binary alloy exist whose solid phase at equilibrium contains 36 mass % Ni and whose liquid phase contains 20 mass % Ni? Explain. Refer to Figure 5.3.

5.8. A silver–20 mass % copper alloy is slowly heated from room temperature.
   (a) State the temperature at which a liquid phase starts to form.
   (b) State the composition of this liquid phase.
   (c) At which temperature is the alloy completely liquefied?
   (d) Give the composition of the solid just before complete melting has occurred.

Suggestions for Further Study

D.R. Askeland, *The Science and Engineering of Materials*, 3rd Edition, PWS Publishing Co., Boston (1994).

W.D. Callister, Jr., *Materials Science and Engineering*, 4th Edition, Wiley, New York (1997).

P. Haasen, *Physikalische Metallkunde*, 3rd Edition, Springer-Verlag, New York (1994).

E. Hornbogen, *Werkstoffe*, 6th Edition, Springer-Verlag, Heidelberg, New York (1994).

M. Ohring, *Engineering Materials Science*, Academic, New York (1995).

R. Stevens, *Zirconia and Zirconia Ceramics*, Magnesium Electron Ltd., UK (1986).

6

Atoms in Motion

6.1 • Lattice Defects and Diffusion

So far, when discussing the properties of materials we tacitly assumed that the atoms of solids remain essentially stationary. From time to time we implied, however, that the behavior of solids is affected by the thermally induced vibrations of atoms. The changes in properties increase even more when atoms migrate through the lattice and take new positions. In order to gain a deeper insight into many mechanical properties, we therefore need to study the "dynamic case." It will become obvious during our endeavor that the motion of atoms through solids involves less effort (energy) when open spaces are present in a lattice, as encountered, for example, by empty lattice sites. Thus, we commence with this phenomenon.

6.1.1 Lattice Defects

We have repeatedly pointed out in previous chapters that an ideal lattice is rarely found under actual conditions, that is, a lattice in which all atoms are regularly and periodically arranged over large distances. This is particularly true at high temperatures, where a substantial amount of atoms frequently and randomly change their positions leaving behind empty lattice sites, called *vacancies*. Even at room temperature, at which thermal motion of atoms is small, a fair number of lattice defects may still be found. The number of vacancies per unit volume, n_v, increases exponentially with the absolute temperature, T, according to an equation whose generic type is commonly attributed to **Arrhenius**:[1]

$$n_v = n_s \exp\left(-\frac{E_f}{k_B T}\right), \qquad (6.1)$$

[1]Svante August Arrhenius (1859–1927), Swedish chemist and founder of modern physical chemistry, received in 1903, as the first Swede, the Nobel prize in chemistry. The Arrhenius equation was originally formulated by J.J. Hood based on experiments, but Arrhenius showed that it is applicable to almost all kinds of reactions and provided a theoretical foundation for it.

where n_s is the number of regular lattice sites per unit volume, k_B is the Boltzmann constant (see Appendix II), and E_f is the energy that is needed to form a vacant lattice site in a perfect crystal.

As an example, at room temperature, n_v for copper is about 10^8 vacancies per cm^3, which is equivalent to one vacancy for every 10^{15} lattice atoms. If copper is held instead near its melting point, the vacancy concentration is about 10^{19} cm^{-3}, or one vacancy for every 10,000 lattice atoms. It is possible to increase the number of vacancies at room temperature by quenching a material from high temperatures to the ambient, that is, by freezing-in the high temperature disorder, or to some degree also by plastic deformation.

Other treatments by which a large number of vacancies can be introduced into a solid involve its bombardment with neutrons or other high energetic particles as they exist, for example, in nuclear reactors (*radiation damage*) or by ion implantation. These high energetic particles knock out a cascade of lattice atoms from their positions and deposit them between regular lattice sites (see below). It has been estimated that each fast neutron may create between 100 and 200 vacancies. At the endpoint of a primary particle, a *depleted zone* about 1 nm in diameter (several atomic distances) may be formed which is characterized by a large number of vacancies.

Among other point defects are the **interstitials**. They involve foreign, often smaller, atoms (such as carbon, nitrogen, hydrogen, oxygen) which are squeezed in between regular lattice sites. The less common **self-interstitials** (sometimes, and probably not correctly, called *interstitialcies*) are atoms of the *same* species as the matrix that occupy interlattice positions. Self-interstitials cause a substantial distortion of the lattice. In a *dumbbell*, two equivalent atoms share one regular lattice site. **Frenkel defects** are vacancy/interstitial pairs. **Schottky defects** are formed in ionic crystals when, for example, an anion as well as a cation of the same absolute valency are missing (to preserve charge neutrality). Dislocations are one-dimensional defects (Figure 3.20). Two-dimensional defects are formed by grain boundaries (Figure 3.15) and free surfaces at which the continuity of the lattice and therefore the atomic bonding are disturbed. We shall elaborate on these defects when the need arises.

6.1.2 Diffusion Mechanisms

Diffusion by Vacancies

Vacancies provide, to a large extent, the basis for *diffusion*, that is, the movement of atoms in materials. Specifically, an atom may move into an empty lattice site. Concomitantly, a vacancy migrates in the opposite direction, as depicted in Figure 6.1. The prerequisite for the jump of an atom into a vacancy is, however, that the atom possesses enough energy (for example, thermal energy) to squeeze by its neighbors and thus causes the lattice to expand momentarily and locally, involving what is called **elastic strain**

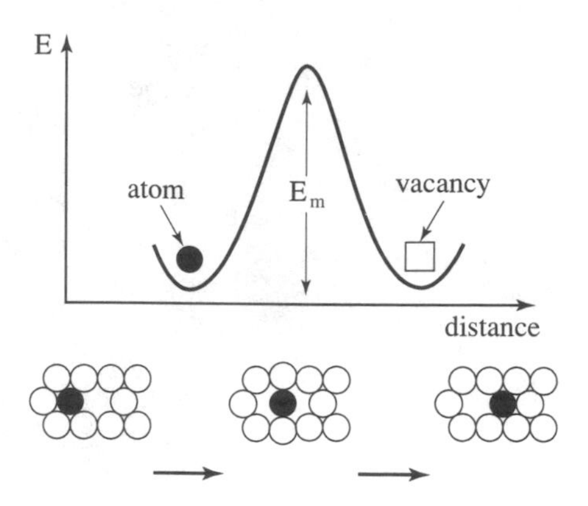

FIGURE 6.1. Schematic representation of the diffusion of an atom from its former position into a vacant lattice site. An activation energy for motion, E_m, has to be applied which causes a momentary and local expansion of the lattice to make room for the passage of the atom. This two-dimensional representation shows only part of the situation. Atoms above and below the depicted plane may contribute likewise to diffusion.

energy. The necessary energy of motion, E_m, to facilitate this expansion is known as the **activation energy** *for vacancy motion*, which is schematically represented by an energy barrier shown in Figure 6.1. E_m is in the vicinity of 1eV. The average *thermal* (kinetic) energy of a particle, E_{th}, at the temperatures of interest, however, is only between 0.05 to 0.1 eV, which can be calculated by making use of an equation that is borrowed from the kinetic theory of particles (see textbooks on thermodynamics):

$$E_{th} = \frac{3}{2} k_B T. \tag{6.2}$$

This entails that for an atom to jump over an energy barrier, large fluctuations in energy need to take place until eventually enough energy has been "pooled together" in a small volume. Diffusion is therefore a statistical process.

A second prerequisite for the diffusion of an atom by this mechanism is, of course, that one or more vacancies are present in neighboring sites of the atom; see Eq. (6.1). All taken, the *activation energy for atomic diffusion*, Q, is the sum of E_f and E_m. Specifically, the activation energy for diffusion for many elements is in the vicinity of 2 eV; see Table 6.1

Interstitial Diffusion

If atoms occupy interstitial lattice positions (see above), they may easily diffuse by jumping from one interstitial site to the next without involving vacancies. Interstitial sites in FCC lattices are, for example, the center of a cube or the midpoints between two corner atoms. Similarly as for vacancy diffusion, the adjacent matrix must slightly and temporarily move apart to let an interstitial atom squeeze through. The atom is then said to have dif-

TABLE 6.1 Selected diffusion constants (volume diffusion)

Mechanism	Solute	Host material	$D_0\left[\frac{m^2}{s}\right]$	Q [eV]
Self diffusion	Cu	Cu	7.8×10^{-5}	2.18
	Al	Al	1.7×10^{-5}	1.40
	Fe	α-Fe	2.0×10^{-4}	2.49
	Si	Si	32×10^{-4}	4.25
Interstitial diffusion	C	α-Fe (BCC)	6.2×10^{-7}	0.83
	C	γ-Fe (FCC)	1.0×10^{-5}	1.40
Interdiffusion	Zn	Cu	3.4×10^{-5}	1.98
	Cu	Al	6.5×10^{-5}	1.40
	Cu	Ni	2.7×10^{-5}	2.64
	Ni	Cu	2.7×10^{-4}	2.51
	Al	Si	8.0×10^{-4}	3.47

fused by an interstitial mechanism. This mechanism is quite common for the diffusion of carbon in iron or hydrogen in metals but can also be observed in nonmetallic solids in which the diffusing interstitial atoms do not distort the lattice too much. The activation energy for interstitial diffusion is generally lower than that for diffusion by a vacancy mechanism (see Table 6.1), particularly if the radius of the interstitial atoms is small compared to that of the matrix atoms. Another contributing factor is that the number of empty interstitial sites is generally larger than the number of vacancies. In other words, E_f (see above) is zero in this case.

Interstitialcy Mechanism

If the interstitial atom is of the same species as the matrix, or if a foreign atom is of similar size compared to the matrix, then the diffusion takes place by pushing one of the nearest, regular lattice atoms into an interstitial position. As a result, the former interstitial atom occupies the regular lattice site that was previously populated by the now displaced atom. Examples of this mechanism have been observed for copper in iron or silver in AgBr.

Other Diffusion Mechanisms

Diffusion by an *interchange* mechanism, that is, the simultaneous exchange of lattice sites involving two or more atoms, is possible but energetically not favorable. Another occasionally observed mechanism, the *ring exchange*, may occur in substitutional, body-centered cubic solid solutions that are less densely packed. In this case, four atoms are involved which jump synchronously, one position at a time, around a circle. It has been calculated by Zener that this mode requires less lattice distortions and, thus, less energy than a direct interchange.

Self-Diffusion and Volume Diffusion

Diffusion involving the jump of atoms within a material consisting of only *one* element is called *self-diffusion*. (Self-diffusion can be studied by observing the motion of radioactive tracer atoms, that is, isotopes of the same element as the nonradioactive host substance.) Diffusion within the bulk of materials is called *volume diffusion*.

Grain Boundary Diffusion

Grain boundaries are characterized by a more open structure caused by the lower packing at places where two grains meet. They can be represented by a planar channel approximately two atoms wide, as schematically depicted in Figure 6.2. Grain boundaries therefore provide a preferred path for diffusion. The respective mechanism is appropriately called *grain boundary diffusion*. It generally has an activation energy of only one-half of that found for volume diffusion since the energy of formation of vacancies E_f (see above) is close to zero. This amounts to a diffusion rate that may be many orders of magnitude larger than in the bulk, depending on the temperature. However, grain boundaries represent only a small part of the crystal volume, so that the contribution of grain boundary diffusion, at least at high temperatures, is quite small. As a rule of thumb, volume diffusion is predominant at temperatures above one-half of the melting temperature, T_m, of the material, whereas grain boundary diffusion predominates below 0.5 T_m.

Surface Diffusion

Further, free surfaces provide an even easier path for migrating atoms. This results in an activation energy for *surface diffusion* that is again approximately only one-half of that for grain bound-

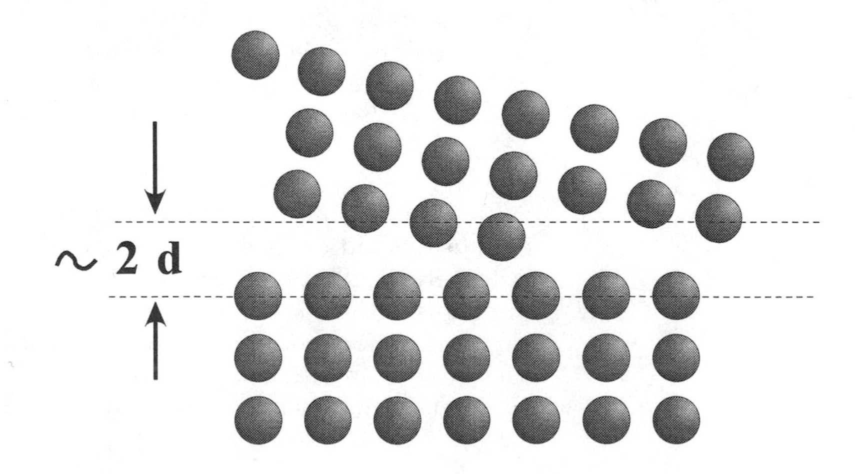

FIGURE 6.2. Schematic representation of a planar diffusion channel between two grains. (Grain boundary diffusion.)

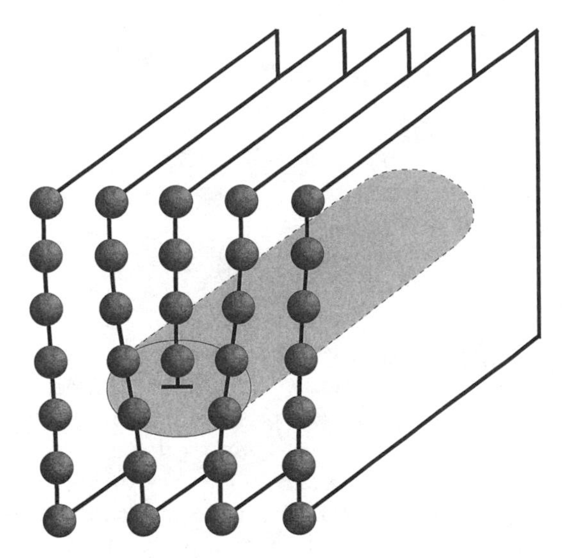

FIGURE 6.3. Schematic representation of a diffusion channel caused by an edge dislocation. (Dislocation-core diffusion.)

ary diffusion. One should keep in mind, however, that, except for thin films, etc., the surface area comprises only an extremely small fraction of the total number of atoms of a solid. Moreover, surfaces are often covered by oxides or other layers which have been deliberately applied or which have been formed by contact with the environment. Thus, surface diffusion represents generally only a small fraction of the total diffusion.

Dislocation-Core Diffusion

Finally, a dislocation core may provide a two-dimensional channel for diffusion as shown in Figure 6.3. The cross-sectional area of this core is about $4d^2$, where d is the atomic diameter. Very appropriately, the mechanism is called *dislocation-core diffusion* or *pipe diffusion*.

6.1.3 Rate Equation

The *number of jumps per second* which atoms perform into a neighboring lattice site, that is, the *rate* or *frequency for movement*, f, is given again by an *Arrhenius-type equation*:

$$f = f_0 \exp\left(-\frac{Q}{k_B T}\right), \tag{6.3}$$

where f_0 is a constant that depends on the number of equivalent neighboring sites and on the vibrational frequency of atoms (about 10^{13} s^{-1}). Q is again an activation energy for the process in question.

We see from Eq. (6.3) that the jump rate is strongly temperature-dependent. As an example, one finds for diffusion of carbon atoms in iron at room temperature ($Q = 0.83$ eV) about one jump

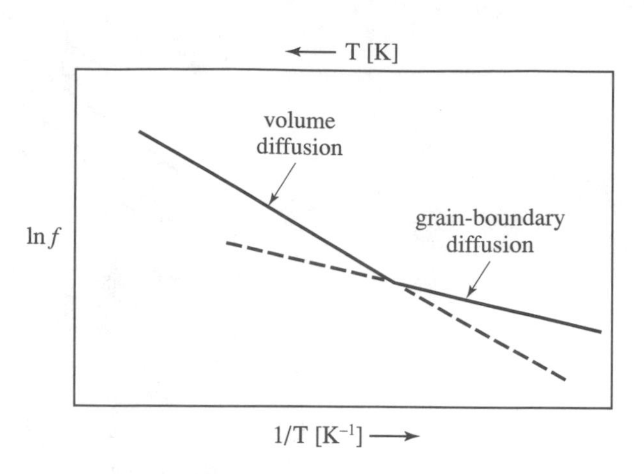

FIGURE 6.4. Schematic representation of an Arrhenius diagram. Generally a logarithmic scale (base 10) and *not* an ln scale is used. The adjustment from ln to log is a factor of 2.3. The difference between volume diffusion and grain boundary diffusion is explained in the text. The slopes represent the respective activation energies.

in every 25 seconds. At the melting point of iron (1538°C) the jump rate dramatically increases to 2×10^{11} per second.

6.1.4 Arrhenius Diagrams

Arrhenius equations are generally characterized by an exponential term that contains an activation energy for the process involved as well as the reciprocal of the absolute temperature. It is quite customary to take the natural logarithm of an Arrhenius equation. For example, taking the natural logarithm of Eq. (6.3) yields:

$$\ln f = \ln f_0 - \left(\frac{Q}{k_B}\right)\frac{1}{T}. \tag{6.4}$$

This expression has the form of an equation for a straight line, which is generically written as:

$$y = b + mx. \tag{6.5}$$

Staying with the just-presented example, one then plots the experimentally obtained rate using a logarithmic scale versus $1/T$ as depicted in Figure 6.4. The (negative) slope, m, of the straight line in an Arrhenius diagram equals Q/k_B, from which the activation energy can be calculated. The intersect of the straight line with the y-axis yields the constant f_0. This procedure is widely used by scientists to calculate activation energies from a series of experimental results taken at a range of temperatures.

6.1.5 Directional Diffusion

Self-diffusion is random, that is, one cannot predict in which direction a given lattice atom will jump if it is surrounded by two or more equivalent vacancies. Indeed, an individual atom often migrates in a haphazard zig-zag path. In order that a bias in the direction of the motion takes place, a *driving force* is needed. Driving forces are, for example, provided by concentration gradients in an alloy (that is, by regions in which one species is more abundant compared to another species). Directional diffusion can also occur as a consequence of a strong electric current (electromigration) or a temperature gradient (thermomigration).

We learned already in Chapter 5 that concentration gradients may occur during solidification of materials (coring). These concentration gradients need to be eliminated if a homogeneous equilibrium structure is wanted. The mechanism by which homogenization can be accomplished makes use of the just-discussed drift of atoms down a concentration gradient. Diffusion also plays a role in age hardening, surface oxidation, heat treatments, sintering, doping in microelectronic circuits, diffusion bonding, grain growth, and many other applications. Thus, we need to study this process in some detail.

6.1.6 Steady-State Diffusion

Fick's first law describes the diffusion of atoms driven by a concentration gradient, $\partial C/\partial x$, through a cross-sectional area, A, and in a given time interval, t. The concentration, C, is given, for example, in atoms per m^3. The pertinent equation was derived in 1855 by A. Fick and reads for one-dimensional atom flow:

$$J = -D\,\frac{\partial C}{\partial x}, \tag{6.6}$$

where J is called the *flux*:

$$J = \frac{M}{At}, \tag{6.7}$$

measured in atoms per m^2 and per second (see Figure 6.5) and D is the diffusion coefficient or diffusivity (given in m^2/s). M is defined as mass or, equivalently, as the number of atoms. The negative sign indicates that the atom flux occurs towards lower concentrations, that is, in the downhill direction. The diffusivity depends, as expected, on the absolute temperature, T, and on an activation energy, Q, according to an Arrhenius-type equation:

$$D = D_0 \exp\left(-\frac{Q}{k_B T}\right), \tag{6.8}$$

where D_0 is called the (temperature-independent) pre-exponential diffusion constant (given in m^2/s). The latter is tabulated in diffusion handbooks for many combinations of elements. A selection of diffusion constants is listed in Table 6.1. There exists a connection between the diffusion coefficient, D, and the jump

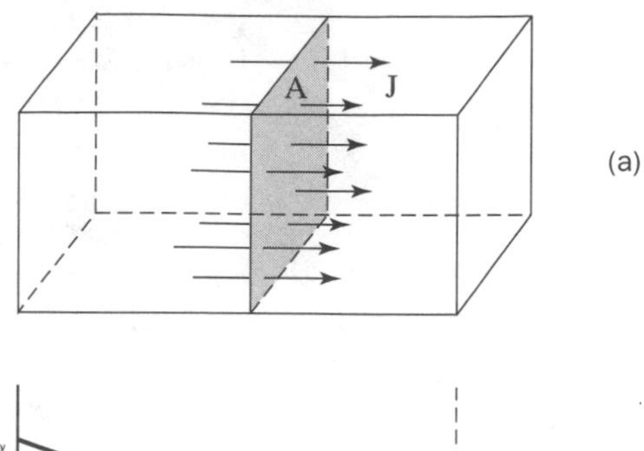

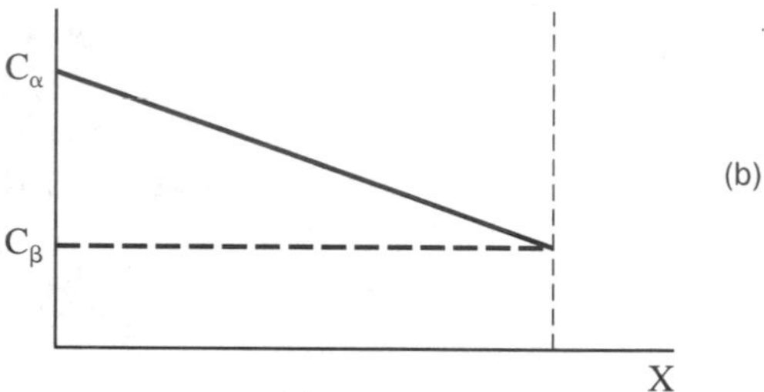

FIGURE 6.5. (a) Steady-state diffusion through a slab. (b) Linear concentration gradient which stays constant with time by constantly supplying solute atoms on the left and removing the same number of atoms on the right side of the slab. C_α and C_β are two assumed surface concentrations, where $C_\alpha > C_\beta$.

frequency of the atoms, f, which we introduced above [Equation (6.3)]. This interrelationship reads:

$$D = \frac{1}{6}\,\lambda^2 f, \tag{6.9}$$

where λ is the diffusion jump distance which is, in isotropic systems, identical with the interatomic distance.

Fick's first law [Eq. (6.6)] may be used to describe *steady-state flow*. One assumes for this case an infinite source and an infinite sink, respectively, on the opposite ends of a plate of metal that causes a constant flow of solute atoms through a given area. Steady-state flow is observed, for example, when gases (such as hydrogen or oxygen) diffuse through metals as a consequence of a constant (but different) gas pressure on each side of a plate. This requires a supply of gas atoms on one side and a removal of the same amount of gas atoms on the other side. As an example, hydrogen-filled gas tanks for fuel-cell-propelled automobiles are empty after about two months. (Hand tools in university labs seem to disappear by a similar mechanism.)

6.1.7 Nonsteady-State Diffusion

Fick's second law deals with the common case for which the concentration gradient of a diffusing species, A, in the host material, B, changes gradually with time (Figure 6.6). The *nonsteady-state* (or dynamic) case is governed by the partial differential equation:

$$\frac{\partial C}{\partial t} = D\,\frac{\partial^2 C}{\partial x^2}, \tag{6.10}$$

which can be solved whenever a specific set of boundary conditions is known. It should be noted that the diffusion coefficient in Eq. (6.10) has only a constant value if one considers self-diffusion or possibly for very dilute systems. In general, however, D varies with concentration so that Eq. (6.10) must be written in more general terms as:

$$\frac{\partial C}{\partial t} = \frac{\partial}{\partial x}\left(D\,\frac{\partial C}{\partial x}\right). \tag{6.10a}$$

The solution of Eq. (6.10), assuming D to be constant with concentration and for "infinitely long" samples (i.e., rods which are long enough so that the composition does not change at the outer ends), is:

$$\frac{C_i - C_x}{C_i - C_0} = erf\left(\frac{x}{2\sqrt{Dt}}\right). \tag{6.11}$$

In other words the solution (6.11) is only valid when the length of a sample is larger than $10\sqrt{Dt}$. The parameters in Eq. (6.11) are as follows: C_x is the concentration of the solute at the distance x; C_i is the constant concentration of the solute at the interface dividing materials A and B after some time, t; and C_0 is the initial solute concentration in material B. The initial solute

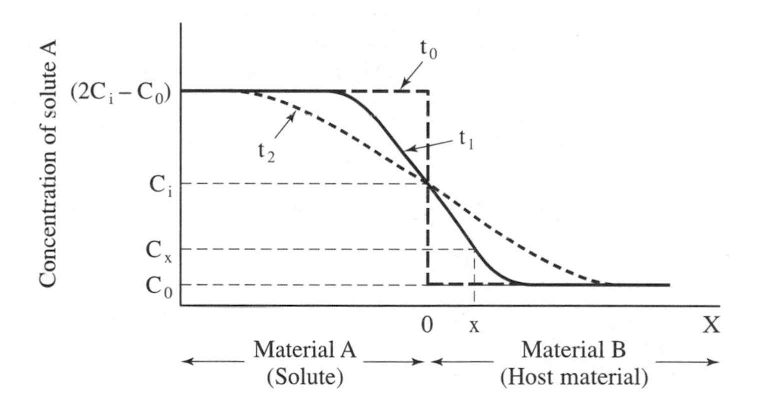

FIGURE 6.6. Concentration profiles (called also penetration curves) for nonsteady-state diffusion of material A into material B for three different times. The concentration of the solute in material B at the distance x is named C_x. The solute concentration at $X = 0$, that is, at the interface between materials A and B, is C_i. The original solute concentration in material B (or at $X = \infty$) is C_0. A mirror image of the diffusion of B into A can be drawn if the mutual diffusivities are identical. This is omitted for clarity.

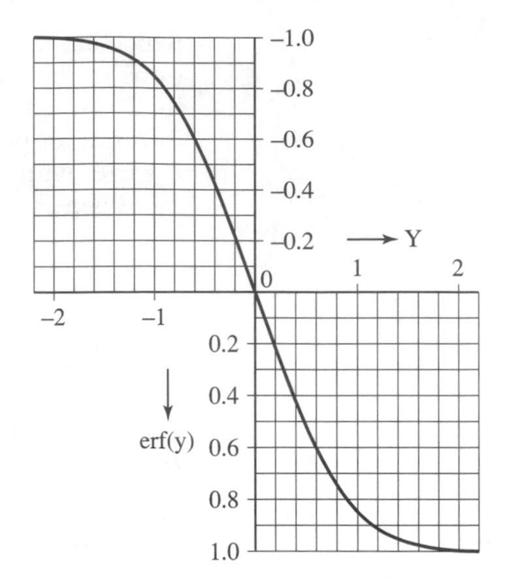

FIGURE 6.7. The Gaussian error function *erf*(*y*) as a function of *y*. Compare to Figure 6.6.

concentration in material A is $2C_i$–C_0; see Figure 6.6. The right-hand side in Eq. (6.11) is called the *probability integral* or *Gaussian error function,* which is defined by:

$$erf(y) = \frac{2}{\sqrt{\pi}} \int_0^y e^{-\phi^2} d\phi. \qquad (6.12)$$

It is tabulated in handbooks similarly as trigonometric or other functions and is depicted in Figure 6.7. It can be observed that the error function (Figure 6.7) reproduces the concentration profile of Figure 6.6 quite well. (See also Problem 6.1.) A reasonable estimate for the distance, X, stating how far solute atoms may diffuse into a matrix during a time interval, t, can be obtained from:

$$X = 2\sqrt{Dt}. \qquad (6.13)$$

6.1.8 Interdiffusion

The boundary condition for which Eq. (6.10) was solved stipulates, as already mentioned, that the initial compositions of materials A and B do not change at the free ends. Now, if the diffusion process is allowed to take place for long times, and if it is additionally conducted at high temperatures, a complete mixing of the two materials A and B will eventually take place. In other words, a uniform concentration of both components is then found along the entire couple, as indicated in Figure 6.6 by the horizontal line marked C_i. In this case, which is called complete *interdiffusion*, Eq. (6.11) is no longer a valid solution of Fick's second law. Moreover, Eq. (6.11) loses its validity long before a uniform composition is attained.

6.1.9 Kirkendall Shift

So far, we have considered only the diffusion of one component (say, material A) into another material (say, B). If both elements have the same diffusivity in each other (that is, if A diffuses with the same velocity into B as B diffuses into A), then a mirror image of the diffusion profiles shown in Figure 6.6 can be drawn for the other component. In many cases, however, this condition is not fulfilled. For example, copper diffuses with different velocity into nickel than vice versa (Table 6.1). Let us assume that element A diffuses faster into B than B into A. As a consequence, more A atoms cross the initial interface between the two elements in one direction than B atoms into the opposite direction. The result is that the initial interface shifts in the direction of element A.

This can be best observed, as done by Kirkendall, by inserting inert markers (usually fine wires of metals having a high melting point) between the two bars of metals A and B before joining them. It is then observed that these markers move into the direction of the A metal upon prolonged heating of this diffusion couple, as schematically depicted in Figure 6.8. The shift is not very large, so that annealings close to the melting point of the elements lasting for many days are necessary.

After cooling, small sections parallel to the interface may be removed in a lathe which are then chemically analyzed for their compositions. This latter procedure is, incidentally, common for many diffusion experiments. As an alternative, radioactive "tracer elements" are utilized for determining the amount of diffused species rather than the less accurate chemical analysis. An even more advanced technique utilizes the microprobe, which scans an X-ray beam along the specimen whose response signal eventually yields the composition. It might be noted in passing that in cases for which the atom flow of the species under consideration is considerably unbalanced, some porosity in the diffusion zone might develop.

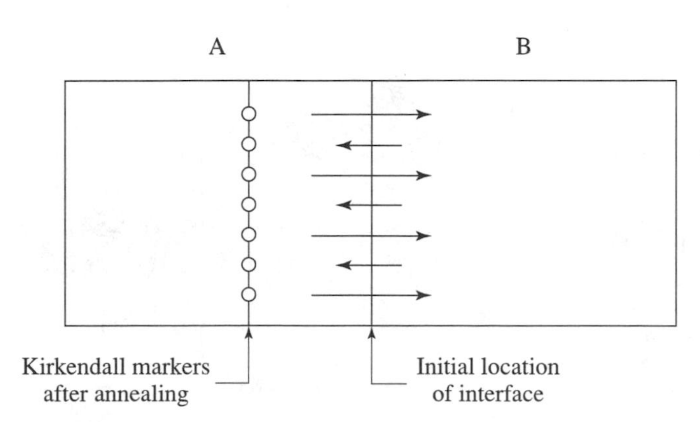

FIGURE 6.8. Schematic representation of the movement of inert markers (inserted at the initial interface) due to faster movement of A atoms into B than vice versa (Kirkendall shift).

6.1.10 Diffusion in Amorphous Solids

So far we tacitly assumed diffusion in crystalline solids only. It goes almost without saying that diffusion also occurs in amorphous solids. Indeed, because of the more open structure in amorphous solids, the atoms may diffuse there with even less effort than in crystals.

6.2 • Diffusion in Ceramics and Polymers

Diffusion phenomena have been mostly studied in metals. Nevertheless, it is quite conceivable that similar mechanisms would also take place in ceramic materials. Thus, the basic theories discussed so far may essentially also apply to nonmetals. However, there are several modifications that need to be taken into consideration when discussing ceramics. First, the ionic and covalent binding forces (Section 3.2) hold the atoms much stronger into their lattice positions so that the diffusion of matrix atoms in ceramics is substantially more difficult. Second, ionic compounds (Figure 3.18) need to maintain their charge neutrality. Thus, any removal of a cation (having, say, one positive charge) would also require the removal of an anion of equal, but negative charge.

We mentioned earlier that such a pair of vacancies having opposite charges is called a *Schottky defect*. This type of disturbance is dominant in alkali halides such as NaCl. (In contrast to this, vacancy-interstitial pairs, that is, Frenkel defects, are dominant in AgCl or AgBr.) The situation becomes even more involved when a matrix ion is replaced by an impurity ion of different valence. If, for example, a divalent Mg^{2+} ion substitutes for a monovalent Na^{+} ion in NaCl, another Na^{+} ion needs to be removed in order to restore charge neutrality, thus creating a cation vacancy

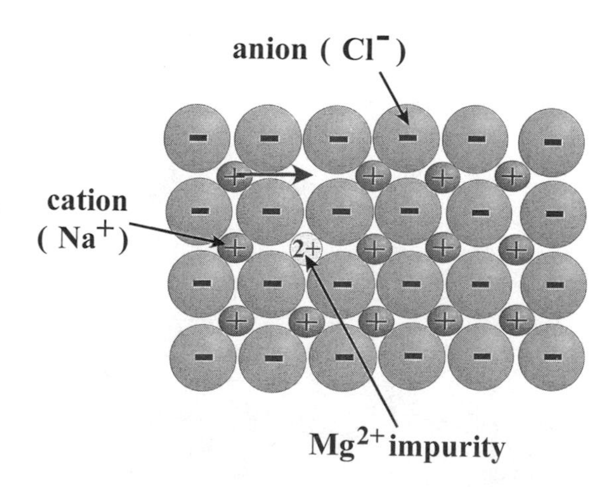

FIGURE 6.9. Schematic representation of a (100) plane of an ionic crystal having the NaCl structure and diffusion of a cation towards a cation vacancy. See also Figure 3.18.

(Figure 6.9). Cations, because of their smaller size (Figure 3.2), diffuse more readily than anions but still must squeeze between two anions to reach a vacant lattice site; see Figure 6.9. It is therefore not surprising that the activation energy for diffusion in ionic crystals is about twice as large as for volume self-diffusion in many metals. In short, self-diffusion in ceramic materials is generally less prevalent than in metals and alloys. However, the forced migration of ions in an electric field plays an important role in ionic conduction. We shall return to this topic in Chapter 11.6.

Diffusion Through Polymers

Diffusion of atoms or gas molecules between the individual polymer chains occurs faster in amorphous polymers compared to that in crystalline polymers, as can be inferred from Figure 3.7. The rate of diffusion through polymers depends additionally on the size of the migrating atoms. This is exploited in desalination plants where a separation between water molecules and dissolved ions is achieved. Diffusion *within* a polymer chain, involving the motion of covalently bound atoms, is, instead, energetically less favored as outlined above for ceramics.

6.3 • Practical Consequences

The previous pages provided some evidence for the fact that atoms are generally in motion, particularly if temperature or diffusion coefficients are high. The diffusion is directional when specific types of gradients are present. These gradients may be caused, for example, by local differences in composition of the constituents, or by local accumulations of empty lattice sites, or by differences in crystal energy. As a result of the motion of atoms, the properties of materials often change with time. This may have detrimental consequences upon the long-term stability of materials that are in service. On the other hand, the motion of atoms can be advantageously exploited for the improvement of the properties of materials. A few examples may illustrate both cases.

6.3.1 Age Hardening

The controlled precipitation of a second phase out of a metastable solid solution yields an increase in hardness as discussed in Section 5.3. The hardening effect depends on the annealing time and annealing temperature. Age hardening is a diffusion-controlled process driven by the instability of the quenched solid solution. As an example, some copper atoms in an Al–4% Cu alloy migrate through the matrix and eventually form copper monolayers (platelets) on {100} planes called GP-I Zones, as discussed in Section 5.3. In essence, the precipitated GP zones have a lower energy state compared to the metastable solid solution. In other words, the diffusion of copper atoms to {100} planes and the formation of GP zones are driven by a gradient in the free energy.

6.3.2 Grain Growth

The growth of grains is likewise driven by the tendency of materials to assume the lowest possible energy state. Specifically, a large number of small grains and, thus, a large amount of grain boundary area represent a comparatively high energy state because the atoms in or near grain boundaries are not efficiently packed, leaving a substantial amount of open space. In contrast, a few large grains (or a single crystal) are in a lower energy state. In order to reduce the free energy, crystal structures tend to increase their grain size. The mechanism by which some grains grow at the expense of others involves the slight shift of atoms situated near grain boundaries from one grain to the next, thus adding atom by atom to the planes of neighboring grains.

As pointed out in Section 5.5, the strength of materials decreases when only a few large grains are present. Thus, a large-grained microstructure is generally undesirable. (There are some exceptions where large grains or even single crystals are wanted, such as for turbine blades and at high temperatures to avoid creep or in microelectronic devices.) To prevent grain growth, impurity particles (which are selected so that they are essentially insoluble in the matrix) are occasionally added to metals or alloys. They reduce the grain boundary area and therefore the total grain boundary energy, thus reducing grain growth and, consequently, preventing a decrease in strength. This procedure is called *grain boundary pinning*. As with diffusion-controlled processes, an increase in temperature accelerates the grain growth. This could have unwanted consequences when materials are put in service at high temperatures.

6.3.3 Sintering

Sintering is driven by a thermally activated process. This technique involves the joining together of small (0.1- to 50-μm) particles consisting, for example, of high melting point metals, or of ceramics, or of composites. The powdered constituents are pressed into forms and then heated at high temperatures for extended times, causing the surface atoms to diffuse into the empty spaces between the particles. Concomitantly, vacancies diffuse into the grain boundaries and are annihilated there. This way the pores eventually close up and the work piece shrinks, that is, it becomes dense. The powders are obtained either by grinding brittle materials (pulverization), by chemical conversion, or by directing a pressured gas stream toward a thin stream of liquid metal ("atomization").

6.3.4 Annealing

Annealing of metals and alloys is an important procedure that is applied, for example, to cold-worked metals and alloys to restore the ductility which the material had before plastic deformation. Thus, it makes materials susceptible for renewed deformation and shaping. Annealing of cold-worked materials at *moderate temperatures*, called *recovery* or *stress-relief annealing*, causes the tangled dislocations to move. They eventually rearrange and form subgrains. The deformation-induced stress is reduced without substantially changing the number of dislocations. Thus, the strength of stress-relieved materials is almost maintained. The result is referred to as a *polygonized* structure. (Even though recovery is a thermally activated process, it is not classified as diffusion-controlled.)

It should be briefly added at this point that cold-working reduces the electrical conductivity of metals (see Part II). The decrease in conductivity is, however, restored by recovery annealings. This is of great importance to industries which produce wires for electrical power transmission, because wire drawing causes high strength and moderate annealing reinstates high conductivity.

If the cold-worked material is annealed at approximately 0.4 times the absolute melting temperature (T_m), new grains nucleate and grow. This process is appropriately called *recrystallization*. It restores the original high ductility, low strength, and relatively low dislocation density. The reduction in strength thus obtained could have detrimental consequences when a material is put in service at high temperatures. Recrystallization is driven by the tendency to completely eliminate the deformation-induced strain energy.

The just-mentioned recrystallization temperature depends on the amount of cold-work. Specifically, a large amount of deformation reduces the recrystallization temperature because of the larger amount of strain energy which was introduced by the work hardening process. On the other hand, small amounts of cold-work (several percent) do not lead to appreciable recrystallization upon heating within a reasonable amount of time.

Metals and alloys that are heat-treated at temperatures above 0.4 T_m eventually tend to undergo further *grain growth*. We have already discussed previously the implications of large-grained materials on the mechanical properties and explained there why large-grained materials have a lower strength than fine-grained materials.

6.3.5 Creep

We briefly mentioned in Chapter 2 that materials, when held at high temperatures, generally undergo progressive plastic deformation while under a stress level that is well below the room temperature yield strength. This mechanism may be a diffusion-controlled process that involves, among others, the climbing of edge dislocations. It is called *dislocation creep*. In Chapter 3, we talked extensively about dislocations and explained there that the slip of dislocations is the principle mechanism for plastic deformation under stress. We also elucidated that eventually the movement of dislocations may be blocked by precipitates or other obstacles. Now, if the temperature is above 0.3 times the absolute melting temperature and a metal is under stress, the pinned dislocations may be unlocked. Specifically, the stress provides a driving force that allows the atoms at the bottom of the extra half-plane of an edge dislocation to diffuse away and thus cause the dislocation to climb as depicted in Figure 6.10. (Alternatively, atoms may diffuse to the bottom of an edge dislocation, causing the "climbing" to be in the opposite direction.) Eventually the pinned dislocation has climbed away from the sphere of influence of the pinning site. The dislocation may then undergo further slip until the next obstacle is encountered. This repeated climbing and gliding leads to continuous plastic deformation, that is, to creep.

The creep rate at a constant applied stress can best be visualized by plotting the creep strain, ϵ, as a function of creep time, t, as shown in Figure 6.11. Three regions may be distinguished. Initially, the creep progresses quite rapidly but eventually slows down as the deformation of the material becomes more difficult due to work hardening, as discussed in Section 3.4. This region is called **primary** or **transient creep**. In a second range steady-state creep occurs when, as an average, the same amount of dislocations climb away from obstacles as dislocations are blocked on obstacles. In this case the creep rate is constant, as seen in Fig-

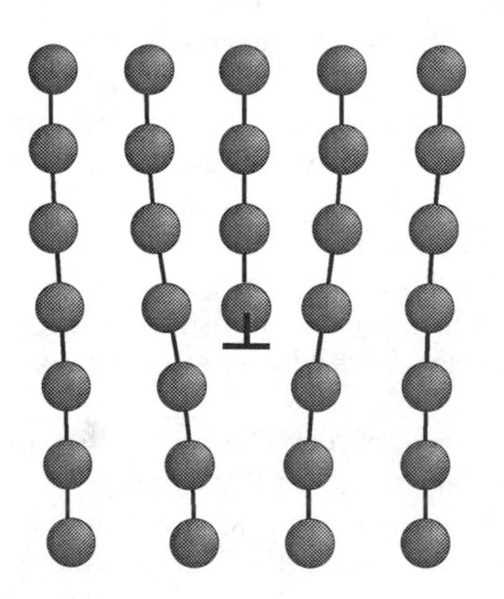
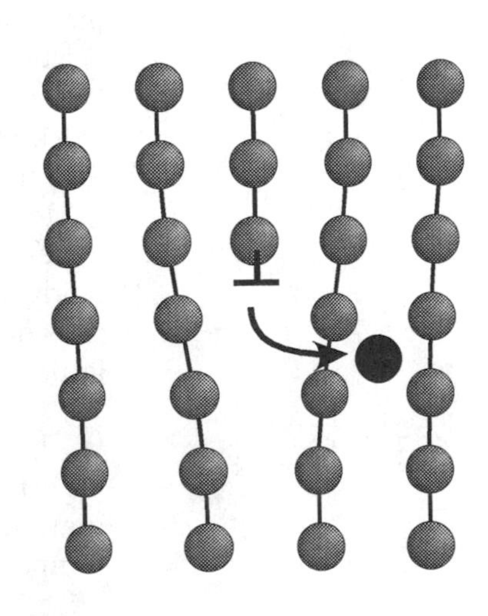

FIGURE 6.10. Climbing of a dislocation at high temperatures caused by diffusion of atoms away from the bottom of the extra half-plane of an edge dislocation. The moving atoms could either fill vacancies or jump into interstitial positions. (The dislocation moves in the opposite direction when atoms are attached to it.)

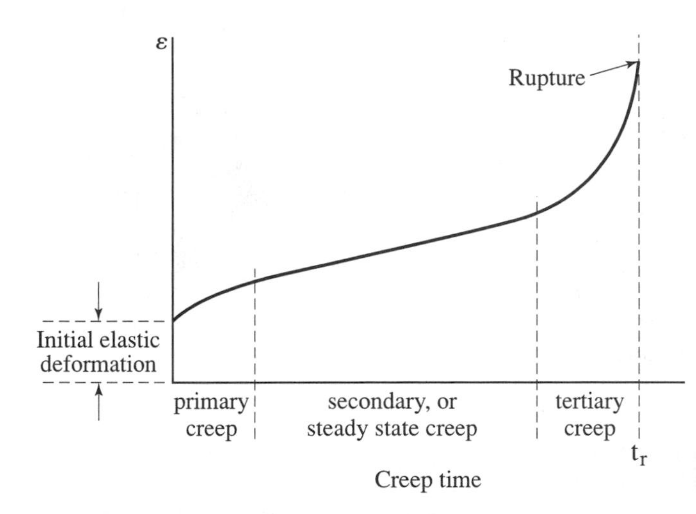

FIGURE 6.11. Schematic representation of a creep curve as typically encountered for single-phase materials in which the creep strain, ϵ is depicted as a function of creep time, t during which a constant stress is applied to the solid. The slope in the secondary range is termed the steady-state creep rate, $\left(\frac{d\epsilon}{dt}\right)_s = \dot{\epsilon}_s$.

ure 6.11. The **steady-state creep** rate, $\dot{\epsilon}_s = (d\epsilon/dt)_s$, obeys as usual an Arrhenius-type relation, which reads in the present case:

$$\dot{\epsilon}_s = A\ \sigma^n \exp\left(-\frac{Q}{k_B T}\right), \tag{6.14}$$

where A is the creep constant, σ is the stress [Eq. (2.1)], n is another constant called the creep exponent, which varies between 3 and 8, and Q is the activation energy for creep (having about the value of that for self-diffusion). This type of creep is called **power law creep**.

After even longer times, the creep rate accelerates again (tertiary creep region), which leads to void formation at grain boundaries and possibly necking of the work piece [Figure 2.5(a)]. Eventually the voids join, causing fracture due to separation at grain boundaries. Likewise, internal cracks or cavities may join. At this point, the *rupture lifetime*, t_r, has been reached.

At low stress levels and very high temperatures, an alternative mechanism may cause creep. It involves the force-induced elongation of grains in one direction (and shrinkage in the other) due to the migration of atoms or vacancies between grain facets; see Figure 6.12. Again, diffusion plays a decisive role which allows the motion of atoms within a grain. It is called **diffusion creep** (or **Nabarro–Herring creep**.) The creep rate is in this case proportional to the stress, σ, and inversely proportional to the square of the grain size, d. The activation energy, Q, is about that for self-diffusion; see Table 6.1. One finds:

$$\dot{\epsilon} = \frac{d\epsilon}{dt} = \frac{B\sigma}{d^2} \exp\left(-\frac{Q}{k_B T}\right), \tag{6.15}$$

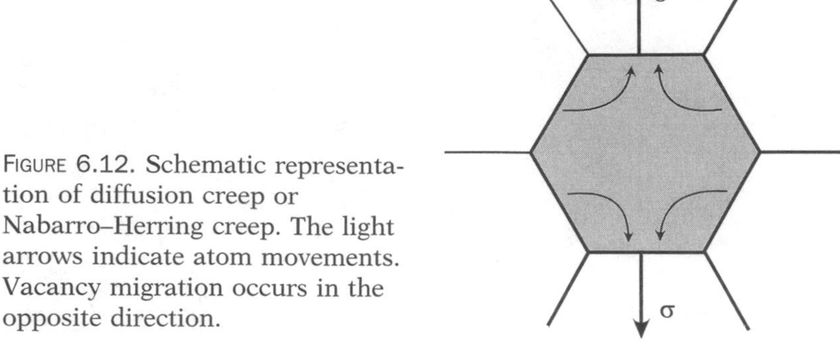

FIGURE 6.12. Schematic representation of diffusion creep or Nabarro–Herring creep. The light arrows indicate atom movements. Vacancy migration occurs in the opposite direction.

where B is a constant. (The other parameters have their usual meanings.)

In a further mechanism which involves creep deformation and which occurs at somewhat lower temperatures, vacancies are thought to migrate along the grain boundaries (**Cobble creep**). The activation energy is now that for grain boundary diffusion.

6.3.6 Creep in Ceramic Materials

Ceramics, such as metal oxides, carbides, halides, etc., are inherently more resistant to creep than other materials. First, because of their relatively high melting points, they start to creep at higher temperatures than metals. Second, ceramics start to creep above 0.4 to 0.5 T_m. This is of considerable importance for, say, gas turbine blades made of ceramics where local temperatures in the vicinity of 1200°C are not uncommon. Third, since the atoms in ceramics are bonded predominantly by strong ionic or covalent forces (Section 3.2), any diffusion in ceramic materials is extremely small and generally only possible if, for example, a cation vacancy exists in which a cation may jump. (Cations are smaller in size, and therefore diffuse faster than anions as explained in Section 6.2.) Because of this smaller diffusion rate, dislocation climb (Figure 6.10) is less likely to occur. Ceramics creep instead mainly by grain boundary sliding, coalescence of microcracks, and due to porosity, if present.

Creep in Glasses

On the other hand, the creep rates of **glasses** are linked to their viscosity. Specifically, glasses become more pliable at high temperatures at which, for example, silicate (SiO_2) rings, chains, or islands move past each other when a force is applied to the material (see Figure 15.3). An Arrhenius-type equation exists which relates the viscosity, η, of a glass to an activation energy, Q_η:

$$\eta = \eta_0 \exp\left(\frac{Q_\eta}{k_B T}\right). \quad (6.16)$$

The activation energy depends on the ease with which atom groups slide past each other. Certain additions to glass, such as alkalis, lead, calcium, etc. (called *modifiers*), break up the glass network structure to some degree, thus reducing its strength and viscosity and causing an enhancement of the creep rate at a given temperature. See also Chapter 15.

6.3.7 Polymers

Polymers creep above the glass transition temperature at which the material is in a rubbery state. Below this temperature polymers are hard and sometimes brittle, and essentially do not creep. We return to this subject in Section 16.4.

Summary

In summary, some materials, such as ceramics, diamond, and tungsten, creep at rather high temperatures due to their high T_m and their specific crystal structure, whereas others, such as lead, tin, and many polymers (nylon, polypropylene, etc.), already creep at room temperature, for example, due to gravity. Incidentally, *glaciers* (ice) also move due to creep.

6.3.8 Superplasticity

An interesting phenomenon, which involves Nabarro–Herring creep, is superplasticity: As already explained above (Figure 6.12), the strain under tension causes under certain circumstances a stress-biased diffusion of atoms from grain boundaries lying essentially parallel to the stress direction to other boundaries perpendicular to the stress. As a result some alloys have been observed to elongate enormously. The present world record is held by Higashi in Japan at 8000 % elongation! Superplasticity is defined to be the ability of polycrystalline materials to exhibit, in a generally isotropic manner, very high tensile elongations prior to fracture. Several prerequisites for this to occur are necessary:

- The metal must have a very fine grain structure (i.e., grain sizes in the micrometer range).
- The alloy needs to be strained at high temperatures (about 0.6 T_m).
- The strain must be applied at very slow rates (typically 0.01

sec^{-1}) so that the diffusion of atoms can keep pace with the employed strain rate.
- The grains should easily slide and roll over one another under stress.

Examples for superplastic alloys are Al with 6 mass% Cu and 0.5 mass% Zr (Trade name SUPRAL), Bi-Sn, Al-Cu eutectic, Ti-6%Al-4%V, and Zn-23%Al.

Applications of superplasticity include the plastic deformation of certain consumer parts at modest stresses and by using inexpensive die materials, making this technique suitable for small production runs.

It should be added that ultrafine-grained ceramics can also be superplastically deformed by the Nabarro–Herring mechanism. Finally, as we will discuss in Chapter 15, glass may be drawn out rapidly in an appropriate temperature range into very long and fine fibers and thus, show likewise superplasticity.

6.3.9 Electromigration

Finally, a few words should be added about the diffusion of metal ions driven by strong electric fields. This mechanism is called *electromigration* and is of great concern to the electronics industry. It involves the forced motion of, for example, aluminum ions in aluminum thin film stripes that interconnect the individual transistors in microelectronic circuits. As a consequence of the unidirectional motion of the Al ions from the cathode to the anode, voids near the cathode appear which lead to failure of the device. The preferred paths for the electromigrating aluminum ions are the grain boundaries. This is understandable when it is recognized that the activation energy for diffusion in grain boundaries is only one-half of that for volume diffusion. Numerous attempts have been undertaken in the past to alleviate electromigration in thin films. Among them are the inoculation of grain boundaries by foreign atoms (such as copper) to block some of the grain boundary diffusion, orienting the grain boundaries perpendicular to the electric field (creating a so-called bamboo structure), or utilizing better suited metallization materials, such as gold or copper which have a higher activation energy than aluminum.

6.4 • Closing Remarks

The previous chapters have taught us that the arrangement of atoms or, better, their lack of regular arrangement, together with the motion of atoms and their mutual interactions, play a key role in understanding some fundamental mechanical properties of materials. The motion of atoms is enhanced by thermal activation. This thermal activation is often governed by an exponential law that contains the absolute temperature and an activation energy as variables, as exemplified by a number of Arrhenius-type equations.

Problems

6.1. Calculate C_x for $x = -\infty, x = 0$, and $x = +\infty$ using Eq. (6.11) and compare the results with Figure 6.6.

6.2. Calculate the concentrations C_x for $t \to \infty$ and $t \to 0$ using Eq. (6.11) and compare the results with Figure 6.6. (*Hint*: For $t \to 0$, use positive x as well as negative x values.)

6.3. Calculate the ratio between the diffusion rates of grain boundary diffusion and volume diffusion assuming an activation energy for grain boundary diffusion to be one-half of that for volume diffusion. Take $T = 500°C$; $Q_v = 2$ eV; $(f_0)_v = 6 \times 10^{15}$ $1/s$ and $(f_0)_g = 1.5 \times 10^{11}$ $1/s$.

6.4. In an aluminum alloy, the atomic diffusion constant $D = D_0 \exp(-Q/k_BT)$ is given by $D_0 = 10$ cm^2/s and $Q = 1.3$ eV. The vacancies in this material are thought to have a formation energy, E_f, of 0.7 eV. For an observable microscopic atomic diffusion, D should be larger than 10^{-18} cm^2/s.

(a) Give the atomic diffusion constant for $T = 300$ K (room temperature). Is its value sufficient for observable diffusion?

(b) The alloy is homogenized at 750 K followed by a quench into water. What is the vacancy concentration at this temperature? (Use the room temperature atomic mass and density for Al.)

(c) During quenching, 70% of these vacancies anneal out; the rest are quenched-in and are present at 300 K. What is their concentration?

(d) What happens to these vacancies? Do they anneal out at 300 K? Determine the diffusion constant of the vacancies assuming $D_{vo} = 10$ cm^2/s. (*Hint*: Note that the diffusion constant for vacancies, D_v, equals $D_{vo} \exp-[(Q - E_f)/k_BT]$.) Is there an observable vacancy diffusion?

(e) What happens to the atomic diffusion constant at 300 K during annealing out of these vacancies?

6.5. Calculate the approximate time that is needed for diffusing carbon into steel to a depth of 1 mm when the diffusion coefficient at the temperature of heat treatment is 2×10^{-7} cm^2/s.

6.6. Problem 6.5 above and the rule-of-thumb equation (6.13) do not specify the concentration C_x at the point X. The following exercise alleviates this shortcoming for some specific cases: Derive an equation similar to (6.13) from (6.11) by assuming (a) $C_x = C_0$ ("far field value"), and (b) $C_x = (C_i + C_0)/2$ ("effective diffusion depth" which is the "average value between surface and final concentrations"). (c) For what value of C_x does one obtain (6.13)?

Suggestions for Further Study

R.T. DeHoff, *Thermodynamics in Materials Science*, McGraw-Hill, New York (1993).

R. Haase, *Thermodynamics of Irreversible Processes*, Dover, New York (1990).

R.E. Hummel, "Electromigration and Related Failure Mechanisms in Integrated Circuit Interconnects," *International Materials Reviews*, **39** (1994), p. 97.

H. Mehrer, *Diffusion in Metals and Binary Alloys*, in: Diffusion in Condensed Matter, P. Heitjans and J. Kärger Eds., Springer-Verlag, Heidelberg, New York (2003).

P. Shewmon, *Diffusion in Solids*, McGraw-Hill, New York (1963).

7

The Iron Age

Historians claim that the Iron Age began between 1500 and 1000 B.C. (at least in some parts of the world). This does not mean that iron was unknown to man before that time; quite the contrary is the case. Meteoric iron (which has a large nickel content) must have been used by prehistoric people as early as 4000 B.C. They made tools and weapons from it by shaping and hammering. It is thus quite understandable that in some ancient languages the word for iron meant "metal from the sky". Naturally, the supply of meteoric iron was limited. Thus, stone, copper, and bronze were the materials of choice at least until the second millennium B.C. There were, however, some important uses for iron ores during the Bronze Age and also during the Chalcolithic period. As explained already in Chapter 1, copper needs a fluxing agent for the smelting process when using malachite. For this, iron oxide was utilized, which was known to react during smelting with the unwanted sand particles that are part of malachite. Eventually, a slag was formed which could be easily separated from the copper after the melt had cooled down.

It has been frequently debated and asked by scholars in which way early man might have produced iron, utilizing terrestrial sources in particular, since the melting point of iron is 1538°C. This temperature was essentially unachievable during that period, at least in the western (or middle eastern) part of the world. The answer can probably be found by considering the abovementioned slag, large amounts of which have been found in areas at which major copper smelting operations were conducted. This slag was observed to contain some reduced iron, but in a porous condition, which is today known by the name of **sponge iron** or **bloom**. When bloom is repeatedly hammered at high temperatures, the slag can be eventually removed and the iron is compacted. In this way nearly pure iron is obtained. The end product is known among today's metallurgists by the name of **wrought iron**. Therefore, it can be reasonably assumed that iron production took its way from reducing iron ore into spongy bloom, which is a process that needs a lower temperature (about 1000°C) than melting pure iron. In other words, the temperature was never high enough to yield a liquid product. Bloom was then eventually hammered into wrought iron. However, pure iron is quite soft; actually, it is softer than bronze, as can be seen from Figure 7.1. Additionally, pure iron corrodes readily when exposed to air of high humidity. As a consequence, pure iron must have been of little interest to early man, at least until a major discovery was made to produce *"good iron"*, as it was named in old records.

The discovery of good iron is credited by many archaeometallurgists to the Hittites, or probably to subjects of the Hittites (called the Chalybes) who lived for some time in the Anatolia-Mesopotamia region which is today Turkey. The Hittites conquered large areas of the Mediterranean, such as Assyria, Babylon, and Northern Palestine. They challenged the Egyptians and the Syrians. The system of government of the Hittites is said to have been more advanced than that of many of their neighbors, and their legal system emphasized compensation for wrongdoings rather than punishment. The Hittite language (belonging to the indogermanic languages) was recorded in hieroglyphics or in cuneiform (a system of syllabic notations, borrowed from the Mesopotamians), and their international correspondence was written in the Akkadian tongue. Legend has it that their successful weapons consisted of swords, spears, and arrows made of iron which pierced through the bronze shields of their enemies. But their light and fast chariots certainly must have like-

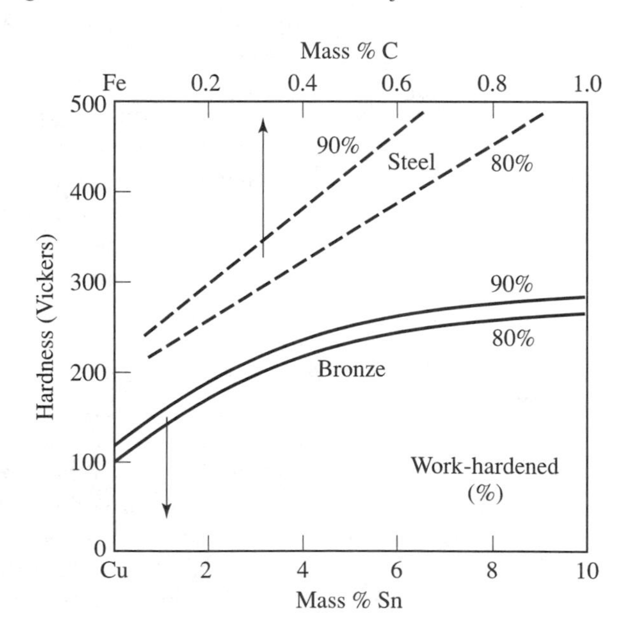

FIGURE 7.1. Hardness of various steels and bronzes as a function of composition and degree of work hardening. The work hardening is given in percent reduction of area (see also Table 2.1).

wise contributed to their victories. The secret of making good iron is said to have been kept by the Hittites for two hundred years, that is, from about 1400 B.C. to about 1200 B.C.

Good iron was produced by applying repeated cycles of heating a piece of bloom in a charcoal furnace near 1200°C for softening purposes and by subsequently hammering to remove the slag and to compact it. During the heat treatment, the bloom and eventually the iron was frequently exposed to the carbon monoxide gas of the burning charcoal. As has been explained in Chapter 6, this procedure is supportive of carbon diffusion into the surface of iron. As a result, an iron–carbon alloy is formed (called *steel*) which is substantially harder than bronze (e.g., Cu–10% Sn), even if the carbon content is only about 0.5% as seen in Figure 7.1. (Steel is defined to be iron that contains up to 2.11 mass percent carbon.) The steel of antiquity had about 0.3 to 0.6% carbon. Cold-working adds additional strength (Figure 7.1). Iron Age man must have also observed that limiting the carbonization to only the surface (such as the edge of a blade or the tip of a tool) combined high hardness of the surface with good ductility in the interior. Examples of "selective steeling" whose end product is iron with 1.5% carbon on the surface have been found from time periods as early as 1200 B.C. This process corresponds essentially to modern *case hardened iron*.

The role of carbon on the hardness of iron and steel was, however, not recognized for a long time. Indeed, the Greek philosopher and scientist Aristotle (384–322 B.C.), among others, believed (contrary to the truth) that steel was a purer form of iron due to the "purifying effect of charcoal fire." It was not before 1774, when S. Rinman, a Swedish metallurgist, discovered a "graphite-like residue" when cast iron was dissolved in acid. Seven years later, Bergman and Gadolin finally reported the different amounts of carbon in various irons and steels.

There were two more discoveries which were probably made during the first millennium B.C. that improved the quality of carbonized iron even further. One of them (interestingly enough, described in Homer's *Odyssey*) involves *quenching*, that is, a rapid cooling of a red-hot piece of carbonized iron into cold water. This procedure hardens the work piece considerably more, sometimes even to the extent of brittleness. As a result, quenched swords, tools, and other utensils may have cracked or even shattered.

The other discovery which was made during the end of the first millennium B.C. entailed a short-time reheating of a previously quenched piece of steel to about 600°C. This procedure, which is known today as *tempering*, restores some ductility and relieves the brittleness at the expense of some loss in hardness. We shall provide the scientific explanations for all these processes in Chapter 8.

After a relatively short time of Hittite dominance over the Mediterranean region (from about 1900 B.C.), some European tribes, vaguely called in the literature the "Sea Peoples," overran the Middle East in about 1200 B.C., destroying almost everything on their way. This caused the Hittites to vanish almost into oblivion. The destruction of the Hittite empire probably caused the scattering of their subjects and with them their iron-making skills. It can be observed that, after this dispersal of Hittite metal artisans, iron production was eventually conducted almost everywhere in the western part of the world. For example, iron making was practiced by the Celtic tribes (from about 500 B.C.), who lived in Europe between the Mediterranean and the Baltic Seas and from the Atlantic Ocean to the Black Sea. They put iron rims on the wheels of their chariots, fitted their horses with shoes, produced iron plowshares, and invented the chain armor (see Plate 7.5). Despite their achievements, the distribution of iron utensils was not widely spread and was probably limited to the upper class, mostly because of its labor-intensive production from iron bloom.

Before our attention is directed to iron smelting in the Far East, we need to discuss the important question concerning the reasons why the Bronze Age people abandoned their well-established technology and turned to a new material, that is, to iron. Certainly, iron ore was more abundant than copper or tin. Actually, 5% of the earth's crust consists of iron whereas the abundance of copper and tin on the earth's crust is only 50 and 3 parts per million, respectively. Additionally, iron was often available on the surface of the earth, which did not necessitate underground mining. But, as discussed above, the high melting temperature of iron was virtually inaccessible in the Mediterranean basin during the second millennium B.C. The apparent reason for turning away from bronze was something much graver, namely, the interruption of trade routes probably by the above-mentioned Sea Peoples. As a consequence, the supply of tin, wherever it may have come from, was cut. When new bronze articles were wanted one had to rely on "recycled" bronze, that is, on bronze which was obtained by melting down earlier goods. Thus, 2000 years of bronze technology came to a halt in the Middle East in a relatively short span of time for lack of raw materials.

The same conditions naturally did not apply to the Far East. Thus, the Bronze Age lasted there somewhat longer. Eventually, however, iron making came to China most likely from the West between 1000 and 650 B.C. as a result of the above-mentioned dispersion of Hittite metal artisans and their know-how. At the beginning, the Chinese most probably applied the Western technique of converting bloom into wrought iron with subsequent carbonizing and possibly quenching and tempering. Very soon, however, Chinese iron makers went their own way by utilizing much larger and more powerful, horizontally operated, double-acting box bellows. They were driven by animals or water wheels and probably also by several humans. (This application of large amounts of forced air was reinvented in the West in the fifteenth century and was then called the **blast furnace**.) Most important, however, the Chinese also increased their carbon monoxide content by enlarging their furnaces and substantially increasing the amount of charcoal fuel. As a result of both improvements, significant amounts of carbon diffused into the iron. This, in turn, decreased the melting point of the resulting charge to as low as 1148°C. We have learned already in Chapter 5 that the melting temperature of metals is often reduced by alloying, that is, by adding a second constituent to a substance. (In the present case, the lowest melting point is obtained for the eutectic composition, which involves iron with 4.3% C, as we shall see in Chapter 8.) As a result of this new technology, iron could be cast similarly as bronze. Today, crude cast iron taken directly from the furnace is called **pig iron** because a row of parallel molds is said to resemble little piglets drinking on their mother.

Iron that contains large amounts of carbon is quite hard, but it is also brittle. The material is therefore almost worthless for tools and weapons because it cracks or shatters easily when a blow is applied to it. Thus, cast iron requires an additional treatment. This new treatment was probably introduced by the Chinese at about 500 B.C. It consisted of removing some of the excess carbon from the surface of high carbon iron. This eventually yielded a steel jacket that has similar properties as the steel that western people had produced when carbonizing wrought iron. To accomplish the reduction of carbon a piece made of cast iron was heated at temperatures between 800 and 900°C in the presence of air. The oxygen in the air combines with some of the carbon and forms carbon monoxide gas, which is allowed to escape. In essence, both the Chinese and the Mediterranean people eventually achieved a similar product but arrived at it from opposite directions. The main advantage of the Chinese technology was, however, that the Chinese could shape their products by casting, which allowed easy mass production, whereas the Western world had to shape and carbonize their goods individually by hammering.

Not enough. There was still another development that arose from China in the first century A.D. It involved the stirring of carbon-rich iron to allow the oxygen from the air to react with the carbon of the melt. As a consequence, the carbon content was already reduced in the melt to the extent that it yielded steel. This process, which is called *puddling* today, was rediscovered in England in 1784. During the latter part of the Han dynasty (202 B.C.–A.D. 220), immense industrial complexes were operating near Zheng-Zhou and other places containing several huge furnaces (4×3 m in area and about 3 m high) which might have produced several tons of iron per day. Moreover, the Chinese already used the technique of *stack casting*, that is, they arranged several molds on top of and next to each other, which enabled them to cast up to 120 articles at the same time. The large-scale production of plowshares, hoes, cart bearings, and harness buckles made the output quite cost-effective. It thus allowed a wide distribution of tools for working the fields and digging irrigation channels, which in turn might have led to a larger production of agricultural crops and probably even to an increase in population. In other words, the change that was brought about by the introduction of iron and steel slowly revolutionized the way the Chinese and probably other peoples lived and worked. Nomads have no need for agricultural tools, but the availability of these tools probably contributed to the settlement of some nomads. And finally, iron paved the way from agriculture to industry. The potential for China to become a world-wide industrial power already 2000 years ago was laid by these inventions and by their large-scale exploitation. But Chinese bureaucracy upheld by Confucian-trained civil servants apparently stifled new ideas and the expansion of major trade beyond the boundaries of China.

The same revolutionizing developments that were caused by the use of iron were eventually also seen in other parts of the world. Goods made of iron were traded virtually everywhere. Iron axes reduced the amount of forests to provide fuel and to clear land for feeding more people. Weapons made of iron or steel (see for example Plate 7.2) unfortunately provided the means to conquer and often destroy other civilizations. Knights wore armors of iron. Indeed, iron was and still is, in many cultures, the symbol for strength, power, and will.

One particular amazing demonstration of iron workmanship is the famous iron pillar next to the Qutub Minar tower on the outskirts of Delhi in India. The pillar made of forged iron is seven meters tall and has a purity of about 99.2%, containing only small amounts of sulphur (0.08%), phosphorus (0.11%), silicon (0.46%), and carbon (0.08%). It dates back to the fourth century A.D. and probably was manufactured by heating and hammering together a large number of small iron pieces. Most amazingly, however, the iron pillar has not experienced any corrosion during the 1500 years in which it has been exposed to air. It is speculated that the lack of rusting is due to a combination of climatic factors, high P and S contents, and a large heat capacity. Another, larger iron pillar was found at Sarnath, which was produced between 300 and 200 B.C. and before it broke was almost 14 m tall.

The time at which iron was first smelted in India is not exactly known. However, iron production is mentioned in the *Rigveda*, which is the oldest known Hindu religious book. Conservative estimates place its origin around 1200 B.C. Other sources claim that iron smelting in India did not commence before 600 B.C.

Another specialty of India that was produced and sold virtually across the entire continent from the first millennium B.C. until the middle ages was the so-called **wootz steel** which was later named **Damascus steel**. It is said that it was the raw material for the best swords and daggers of that time (see Plates 7.3 and 7.4). Wootz steel was made by placing small pieces of wrought iron or sponge iron (see above) together with some wood chips and leaves in small clay crucibles which were sealed with a clay lid and then heated in air-blast–enhanced fires. This enabled the carbon from the plant material to evenly penetrate the iron, thus providing an essentially homogeneous iron–carbon steel. (In contrast to this, the Mediterranean or Chinese technologies allowed only a portion of the work piece to be steel, as described above. The other parts were either wrought iron or cast iron.) The Indians succeeded in keeping their technique a secret until the seventh century A.D., after which the Syrians near Damascus and the Spaniards near Toledo came up with their own versions. The Damascus swords of later times were produced by joining and folding through hammer-welding alternate bars of iron and steel. Application of a dilute acid colored the steel sections leaving the iron relatively bright. Wootz steel was also reinvented in England during the eighteenth century.

When reporting about the advances which major civilizations contributed to the art of iron and steel making, it is often overlooked that less known peoples were also able to prepare iron goods. Their contributions might not have been as spectacular as those described above, but they still had some local impact on the lives in certain societies. Among them were the Haya people, who lived near the shores of Lake Victoria, which is part of today's Tanzania in East Africa. Their folktales are full of stories about iron making, and the vocabulary with which they are told is enriched with reproductive symbolism. (A PBS documentary entitled, "The Tree of Iron," witnesses to this effect.)

During the Middle Ages the knowledge of metallurgy in general and iron and steel making in particular precipitated into written documents. Among them were books like *De re metallica* by the German extractive metallurgist and miner Georgius Agricola, who summarized in 1556 the then-available knowledge on smelting, refining, and analytical methods as well as on prospecting and concentration of ores. Another book by the Italian metalworker Vannoccio Biringuccio, entitled *De la pirotechnia*, reported in much detail on smelting, analytical methods, casting, molding, core making, and foundry practices. Not all documents of that time were, however, of the same authority as the two works just mentioned. For example, the book *Von Stahel und Eysen* (*On Steel and Iron*), which was published in Nuremberg (Germany) in 1532 by an alchemist, is quite an entertaining read in the twentieth century. A few samples may illustrate this. One finds the following recipe in this volume under the heading "How iron is to be hardened and some of the hardness drawn again":

> Take the stems and leaves of vervain, crush them, and press the juice through a cloth. Pour the juice into a glass vessel and lay it aside. When you wish to harden a piece of iron, add an equal amount of a man's urine and some of the juice obtained from the little worms known as cockchafer grubs. Do not let the iron become too hot but only moderately so; thrust it into the mixture as far as it is to be hardened. Let the heat dissipate by itself until the iron shows gold-colored flecks, then cool it completely in the aforesaid water. If it becomes very blue, it is still too soft.

A good procedure for tempering might have been:

> Take clarified honey, fresh urine of a he-goat, alum, borax, olive oil, and salt; mix everything well together and quench therein.

Actually, the nitrogen in urea (H_2NCONH_2) probably led to nitrated, "case-hardened" iron. Here is another useful recipe:

> How To Draw The Hardness of Iron: Let the human blood stand until water forms on top. Strain off this water and keep it. Then hold the hardened tools over a fire until they have become hot and subsequently brush them with a feather soaked in this water; they will devour the water and become soft.

A splendid method of hardening could have been:

> Take varnish, dragon's blood, horn scrapings, half as much salt, juice made from earthworms, radish juice, tallow, and vervain and quench therein. It is also very advantageous in hardening if a piece that is to be hardened is first thoroughly cleaned and well polished.

(The latter recipe must have had restricted use because of the limited availability of dragon's blood.) Those who will not have instant success with these instructions may take comfort when reading in that book: "If fault should be found with some of the recipes, pray do not reject the whole book. Perhaps the fault lies in the user himself, because he did not follow the instructions correctly. All arts require practice and long experience, and their mastery is only gradually acquired."

Legend tells us about a medieval, germanic weapon smith, called Wieland, whose swords were unsurpassed in strength and sharpness. His secret recipe supposedly involved the filing of a forged piece of iron into a coarse powder that was fed to his chicken. He then separated the iron from the feces with a magnet. After seven passes, a superb sword was eventually forged from this material, which won a critical contest by slicing a competitor, who was wearing his armor, in half. A metallurgist working for a German company tried this procedure in the 1930s and found that the chickens reduced the carbon content of the iron in their digestive system while enriching the iron with nitrogen. This made the steel stronger, as scientific studies have demonstrated.

Today, iron is reduced from its ore in massive **blast furnaces**, up to 30 meters high, in which the preheated air is blast-injected through nozzles (*tuyères*) near the bottom of the furnace to obtain the necessary high temperatures. They are fueled by the more efficient *coke* (derived from coal) and fluxed by limestone. One blast furnace yields in excess of 4000 tons of pig iron per day. Elimination of very fine particles in the raw material (called *burden*), which may restrict the flow of gas, has resulted in productivity increases of as much as 100%. In countries with cheap electricity combined with easy access to raw materials, *electric furnace smelting* is used to produce pig iron. Another technique, called **direct reduction processes**, involves rotary kilns in which a mixture of pure, dry iron ore, a reducing substance, and limestone (as flux) is heated to about 1000°C. Reducing agents are, for example, coal, coke, graphite, fuel oil, or hydrogen. The process of making low-carbon–iron (or steel) directly from ore is quite attractive compared to a two-stage process in which high-carbon pig iron is produced first and then later is purified to steel. However, for the direct process the ore must be very rich, finely divided, and intimately mixed with the reducing agent in correct proportions. Thus, only 2% of the iron and steel produced in the world today are made in this way. The end product is nearly pure sponge iron or bloom as historically known for millennia. High purity sponge iron is also used in the chemical industry as a strong reducing agent and for powder metallurgy.

For modern *steel making* from pig iron mainly three procedures are utilized. Briefly, the **basic oxygen process** (BOP), which was invented in 1952 in Austria, involves blowing oxygen into the molten iron (which may contain up to 25% scrap steel) either from the top by means of a retractable lance or from the bottom through tuyères. The oxygen combines with the carbon and other impurities, thus reducing the carbon content to form steel (i.e., iron with less than 2.11% C). The BOP techniques are fast and cost-effective but do not provide an exact chemical composition of the steel. Further, the possibility for recycling of scrap steel is limited.

The **open-hearth process** uses a shallow, swimming pool–shaped (about 27 m × 9 m), fire brick-lined furnace in which air (heated essentially by oil burners) is blown horizontally over the surface of the melt. An oxygen lance from the top speeds up the carbon reduction. Since this technique is slow and utilizes oil for heating the air, and further, since it produces large volumes of polluting waste gases (carbon monoxide), its application has steadily declined to about 5% in the United States. On the positive side, considerable amounts of scrap iron (usually up to 50%) can be used. The principle features of this process were developed in 1864/68 by Siemens and Martin.

For melting and refining primarily steel scraps, **electric arc furnaces** are common and cost-effective. Again, oxygen is injected during the process. The heat is generated involving principally the electrical resistance between carbon electrodes and the ingot.

The **Bessemer converter**, which came into wide use in the middle of the nineteenth century, has been virtually replaced by the above-mentioned steel making techniques because of serious disadvantages of its products. Specifically, it has been found that the lack of ductility and workability of Bessemer steel was due to small percentages of nitrogen (about 0.015%!) that prevented its deep drawing into sheet metal. The Bessemer process involved air that was blown upward through molten pig iron held in a refractory-lined, pear-shaped vessel. Two types of linings were used. They consisted either of *basic* bricks for phosphorus-rich and silicon- poor pig iron (fluxed with limestone to remove the phosphorus which makes the iron brittle and thus useless) or acid lining for Si-rich/P-poor raw iron.

The large consumption of charcoal as fuel and as a reducing agent for iron smelting eventually took its toll on the forests of industrialized countries. Charcoal production involves the incomplete burning of wood which is stacked in a pile and covered with earth and leaves. This process drives off the volatile constituents in the wood that would otherwise contaminate the iron and thus may compromise the properties of steel. In order to preserve her forests and ship-building capability, England's Parliament in 1584 severely restricted the cutting of timber for charcoal production. To alleviate the resulting charcoal shortage, an alternative reducing agent and fuel was sought and eventually found in the seventeenth century. This substance is *coke*, which is obtained by heating soft coal powder in an airtight oven for the purpose of driving off the volatile impurities of coal. The result is 88% carbon.

The long history of iron and steel making is characterized by secrets, emotions, and drives for dominance. Wars have been regrettably fought and won *with* iron and steel, and wars have been fought *because* of iron and steel. For centuries apprentices and students have been trained and educated in the art and science of iron making. Entire research centers have been devoted solely for the purpose of better understanding and improving iron and steel. Societies and journals are exclusively devoted to iron and steel. Price wars are fought to increase the sales of steel, and subsidies are paid by some governments in order to keep local steel production alive and competitive. Steel is a commodity that is traded on international financial exchanges. Transportation systems (mostly railways) have been built to ship the raw materials (iron ore and coal) and finished products where needed.

Despite this, the relative role that steel plays in our daily life is slowly diminishing. This does not mean that overall steel production is reduced; quite the contrary is true. The world crude steel production from June 1994 to June 2001 still increased by about 10%, the world pig iron production increased also by 10%, and only the amount of direct reduced iron stayed pretty constant. (**Crude steel** is steel of variable carbon content that is destined for reprocessing.) Figures 7.2 and 7.3 provide a short summary for the world iron and steel production as published by the International Iron and Steel Institute. The reason for the relatively slow increase in iron and steel production despite an ever increasing demand by developing countries can be found in the fact that new materials, in particular plastics, are replacing traditional steel applications. Additionally, aluminum likewise is gaining increased market shares, mainly because of its light weight and its corrosion resistance.

In Chapter 18 we shall compare the production figures of iron with those of other materials such as wood, cement, polymers, and aluminum, and discuss how the usage of these materials changed in the past 20 years. We shall also look at the price development of some materials in the past 100 years and contemplate on the remaining reserves of important minerals.

Before we close this chapter it might be worthwhile to consider where iron and iron ores can be found. Iron in its various forms is the fourth most abundant element behind oxygen, silicon, and aluminum. We mentioned earlier that 5% of the earth's crust consists of iron. Interestingly enough, 14.3% of the *moon's* surface appears to contain iron, as mineral samples collected in the Sea of Tranquility seem to indicate. The inner core of the earth (about 2400 km in diameter) consists mostly of solid iron. The outer core of the earth (some 2200 km thick) contains liquid iron and nickel. It is believed that 4.5 billion years ago a cloud of dust condensed into planet earth and became liquid from radioactive decay and

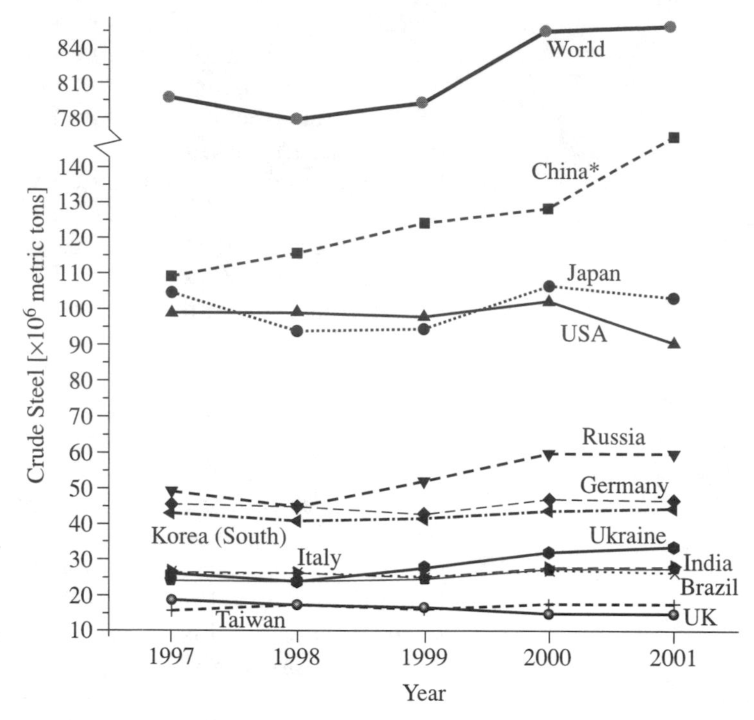

Figure 7.2. World production of **crude steel** by country and total world production [$\times 10^6$ metric tons]. Note the differences in Figure 7.3. For steel prices, see Fig. 18.1. *As reported by Chinese State Statistical Bureau that the Chinese Government considers as official statistical data. *Source*: International and US Iron and Steel Institute.

possibly from meteoric impacts. During cooling, the heavier materials such as iron or nickel "sank" to the center.

Iron assumes a rather extraordinary place among all elements. Light elements are created in hot stars by nuclear *fusion* of still lighter elements under release of energy. For example, four hydrogen nuclei fuse together in our sun into a helium nucleus (under release of two positrons), thus providing the energy that supports life on earth. Now, this energy-releasing fusion of light nuclei can only produce elements up to iron. Elements heavier than iron are instead created by nuclear *fission*, that is, by splitting heavy nuclei into lighter ones, again under release of energy. (The fission of uranium 235 is a common example of this process.) As a result of these two rather contrary reactions, one observes that iron has energetically the most stable nucleus of all elements. This explains its relative abundance.

The major terrestrial iron deposits on the earth's crust are distributed in the *iron belt* along the northern temperate zone extending from North America to Europe, Russia, and North China (Figure 7.4). By comparison, the South American, African, Indian, and Australian deposits are of lesser significance. (Incidentally, the major coal deposits follow the iron belt throughout the northern continents.) Whole mountains may consist essentially of iron ore,

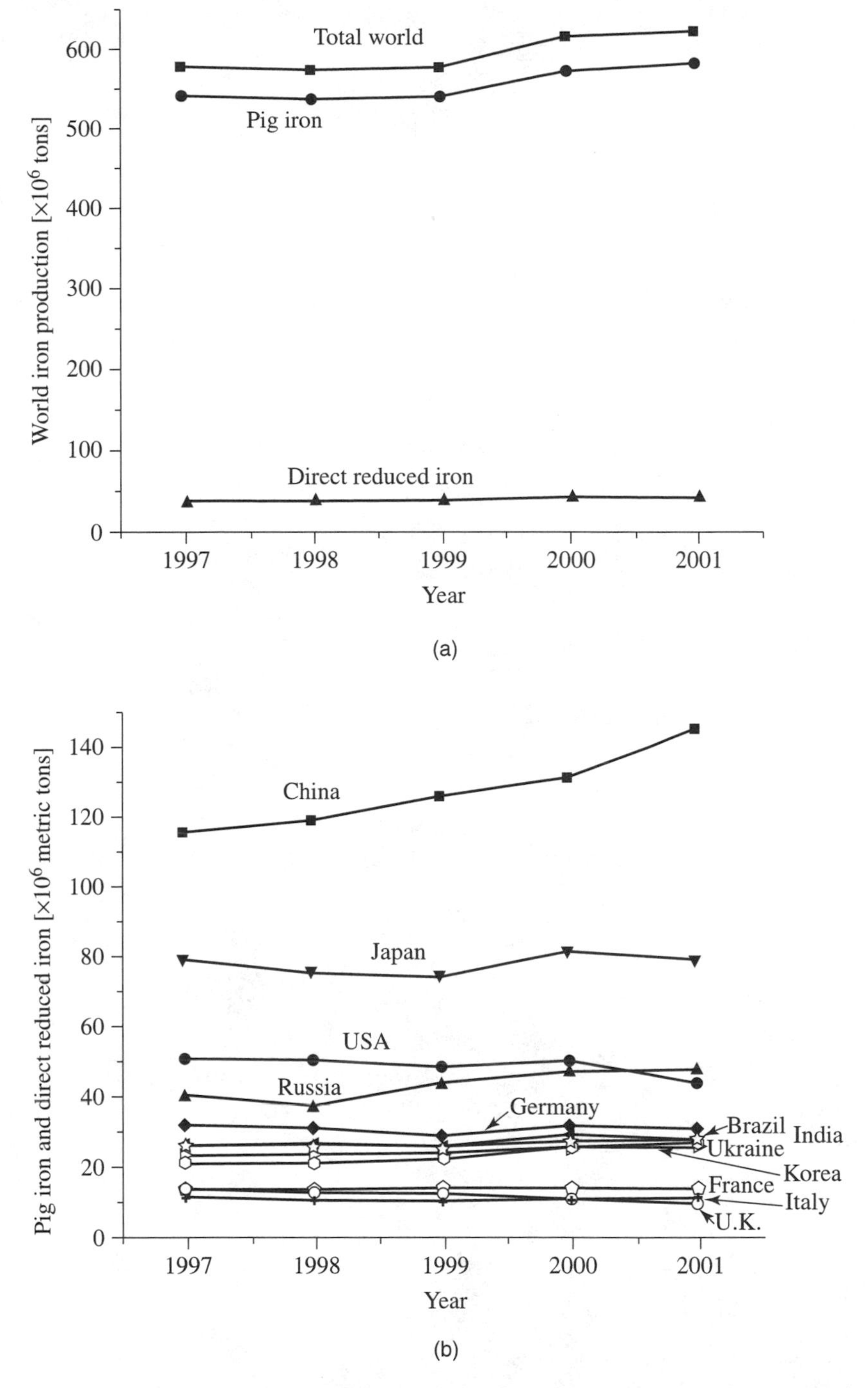

Figure 7.3. World iron production (a) Total, and divided into pig iron and direct reduced iron. (b) Combined for different countries. [$\times 10^6$ metric tons]. *Source*: American Iron and Steel Institute. *Notes*: The European Common Market produced in 2001 more than 87×10^6 metric tons of pig iron and direct reduced iron. Not all of the European countries are listed in the figure. The data from China are as reported by the State Statistical Bureau that the Chinese government considers as official statistical data.

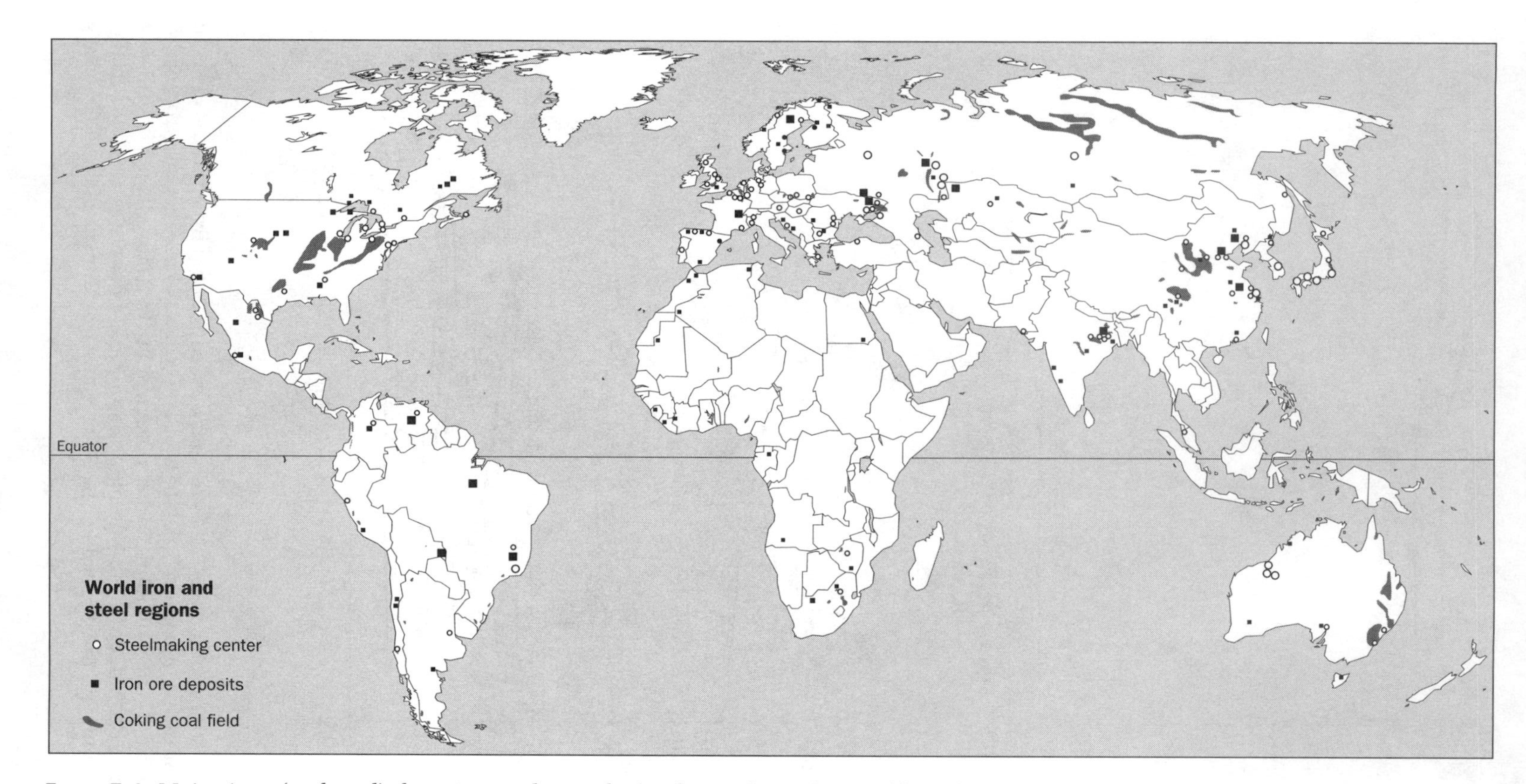

FIGURE 7.4. Major iron (and coal) deposits on the earth. (Redrawn from the *World Book Encyclopedia*, © 1997 World Book, Inc. By permission of publisher.)

such as the one near Kirunavara (Sweden) or the Iron Mountain in Styria (Austria). Blocks of pure iron weighing several tons are found on a small island named Disko near Western Greenland. *Cosmic iron* does not only impact the earth in the form of meteorites, but it also descends continuously (peaking in autumn) as cosmic dust. We shall elucidate in a moment that this constant influx of fine-dispersed iron has some nourishing effects on terrestrial life. It is assumed that cosmic iron originates from the constellation of *Scorpio*. Spectroscopic analysis of the sun's corona has revealed that similar iron quantities exist in the sun as on earth.

Last but not least, iron plays a vital role in *hemoglobin*, that is, for the red blood cells, without which we could not breathe air. Iron is also known to help the body to properly produce heat. Further, it strengthens the forces of consciousness. (Anemic persons lacking sufficient iron are often unable to keep sufficiently warm and may suffer from a lack of resolution.) In chlorophyll, the iron acts as a catalyst without which photosynthesis does not work. Indeed, all plants contain iron in their ash. Particularly in anise, stinging nettle, spinach, and the fruit of water chestnuts, iron is quite abundant. Without iron, the rocks and soils would not be brown, green, yellow, or red; the plants would not be green; and man would have no complexion.

The most important iron ores are oxides (hematite, magnetite), hydroxides (limonite), and carbonates (siderite); see Plate 7.1. Another iron ore is pyrite (FeS_2), which crystallizes mostly in cubes but also in pentagonal dodecahedrons. Pyrite always contains some gold and sometimes traces of copper. Magnetic pyrite (pyrrhotite), a different ore, contains about equal amounts of iron and sulphur atoms and crystallizes hexagonally. It usually contains some cobalt and nickel. Iron in connection with oxygen has two possible valencies: In its bivalent form it is called *ferrous iron*, whereas the trivalent compound is named *ferric iron*. At a lower degree of oxidation, iron is alkaline; at a higher degree, it is acid. Thus, magnetite, i.e., $FeO \cdot Fe_2O_3$ (see Plate 7.1), is a salt of ferric acid. Magnetite is ferrimagnetic and is therefore the material from which magnetic compasses were first made. We return to this topic in Chapter 12. The British coined the word *lodestone*[1] for magnetite which meant to say that this mineral points the way.

Iron has played a significant role in ancient mythology as the symbol of strength, resolve, and will. Numerous German fairy tales or the Finnish epic, *The Kalevala*, may serve as examples. The iron smith was usually portrayed as a well-respected, power-

---

[1]lode (old English) = to lead, to guide

ful hero in folklore. An oath taken over the anvil of a blacksmith was sacred and could never be broken. Thus, it was fashionable for eloping couples from England to get married by a blacksmith at Gretna Green, just across the border in Scotland. On the other hand, it has become the destiny of iron, more than any other metal, to serve war and destruction.

The following chapter presents the scientific basis for understanding the multitude of mechanical properties of iron and steel.

Suggestions for Further Study

R. Balasubramaniam, *Delhi Iron Pillar: New Insights*, Indian Institute of Advanced Studies, Shimla and Aryan Books International, New Delhi (2002).

S. Das, *The Economic History of Ancient India*, Howrah, Calcutta (1925).

J.G. Macqueen, *The Hittites and Their Contemporaries in Asia Minor*, Thames and Hudson, London (1986).

V.C. Pigott, "Iron Versus Bronze," *Journal of Metals*, August 1992, pp. 42ff.

R. Raymond, *Out of the Fiery Furnace—The Impact of Metal on the History of Mankind*, The Pennsylvania State University Press, University Park, PA (1986).

P.R. Schmidt and S.T. Childs, "Ancient African Iron Production," *American Scientist*, **83** (1995), p. 524.

R.F. Tylecote, *A History of Metallurgy*, The Metals Society, London (1976).

J.C. Waldbaum, *The Coming of the Age of Iron*, T.A. Wertime and J.D. Muhly, Editors, Yale University Press, New Haven, CT (1980).

8

Iron and Steel

8.1 • Phases and Microconstituents

A deeper understanding of the diverse properties of iron and various steels is gained by inspecting the iron–carbon phase diagram. Actually, for the present purposes only the portion up to 6.67% C is of interest; see Figure 8.1.

The various phases are known by specific names, such as the hard and brittle intermetallic phase Fe_3C (6.67% C), which is called iron carbide or *cementite*; the FCC, non(ferro)magnetic, γ-phase named *austenite*, and the BCC α-phase known as *ferrite*. Further, a high-temperature BCC phase called δ-*ferrite* and the eutectoid phase mixture (α + Fe_3C) named *pearlite*. Not enough. Two more microconstituents known as *bainite* and *martensite*, respectively, exist which are formed by specific heat treatments. The latter will be discussed in Section 8.3. These names came into existence either because of their properties or appearance under the microscope (such as cementite and pearlite) or to commemorate certain scientists who devoted their lives to the study of these microconstituents (such as Sir W.C. Roberts–Austen, English Metallurgist, 1843–1902; A. Martens, German Engineer, 1850–1914; and E.C. Bain, American Metallurgist).

Several three-phase reactions are evident from Figure 8.1. The eutectic reaction at Fe–4.3% C lowers the melting temperature of iron to 1148°C, as mentioned in Chapter 7. Further, a *eutectoid* reaction ($\gamma \rightarrow \alpha + Fe_3C$) at 727°C and a *peritectic* reaction at 1495°C need to be emphasized. Finally, two *allotropic* transformations during cooling from δ-ferrite to austenite and from there to ferrite take place. The α-, γ-, and δ-phases consist of solid solutions in which the carbon is *interstitially* dissolved in iron.

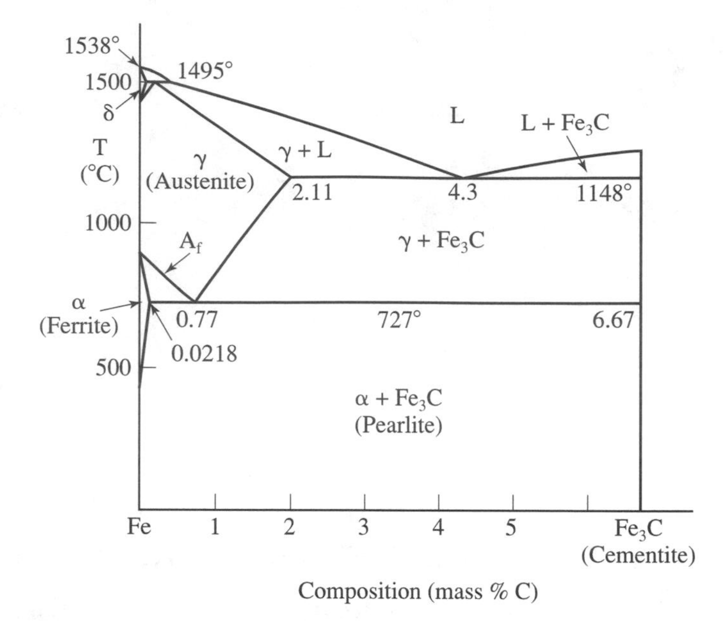

FIGURE 8.1. Portion of the iron–carbon phase diagram. (Actually, this section is known by the name *Fe-Fe_3C phase diagram*.) A_f is the highest temperature at which ferrite can form. As before, the mass percent of solute addition is used (formerly called weight percent).

8.2 • Hardening Mechanisms

Eutectoid Steel

Several hardening mechanisms take place. First, the α-, γ-, and δ-phases are solid-solution strengthened as discussed in Section 5.1. Second, pearlite involves dispersion strengthening caused by the interaction of hard and brittle cementite with the relatively soft and ductile ferrite (Section 5.4). More specifically, the α- and Fe_3C phases grow in the form of thin plates or lamellae, similarly as in eutectic reactions and as schematically depicted in Figure 8.2. However, the plates are much thinner for pearlite than in a eutectic structure, which is necessitated by the shorter diffusion lengths encountered at lower temperatures. In short, the primary reason why **eutectoid steel** (iron with 0.77 mass % C) is harder than pure iron or ferrite is because of the dispersion of hard cementite in soft ferrite in the form of plate-shaped pearlite, as shown in Figure 8.2.

Hypoeutectoid Steel

The above statements need some fine tuning. For **hypoeutectoid** compositions (below 0.77% C; see Section 5.2.2) the ferrite is the primary and continuous phase which, upon cooling from the γ-field, nucleates and grows at the grain boundaries of austenite. In other words, the α-phase quasi-coats the grain boundaries of austenite. Below 727°C, the pearlite finally precipitates in the remaining γ-phase by a eutectoid reaction. It is thus surrounded

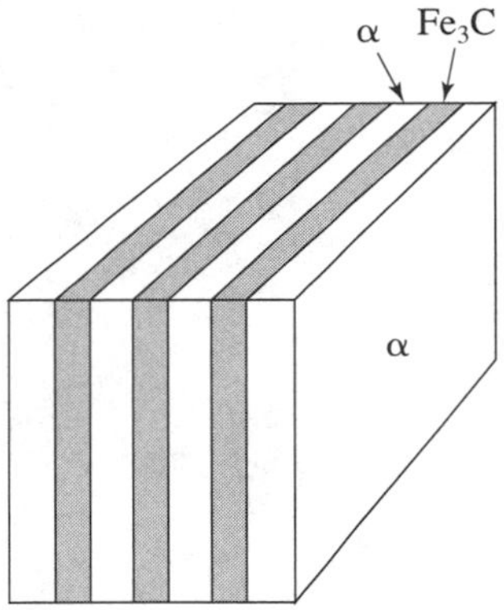

FIGURE 8.2. Schematic representation of a lamellar (plate-like) microstructure of steel called pearlite obtained by cooling a eutectoid iron–carbon alloy from austenite to below 727°C. Pearlite is a mixture of α and Fe_3C. Compare to Figure 5.9.

by primary α, as schematically depicted in Figure 8.3(a). The resulting steel is hard but still ductile due to the continuous and soft ferrite. The strength of hypoeutectoid steels initially increases with rising carbon content, but eventually levels off near the eutectoid composition.

There are some more mechanisms that may further increase the hardness of hypoeutectoid steel. We learned in Section 5.3 that a large number of small particles pose an enhanced chance for blocking the moving dislocations. This causes an increase in strength compared to the action of only a few but large particles. The same is true for the number and size of pearlite domains or *"colonies"*. The number of pearlite colonies can be increased by providing small austenitic grains to begin with on whose boundaries the pearlite eventually nucleates. Specifically, the hardness

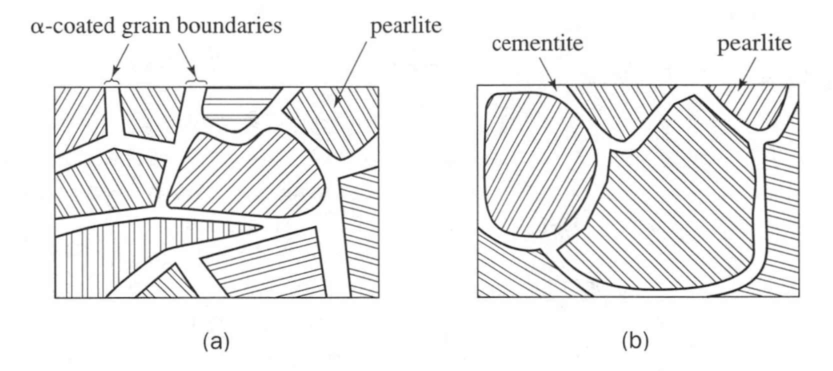

FIGURE 8.3. Schematic representation of (a) a *hypo*eutectoid microstructure of steel at room temperature containing primary α and pearlite microconstituents (the latter consisting of two phases, i.e., α and Fe_3C); (b) a *hyper*eutectoid microstructure of steel. Note that the primary phases in both cases have "coated" the former grain boundaries of the austenite.

is increased by annealing the steel in the γ-field slightly (e.g., 25°C) above A_f to prevent grain growth. This is called an *austenitizing treatment*. Alternatively, grain refiners can be used. Still another technique for increasing the strength is to produce a finer pearlite, that is, by reducing the size of the individual plates (Figure 8.2). This can be accomplished by increasing the cooling rate (called *normalizing*), for example, by air-cooling the work piece. (On the other hand, slow cooling in a furnace, called *full annealing*, yields coarse pearlite, that is, steel with less strength.)

Hypereutectoid Steel

The situation is somewhat different for **hypereutectoid** steels (iron with carbon concentrations between 0.77 and 2.11% C). In this case, the primary constituent is the hard and brittle cementite which nucleates on the grain boundaries of austenite upon cooling. These cementite nuclei grow and eventually join each other, thus forming a continuous Fe_3C microconstituent. Upon further cooling below 727°C, the pearlite precipitates out of the remaining γ microconstituent. This results in pearlite particles (colonies) that are dispersed in a continuous cementite; Figure 8.3 (b). The resulting steel is therefore brittle. To improve the ductility one would have to anneal the steel for an extended time just slightly above or below the eutectoid temperature. This produces rounded discontinuous cementite due to the tendency of elongated constituents to reduce their surface energy (i.e., their boundary area), thus eventually forming spherical particles. In other words, the extended heat treatment near the eutectoid temperature yields spherical Fe_3C particles in a ferrite matrix. This process, called *spheroidizing*, improves the machinability of hypereutectoid steel.

8.3 • Heat Treatments

TTT Diagrams

We learned in Chapter 7 that earlier civilizations had an intuitive knowledge of the fact that certain heat treatments such as annealing, quenching, and tempering would alter and improve the mechanical properties of steel. We shall now provide the scientific basis for understanding these treatments. For this a **time–temperature–transformation (TTT)** diagram needs to be presented as depicted in Figure 8.4. Let us consider a few specific cases.

(a) By quenching a *eutectoid steel* from above 727°C, that is, from the austenite region, to a temperature slightly below 727°C (indicated by the arrow "a" in Figure 8.4), only little undercooling of the austenite takes place. The driving force for ferrite and cementite nucleation is therefore small. As a consequence, the time span is relatively long until ferrite and cementite nuclei start to form at the grain boundaries of austenite. The time at which the pearlite begins

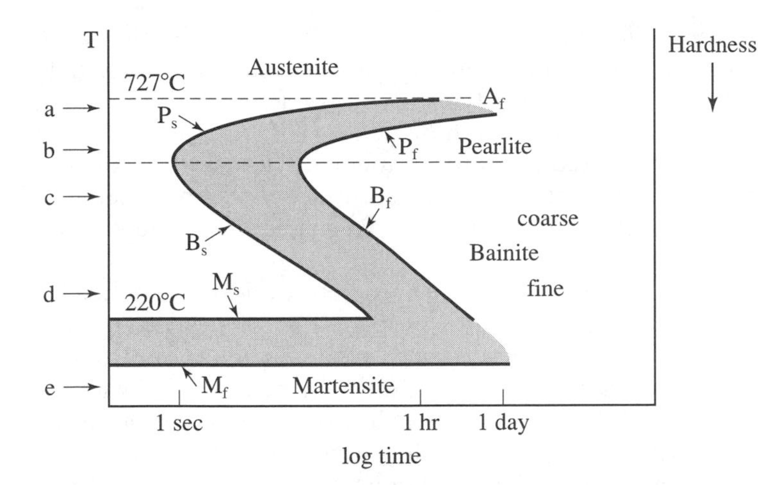

FIGURE 8.4. Schematic representation of a TTT diagram for eutectoid steel. The annealing temperatures (a) through (e) refer to specific cases as described in the text. Note that the hardness scale on the right points downward.

to nucleate is called the *pearlite start time*, or abbreviated P_s. Upon holding the work piece further at the same temperature, the nuclei grow in size until all austenite has been eventually transformed into ferrite and cementite platelets, that is, into pearlite. This has occurred at the *pearlite finish time*, P_f. Since the transformation temperature is quite high, the diffusion is fast and the diffusion distances may be long. For this reason, and because the density of the nuclei was small, the pearlite is coarse and the hardness of the work piece is relatively low; see Figure 8.4. In summary: A small temperature difference during quenching causes little undercooling which yields only a small number of nuclei. As a consequence the pearlite is coarse and the hardness is relatively small.

(b) The situation is somewhat different if austenitic steel is quenched to a lower temperature, as indicated by "b" in Figure 8.4. The undercooling is now larger, which causes a shorter nucleation time. Moreover, one encounters shorter diffusion distances and a larger number of nuclei due to the lower temperature. As a consequence, the pearlite is finer and thus harder. The time until the entire transformation is completed is relatively short, as can be deduced from Figure 8.4.

(c) If the temperature to which austenitic, eutectoid steel is quenched is reduced even further, the interplay between an enhanced tendency toward nucleation and a reduced drive for diffusion causes the cementite to precipitate in microscopically small, elongated particles (needles) that are imbedded in a ferrite matrix. This new microconstituent has been named *bainite*, and the respective times for the start and finish of the transformation have been designated as B_s and B_f. Heat treatments just below the "nose" of the TTT curves (e.g., "c" in Figure 8.4) produce *upper* or *coarse bainite*. Bainite is harder than pearlite, and the presence of a ferrite matrix causes the steel to be ductile and tough.

(d) *Lower* or *fine bainite* is even harder (but less ductile) than coarse bainite due to its larger number of small cementite particles. This microconstituent forms by reducing the quenching temperature even further, as indicated by "d" in Figure 8.4. The long times for heat treatment to complete the transformation are, however, often prohibitive for proceeding on this avenue, particularly since other treatments can be applied to achieve similar results; see below.

(e) If austenitic, eutectoid steel is very rapidly quenched to room temperature (to prevent the formation of pearlite or bainite), a very hard and brittle, body-centered tetragonal (BCT) structure, called *martensite*, is instantly formed. No diffusion of atoms is involved. Instead, a slight shift of the location of atoms takes place. This allows the transformation from FCC to BCT to occur with nearly the velocity of sound. Indeed, needle-shaped microconstituents can be observed in the electron microscope to shoot out from the matrix. The reason for the increased hardness and the greatly reduced ductility is that BCT has no close-packed planes on which dislocations can easily move. Another cause is the large c/a ratio, which distorts the lattice and leads to substantial twinning. The hardness of steel martensite increases with rising carbon content, leveling off near 0.6% C.

When austenitic steel is quenched to temperatures between the M_s and M_f temperatures (see Figure 8.4) only a portion of the austenite is transformed into martensite. Specifically, the amount of martensite, and thus the hardness, increase with decreasing temperature. Prolonging the annealing time at a given temperature does not change the amount of martensite, as can be deduced from Figure 8.4.

The quenching medium has an influence on the martensitic transformation. It affects the rate at which a work piece is cooled from austenite to below the M_f temperature without allowing pearlite or bainite microconstituents to form. As an example, the cooling rate in brine is five times faster than in oil and two times faster than in plain water. The quench rate can be even doubled by stirring the medium. (The severity of a quench is determined by the *H-coefficient* of the medium.)

Further, the shape and size of a piece to be heat-treated influences the rate of transformation and thus its hardness. For example, if a thick part is quenched from austenite, the surface is affected more severely than the interior. This may cause a more complete martensitic transformation on the outside compared to the interior, and may thus result in quench cracks due to residual stresses. Moreover, a large mass as a whole may not be effectively quenched because of a lack of efficient heat removal.

Martensitic steel is essentially too brittle to be used for most engineering applications. Thus, a subsequent heat treatment, called *tempering*, needs to be applied. This causes the precipitation of equilibrium ferrite in which very fine cementite particles are dispersed. The result is an increase in ductility at the expense of hardness. Tempering between 450 and 600°C is typical. Considerable skill and experience are involved when performing quenching and tempering. Because of the importance of these heat treatments, many metal shops have wall charts that provide guidance for the proper procedures which allow one to obtain specific mechanical properties.

It should be noted in passing that diffusionless phase transformations (i.e., martensitic transformations) are also observed in other alloys or substances. Among them are martensitic transformations in certain copper–zinc alloys, in cobalt, or many polymorphic ceramic materials. Some alloys (such as NiTi, Cu-Al-Ni, Au-Cd, Fe-Mn-Si, Mn-Cu, Ag-Cd, or Cu-Zn-Al) which have undergone a thermo-mechanical treatment that yields a martensitic structure possess a *shape memory effect*. After deformation of these alloys, the original shape can be restored by a proper heat treatment which returns the stress-induced martensite into the original austenite. Some materials also change their shape upon recooling. They are called two-way shape memory alloys in contrast to one-way alloys which change only when heated. Only those materials that exert a significant force upon shape change are of commercial interest, such as Ni-Ti and the copper-based alloys. (An Italian entrepreneur exploited this effect to create a smart shirt that automatically rolls up its sleeves at elevated temperatures and that can be smoothed out by activating a hair dryer.)

The TTT diagrams for **noneutectoid** steels need to be modified somewhat to allow for the austenite-containing two-phase regions (i.e., $\gamma + \alpha$ or $\gamma + Fe_3C$); see Figure 8.1. Let us consider, for example, a *hypo*eutectoid steel. To accommodate for the transformation from γ to $(\alpha + \gamma)$ and from there to α + pearlite, etc., an additional line has to be inserted beginning at the nose of the TTT diagram and reaching to higher temperatures. It represents the ferrite start temperature F_S; see Figure 8.5. Let us consider again a few specific cases.

(a) Quenching a hypoeutectoid steel from above A_f (i.e., the highest temperature at which ferrite can form) to a temperature between A_f and the eutectoid temperature results in a mixture of γ and primary α; see Figures 8.1 and 8.5. Once formed, the amount of ferrite does not change any further when extending the annealing time; see "a" in Figure 8.5.

(b) Austenitizing and quenching a hypoeutectoid steel to a temperature slightly above the nose in a TTT diagram yields relatively quickly a mixture of γ and primary α. The remaining

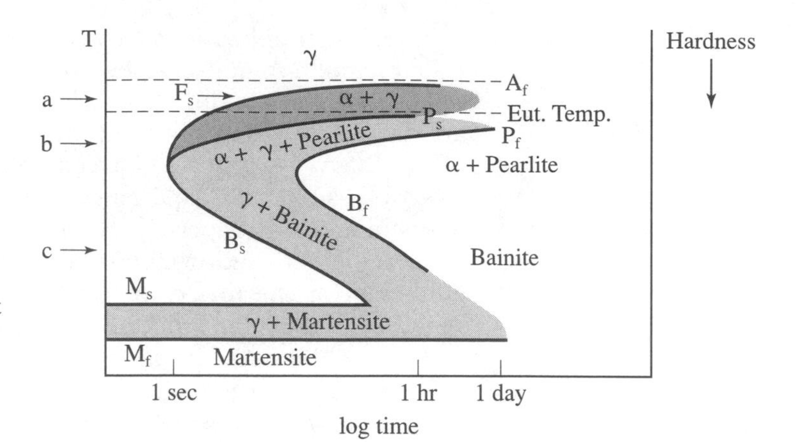

FIGURE 8.5. Schematic representation of a TTT diagram for a hypoeutectoid plain carbon steel. A_f is the highest temperature at which ferrite can form; see Figure 8.1. F_s is the ferrite start temperature.

austenite eventually transforms into pearlite upon some further isothermal annealing. The transformation into pearlite is completed at the pearlite finish temperature, P_f; see "b" in Figure 8.5.

(c) Finally, quenching and holding the same steel to a temperature just below the nose yields, after crossing the B_f line, only bainite, which has, in contrast to pearlite, no fixed composition.

Similar TTT diagrams as in Figure 8.5 are found for **hypereutectoid** plain carbon steels. The differences are an $Fe_3C + \gamma$ field (instead of the $\alpha + \gamma$ field) and a cementite start curve, C_s (instead of the ferrite start line, F_s).

The martensitic transformations for hypo- and hypereutectoid steels behave quite similar as outlined above. However, the M_s and M_f temperatures depend on the carbon content, as shown in Figure 8.6. Unfortunately, the M_f temperature cannot be clearly determined by visual inspection only. Other techniques, such as resistivity or X-ray diffraction measurements, need to be applied to obtain a reliable value. Further, the martensitic transforma-

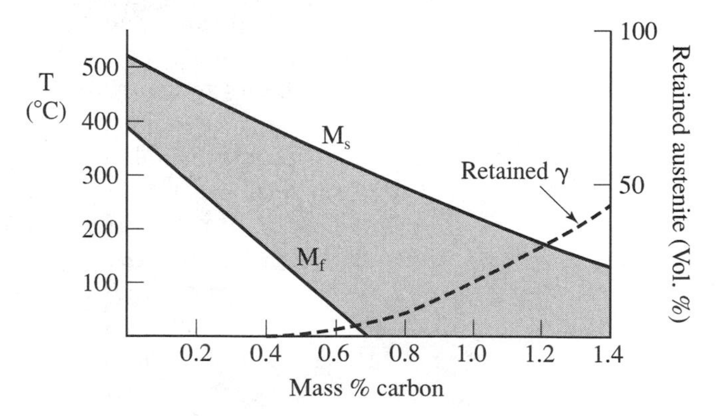

FIGURE 8.6. Schematic representation of the influence of carbon concentration on the M_s and M_f temperatures in steel and on the amount of retained austenite (given in volume percent).

tion is seldom entirely completed even at very low temperatures. This results in some retained austenite, as indicated in Figure 8.6. Retained austenite can be of concern after tempering in that it may lead to some brittleness due to its transformation to martensite upon tempering and cooling to room temperature.

8.4 • Alloyed Steels

Alloying elements, such as Mn, Si, Ni, Cu, Mo, and V, are often added to steel in various quantities (often well under 1%) in order to favorably alter its properties. These additional constituents generally shift the nose in a TTT diagram to longer times. As a consequence, no pearlite or bainite is inadvertently formed upon quenching, and the martensitic transformation can be brought to completion even in large work pieces despite the fact that the cooling rate might have been relatively slow. This feature is referred to as *hardenability* and expresses the ease with which martensite is formed upon quenching.

A second effect that alloying elements provide is a *shift of the eutectoid composition* to lower carbon concentrations. One mass % of molybdenum, for example, reduces the eutectoid composition of plain carbon steel from 0.77 to 0.4% C. This has some influence on the primary microconstituent which is formed upon cooling. Specifically, a reduction of the eutectoid composition might lead to primary cementite instead of primary ferrite in a given plain-carbon–steel.

Third, some constituents (such as Mn and Ni) considerably decrease the *eutectoid temperature*, T_E, and the temperatures at which ferrite and cementite are first formed (Figure 8.1). For example, 5% Ni decreases T_E of plain carbon steel by about 70°C. Other elements, such as Cr, W, and Mo, *increase* the eutectoid temperature instead. These changes have to be considered when austenitizing treatments are conducted. Fourth, the *martensitic start and finish temperatures* are reduced by alloying elements. Moreover, the entire TTT diagram might undergo some variations. Fifth, the time needed for *tempering* is generally diminished by alloying.

Stainless Steel

Sixth, and nearly most importantly, appreciable additions of chromium to iron (at least 12%) yield corrosion-resistant steels called **stainless steels**. They derive this property from a protective layer of chromium oxide which forms on the free surface. However, with rising Cr concentrations the amount of austenite decreases in either binary Fe–Cr or some chromium-containing iron–carbon steels which cause the ferrite to be the dominant microconstituent. The strengthening mechanism in ferrite stainless steels (up to 30% Cr and less than 0.12% C) is then by solid-solution hardening. Still, martensitic stainless steels (for knives or ball bearings) and austenitic stainless steels (containing Ni and Cr) are also frequently manufactured and used if the high price can justify this. For details we refer again to the above-mentioned handbooks.

8.5 • Cast Irons

Cast iron is the principal material which the old Chinese manufactured 3000 years ago as a result of increasing the carbon content during the smelting of iron ore. This enabled them to reduce the melting temperature possibly to as low as 1150°C, thus gaining liquid metal that could be effectively cast (see Chapter 7). Further, raw cast iron (today called pig iron) is the material that flows out of a blast furnace, as likewise explained in Chapter 7. Specifically, iron having a carbon content above 2.11 mass percent is generally referred to as cast iron even though carbon concentrations between 2.5 and 4.5% are more typical for practical applications. (See in this context the carbon concentrations of common iron-containing materials shown on the bottom of Figure 8.7.)

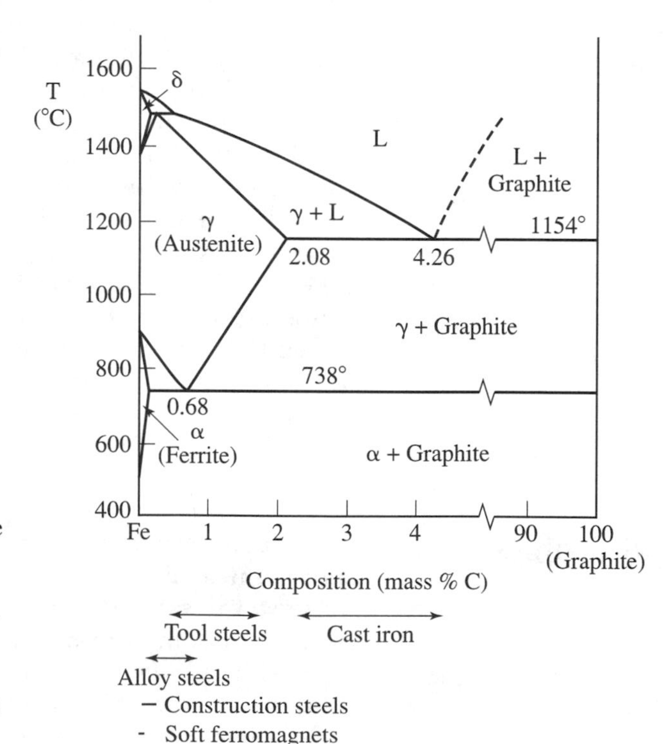

FIGURE 8.7. The equilibrium iron–carbon phase diagram showing graphite as the stable phase. Compare to Figure 8.1, in which the metastable intermetallic phase Fe_3C is prominently included. The dashed line marks the solubility of graphite in liquid Fe. Note the carbon contents of common iron-containing materials, which are shown below the phase diagram.

When discussing steels, we worked with the iron–cementite phase diagram. Now, cementite is actually a metastable compound which, when brought to its equilibrium state, decomposes into ferrite and graphite. (Graphite is one of the modifications of carbon, diamond being another.) The tendency for this dissociation increases for larger carbon concentrations and for slow cooling rates.

Further, small additions of silicon (in the amount of about 2%) promote graphite formation. Thus, cast iron is generally a somewhat different material than steel. It has important new mechanical properties, as we shall see momentarily. Actually, even the phase diagram is slightly modified, as can be observed by comparing Figure 8.1 with Figure 8.7. The main features of the iron–graphite phase diagram are the absence of the cementite intermetallic phase, a slightly higher eutect*oid* temperature ($\Delta = 11°C$), a slightly higher eutect*ic* temperature ($\Delta = 6°C$), and an insignificant shift of the eutectic and eutectoid compositions to smaller carbon contents. Figure 8.7 also indicates, that below the eutectoid temperature, ferrite and graphite are intermixed, as is common in any two-phase region.

Gray Cast Iron

Several types of cast irons are in use. Their names have been derived from their appearances. Among them, **gray cast iron** is utilized extensively, mainly because it is the least expensive metallic material of all, but also because it has the ability to effectively damp mechanical vibrations. For these reasons, gray cast iron is frequently used for the body of heavy machines in which the vibrations need to be reduced. A few more favorable properties are found in gray cast iron: It flows easily when in the liquid state, allowing castings of intricate shapes. Further, it shrinks little after casting and has a high resistance toward wear. However, gray cast iron is hard and brittle and relatively weak in tension. The ductility is low, and it shatters easily when exposed to a blow. Thus, tools, especially hammers, should never be made out of gray cast iron, even though cheap tools are occasionally manufactured in this way.

Gray cast iron contains generally between 1 and 3 mass percent *silicon*. This causes the graphite to precipitate in the form of interconnected flakes, also called *eutectic cells* or clusters (quite similar in appearance to potato chips), imbedded in a ferrite matrix, see Figure 8.8(a). The flakes have nucleated on those points where they are interconnected. However, if the Si concentration is lowered or the cooling rate is increased, the decomposition of cementite into graphite and ferrite is incomplete and the graphite is then surrounded by cementite.

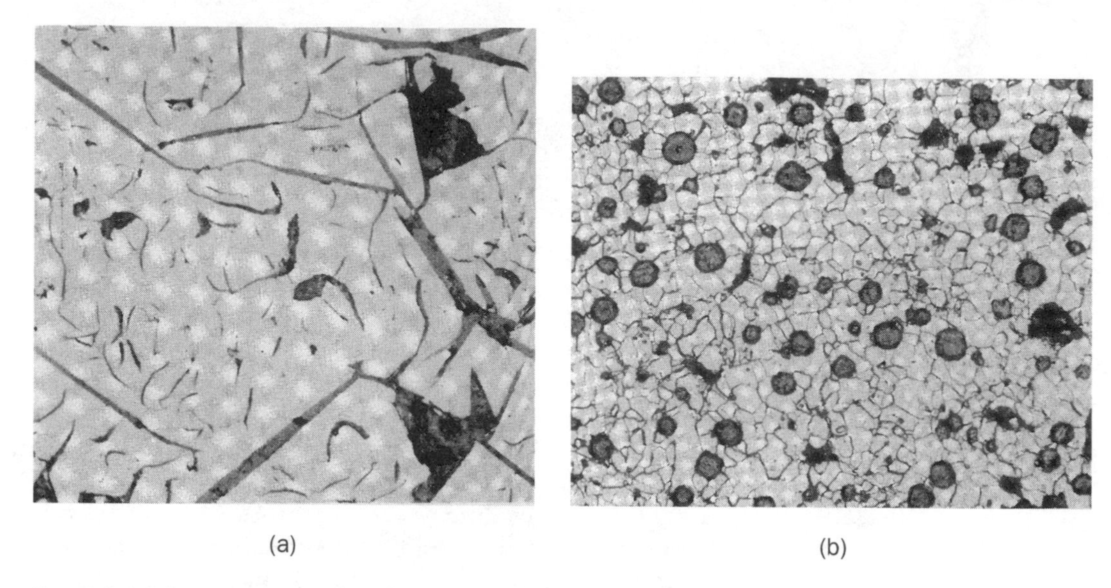

(a) (b)

FIGURE 8.8. Photomicrographs of (a) graphite flakes in gray cast iron (as polished, not etched, 100×), and (b) nodular or ductile cast iron (annealed for 6 hr at 788°C and furnace cooled, 100× 3% nitel etch). Reprinted with permission from Metals Handbook, 8th Edition, Vol. 7 (1972), ASM International, Materials Park, OH, Figures 647 and 709, respectively, pages 82 and 89, respectively.

Nodular Cast Iron

The ductility and the strength of cast iron can be improved considerably (approaching properties known for steel) by adding only a small amount (less than 1%!) of magnesium and/or cesium. This causes the graphite to precipitate in the form of spherical particles (nodules) that are imbedded in pearlite [Figure 8.8 (b)]. Moreover, if this material, called **nodular or ductile cast iron**, is heated for several hours near 700°C, or is slowly cooled, a ferrite matrix will result which increases the ductility. The resulting material is heavily used in the automotive and machine industries, such as for crankshafts, gears, and valves.

White Cast Iron

Another type of cast iron, called **white cast iron**, contains less than 1% Si and forms by rapid cooling. As we know from above, this causes the retention of cementite as a matrix and thus a very hard and brittle material. If thick pieces of white cast iron are appropriately quenched, only the surface contains the hard phase whereas the interior, because of the slower cooling rate, transforms into gray cast iron. This combination which contains a hard and therefore wear-resistant surface is used, for example, for rollers in rolling mills.

Malleable Iron

The ductility of white cast iron can be improved substantially by heating it between 800 and 900°C in an inert atmosphere (to prevent oxidation). As the reader might have expected, this causes the decomposition of cementite into graphite clusters in a ferrite (or pearlite) matrix, depending on the cooling rate. The result is a high strength and fair ductility. The material is therefore called **malleable iron**. It is used again in the automotive industry for transmission gears or valves, but also for other heavy-duty applications.

8.6 • Closing Remarks

The reader who has been confronted with the information contained in this chapter for the first time might have felt occasionally overwhelmed (and probably at times even bored) by the multitude of heat treatments, compositions, and microconstituents that are employed in today's iron and steel technology. Still, this chapter could only scratch the surface of the wealth of information which is additionally available. Entire handbooks are devoted to this very topic and are designed to guide the practitioner through the apparent jungle of alloys, heat treatments, and mechanical properties. The foregoing pages may have helped to outline the basic principles which are involved in iron and steel making. Knowing these elements increases our respect and admiration for former generations which were able to master essential parts of the technology thousands of years ago. The basic principles remain mostly the same. However, the understanding of the mechanisms involved has increased dramatically over the past 150 years, which has led to new compositions and procedures.

Problems

8.1. Give the amount of austenite and cementite (in percent) which are in a solid eutectic iron–carbon alloy at the eutectic temperature.

8.2. Calculate the carbon content of a steel that contains 90.3 mass % austenite and 9.7 mass % cementite at 1000°C.

8.3. A steel having a carbon concentration of 0.65% contains, when cooled from austenite, certain amounts of primary ferrite and pearlite. Calculate the amounts (in percent) of these two microconstituents at the eutectoid temperature. What are the carbon contents of these two microconstituents?

8.4. Carbon atoms can best be interstitially accommodated in FCC γ-iron by occupying the site halfway between two nearest corner atoms. Calculate the size of this open interstitial site by knowing that the radius of a γ-iron atom is 1.29 Å. Now, the radius of a carbon atom is 0.71 Å. How can this carbon atom fit into the open space?

8.5. Perform the same calculation as in Problem 8.4 above by considering BCC α-iron that has an atomic radius of 1.24 Å. [*Hint:* Take as the coordinates of the interstitial site in BCC 1/2, 0, 1/4 (tetrahedral site). Alternately, take as coordinates of the interstitial site 1/2, 1/2, 0 (octahedral site). Which one is larger?]

8.6. Consider the two interstitial sites in Problems 8.4 and 8.5 for γ-iron and α-iron, respectively. Which of the two interstitial sites is larger? What do you conclude with respect to the solubility of carbon atoms in α-iron compared to γ-iron?

8.7. Why are the atomic radii for α-iron and γ-iron different?

8.8. It has been observed by metallurgists at the Society of Naval Architects and Marine Engineers (Jersey City, USA) that the rivets of the Titanic contained three times more slag than those of modern wrought iron. (Slag is the glassy residue left over from smelting metallic ores, see Chapter 7). Discuss the possible influence of brittle slag on the ductility of wrought iron and how it could have compromised the hull of the Titanic.

Suggestions for Further Study

R.W.K. Honeycombe, *Steels: Microstructure and Properties*, 2nd Edition, Arnold, London (1996).

Metals Handbook, Vol. 1, Properties and Selection: Irons and Steels and High-Performance Alloys, ASM International, Materials Park, OH (1990).

Metals Handbook, Vol. 7, 8th Edition, R.F. Mehl, Editor, *Atlas of Microstructures of Industrial Alloys*, ASM International, Materials Park, OH (1972).

The Making, Shaping, and Treating of Steel, 10th Edition, AISE, Pittsburgh, PA (1985).

C.F. Walton and T.F. Opav, Editors, *Iron Castings Handbook*, Iron Castings Society, Des Plaines, IL (1981).

Worldwide Guide to Equivalent Irons and Steels, 2nd Edition, ASM International, Materials Park, OH (1987).

LEGENDS FOR COLOR PLATES

Plate 1.1. Stone Age knife, arrow points, and spear tips (projectile points) found in Florida, USA. *Top row*, from, left to right: (1) Clovis point from Early Paleo-Indian period, 10,000–8000 B.C.; (2) scraper knife, Late Paleo-Indian period, 9000–8000 B.C.; (3) Bolen beveled point, Late Paleo-Indian period, 8000–7000 B.C.. *Second row:* (4) Variant of a Wacissa point, Dalton Late period, 7000–6000 B.C.; (5) Kirk serrated point, early Preceramic Archaic period, 6000–5000 B.C.; (6) preform (never finished) of an Archaic stemmed point, 6000–2000 B.C.; (7) Florida Archaic stemmed point, 3000–2000 B.C. *Bottom row*: (8) Citrus point from Florida Transitional period, 1200–500 B.C.; (9) Leon point, AD 200–1250; (10) Pinellas point, AD 1250–1600. (Courtesy of the Collections of the Anthropology Department of the Florida Museum of Natural History.)

Plate 1.2. Obsidian, a dark volcanic glass. Note the sharp edges and its translucency in the thinner parts.

Plate 1.3. Copper minerals: [14] azurite (basic cupric carbonate); [15] boronite (iron copper sulfide); [16] chalcopyrite (copper iron sulfide); [17] malachite (basic copper carbonate).

Plate 1.4. Grave from the Copper Age cemetery at Varna on the Black Sea coast of Bulgaria (about 4300 B.C.). Gold ornaments, copper objects, flint, and copper weapons suggest a high-ranking individual. (Courtesy of the Arheologicheski Muzei, Varna, Bulgaria.)

Plate 1.5. The Iceman's axe, 60.8 cm long, consists of a wooden haft, a leather or skin binding and a copper blade. For the hafting fork, not visible under the lashing, the carver chose the fork of a branch. For this photograph, the blade was pulled out slightly from its fixing. (Approximate age: 3300 B.C.) (Courtesy of the Römisch-Germanisches Zentralmuseum, Mainz, Germany.)

Plate 1.6. The art of working copper goods in Arabic countries probably remains about the same as it existed a thousand years ago.

Plate 1.7. Cylinder seal stamp of the Akkadian period (2370–2130 B.C.) depicting a goddess and a god (recognizable by their "horns of divinity") which stand atop winged felines with plume-like tails. The motifs were carved with a copper or bronze bow drill, utilizing abrasive powders, into black and green serpentine (a decorative stone consisting of $Mg_3Si_2O_5(OH)_4$). Cylinder seal stamps were used to impress signatures of authority on business contracts or other documents written on clay tablets while wet. The hole in the center enabled the owner to string a cord through the stamp and wear it around his neck. The right picture shows an impression on clay (Courtesy of the Habib Anavian Collection, New York.)

Plate 4.1. Some minerals used during the Bronze Age: [39] tin mineral, cassiterite (tin oxide); [7] arsenic mineral, oripment (arsenic trisulfide); [25] lead mineral, Galena (lead sulfide); [45] zinc mineral, sphalerite (zinc sulfide).

Plate 4.2. Bronze helmet, southwest Iran, Elamite style, ca. 1300 B.C., ornamented with four deities, decorated with gold and silver. (Courtesy of The Metropolitan Museum of Art, New York. Fletcher Fund, 1963, Accession Nr. 63.74 ANE.)

Plate 4.3. Ritual bronze vessel with ram's head decoration; about 1300 B.C., Shang Dynasty, China. These vessels were used for ritual sacrifices to the ancestors. The decoration with animal heads was done because the Chinese believed that animals would communicate with the ancestors in the spirit world. (Courtesy of the British Museum, London, Accession Nr. PS174465.)

Plate 7.1. Some iron minerals: [20] hematite (ferric oxide, Fe_2O_3); [21] limonite (ferric oxide hydroxide, $Fe_2O_3 \cdot H_2O$; [22] magnetite (ferrous ferric oxide, $FeO \cdot Fe_2O_3$); [23] siderite (ferrous carbonate, $FeCO_3$).

Plate 7.2. Wrought iron dagger, Iran, Luristan, ca. 700 B.C. Length: 49.4 cm. (Courtesy of The Metropolitan Museum of Art, New York. Accession Nr. 61.62 ANE.)

Plate 7.3. Damascene sword, iron, inlaid with gold, Indian, late eighteenth century. (Courtesy of The Metropolitan Museum of Art, New York. Accession Nr. 36.25.1303.)

Plate 7.4. Detail of Plate 7.3. Sword blade, iron hilt, inlaid with gold damascene decoration, Indian, late eighteenth century. (Courtesy of The Metropolitan Museum of Art, New York. Bequest of George C. Stone, 1935. Accession Nr. 36.25.1303.)

Plate 7.5. Iron, bronze, and ceramic utensils found in Celtic graves of the so-called Hallstatt people (about 800–400 B.C., i.e., late Bronze and early Iron Age in Central and Western Europe). The term "Hallstatt culture" is derived from prehistoric cemeteries and salt mines near the village of Hallstatt in upper Austria where Celtic artifacts were first identified. The art of iron making was probably brought to the Celts by the Etruscans (Northern Italy).

Plate 15.1. Goblet earthenware with painted decoration (goats, giraffes, sheep). Amaratian white line ware, Egypt, pre-Dynastic, Nagada I, 4000–3500 B.C. Height: 25.2 cm. The clay is of plain red ware, fired buff to reddish-brown. The vessel was covered with a thin, hard, reddish-brown slip, burnished, and polished. Decoration in thick, creamy-white to yellowish-buff paint. (Courtesy Ashmolean Museum, Oxford. Accession Nr. 1895.482.)

Plate 15.2. Egyptian amphoriskos, about 1400–1360 B.C. New Kingdom, Eighteenth Dynasty. Nearly opaque, deep blue glass matrix, opaque white, yellow, light blue, and turquoise glasses, core-formed, trail-decorated, and tooled. Height: 11.8 cm. (Courtesy of the Corning Museum of Glass, Corning, NY. Accession Nr. 50.1.1.)

Plate 15.3. Ribbon glass bowl, first century B.C.–first century A.D. Roman Empire, probably Italy. Multicolored glasses in preformed rods, fused in or on a mold, polished. Ribbon-glass technique. Diameter: 13.5 cm. (Courtesy of the Corning Museum of Glass, Corning, NY. Accession Nr. 66.1.214.)

Plate 15.4. Greek water jar (*hydria* or *kalpis*). Red figures with some gilding. Attic, by the Meidias Painter, 420–410 B.C. Height: 52 cm. Black glaze was applied which was scratched away at the places of figures to reveal the red of the clay beneath. (Courtesy of the British Museum, London. Accession Nr. E224.)

Plate 15.5. Glazed brick relief from the Royal Processional Street, Babylon, reign of Nebuchadnezzar II, about 604–561 B.C. Author's own photograph. Wall is on display at the Pergamon Museum, Berlin, Germany.

Plate 15.6. Ceramic life-size warriors, excavated near Xi'ang, China, in the tomb of Qin Shi Huang (259–210 B.C.), the first emperor of China. The terra cotta sculptures (about 6,000) stood between 175 and 198 cm tall and were manufactured by combining the use of molds and hand-working. The heads were created separately. The textured gray clay shrunk 18% during firing. The sculptures were originally painted. They were created to spare the live burial of the warriors after the death of the emperor.

Plate 15.7. Tin-glazed earthenware (*majolica*) formerly called *Orvieto pottery*. High-temperature green and manganese purple, Italy, late fourteenth century. Height: 28 cm. (Courtesy of the Musée du Louvre, Paris, France. No. OA7394.7398.)

Plate 15.8. Coffee pot, Böttger stoneware, enamel colors. The outline is derived from Turkey or Persia and is one of the earliest products of the Meissen Factory (about 1712). Note the Oriental details such as the dragon's head at the base of the spout. Height: 15.2 cm. (Courtesy of the Schloßmuseum Schloß Friedenstein, Gotha, Germany.)

Plate 15.9. Dish Faïence, France, eighteenth century, an elegant example of the *décor rayonnant* of the golden age of Rouen. The central "rose" or "wheel" is made up of foliated scrollwork in trellised or diapered sections. (Réunion des Musés Nationaux/Art Resource, NY)

Plate 15.10. Unglazed, incised, and burnished water bottle recovered from the Picnic Mound in Hillsborough County, Florida, USA. "Safety Harbor Period," A.D. 900–1500. (Courtesy of the Collections of the Anthropology Department of the Florida Museum of Natural History. FLMNH Cat. No. 76661.)

Plate 15.11. *Cliff Palace* is one of the hundreds of cliff dwellings which sheltered typically about 350–400 Pueblo Indians in Mesa Verde, Colorado, USA. They were built in the 1200s and abandoned by A.D. 1300, probably due to a drought. The stones for the walls were secured by mud mortar and covered with mud plaster. The structures are protected from the destructive forces of the weather by the cliff overhangs.

Plate 16.1. Unreeling of silk threads (fine line of fiber) from cocoons in a Chinese silk factory.

Plate 16.2. Reproduction of a paper-like sheet similar to those produced by the ancient Egyptians. "Cyperus papyrus" fibers were placed side by side and crossed at right angles with another identical layer, all of which was subsequently dampened and pressed.

Plate 16.3. Wooden fence post which has been partially decayed over an extended period of time. Note the different degrees of deterioration of the *spring wood* (low density) compared to the *summer wood* (darker, denser, more resistant) rings.

Plate 17.1. Gold single crystals from Rosia Montana in Transylvania/Rumania. Length of the largest crystal: 15mm (Courtesy of W. Lieber, Heidelberg, Germany.)

Plate 17.2. Gold cup from Sicily, seventh century B.C. The relief decoration consists of a slow, stately procession of six bulls. Siculo-Greek work. (Courtesy of the British Museum, London. Accession Nr. PS207800.)

Plate 17.3. Etruscan gold necklace with heads of river-gods, sirenes, flowers, buds, and scarabs. Also shown is an ear stud, richly decorated with filigree granulation and inlay. Sixth century B.C. (Courtesy of the British Museum, London. Accession Nr. PS036193.)

Plate 17.4. Greek gold necklace composed of interlocking rosettes and other ornaments from which hang flower buds and female heads. Filigree decoration; fourteenth century B.C. (Courtesy of the British Museum, London. Accession Nr. PS268530.)

Plate 17.5. Gold plates, 5–4 millennium B.C. (Courtesy of the Archeologicheski Muzei, Varna, Bulgaria.)

PLATE 1.1. Stone-Age knife, arrow, and spear tips.

PLATE 1.2. Obsidian.

PLATE 1.3. Copper minerals.

Plate 1.4. Copper-Age grave site.

Plate 1.5. Copper axe.

Plate 1.6. Copper working.

Plate 1.7. Cylinder seal stamp.

Plate 4.1. Bronze minerals.

Plate 4.2. Bronze helmet.

Plate 4.3. Bronze ritual vessel.

PLATE 7.1. Iron minerals.

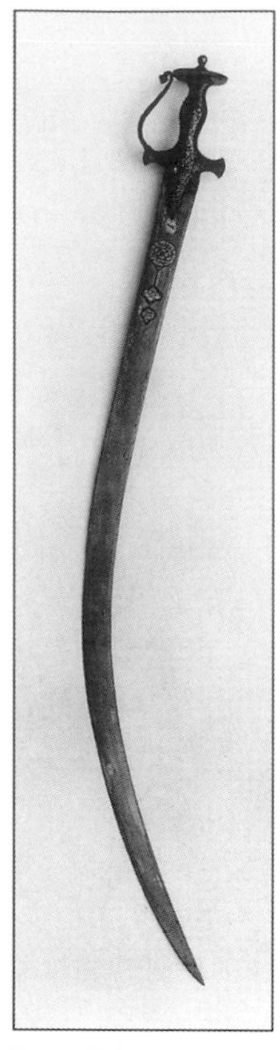

PLATE 7.3. Damascene sword.

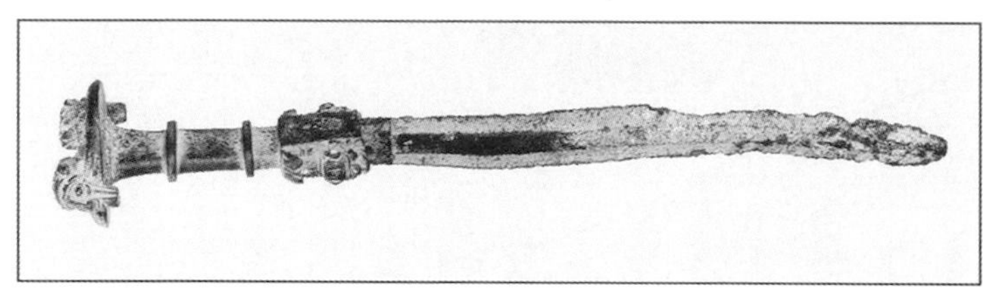

PLATE 7.2. Iron dagger.

PLATE 7.4. Damascene sword (detail).

PLATE 7.5. Iron-Age utensils (Celtic).

Plate 15.1. Earthenware goblet.

Plate 15.2. Glass Amphoriskos (Egypt).

Plate 15.3. Ribbon glass bowl (Roman).

Plate 15.4. Glazed water jar (Greece).

Plate 15.5. Glazed brick relief (Babylon).

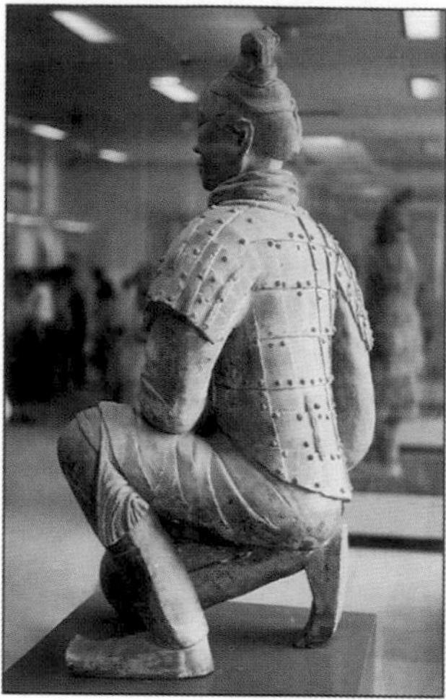

PLATE 15.6. Ceramic warriors (China).

PLATE 15.7. Tin-glazed earthenware (Majolica).

PLATE 15.8. Stoneware (Germany).

PLATE 15.9. Dish Faïence (France).

PLATE 15.10. Burnished water bottle (North America).

PLATE 15.11. Cliff dwellings (North America).

PLATE 16.1. Silk unreeling (China).

PLATE 16.2. Papyrus (Egypt).

PLATE 16.3. Growth rings in wood.

Plate 17.1. Gold crystals.

Plate 17.2. Gold cup.

Plate 17.3. Gold necklace (Etruscan).

Plate 17.4. Gold necklace (Greece).

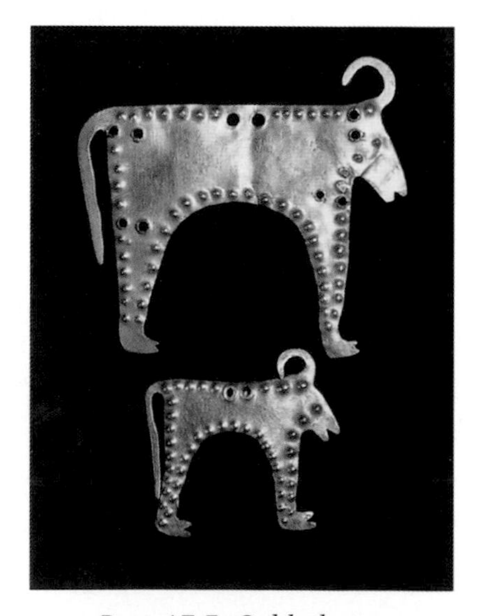

Plate 17.5. Gold plates.

9

Degradation of Materials (Corrosion)

9.1 • Corrosion Mechanisms

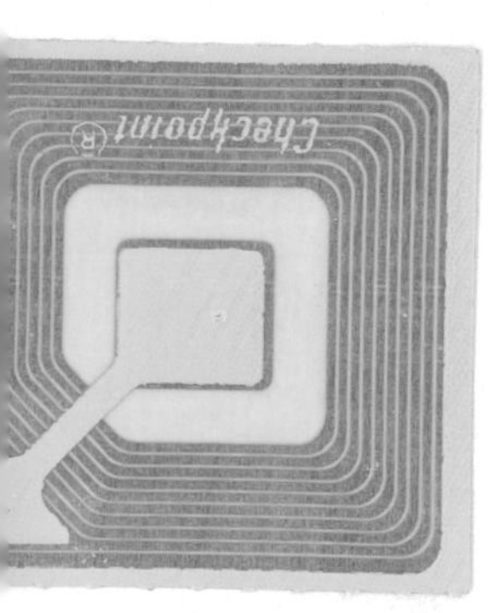

Despite the numerous useful properties of iron and steel and the cultural changes that came along with the introduction of iron, it has to be kept in mind that iron and steel are plagued by a grave detriment. This detriment is **rusting**, also often referred to by the generic and more accommodating names *corrosion* or **environmental interaction**. Specifically, rusting destroys goods valued at approximately 5% of the gross national product in industrialized countries. Billions of dollars have to be spent annually to replace or repair corrosion-related damages or to prevent corrosion. (About \$250 billion per year in the United States alone.) Moreover, corrosion can weaken the strength of structures made from iron and changes their appearance. It is of little consolation that many other materials such as glass and polymers likewise undergo some form of deterioration. Rusting transforms iron or ferrous alloys into ceramic compounds (e.g., iron into iron oxide or hydrated iron oxide), as we shall elucidate momentarily. Actually, corrosion is a slow form of burning. In short, rusting is a prime destructive mechanism that affects a society which places its trust and investments into iron and steel.

Many industrialists do not see these facts as just stated. For one, they consider the advantage of rusting to be that it creates a "repetitive market" for the goods which have been destroyed. This yields jobs for otherwise unemployed people and dividends for investors. Further, the usage of iron and steel is often an economic factor. Occasionally, replacing relatively inexpensive goods made of steel may be less costly than building equipment out of more durable materials that may cost several times more.

Third, it is often emphasized that many goods fail by other modes, e.g., by fatigue or wear, before major corrosion takes place. Fourth, corrosion may return industrial products back to the environment after, of course, a long time period has elapsed. Finally, some people consider the color of rusting steel quite appealing and advertise it as "weathering steel." Actually, the exteriors of a number of commercial buildings have been covered with "weathering steel." Such a surface is, however, not appropriate for all climate zones because the rust, washed down by the rain, may cause some staining of the surrounding grounds. Regardless, corrosion needs to be understood to stay in control of degradation processes. This shall be attempted in the present chapter.

Oxidation

There are several types of environmental interactions which materials may undergo. Among them is **oxidation**, that is, the formation of a nonmetallic surface film (or scale) which occurs where a metal is exposed to air. Essentially, most metals and alloys experience some form of superficial oxidation in various degrees. Often, these surface films are not necessarily disabling, that is, they create protective layers which shield the underlying material from further attack. A chromium oxide film that forms on stainless steel, as discussed in Chapter 8, or a thin, aluminum oxide film that protects bulk aluminum are examples of this. In some cases, an oxide layer is even most welcome, as in insulating SiO_2 films which readily grow on silicon wafers and thus provide a basis for microminiaturization in the electronics industry.

P–B Ratio

Iron oxide (such as FeO, or Fe_2O_3, etc.) films initially form a protective layer on iron. However, the specific volumes of the oxides are larger than that of metallic iron. This leads, as the thickness of the oxide layer increases, to high compressive stresses in the film and eventually to a flaking from the bulk. As a result, fresh metal is newly exposed to the environment and the oxidation cycle starts again. The ratio between oxide volume and metal volume (both per metal atom) is called the **Pilling–Bedworth (P–B) ratio**. Its values range from less than 1, as for MgO on Mg (leading to a porous film due to tensile stresses in the film), to more than 2, as in the just-mentioned case of iron oxides on iron. A continuous, nonporous, protective, and adherent film is encountered when the volumes of oxide and metal are about the same, that is, if the P–B ratio is between 1 and 2, as for Al, Cr, and Ti oxides.

Moreover, the difference in *thermal expansion coefficients* between oxide and metal may lead to cracks in the oxide films when considerable thermal cyclings are imposed on the material. As a rule, oxides have a smaller expansion coefficient than the respective metals.

Free Energy of Formation

The tendency toward oxidation in gaseous (e.g., oxygen-containing) environments is different for various metals. Specifically, the oxidation is driven by the **free energy of formation**, which depends on the reaction temperature as well as on the species itself. As an example, the free energy of formation for aluminum oxide or titanium oxide has a large negative value that leads to a relatively stable oxide; see Figure 9.1. In contrast to this, copper or nickel oxides have a small driving force toward oxidation and therefore form relatively unstable oxides.

Rate of Oxidation

The **rate of oxidation** also should be considered. It depends on the kind of film that is forming. For example, a porous film (see above) allows a continuous flow of oxygen to the metal surface which, in turn, leads to a linear oxidation rate with time. In contrast, the most *protective* films are known to grow much slower, that is, in general, logarithmically with time. Somewhere in between are the growth rates for nonporous oxide layers, as in iron or copper, where a parabolic time-dependence has been found.

Leaching

Oxidation is only one form of environmental interaction which materials undergo. Some chemical elements are simply dissolved by aqueous solutions. This is called **leaching**. The well-known lead contamination of the drinking water in old Rome by their leaden water pipes may serve as an example. Lead contamination in drinking water, however, has not been eliminated completely in modern days, as indicated by "consumer warning" labels which are packed along with faucets. The reason for this is lead-containing solder joints (for connecting copper pipes) or the use of leaded brass for faucets to facilitate better machining to final shape. (Be-

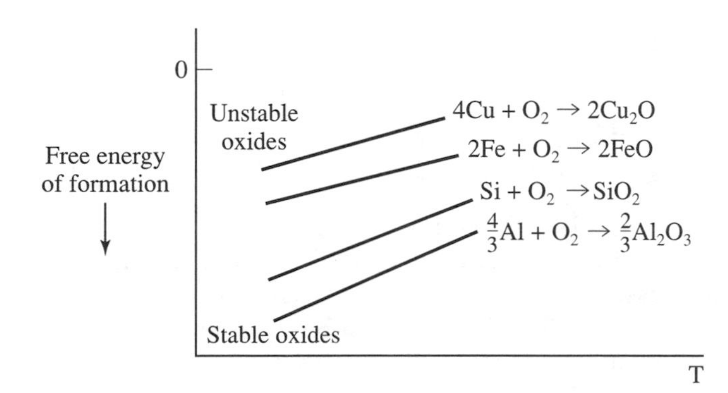

FIGURE 9.1. Schematic representation of the free energy of formation for selected metals as a function of temperature. A small negative free energy of formation means a small driving force toward oxidation and an unstable oxide. (Richardson–Ellingham diagram.)

cause of governmental regulations, manufacturers are now switching to bismuth–copper alloys.) “Soft” water is generally more corrosive (dissolves more trace elements) than unsoftened water. [To avoid lead contamination of drinking water, it is often recommended to run the water for a few seconds before use, particularly if it was standing for some time (e.g., overnight) in the line or faucets. Moreover, it is advisable to avoid drinking or cooking with water drawn from the hot water side of the tap altogether.]

Dealloying

Selective leaching or **dealloying** are terms used when one alloy constituent is preferentially dissolved by a solution. Selective leaching of zinc from brass (called *dezincification*), or selective loss of graphite in buried gray cast iron gas lines (possibly leading to porosity and explosions) may serve as examples.

Corrosion is often more fully described by **electrochemical reactions** in which free electrons are interpreted to be transferred from one chemical species (or from one part of the same species) to another. Specifically, the deterioration of metals and alloys is interpreted to be caused by an interplay between oxidation and reduction processes. This shall be explained in a few examples. During the *oxidation* process, electrons are transferred from, say iron, to another part of iron (or a different metal) according to the reaction equation:

$$Fe \rightarrow Fe^{2+} + 2e^{-} \tag{9.1}$$

or generally for a metal, M:

$$M \rightarrow M^{n+} + ne^{-}, \tag{9.2}$$

where n is the valency of the metal ion or the number of electrons transferred, and e^{-} represents an electron. The site at which oxidation takes place is defined to be the *anode*. During a *reduction* process, the free electrons which may have been generated during oxidation are transferred to another portion of the sample and there become a part of a different chemical species according to:

$$2H^{+} + 2e^{-} \rightarrow H_2 \tag{9.3}$$

or in the case of a metal, M:

$$M^{n+} + ne^{-} \rightarrow M. \tag{9.4}$$

The site where reduction takes place is called the *cathode*. In other words, oxidation and reduction can be considered as mirror-imaged processes. As an example, the dissolution of iron in an acid solution (e.g., HCl) is represented by the above equations (9.1) and (9.3), or, in summary:

$$Fe + 2H^{+} \rightarrow Fe^{2+} + H_2. \tag{9.5}$$

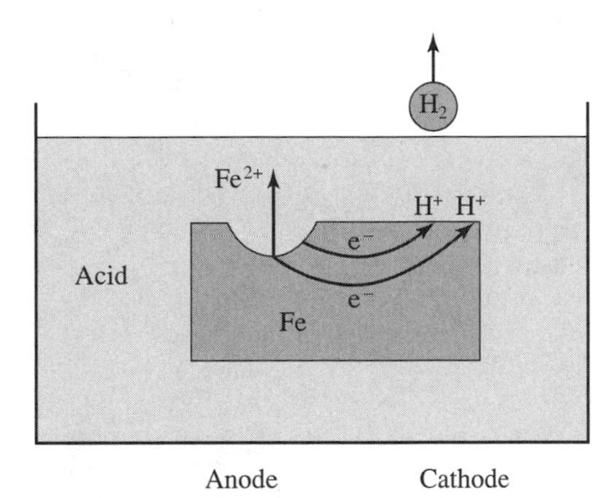

FIGURE 9.2. Schematic representation of the dissolution (corrosion) of iron in an acid solution.

The iron ions thus created transfer into the solution (or may, in other cases, react to form an insoluble compound). The process just described is depicted schematically in Figure 9.2.

Rust forms readily when iron is exposed to damp air (relative humidity > 60%) or oxygen-containing water. Rusting under these conditions occurs in two stages in which iron is, for example, at first oxidized to Fe^{2+} and then to Fe^{3+} according to the following reaction equations:

$$Fe + \tfrac{1}{2}O_2 + H_2O \rightarrow Fe(OII)_2 \tag{9.6}$$

and

$$2Fe(OH)_2 + \tfrac{1}{2}O_2 + H_2O \rightarrow 2Fe(OH)_3. \tag{9.7}$$

Fe $(OH)_3$, that is, hydrated ferric oxide, is insoluble in water and relatively inert, that is, cathodic. As outlined above, $Fe(OH)_3$ is not the only form of "rust." Indeed, other species, such as FeO, Fe_2O_3, FeOOH, and Fe_3O_4, generally qualify for the same name. In these cases, different reaction equations than those shown above apply.

9.2 • Electrochemical Corrosion

Electrochemical corrosion is often studied by making use of two electrochemical half-cells in which each metal is immersed in a one-molar (1-M) solution of its ion. (A 1-M solution contains 1 mole of the species in 1 dm^3 of distilled water. One mole is the atomic mass in grams of the species.) The two half-cells are separated by a semipermeable membrane which prevents interdiffusion of the solutions but allows unhindered electron transfer. Figure 9.3 depicts an example of such an *electrochemical cell* or *galvanic couple* in which a piece of iron is immersed in a solution

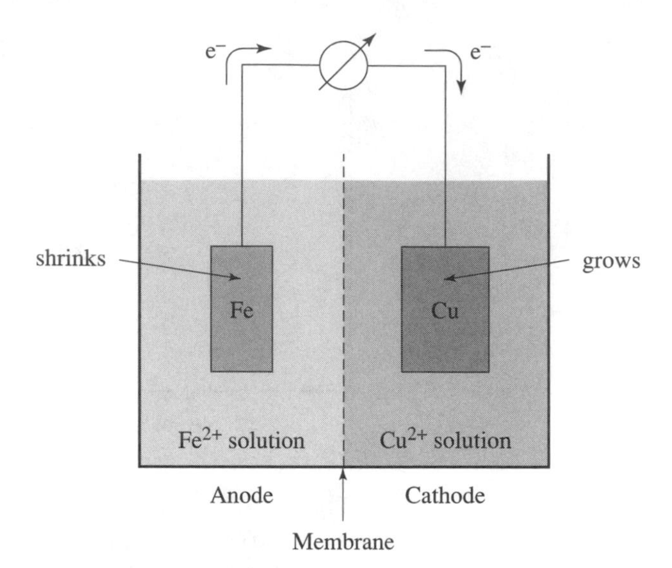

FIGURE 9.3. Schematic representation of two electrochemical half-cells in which two metal electrodes are immersed in 1-M solutions of their ions (*Galvanic couple*).

that contains 1-M Fe^{2+} ions. The other part consists of an analogous copper half-cell. The two metal electrodes are electrically connected through a voltmeter for reasons that will become obvious momentarily. After some time one observes that the iron bar has slightly decreased in size, that is, some of the solid Fe atoms have transferred as Fe^{2+} into the solution. On the other hand, some copper ions have been electroplated onto the copper electrode, which, as a consequence, has increased somewhat in size. The pertinent reaction equations are thus as follows:

$$\mathrm{Fe} \rightarrow \mathrm{Fe}^{2+} + 2e^- \tag{9.8}$$

$$\mathrm{Cu}^{2+} + 2e^- \rightarrow \mathrm{Cu}. \tag{9.9}$$

During the entire process, a voltage (called electromotive force, or emf) of 0.78 V is measured. One concludes from this experimental result that copper is more noble than iron. Indeed, a table can be created which ranks all metals (or their electrode reactions) in decreasing order of inertness to corrosion along with their emf's compared to a reference for which the hydrogen reaction has been arbitrarily chosen; see Table 9.1. (The standard hydrogen reference cell consists of an inert platinum electrode that is "immersed" in 1 atm of flowing hydrogen gas at 25°C.)

The electromotive forces, E_0, listed in Table 9.1 are valid for 1-M solutions and for room temperature (25°C) only. For other conditions, the potential of a system needs to be adjusted by making use of the **Nernst equation**:

$$E = E_0 + \frac{RT}{nF} \ln C_{\mathrm{ion}} = E_0 + \frac{0.0592}{n} \log C_{\mathrm{ion}} [\mathrm{V}], \tag{9.10}$$

TABLE 9.1. Standard emf series for selected elements in 1-M solution at 25°C

Electrode reaction	Electrode potential, or emf E_0 (V)	
$Au^{3+} + 3e^- \rightarrow Au$	+1.42	Cathodic
$Pt^{2+} + 2e^- \rightarrow Pt$	+1.2	
$Ag^+ + 1e^- \rightarrow Ag$	+0.80	
$Cu^{2+} + 2e^- \rightarrow Cu$	+0.34	
$\mathbf{2H^+ + 2e^- \rightarrow H_2}$	**0.000**	
$Sn^{2+} + 2e^- \rightarrow Sn$	−0.14	
$Ni^{2+} + 2e^- \rightarrow Ni$	−0.25	
$Fe^{2+} + 2e^- \rightarrow Fe$	−0.44	
$Cr^{3+} + 3e^- \rightarrow Cr$	−0.74	
$Zn^{2+} + 2e^- \rightarrow Zn$	−0.76	
$Al^{3+} + 3e^- \rightarrow Al$	−1.66	
$Mg^{2+} + 2e^- \rightarrow Mg$	−2.36	Anodic (Oxidation)

where R and F are gas and Faraday constants, respectively (see Appendix II), T is the absolute temperature [set equal to 298.2 K or 25°C in the right side of Eq. (9.10)], n is again the number of electrons transferred, and C_{ion} is the molar ion concentration. As expected, one obtains $E = E_0$ for a 1 M solution. Equation (9.10) shows that a tenfold increase in the ion concentration leads to a rise of the half-cell potential of 59.2 mV for a single electron reaction.

Galvanic Series

It is customary to rank metals and alloys in a simpler way, that is, in terms of their relative reactivities to each other in a given environment. Table 9.2 presents such a list, called the **galvanic series**. It includes some commercial metals and alloys that have been exposed to sea water. No emf's are usually included in a galvanic series.

Galvanic Corrosion

The galvanic series of electrochemical corrosion explains a number of important mechanisms. To start with, when two different metals (such as an iron and a copper pipe) are electrically connected and are exposed to an electrolyte (e.g., ordinary water), the less noble metal (here, the iron) corrodes near the junction (Figure 9.4). This is an example of a mechanism called **galvanic corrosion**. It can be essentially prevented by inserting an insulated coupling (plastic) between the two unlike pipes. (This technique does not prevent galvanic corrosion in the case when, for example, copper is dissolved "upstream" and then plated on the steel pipe "downstream".) Galvanic corrosion also occurs in a

TABLE 9.2. Galvanic series of some commercial metals and alloys exposed to sea water[a]

Platinum	Cathodic
Gold	↓
Graphite	
Titanium	
Stainless steel (passive)	
Copper–nickel alloys	
Bronze	
Copper	
Tin	
Sn–50% Pb solder	
Stainless steel (active)	
Cast iron	
Iron and steel	
Aluminum alloys	
Aluminum	
Galvanized steel	
Zinc	
Magnesium	Anodic

[a]Note that each metal or alloy is anodic with respect to those above it.

marine environment. Examples are steel rivets that have been (improperly) used to join aluminum sheets, or steel ball bearings of sliding doors that sit on an aluminum track. The list of "galvanic corrosion sins" observed in daily life is almost endless and, of course, is not restricted to iron and steel.

Galvanic corrosion is also encountered due to concentration differences of ions or dissolved gases in an electrolyte. The respective mechanism involved has been termed a *concentration cell*. The deterioration often occurs at the site at which the solution is most diluted. For example, the oxygen concentration is lower immediately under a drop of water sitting on an iron plate. In contrast, the oxygen concentration at the periphery of the water drop is much higher due to its increased exposure to the atmosphere. As a consequence, an iron plate corrodes under the *center of* the water drop and forms what is called a *corrosion pit*; see Figure 9.5(a). Similar observations are made immediately below the waterline

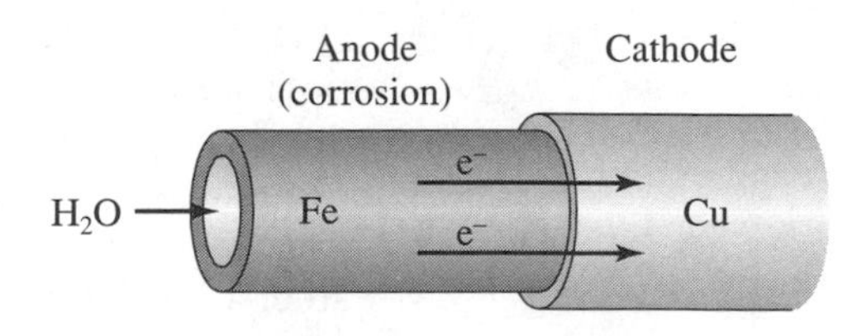

FIGURE 9.4. Galvanic corrosion of two dissimilar metals (water pipes) (see Table 9.1).

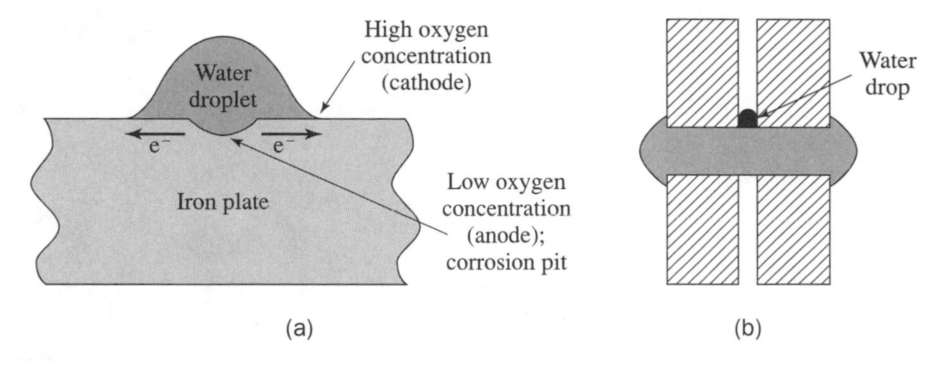

FIGURE 9.5. Schematic representation of concentration cells due to an oxygen concentration gradient (a) under a water drop and (b) in a crevice.

of ships, which is, for more than one reason, the preferred site for deterioration (*waterline corrosion*). Finally, crevices between two metal plates which have been mechanically joined may trap water that deprives the affected area from atmospheric oxygen and thus leads to *crevice corrosion* there; Figure 9.5(b). Crevice corrosion can be prevented by, for example, closing the ends of the crevice with a weld using a material that is similar to the plates.

Cathodic Protection

On the positive side, galvanic corrosion can be put to good use for preventing deterioration of buried steel pipes. For this, the metal to be protected needs to be supplied with electrons that force the pipe to become cathodic. Specifically, a less noble metal (e.g., Mg, Al, Zn; see Table 9.2) is wired to a steel pipe, while both are imbedded in a suitable back-fill of moist soil, Figure 9.6(a). The magnesium (or Mg–Zn alloy) is then slowly consumed and eventually needs to be replaced.

Understandably, the less noble metal is called the *sacrificial anode* and the mechanism just described is known by the name **cathodic protection**. Cathodic protection is also employed for ships, tanks, and hot-water heaters, to name a few examples. Another method to provide the metal to be protected with electrons is by connecting it to a direct-current power source (e.g., a solar photovoltaic cell) as depicted in Figure 9.6(b). If unprotected, the corrosion of buried steel pipes would be caused either by *long line currents* that flow through the pipes from areas of one type of soil to another one, or by *short line currents* from the bottom to the top of the pipe, again caused by different soil species or by differential aeration of the soil.

A further useful application of galvanic corrosion is made in alkaline batteries in which the emf, generated between two dissimilar metals, is utilized until the less noble metal has been consumed.

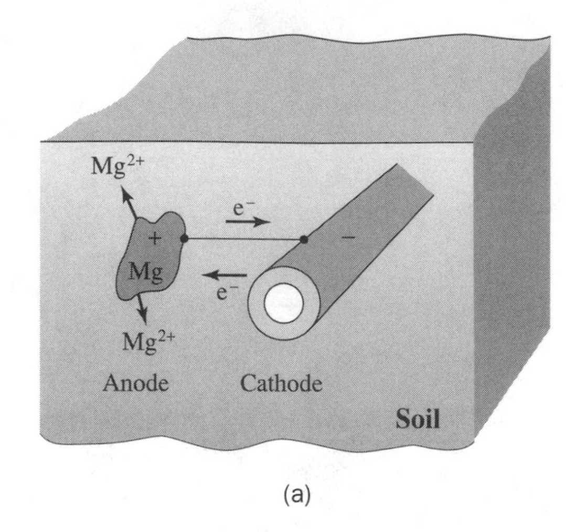

(a)

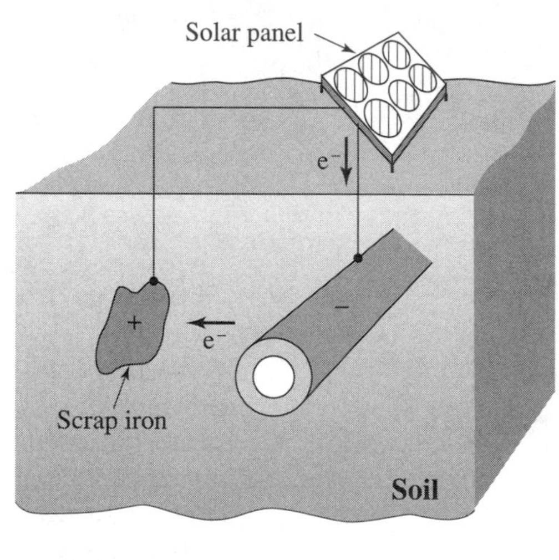

(b)

FIGURE 9.6. (a) Sacrificial anode which is wired to a steel pipeline to provide cathodic protection. (b) A photovoltaic device supplies electrons to a steel pipe for cathodic protection. The connecting wires need to be electrically insulated!

Faraday Equation

The question then arises, how much metal is electrochemically removed by corrosion (or deposited) in a given time interval, t, while applying a certain current, I? M. **Faraday** (see Chapter 10) empirically arrived at the following **equation**:

$$W = \frac{q \cdot M}{n \cdot F} = \frac{I \cdot t \cdot M}{n \cdot F}, \tag{9.11}$$

where W is the liberated mass, q is the electric charge, M is the atomic mass of the metal, F is the Faraday constant (see Appendix II), and n is the valence of the metal ion. We shall return to this topic in Chapter 10.

9.3 • Practical Consequences

Finally, we shall briefly summarize the corrosion properties of some selected materials and designs. This list contains observations which have been accumulated over a time span of many decades and belong to the "must-know" repertoire of the corrosion-conscious engineer.

- Low-alloy steel (up to 2% of total alloy content) has a much better resistance to atmospheric corrosion than plain carbon steel.
- Stainless steels (iron containing at least 12% chromium) derive their corrosion resistance from the high reactiveness of Cr and its alloys (Table 9.1), which leads to a protective (i.e., insulating) coating or Cr-containing corrosion products and thus to an interruption of the corrosion current. This mechanism is called *passivation*. (Note in this context the ranking of "active" stainless steel compared to "passive" stainless steel in Table 9.2.)
- Iron can be passivated by immersing it momentarily into highly concentrated nitric acid, which causes a thin film of iron hydroxide to form. Iron, protected this way, is then no longer attacked by nitric acid of lower concentrations.
- Martensitic stainless steels usually display a better corrosion resistance when in the hardened state compared to its annealed condition.
- Cold worked metals are often more severely attacked by corrosion than annealed metals.
- A two-phase alloy is generally more severely attacked by corrosion than a single-phase alloy or a pure metal. This is due to the possible presence of a *composition cell,* where one phase may be cathodic to the other. As an example, cementite is cathodic with respect to ferrite. This leads to microcorrosion cells within the sample consisting of α and Fe_3C plates as shown in Figure 9.7(a). Under these circumstances, the α tends to be sacrificed by protecting the Fe_3C.
- Precipitation of a second phase or segregation at grain boundaries may cause a galvanic cell and leads to *intergranular corrosion*. As an example, chromium carbide particles may precipitate at high temperatures (due to welding or heat treatments) along the grain boundaries of stainless steel. This causes a chromium depletion in the vicinity of these grain boundaries. Again, a concentration cell is formed which may lead to corrosion at the grain boundaries [Figure 9.7(b)].
- *Stress corrosion cracking* occurs in specific metals containing regions that have different stress levels and are exposed to spe-

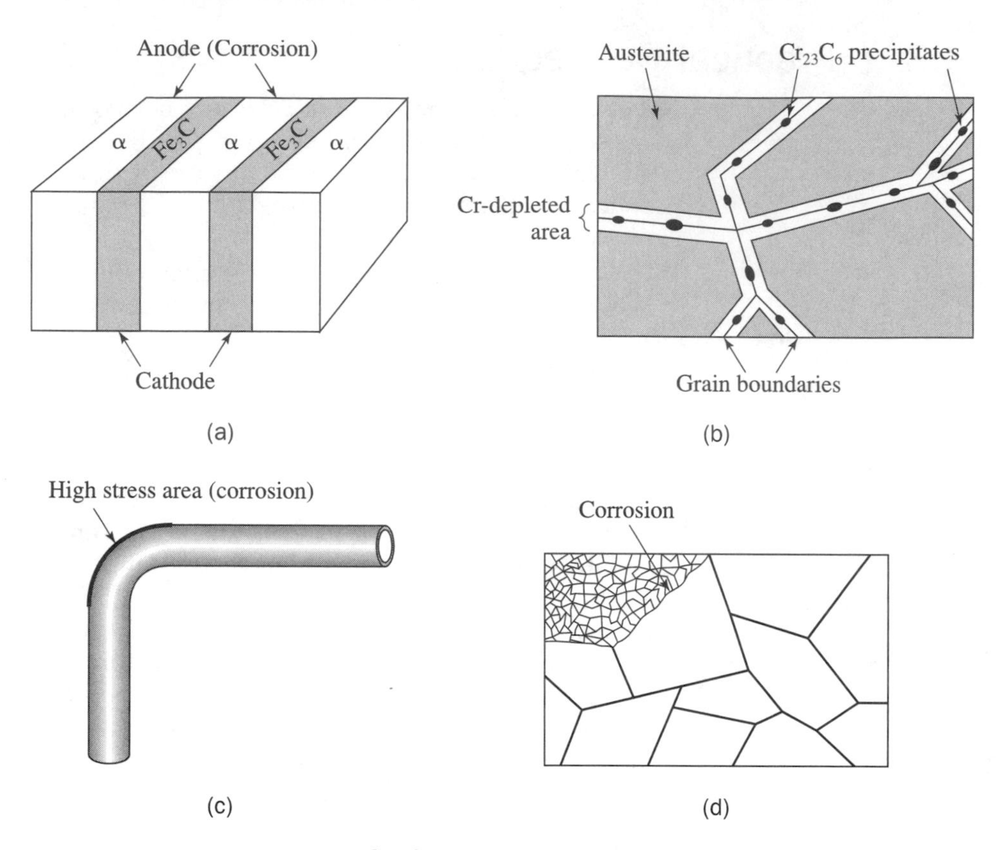

FIGURE 9.7. Examples for corrosion. (a) Composition cell in a two-phase alloy (here, ferrite and cementite platelets) (see Figure 8.2). (b) Intergranular corrosion due to a concentration gradient (concentration cell). (c) Stress corrosion cracking in high-stress (bent) areas of a metal pipe. (d) Grain-size gradients may cause corrosion.

cific environments. These stresses may be residual from bending, or from rapid temperature changes (which cause uneven contraction in various areas), or from different expansion coefficients in two-phase alloys. The high stress areas act as anodes and thus corrode [Figure 9.7(c)].

- Areas containing a high density of grain boundaries (i.e., fine-grained materials) are anodic compared to coarse-grained regions. Thus, corrosion may occur in fine-grained areas if a large variation in grain size is encountered [Figure 9.7(d)].
- We have discussed or implied above several avenues for preventing corrosion. They shall not be repeated here. Another group of corrosion aversions has, however, not yet been mentioned, namely, the coating of metals by paint, enamel (glass),

or polymers. Moreover, steel pipes, nails, or roofing material are often coated with zinc, leading to *galvanized steel*. One of these processes, involving molten Zn, called *hot dip zinc galvanizing*, consumes about 3 million tons of Zn per year in the United States. The protection is still maintained even when the zinc surface is scratched, since zinc is less noble than iron and therefore behaves as a sacrificial anode. The same cannot be said for tin-coated iron (used for cans), in which case the iron is attacked by the electrolyte (see Table 9.2).

- The protective film on metals may be mechanically disturbed in certain parts of machines such as pumps, valves, centrifuges, elbows in pipes, impellers, or agitated tanks due to fluid flow. This leads to accelerated corrosion in combination with erosion and is thus appropriately named *erosion-corrosion*.
- In specific cases, recirculated water (electrolyte) can be treated with chemicals, called *inhibitors*, which mitigate corrosion in various ways. For example, the trace elements preferentially migrate to the anode and increase the electrical resistance there, thus decreasing the corrosion current. As an example, chromate, phosphate, and borate salts are used among other constituents for this in automobile radiators.

All taken, the corrosion of iron, steel, and other metals and alloys is a serious problem. However, when properly understood and acted upon, it can be avoided or brought under reasonable control. Specifically, 15% of corrosion costs are preventable by application of proper methods. Naturally, there is always an economic side for not exercising corrosion prevention as outlined at the beginning of this chapter.

9.4 • Degradation of Polymers and Glass

At the end of this chapter on environmental interactions of materials, some brief remarks on the deterioration of polymers and glass are in place. **Polymers** are generally considered to be more resistant toward acids and alkaline solutions than most metals. Still, strong acids attack, for example, nylon and polyesters. Strong alkalis affect polymethyl methacrylate or polyesters. Nylon absorbs substantial amounts of water if exposed to it for sufficiently long periods of time (1.5% in 24 h). This can cause a reduction in stiffness and hardness. Some polymers such as polyethylene [Figure 3.16(a)], polypropylene, and polyesters are attacked by oxygen and ozone. Other polymers are assaulted by organic solvents. As a matter of fact, the likelihood for dissolution is understandably larger the greater the similarities between the chemical structures of the solvent and the polymer. For example, hydrocarbon rubbers readily absorb hydrocarbon liquids such as gasoline. The mechanism by which polymers are mainly affected is by squeezing solute molecules between the polymer molecules or between the individual polymer chains (see Figures 3.7, 3.16, and 3.17). As a result, the macromolecules are forced apart or the chains are separated and increase in size. This mechanism is called *swelling*. It causes the polymers to become softer and more ductile. In general, the swelling is less severe at lower temperatures and if the polymer has a high degree of cross-linking and crystallinity [Figure 3.17(b)].

Some polymers, for example, polyethylene [Figure 3.16(a)] degrade when exposed to high temperatures in an oxygen atmosphere and become brittle. Other polymers deteriorate under ultraviolet radiation from the sun (or by X-, β-, or γ-rays). Examples include polyvinylchloride [Figure 3.16(b)] and polystyrene. As a result of these high-energetic radiations, some orbital electrons of the involved atoms are removed and the atoms become ionized. The covalent bonds (Figure 3.4) are then broken, which leads to cross-linking or scissions of bonds. Additives that counteract this deterioration are called stabilizers. Alternately, polymers using silicon backbones rather than carbon backbones (Figure 3.16) may be used since they are more resistant to oxidation.

Glass also undergoes some form of corrosion. For example, silica, that is, SiO_2, is readily attacked by alkaline solutions. This is important when knowing that common soda-lime glass contains up to 15% sodium, as will be explained in Chapter 15. Now, any type of aqueous agents, including the moisture of the atmosphere, leaches some sodium ions out of the glass surface and with it forms sodium hydroxide, which, because of its alkalinity, attacks the silica. Moreover, the etching process, which is part of the attack, causes ultramicroscopically small fissures on the surface of glass that eventually have severe effects on the mechanical strength of a glass product. Likewise, carbon dioxide in conjunction with moisture leads to a deterioration of glass. Fortunately, these processes are quite slow and rarely affect the useful lifetime of common glass windows, except in marine environments.

Certain remedies are possible: For example, boric and aluminum oxides can be added to silica to obtain a glass which substantially resists the attack by even hot alkaline solutions. Further, glass containing large amounts of alumina (Al_2O_3) or alkaline-earth oxides is resistant to hot sodium vapor, which destroys ordinary soda-lime glass. This is particularly important for sodium high-pressure light bulbs.

Problems

9.1. Confirm the numerical factor in Eq. (9.10) by inserting the proper constants from the Appendix.

9.2. Calculate the electromotive force from a silver half-cell whose electrolyte contains 1 g of Ag^+ in 1 dm^3 of water.

9.3. Calculate the Cu^{2+}–ion concentration in an electrolyte of an electrochemical half-cell for which the electrode potential has been measured to be 300 mV. How many grams of copper had to be dissolved in 1000 g of water to achieve this concentration?

9.4. Calculate the P–B ratio for cobalt oxide when the density of metallic cobalt is 8.9 g/cm^3 and that of CoO is 6.45 g/cm^3. Is the oxide film continuous or porous?

9.5. Calculate the P–B ratio for silver oxide which is formed through 2 Ag + 1/2 $O_2 \rightarrow Ag_2O$. Take $\delta_{Ag} = 10.5$ g/cm^3 and $\delta_{Ag_2O} = 7.14$ g/cm^3. (*Hint:* Note that there are two metal atoms involved!)

9.6. How much material (in g) is electrochemically removed from a copper electrode when a current of 5 A flows for 0.5 h through an appropriate electrochemical half-cell?

9.7. Calculate the time that is required to electroplate a 1 mm thick gold layer on a 2×2 cm^2 surface when a current of 5 A is allowed to pass through the electrolyte.

Suggestions for Further Study

D.E. Clark and B.K. Zoitos, Editors, *Corrosion of Glass, Ceramics, and Ceramic Superconductors*, Noyes Publications, Park Ridge, NJ (1992).

M.G. Fontana and N.D. Greene, *Corrosion Engineering*, 2nd Edition, McGraw-Hill, New York (1978).

H. Kaesche, *Die Korrosion der Metalle*, 3rd Edition, Springer-Verlag, Berlin (1990).

E.D. Verink, Jr., *Corrosion Testing Made Easy: The Basics*, NACE International, Houston, TX (1994).

Note: Also recommended are general textbooks on materials science.

PART II

ELECTRONIC PROPERTIES OF MATERIALS

10

The Age of Electronic Materials

Stone Age—Bronze Age—Iron Age—what's next? Some individuals have called the present era the *space age* or the *atomic age*. However, space exploration and nuclear reactors, to mention only two major examples, have only little impact on our everyday lives. Instead, electrical and electronic devices (such as radio, television, telephone, refrigerator, computers, electric light, CD players, electromotors, etc.) permeate our daily life to a large extent. Life without electronics would be nearly unthinkable in many parts of the world. The present era could, therefore, be called the age of electricity. However, electricity needs a medium in which to manifest itself and to be placed in service. For this reason, and because previous eras have been named after the material that had the largest impact on the lives of mankind, the present time may best be characterized by the name **Electronic Materials Age**.

We are almost constantly in contact with electronic materials, such as conductors, insulators, semiconductors, (ferro)magnetic materials, optically transparent matter, and opaque substances. The useful properties of these materials are governed and are characterized by *electrons*. In fact, the terms *electronic materials* and *electronic properties* should be understood in the widest possible sense, meaning to include all phenomena in which electrons participate in an active (dynamic) role. This is certainly the case for electrical, magnetic, optical, and even many thermal phenomena. In contrast to this, mechanical properties can be mainly interpreted by taking the interactions of *atoms* into account, as explained in previous chapters.

Electrical Phenomena

The first observations involving **electrical phenomena** in materials probably began when *static electricity* was discovered. (Lightning, of course, preceded these experiments, but this could not be controlled by man.) Around 600 B.C., Thales of Miletus, a Greek philosopher, realized that a piece of amber, having been rubbed with a piece of cloth, attracted feathers and other light particles. Very appropriately, the word "electricity" was later coined by utilizing the Greek word *electron*, which means *amber*. It was apparently not before 2300 years later that man again became seriously interested in electrical phenomena. In 1729, Stephen Gray (a British Chemist) found that some substances conducted the "effluvium" of electricity whereas others did not. In 1733, C.F. Du Fay (a French scientist) postulated the existence of two types of electricity, which he termed glass (or vitreous) electricity and amber (or resinous) electricity, depending on which material was rubbed. Benjamin Franklin[1] later designated to them the plus and the minus sign, implying that one type of electricity would cancel the other. His ideas were based on his famous kite experiments in 1752 in which he demonstrated "the sameness of electrical matter with that of lightning." This classification was expanded almost 100 years later to include five kinds of electricity, namely, frictional, galvanic (animal), voltaic, magnetic (by induction), and thermal.

Magnetism

Magnetism (or, more precisely, ferro- or ferrimagnetism), that is, the mutual attraction of two pieces of iron or iron ore, was likewise already known to the antique world. The term "magnetism" is said to have been derived from a region in Turkey (or northern Greece?), called Magnesia, which had plenty of iron ore. Now, iron does not immediately attract another piece of iron. For this, at least one of the pieces has to be *magnetized*, that is, simply said, its internal "elementary magnets" need to be aligned in parallel. Magnetizing causes no problem in modern days. One merely places a piece of iron into a wire coil through which a direct current is passed for a short time. (This was discovered by the Danish physicist Hans Christian *Oersted* at the beginning of the 19th century.) But how did the ancients do it? There may have been at least two or three possibilities. First, a bolt of lightning could have caused a magnetic field large enough to magnetize a piece of iron or iron ore. Once one magnet had been produced and identified, more magnets could have been obtained by rubbing virgin pieces of iron with the first magnet. There could have been an-

[1]1706–1790, American publisher, scientist, and diplomat.

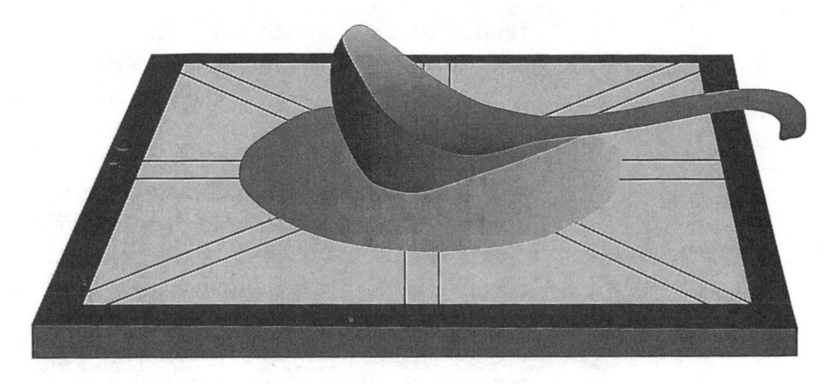

FIGURE 10.1. Depiction of an ancient Chinese compass called a *sinan*. (or *Zhe' nan*) *Zhe* = point; *nan* = south. The spoon-shaped device was carved out of a lodestone and rested on a polished bronze plate. The rounded bottom swiveled on the "earth plate" until the spoon handle pointed to the south.

other possibility. It is known that if a piece of iron is repeatedly hit very hard, its "elementary magnets" will be "shaken loose" and will align in the direction of the earth's magnetic field (which is quite weak, i.e., only about half a gauss). An iron hammer, for example, is *north magnetic* on its face of impact in the northern hemisphere. Could it have been that a piece of iron was used as a hammer and thus became a permanent magnet? A third possibility is that iron- or nickel-containing meteorites responded with an alignment of their "elementary magnets" in an electromagnetic field during their immersion into the earth's atmosphere.

One of the major applications of magnetism was the *compass* which is said to have been invented independently in China (before A.D. 1100, possibly before 1040) and in Western Europe (about A.D. 1187). Other sources emphasize that the Chinese, as early as A.D. 80 (or even earlier), had a device called a *sinan*, which consists of a piece of iron ore carved (by a jade cutter) into the shape of a ladle; see Figure 10.1. When placed on a polished plate of bronze, called the "earth plate," the spoon swiveled until the handle pointed to the south which was considered by the Chinese rulers to be the imperial direction toward which all seats had to face. The ladle resembles the Big Dipper (or great bear) whose pointer stars point to the Polaris or North Star. Another device, the *iron fish compass*, described in A.D. 1044 in a Chinese book was fabricated by allowing molten iron rods to solidify in the north-south direction that is, in the earth magnetic field which induces permanent magnetism in the metal (thermo remanence). The fish-shaped leaf was placed on water where it floated on the surface while the fish's head pointed to the south. A Chinese book printed in 1325 describes a wooden turtle, containing a loadstone and a needle as its tail, pointing to the south. There are no reports that the Chinese used these devices for navigation probably because China was a land-based culture. They were probably used instead to align the edges of pyramids, etc., along the north-south axis or as described above.

In the western world, on the other hand, the first mention of a compass was by an English Augustian monk (Alexander Neckam, 1157–1217) in his book entitled "*De Naturis Rerum*." There is also a document by an Arab writer who, in 1242, reports that a magnetic needle floating on water on top of a wooden splinter points to the north star. The bishop of Acre, Jaques Vitry, wrote in 1218 that the compass is a necessary instrument for navigation on the seas. Around 1300 the south Italian mariners of *Amalfi* are said to have perfected to some degree the compass from a needle floating on water to a round box (called later a "*bussola*") in which a compass card with a wind rose, divided into 32 points, is attached to the rotating needle. During the 15th century it was realized that the compass needle does not point to true north but assumes an angle, called variation (or declination), with the meridian. Magnetism is also mentioned in poetic works such as the *Divine Comedy* by Dante (written between 1310 and 1314) or in *La Bible* by the French monk Guyot de Provins (written about 1206). Magnetism was (and occasionally still is today) considered as a repellent against witchcraft and most anything, to heal madness and insomnia, and as an antidote against poison.

The modern compass consists quite similarly to the bussola of a pivoted bar magnet whose tip, which points to the general direction of geographic north, is called the "north-seeking pole" or simply the *north pole*. The bowl is suspended in gimbals, that is, in rings, pivoted at right angles to each other so that the compass is always level. Around 1500, the term *lodestone* appears in the literature when referring to magnetized iron ore, that is, iron oxide, particularly when used in a compass. This word is derived from the old English word *lode*, which means to *lead* or to *guide*.

Optical Phenomena

The study of **optical** phenomena likewise goes back to antiquity. Interestingly enough, there used to be an intense debate whether in vision something moves from an object to the eye or whether something reaches out from the eye to an object. In other words, the discussions revolved around the question of whether vision is an active or a passive process. Specifically, Pythagoras, a Greek philosopher and mathematician (living during the 6th century

B.C.), believed that light acts like feelers and travels from the eyes to an object and that the sensation of vision occurs when these rays touch that object. *Euclid*, a Greek mathematician, recognized at about 300 B.C. that light propagates in a straight line. Further, he related that the angle of *reflection* equals the angle of incidence when light is impinging the surface between two different media. Even though *refraction* was also known and observed in the antique world, it was not before 1821 when *W. Snell*, a Dutchman, formulated its mathematical relationship. (Refraction is the change in the direction of propagation when light passes the interface between two media having different optical densities.)

Optical materials, particularly glasses, became of prime importance once the refractive power of transparent materials was discovered. This found applications in magnifying glasses and notably in telescopes. Plane and convex mirrors, as well as convex and concave lenses, were known to the Greeks and the Chinese. Their knowledge probably went back to a common source in Mesopotamia, India, or Egypt. There is written evidence that the telescope was invented independently many times before Galileo built his version in 1609. He observed with it the craters of the moon, the satellites of Jupiter, and the orbiting of Venus around the sun, thus shattering the Ptolemaic theory (≈ A.D.150). As we shall describe in Chapter 15, glass was known to the Egyptians as early as 3500 B.C., and crude lenses have been unearthed in Crete and Asia Minor that are believed to date from 2000 B.C.

Modern optical devices include lasers, optical telecommunication, optical data storage (compact disk), and possibly, in the near future, the optical computer.

Thermal Phenomena

Heat was considered to be an invisible fluid, called *caloric*, until late into the eighteenth century. It was believed that a hot piece of material contained more caloric than a cold one and that an object would become warmer by transferring caloric into it. In the mid-1800s, Mayer, Helmholtz, and Joule discovered independently that heat is simply a form of energy. They realized that when two bodies have different temperatures, thermal energy is transferred from the hotter to the colder one when the two are brought into contact.

All taken, electrical, magnetic, optical, and thermal phenomena were considered to be unrelated to each other until the eighteenth century and were thought to be governed by their own independent laws. Many brilliant scientists have corrected this view and enhanced our knowledge on this in the past two centuries. Among them were Oersted, Ampère, Volta, Ohm, Coulomb,

Drude, Seebeck, Henry, Maxwell, Thomson, Helmholtz, and Joule. However, one "natural philosopher," as he called himself, stood out. He was probably the greatest systematic experimental genius the world has known. He believed that certain fundamental laws of nature, such as the interrelationships between electrical, magnetic, and optical phenomena, must and can be found if only the proper experiments were conducted. His name was **Michael Faraday**, who lived from 1791 to 1867. He made most of his fundamental discoveries at the British Royal Institution. His life and his accomplishments shall serve as an example of an individual who significantly advanced our understanding of electronic materials during the nineteenth century.

Michael Faraday, at the age of 14, became the apprentice to a bookbinder and bookseller who generously allowed him to read the books which he bound. Michael was particularly fascinated by the chapter on electricity which he found in the *Encyclopedia Britannica*. Inspired by this reading, he experimented in the back room of the book shop with simple, home-built electric devices, in particular with a machine that created electricity by friction. Soon Faraday visited lectures on natural sciences given by John Tatum (costing one shilling per evening, the sum of which was provided to him by his brother, a blacksmith). While attending them, Faraday took detailed lecture notes which he expanded afterwards from memory. A few months later, Faraday was given several tickets to listen to *Sir Humphrey Davy* (a well-known chemist at that time who lectured at the Royal Institution). Faraday again prepared meticulous notes and sent a copy of them to Davy asking him to be allowed to "enter into the service of science" as his assistant. This wish was eventually granted when Faraday was 21 years old. From then on Faraday helped Davy with his chemical research and his lecture experiments, and served him during his extended visit to Europe.

In 1819, Oersted discovered (in connection with a classroom demonstration) that a wire conveying an electric current would deflect a close-by magnetic compass needle; Figure 10.2(a). (Electric currents were commonly produced at that time by *galvanic cells*, that is, by a repetitive arrangement of copper and zinc plates that were immersed in diluted sulfuric acid.) This gave Faraday the idea to invert the experiment, that is, to hold the magnet fixed and let the current-carrying wire rotate around the magnet. The device which Faraday designed and eventually built was, in principle, the first **electric motor**, that is, the forebearer of all electric motors used today [Figure 10.2(b)]. (Concomitantly, Oersted found that a current-carrying wire would move in a magnetic field. And

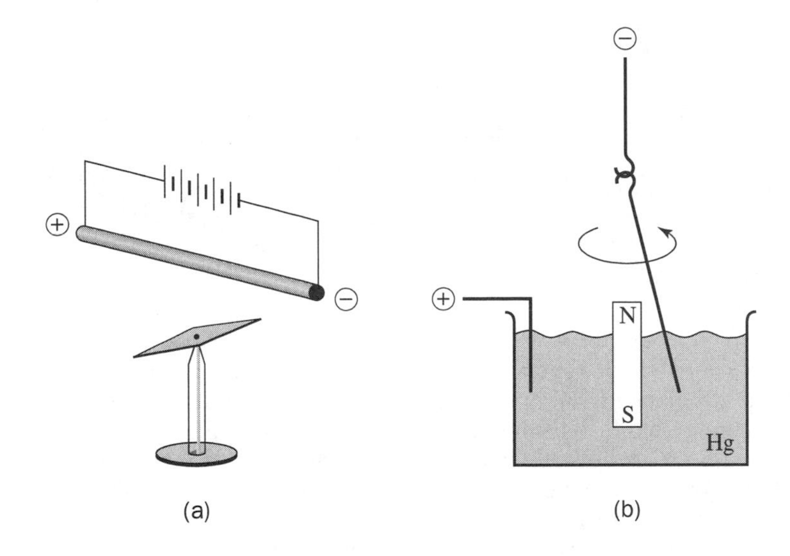

FIGURE 10.2. (a) Oersted's experiment in which a compass needle is deflected by a wire conveying an electrical current (a sophisticated version of this design served later as a *galvanometer*). (b) Schematic representation of Faraday's device in which a current-carrying wire rotates around a permanent magnet. Note that the trough is filled with liquid mercury to provide an electrical contact.

Ampère discovered that two parallel wires through which currents flowed in the same directions attracted each other, whereas they were repelled when the currents flowed in opposite directions.)

Faraday's second and probably greatest single discovery was born out of his conviction that, if electric currents would produce magnetism, the reverse must also occur. Experiments with *stationary* magnets and wires failed. Today we realize that this must be the case; otherwise, a perpetual motion would be created. Finally, in 1831, Faraday demonstrated **electromagnetic induction** in a series of fundamental experiments. The first of them involved an iron ring on which two separate coils were wound as shown in Figure 10.3(a). If the current in the primary circuit was opened and closed, a galvanometer connected to the secondary winding deflected momentarily each time (and in opposite directions, respectively). In a second experiment, Faraday moved a permanent magnet in and out of a wire coil to which a galvanometer was connected. He found deflections in opposite directions depending on which way the magnet was moved; Figure 10.3(b). Finally, a wire or coil that was moved within the field of a horseshoe magnet showed, likewise, deflections on a galvanometer; Figure 10.3(c). Based on these experiments, Faraday created the first dynamo machine consisting of a copper disk which he rotated between the poles of a large permanent magnet. He tapped the induced current from the axis and the edge of that disk.

The common element of these experiments was eventually recognized to be that *any change in magnetic flux induces a pulse of an electrical current in a loop-shaped piece of wire*. Moreover, Fara-

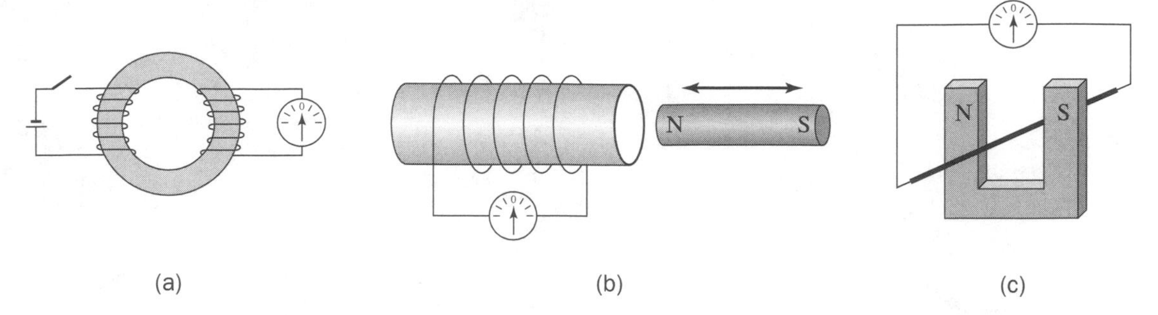

FIGURE 10.3. Schematic representations of Faraday's experiments in which he discovered induction. (a) Soft iron ring on which a primary and a secondary coil of insulated wire is wound. (b) Paper tube on which a wire coil is helically wound. (c) A wire is moved in and out of a magnetic field of a horseshoe magnet. The galvanometer shown is an instrument that measures electric current. It consists, for example, of a compass around which several layers of insulated wire are wound; compare to Figure 10.2(a).

day introduced into physics the concept of magnetic and electric lines of force. This idea, which emphasized a field between magnetic poles, was the major impetus for *Maxwell* to formulate his fundamental equations of electrodynamics.

After having thus established a connection between magnetism and electricity, Faraday felt that a relationship between *light* and magnetism must likewise exist. However, this was harder to prove. He eventually succeeded, in 1844, when he showed that the plane of polarization of light was caused to rotate in a strong magnetic field when the light path was parallel to the direction of the magnetic field; Figure 10.4. The direction of rotation was found to be the same as the direction of current flow in the wire of an electromagnet. This rotation was initially discovered to occur in lead-containing glass (heavy flint glass), which he had created in unrelated research about 20 years earlier. The **Faraday effect** is known today to take place in many liquids, solids, and gases.

In addition, Faraday tried to, but was unsuccessful in finding a possible relationship between *gravity* and electricity, a task that was picked up again by Einstein nearly 100 years later. But even Einstein did not make any progress at this.

Faraday also discovered two quantitative **laws of electrolysis** (i.e., laws which describe the precipitation or liberation of chemical elements on electrodes that are immersed into an electrolyte and to which a voltage is applied). He found, in 1833, after an extended series of laborious experiments, that:

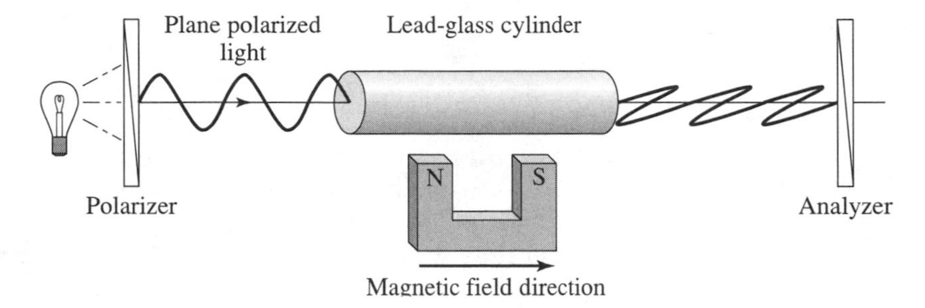

FIGURE 10.4. Schematic representation of the rotation of the plane of polarization of plane polarized light in lead glass, when applying a magnetic field (Faraday effect). In plane polarized light, the electric vector vibrates only in one direction, shown to be the paper plane in the left part of the figure. Polarizer and analyzer are identical devices which allow the light to pass in only one vibrational direction.

1. The amount of material precipitated at an electrode is proportional to the quantity of electricity (current multiplied by time) consumed; see Eq. (9.11).
2. The amount of material precipitated (or liberated) at an electrode by a unit amount of electricity is proportional to its equivalent mass; see Eq. (9.11). (The equivalent mass of a substance is the atomic mass associated with a unit gain or loss of electrons. For example, during the electrolysis of a $MgCl_2$ solution, one unit of electricity, that is, 9.649×10^4 coulombs,[2] or 6.02×10^{23} electrons,[3] deposits 24.312/2 grams of Mg on the negative electrode and liberates 35.453 grams of chlorine gas at the positive electrode; see Appendix IV.)

Furthermore, Faraday introduced the terms anode, cathode, anion, cation, and electrode.

Not enough, Faraday also made contributions to chemistry (liquefaction of chlorine under pressure, isolation of benzene, etc.), and he performed "sponsored research" on stainless steel and on glass. His fundamental studies on nonconducting materials (called *dielectrics*) led to a specific form of recognition: The unit of capacity (that is, the ability to hold an electric charge) was eventually named one *farad*. (The capacitance is one farad when one coulomb of electricity changes the potential between the plates of a capacitor by one volt, see Chapter 11 and Appendix II.) Finally, the *Faraday cage* was named after him. (The Faraday

[2]Faraday constant.
[3]Avogadro constant.

cage is a box of metal screen that shields its interior from electromagnetic radiation.)

Faraday's life was coined by a search for truth and unity. He showed that the above-mentioned five kinds of electricity involved the same principal mechanism. Later he demonstrated that many of the forces of nature are intimately interconnected as described above.

Faraday was also a brilliant public speaker who went to great pains to express his ideas in a clear and simple language. In particular, his lectures on Friday evenings in which he usually showed a number of experiments and his lectures to the youth at Christmas time were quite popular. Regardless of his accomplishments and his fame, Faraday was modest and indifferent to money and honors such as knighthood or becoming the president of the British Royal Society (which he declined to accept). In summary, Michael Faraday was a true scientist and a role model whose impact on society can still be felt today.

Nature of Electrons

The following chapters are mainly concerned with the interactions of electrons with matter. Thus, a brief discussion of the **nature of electrons** is quite in order. We have already mentioned in Section 3.2 that electrons can be considered to be part of an atom which, in an elementary description, orbit the atomic core. Some of these electrons, particularly those in the outermost orbit (i.e., the valence electrons), are often only loosely bound to their nuclei. Therefore, they disassociate with relative ease from their core and then combine to form a "sea" of electrons. These free electrons govern many of the electronic properties of materials, particularly in conductors. In other cases, such as in insulators, the electrons are bound somewhat stronger to their nuclei and thus, under the influence of an alternating external electromagnetic force, may oscillate about their core. This constitutes an electric dipole, as described in Section 3.2. We shall return to this concept when we discuss the electronic properties of dielectric materials.

Now, to our knowledge, nobody has so far seen an electron, even by using the most sophisticated equipment. We experience merely the *actions* of electrons, for example, on a television screen or in an electron microscope. In each of these instances, the electrons seem to manifest themselves in quite a different way, that is, in the first case as a particle and in the latter case as an electron wave. Accordingly, we shall use the terms "wave" and "particle" as convenient means to describe the different aspects of the properties of electrons. This is called the "duality" of the manifestations of electrons. A more complete description of the wave–particle duality of electrons and a quantum-mechanical treatment of electron waves can be found, for example, in the book, *Electronic Properties of Materials*.[4]

Discovery of the Electron

At the end of this chapter we may ponder the question of *when* the electron, as we know and understand it today, was actually discovered. The *particle nature* of electrons was suggested in 1897 by the British physicist J. J. Thomson who experimented with "cathode rays" at the Cavendish Laboratory of Cambridge University. These cathode rays were known to consist of an invisible radiation that emanated from a negative electrode (called a cathode) which was sealed through the walls of an evacuated glass tube that also contained at the opposite wall a second, positively charged electrode; see Figure 10.5.

It was likewise known at the end of the nineteenth century that cathode rays travelled in straight lines and produced a glow when they struck glass or some other materials. J. J. Thomson noticed that the path of these rays could be deflected by magnetic or electric fields, and that cathode rays traveled slower than light and transported negative electricity. In order to settle the lingering question of whether cathode rays were "vibrations of the ether" or instead "streams of particles," he promulgated a bold hypothesis, suggesting that cathode rays were "charged corpuscles which are miniscule constituents of the atom." This proposition—that an atom should consist of more than one particle—

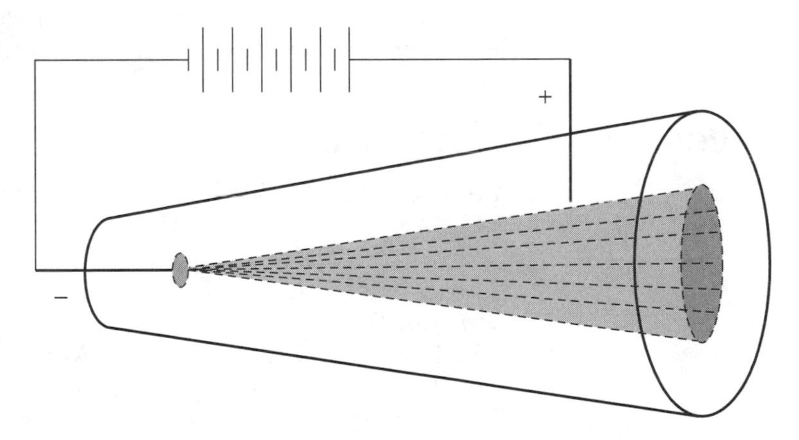

FIGURE 10.5. Schematic representation of a cathode ray tube.

[4]R.E. Hummel, *Electronic Properties of Materials*, 3rd Edition, Springer-Verlag, New York (2001).

was startling for most people at the time. Indeed, atoms were considered since antiquity to be indivisible, that is, the most fundamental building blocks of matter (see Chapter 3.1). The charge of these "corpuscles" was found to be the same (but negative) as that carried by hydrogen ions during electrolysis (about 10^{-19}C). Further, the mass of these corpuscles turned out to be 1/2000th the mass of the hydrogen atom.

A second hypotheses brought forward by J. J. Thomson, suggesting that the "corpuscles of cathode rays are the *only* constituents of atoms," was eventually proven to be incorrect. Specifically, E. Rutherford, one of Thomson's former students, by using a different kind of paricle beam, concluded that the atom resembled a tiny solar system in which a few electrons orbited around a "massive" positively charged center. Today, one knows that the electron is the lightest stable elementary particle of matter and that it carries the basic charge of electricity.

In 1924, de Broglie, who believed in a unified creation of the universe, introduced the idea that electrons should also possess *wave properties*. In other words, he suggested, based on the hypothesis of a general reciprocity of physical laws, that electrons, similarly as light, should display a wave-particle duality. In 1926, Schrödinger cast this idea in a mathematical form. Eventually, in 1927, Davisson and Germer and independently in 1928, G. P. Thomson (the son of J. J. Thomson), discovered electron diffraction by a crystal which experimentally proved the wave nature of electrons.

In the following chapters we shall utilize either the particle nature or, alternately, the wave nature of electrons as they prove to be convenient in explaining the electronic properties of materials.

Suggestions for Further Study

A.D. Aczel, *The Riddle of the Compass*, Harcourt, New York (2001).

D.C. Cassidy, *Uncertainty, The Life and Science of Werner Heisenberg*, Freeman, New York (1992)

J. Lemmerich, Michael Faraday 1791–1867, *Erforscher der Elektrizität*, Beck, Munich (1991)

J.D. Livingston, *Driving Force, The Natural Magic of Magnets*, Harvard University Press, Cambridge, MA (1996)

T.C. Martin, *The Inventions, Researches and Writings of Nikola Tesla*, 2nd Edition, Barnes and Noble, New York, (1992)

11

Electrical Properties of Materials

One of the principal characteristics of materials is their ability (or lack of ability) to conduct electrical current. Indeed, materials are classified by this property, that is, they are divided into conductors, semiconductors, and nonconductors. (The latter are often called insulators or dielectrics.) The **conductivity**, σ, of different materials at room temperature spans more than 25 orders of magnitude, as depicted in Figure 11.1. Moreover, if one takes the conductivity of superconductors, measured at low temperatures, into consideration, this span extends to 40 orders of magnitude (using an estimated conductivity for superconductors of about 10^{20} $1/\Omega$ cm). This is the largest known variation in a physical property and is only comparable to the ratio between the diameter of the universe (about 10^{26} m) and the radius of an electron (10^{-14} m).

The inverse of the conductivity is called **resistivity**, ρ, that is:

$$\rho = \frac{1}{\sigma}. \tag{11.1}$$

The **resistance**, R of a piece of conducting material is proportional to its resistivity and to its *length*, L, and is inversely proportional to its cross-sectional area, A:

$$R = \frac{L \cdot \rho}{A}. \tag{11.2}$$

The resistance can be easily measured. For this, a direct current is applied to a slab of the material. The current, I, through the sample (in ampères), as well as the voltage drop, V, on two potential probes (in volts) is recorded as depicted in Figure 11.2.

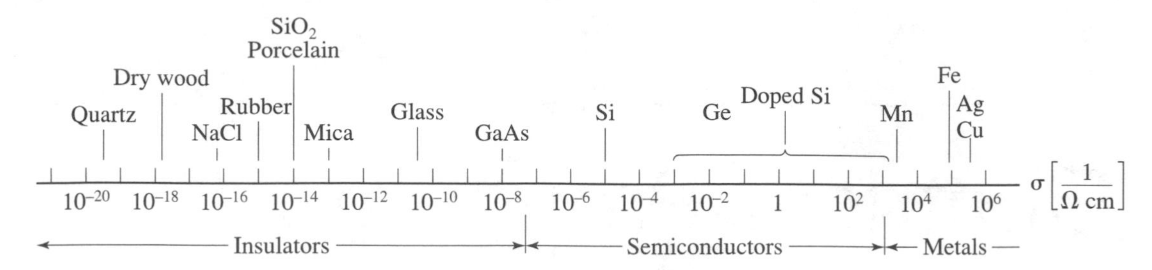

FIGURE 11.1. Room-temperature conductivity of various materials. (Superconductors, having conductivities of many orders of magnitude larger than copper, near 0 K, are not shown. The conductivity of semiconductors varies substantially with temperature and purity.) It is customary in engineering to use the centimeter as the unit of length rather than the meter. We follow this practice. The reciprocal of the ohm (Ω) is defined to be 1 siemens (S); see Appendix II. For conducting polymers, refer to Figure 11.20.

Ohm's Law

The resistance (in ohms) can then be calculated by making use of *Ohm's law*:

$$V = R \cdot I, \tag{11.3}$$

which was empirically found by Georg Simon Ohm (a German physicist) in 1826 relating a large number of experimental observations. Another form of Ohm's law:

$$j = \sigma \cdot \mathscr{E}, \tag{11.4}$$

links current density:

$$j = \frac{I}{A}, \tag{11.5}$$

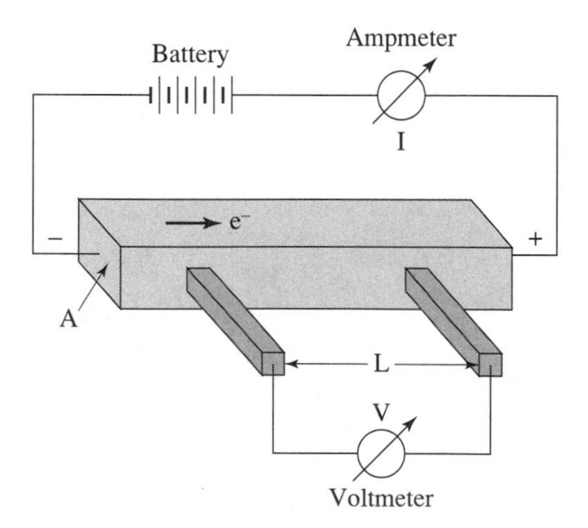

FIGURE 11.2. Schematic representation of an electric circuit to measure the resistance of a conductor.

that is, the current per unit area (A/cm^2), with the conductivity σ (1/Ω cm or siemens per cm) and the electric field strength:

$$\mathscr{E} = \frac{V}{L} \tag{11.6}$$

(V/cm). (We use a script $\mathscr{E}$ for the electric field strength to distinguish it from the energy.)

11.1 • Conductivity and Resistivity of Metals

The resistivity of metals essentially increases linearly with increasing temperature (Figure 11.3) according to the empirical equation:

$$\rho_2 = \rho_1[1 + \alpha(T_2 - T_1)], \tag{11.7}$$

where α is the linear *temperature coefficient of resistivity*, and T_1 and T_2 are two different temperatures. We attempt to explain this behavior. We postulate that the free electrons (see Chapter 10) are accelerated in a metal under the influence of an electric field maintained, for example, by a battery. The drifting electrons can be considered, in a preliminary, classical description, to occasionally collide (that is, electrostatically interact) with certain lattice atoms, thus losing some of their energy. This constitutes the just-discussed resistance. In essence, the drifting electrons are then said to migrate in a zig-zag path through the conductor from the cathode to the anode, as sketched in Figure 11.4. Now, at higher temperatures, the lattice atoms increasingly oscillate about their equilibrium positions due to the supply of thermal energy, thus enhancing the probability for collisions by the drifting electrons. As a consequence, the resistance rises with higher

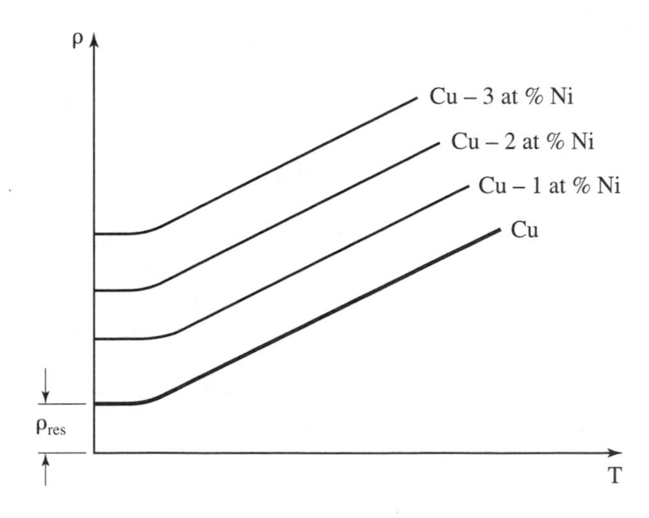

FIGURE 11.3. Schematic representation of the temperature dependence of the resistivity of copper and various copper-nickel alloys. ρ_{res} is the residual resistivity.

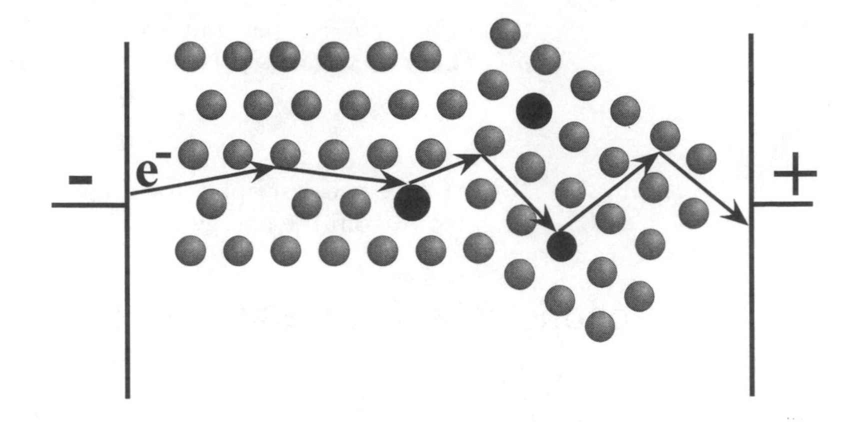

FIGURE 11.4. Schematic representation of an electron path through a conductor (containing vacancies, impurity atoms, and a grain boundary) under the influence of an electric field. This classical description does not completely describe the resistance in materials.

temperatures. At near-zero temperatures, the electrical resistance does not completely vanish, however (except in superconductors). There always remains a *residual resistivity*, ρ_{res} (Figure 11.3), which is thought to be caused by "collisions" of electrons (i.e., by electrostatic interactions) with imperfections in the crystal (such as impurities, vacancies, grain boundaries, or dislocations), as explained in Chapters 3 and 6. The residual resistivity is essentially temperature-independent.

On the other hand, one may describe the electrons to have a wave nature. The matter waves may be thought to be scattered by lattice atoms. Scattering is the dissipation of radiation on small particles in all directions. The atoms absorb the energy of an incoming wave and thus become oscillators. These oscillators in turn re-emit the energy in the form of spherical waves. If two or more atoms are involved, the phase relationship between the individual re-emitted waves has to be taken into consideration. A calculation[1] shows that for a *periodic crystal structure* the individual waves in the forward direction are *in-phase*, and thus interfere constructively. As a result, a wave which propagates through an ideal crystal (having periodically arranged atoms) does not suffer any change in intensity or direction (only its velocity is modified). This mechanism is called *coherent scattering*.

If, however, the scattering centers are not periodically arranged (impurity atoms, vacancies, grain boundaries, thermal vibration of atoms, etc.), the scattered waves have no set phase relationship and the wave is said to be *incoherently scattered*. The energy of incoherently scattered waves is smaller in the forward direction. This energy loss qualitatively explains the resistance. In

[1] L. Brillouin, *Wave Propagation in Periodic Structures*, Dover, New York (1953).

short, the wave picture provides a deeper understanding of the electrical resistance in metals and alloys.

According to a rule proposed by **Matthiessen**, the total resistivity arises from independent mechanisms, as just described, which are additive, i.e.:

$$\rho = \rho_{th} + \rho_{imp} + \rho_{def} = \rho_{th} + \rho_{res}. \tag{11.8}$$

The thermally induced part of the resistivity ρ_{th} is called the *ideal* resistivity, whereas the resistivity that has its origin in impurities (ρ_{imp}) and defects (ρ_{def}) is summed up in the residual resistivity (ρ_{res}). The number of impurity atoms is generally constant in a given metal or alloy. The number of vacancies, or grain boundaries, however, can be changed by various heat treatments. For example, if a metal is annealed at temperatures close to its melting point and then rapidly quenched into water of room temperature, its room temperature resistivity increases noticeably due to quenched-in vacancies, as already explained in Chapter 6. Frequently, this resistance increase diminishes during room temperature aging or annealing at slightly elevated temperatures due to the annihilation of these vacancies. Likewise, work hardening, recrystallization, grain growth, and many other metallurgical processes change the resistivity of metals. As a consequence of this, and due to its simple measurement, the resistivity has been one of the most widely studied properties in materials research.

Free Electrons

The conductivity of metals can be calculated (as P. Drude did at the turn to the 20th century) by simply postulating that the electric force, $e \cdot \mathscr{E}$, provided by an electric field (Figure 11.2), accelerates the electrons (having a charge $-e$) from the cathode to the anode. The drift of the electrons was thought by Drude to be counteracted by collisions with certain atoms as described above. The Newtonian-type equation (force equals mass times acceleration) of this **free electron model**

$$m \frac{dv}{dt} + \gamma v = e \cdot \mathscr{E} \tag{11.9}$$

leads, after a string of mathematical manipulations, to the conductivity:

$$\sigma = \frac{N_f \cdot e^2 \cdot \tau}{m}, \tag{11.10}$$

where v is the drift velocity of the electrons, m is the electron mass, γ is a constant which takes the electron/atom collisions into consideration (called damping strength), $\tau = m/\gamma$ is the average time between two consecutive collisions (called the relaxation time), and N_f is the number of free electrons per cubic meter in the material. We can learn from this equation that semiconductors or insulators which have only a small number of free electrons (or often none at all) display only very small conductivities. (The small number of electrons results from the strong binding forces between electrons and atoms that are common for insulators and semiconductors.) Conversely, metals which contain a large number of free electrons have a large conductivity. Further, the conductivity is large when the average time between two collisions, τ, is large. Obviously, the number of collisions decreases (i.e., τ increases) with decreasing temperature and decreasing number of imperfections.

The above-outlined free electron model, which is relatively simple in its assumptions, describes the electrical behavior of many materials reasonably well. Nevertheless, quantum mechanics provides some important and necessary refinements. One of the refinements teaches us how many of the valence electrons can be considered to be free, that is, how many of them contribute to the conduction process. Equation (11.10) does not provide this distinction. Quantum mechanics of materials is quite involved and requires the solution of the Schrödinger equation, the treatment of which must be left to specialized texts.[2] Its essential results can be summarized, however, in a few words.

Electron Band Model

We know from Section 3.1 that the electrons of isolated atoms (for example in a gas) can be considered to orbit at various distances about their nuclei. These orbits constitute different energies. Specifically, the larger the radius of an orbit, the larger the excitation energy of the electron. This fact is often represented in a somewhat different fashion by stating that the electrons are distributed on different *energy levels*, as schematically shown on the right side of Figure 11.5. Now, these distinct energy levels, which are characteristic for isolated atoms, widen into **energy bands** when atoms approach each other and eventually form a solid as depicted on the left side of Figure 11.5. Quantum mechanics postulates that the electrons can only reside within these bands, but not in the areas outside of them. The allowed energy bands may be noticeably separated from each other. In other cases, depending on the material and the energy, they may partially or completely overlap. In short, each material has its distinct electron energy band structure. Characteristic band structures for the main classes of materials are schematically depicted in Figure 11.6.

---

[2]See, for example, R.E. Hummel, *Electronic Properties of Materials*, 3rd Edition, Springer-Verlag, New York (2001).

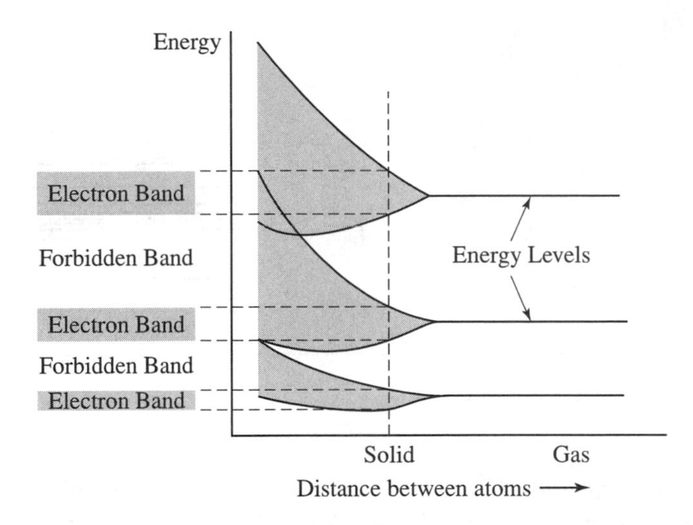

FIGURE 11.5. Schematic representation of energy levels (as for isolated atoms) and widening of these levels into energy bands with decreasing distance between atoms. Energy bands for a specific case are shown at the left of the diagram.

Now, the band structures shown in Figure 11.6 are somewhat simplified. Specifically, band schemes actually possess a fine structure, that is, the individual energy states (i.e., the possibilities for electron occupation) are often denser in the center of a band (Figure 11.7). To account for this, one defines a density of energy states, shortly called the **density of states**, $Z(E)$.

Some of the just-mentioned bands are occupied by electrons while others remain partially or completely empty, similar to a cup that may be only partially filled with water. The degree to which an electron band is filled by electrons is indicated in Figure 11.6 by shading. The highest level of electron filling within a band is called the **Fermi energy**, E_F, which may be compared with the water surface in a cup. (For values of E_F, see Appendix II). We notice in Figure 11.6 that some materials, such as insulators and semiconductors, have completely filled electron bands. (They differ, however, in their distance to the next higher band.)

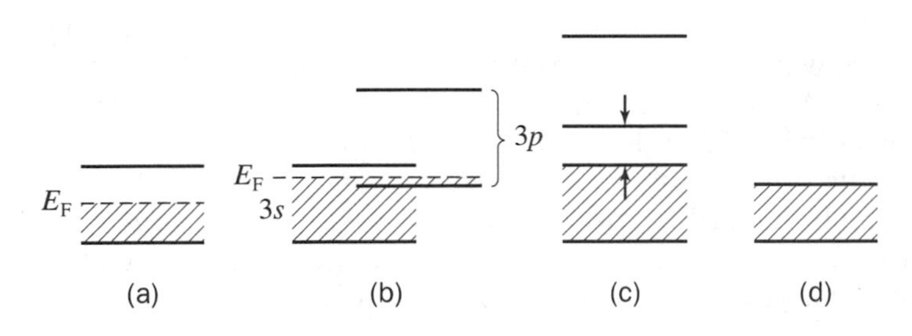

FIGURE 11.6. Simplified representation for energy bands for (a) monovalent metals, (b) bivalent metals, (c) semiconductors, and (d) insulators. For a description of the nomenclature, see Appendix I.

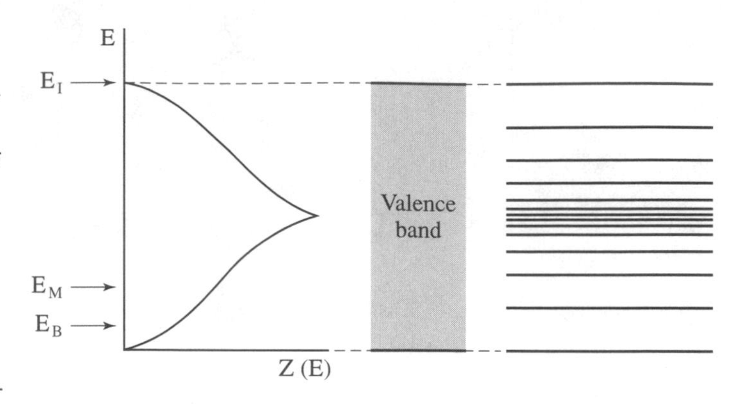

FIGURE 11.7. Schematic representation of the density of electron states $Z(E)$ within an electron energy band. The density of states is essentially identical to the population density $N(E)$ for energies below the Fermi energy, E_F (i.e., for that energy level up to which a band is filled with electrons). Examples of highest electron energies for a monovalent metal (E_M), for a bivalent metal (E_B), and for an insulator (E_I) are indicated.

Metals, on the other hand, are characterized by *partially* filled electron bands. The amount of filling depends on the material, that is, on the electron concentration and the amount of band overlapping.

We may now return to the conductivity. In short, according to quantum theory, only those materials that possess partially filled electron bands are capable of conducting an electric current. Electrons can then be lifted slightly above the Fermi energy into an allowed and unfilled energy state. This permits them to be accelerated by an electric field, thus producing a current. Second, only those electrons that are close to the Fermi energy participate in the electric conduction. (The classical electron theory taught us instead that *all* free electrons would contribute to the current.) Third, the number of electrons near the Fermi energy depends on the density of available electron states (Figure 11.7). The conductivity in quantum mechanical terms yields the following equation:

$$\sigma = \frac{1}{3} e^2 \, v_F^2 \, \tau N(E_F) \tag{11.11}$$

where v_F is the velocity of the electrons at the Fermi energy (called the Fermi velocity) and $N(E_F)$ is the density of filled electron states (called the population density) at the Fermi energy. The population density is proportional to $Z(E)$; both have the unit $J^{-1}m^{-3}$ or $eV^{-1}m^{-3}$. Equation (11.11), in conjunction with Figure 11.7, now provides a more comprehensive picture of electron conduction. Monovalent metals (such as copper, silver, and gold) have partially filled bands, as shown in Figure 11.6(a). Their electron population density near the Fermi energy is high (Figure 11.7), which, according to Eq. (11.11), results in a large conductivity. Bivalent metals, on the other hand, are distinguished by an overlapping of the upper bands and by a small electron concentration near the bottom of the valence band, as shown in Figure 11.6(b). As a consequence, the electron population near the Fermi energy is small (Figure 11.7), which leads to a comparatively low conductivity. Finally, insulators have completely filled (and completely empty) electron bands, which results in a virtually zero population density, as shown in Figure 11.7. Thus, the conductivity in insulators is virtually zero (if one disregards, for example, ionic conduction; see Section 11.6). These explanations are admittedly quite sketchy. The interested reader is referred to the specialized books listed at the end of this chapter.

11.2 • Conduction in Alloys

The residual resistivity of alloys increases with increasing amount of solute content as seen in Figures 11.3 and 11.8. The slopes of the individual ρ versus T lines remain, however, essentially constant (Figure 11.3). Small additions of solute cause a linear shift of the ρ versus T curves to higher resistivity values in accordance with the Matthiessen rule; see Eq. (11.8) and Figure 11.8. Various solute elements might alter the resistivity of the host material to different degrees. This is depicted in Figure 11.8 for silver, which demonstrates that the residual resistivity increases with increasing atomic number of the solute. For its interpretation, one may reasonably assume that the likelihood for interactions between electrons and impurity atoms increases when the solute has a larger atomic size, as is encountered by proceeding from cadmium to antimony.

The resistivity of *two-phase alloys* is, in many instances, the sum of the resistivity of each of the components, taking the volume fractions of each phase into consideration. However, additional factors, such as the crystal structure and the kind of distribution of the phases in each other, also have to be considered.

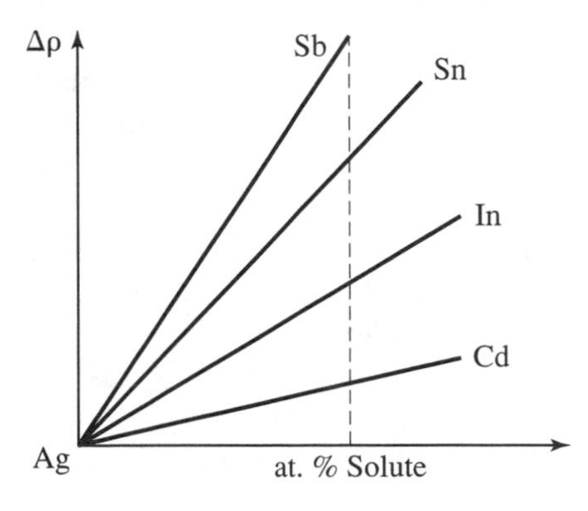

FIGURE 11.8. Resistivity change of various dilute silver alloys (schematic). Solvent and solute are all from the fifth period.

Some alloys, when in the ordered state, that is, when the solute atoms are periodically arranged in the matrix, have a distinctly smaller resistivity compared to the case when the atoms are randomly distributed. Slowly cooled Cu_3Au or CuAu are common examples of ordered structures.

Copper is frequently used for electrical wires because of its high conductivity (Figure 11.1). However, pure or annealed copper has a low strength (Chapter 3). Thus, work hardening (during wire drawing), or dispersion strengthening (by adding less than 1% Al_2O_3), or age hardening (Cu–Be), or solid solution strengthening (by adding small amounts of second constituents such as Zn) may be used for strengthening. The increase in strength occurs, however, at the expense of a reduced conductivity. (The above mechanisms are arranged in decreasing order of conductivity of the copper-containing wire.) The resistance increase in copper inflicted by cold working can be restored to almost its initial value by annealing copper at moderate temperatures (about 300°C). This process, which was introduced in Chapter 6 by the terms *stress relief anneal* or *recovery*, causes the dislocations to rearrange to form a polygonized structure without substantially reducing their number. Thus, the strength of stress-relieved copper essentially is maintained while the conductivity is almost restored to its pre-work hardened state (about 98%).

For other applications, a *high* resistivity is desired, such as for heating elements in furnaces which are made, for example, of nickel–chromium alloys. These alloys need to have a high melting temperature and also a good resistance to oxidation, particularly at high temperatures.

11.3 • Superconductivity

The resistivity in superconductors becomes immeasurably small or virtually zero below a critical temperature, T_c, as shown in Figure 11.9. About 27 elements, numerous alloys, ceramic materials (containing copper oxide), and organic compounds (based, for example, on selenium or sulfur) have been found to possess this property (see Table 11.1). It is estimated that the conductivity of superconductors below T_c is about 10^{20} $1/\Omega$ cm (see Figure 11.1). The transition temperatures where superconductivity starts range from 0.01 K (for tungsten) up to about 125 K (for ceramic superconductors). Of particular interest are materials whose T_c is above 77 K, that is, the boiling point of liquid nitrogen, which is more readily available than other coolants. Among the so-called

TABLE 11.1 Critical temperatures of some superconducting materials

Materials	T_c [K]	Remarks
Tungsten	0.01	—
Mercury	4.15	H.K. Onnes (1911)
Sulfur-based organic superconductor	8	S.S.P. Parkin et al. (1983)
Nb_3Sn and Nb–Ti	9	Bell Labs (1961), Type II
V_3Si	17.1	J.K. Hulm (1953)
Nb_3Ge	23.2	(1973)
La–Ba–Cu–O	40	Bednorz and Müller (1986)
$YBa_2Cu_3O_{7-x}$	92	Wu, Chu, and others (1987)
$RBa_2Cu_3O_{7-x}$	~92	R = Gd, Dy, Ho, Er, Tm, Yb, Lu
$Bi_2Sr_2Ca_2Cu_3O_{10+\delta}$	113	Maeda et al. (1988)
$Tl_2CaBa_2Cu_2O_{10+\delta}$	125	Hermann et al. (1988)
$HgBa_2Ca_2Cu_3O_{8+\delta}$	134	R. Ott et al. (1995)

high-T_c superconductors are the *1-2-3 compounds* such as $YBa_2Cu_3O_{7-x}$ whose molar ratios of rare earth to alkaline earth to copper relate as 1:2:3. Their transition temperatures range from 40 to 134 K. Ceramic superconductors have an orthorhombic, layered, perovskite crystal structure (similar to $BaTiO_3$; see Figure 11.30) which contains two-dimensional sheets and periodic oxygen vacancies. (The superconductivity exists only parallel to these layers, that is, it is anisotropic.) The first superconducting material was found by H.K. Onnes in 1911 in mercury which has a T_c of 4.15 K.

A high magnetic field or a high current density may eliminate superconductivity. In *Type I superconductors*, the annihilation of the superconducting state by a magnetic field, that is, the transi-

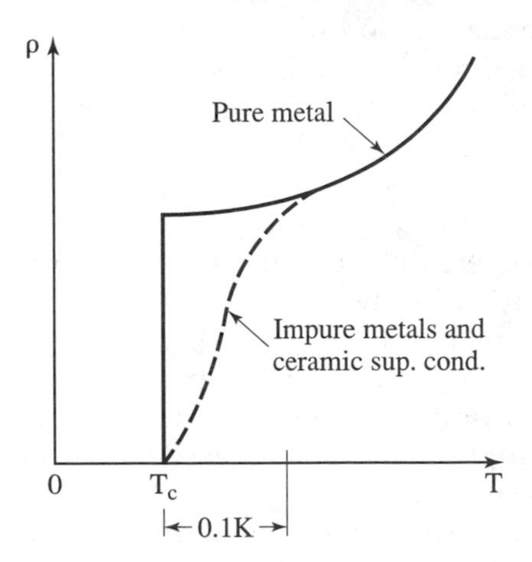

FIGURE 11.9. Schematic representation of the resistivity of pure and compound superconducting materials. T_c is the critical or transition temperature, below which superconductivity commences.

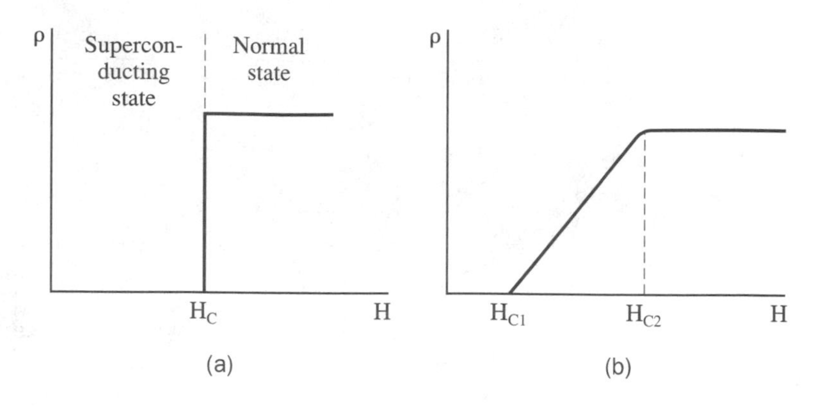

FIGURE 11.10. Schematic representation of the resistivity of (a) Type I (or soft) and (b) Type II (or hard) superconductors in an external magnetic field. The solids behave like normal conductors above H_c and H_{c2} respectively.

tion between superconducting and normal states, occurs sharply; Figure 11.10(a). The critical field strength H_c, above which superconductivity ceases, is relatively low. The destruction of the superconducting state in *Type II superconductors* occurs instead, more gradually, i.e., in a range between H_{c1} and H_{c2}, where H_{c2} is often 100 times larger than Hc_1 [Figure 11.10(b)]. In the interval between H_{c1} and H_{c2}, normal conducting areas, called *vortices*, and superconducting regions are interspersed. The terms "Type I" and "Type II" superconductors are occasionally also used when a distinction between abrupt and gradual transition with respect to temperature is described; see Figure 11.9. In alloys and ceramic superconductors, a temperature spread of about 0.1 K has been found whereas pure gallium drops its resistance within 10^{-5} K.

Type II superconductors are utilized for strong electromagnets employed, for example, in magnetic resonance imaging devices (used in medicine), high-energy particle accelerators, and electric power storage devices. (An electric current induced into a loop consisting of a superconducting wire continues to flow for an extended period of time without significant decay.) Further potential applications are lossless power transmission lines; high-speed levitation trains; faster, more compact computers; and switching devices, called *cryotrons*, which are based on the destruction of the superconducting state in a strong magnetic field. Despite their considerably higher transition temperatures, ceramic superconductors have not yet revolutionized current technologies, mainly because of their still relatively low T_c, their brittleness, their relatively small capability to carry high current densities, and their environmental instability. These obstacles may be overcome eventually, however, by using other materials, for example, compounds based on bismuth, etc., or by producing thin-film superconductors. At present, most superconducting electromagnets are manufactured by using niobium–titanium alloys which are ductile and thus can be drawn into wires.

The quantum-mechanical theory, which explains superconductivity and which was developed in 1957 by Bardeen, Cooper, and Schrieffer, is quite involved and is therefore beyond the scope of this book.

11.4 • Semiconductors

Semiconductors such as silicon or germanium are neither good conductors nor good insulators as seen in Figure 11.1. This may seem to make semiconductors to be of little interest. Their usefulness results, however, from a completely different property, namely, that extremely small amounts of certain impurity elements, which are called *dopants*, remarkably change the electrical behavior of semiconductors. Indeed, semiconductors have been proven in recent years to be the lifeblood of a multibillion dollar industry which prospers essentially from this very feature. Silicon, the major species of semiconducting materials, is today the single most researched element. Silicon is abundant (28% of the earth's crust consists of it); the raw material (SiO_2 or sand) is inexpensive; Si forms a natural, insulating oxide; its heat conduction is reasonable; it is nontoxic; and it is stable against environmental influences.

Intrinsic Semiconductors

The properties of semiconductors are commonly explained by making use of the already introduced electron band structure which is the result of quantum-mechanical considerations. In simple terms, the electrons are depicted to reside in certain allowed energy regions as explained in Section 11.1. Specifically, Figures 11.6(c) and 11.11 depict two electron bands, the lower of which, at 0 K, is completely filled with valence electrons. This band is appropriately called the **valence band**. It is separated by a small gap (about 1.1 eV for Si) from the **conduction band**, which, at 0 K, contains no electrons. Further, quantum mechanics stipulates that electrons essentially are not allowed to reside in the gap between these bands (called the *forbidden band*). Since the filled valence band possesses no allowed empty energy states in which the electrons can be thermally excited (and then accelerated in an electric field), and since the conduction band contains no electrons at all, silicon, at 0 K, is an insulator.

The situation changes decisively once the temperature is raised. In this case, *some* electrons may be thermally excited across the band gap and thus populate the conduction band (Figure 11.11). The number of these electrons is extremely small for statistical reasons. Specifically, about one out of every 10^{13} atoms contributes an electron at room temperature. Nevertheless, this num-

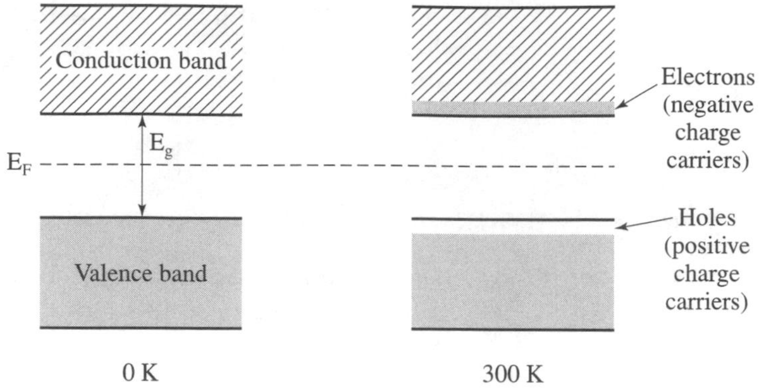

FIGURE 11.11. Simplified band diagrams for an intrinsic semiconductor such as pure silicon at two different temperatures. The dark shading symbolizes electrons.

ber is large enough to cause some conduction, as shown in Figure 11.1. Actually, the number of electrons in the conduction band, N_e, increases exponentially with temperature, T, but also depends, of course, on the size of the gap energy, E_g, according to:

$$N_e = 4.84 \times 10^{15}\, T^{3/2} \exp\left[-\left(\frac{E_g}{2k_B T}\right)\right], \tag{11.12}$$

where k_B is the Boltzmann constant (see Appendix II), and the constant factor in front of $T^{3/2}$ has the unit of $\text{cm}^{-3} \cdot \text{K}^{-3/2}$. The conductivity depends naturally on the number of these electrons but also on their mobility. The latter is defined to be the velocity, v, per unit electric field, $\mathscr{E}$, that is:

$$\mu = \frac{\text{v}}{\mathscr{E}}. \tag{11.13}$$

All taken, the conductivity is

$$\sigma = N_e \cdot \mu \cdot e, \tag{11.14}$$

where e is the charge of an electron (see Appendix II). The mobility of electrons is substantially impaired by interactions with impurity atoms and other lattice imperfections (as well as with vibrating lattice atoms), as explained in Section 11.1. It is for this reason that silicon has to be extremely pure and free of grain boundaries, which requires sophisticated and expensive manufacturing processes called *zone refining* or *Czochralski crucible pulling*.

From the above discussion, particularly from Eq. (11.12), it becomes evident that the conductivity for semiconductors increases with rising temperatures. This is in marked contrast to metals and alloys, for which the conductivity decreases with temperature (see Figure 11.3).

Now, the thermal excitation of some electrons across the band gap has another important consequence. The electrons that have left the valence band leave behind some empty spaces which allow additional conduction to take place in the valence band. Actually, the empty spaces are called *defect electrons* or *electron holes*. These holes may be considered to be positively charged carriers similarly as electrons are defined to be negatively charged carriers. In essence, at elevated temperatures, the thermal energy causes some electrons to be excited from the valence band into the conduction band. They provide there some conduction. Concomitantly, the electron holes which have been left behind in the valence band cause a hole current which is directed in the opposite direction compared to the electron current. The total conductivity, therefore, is a sum of both contributions:

$$\sigma = N_e \mu_e e + N_h \mu_h e, \tag{11.15}$$

where the subscripts e and h refer to electrons and holes, respectively. The process just described is called *intrinsic conduction*, and the material involved is termed an *intrinsic semiconductor* since no foreign elements have been involved.

The Fermi energy of intrinsic semiconductors can be considered to be the average of the electron and the hole Fermi energies and is therefore situated near the center of the gap, as depicted in Figure 11.11.

Extrinsic Semiconductors

The number of electrons in the conduction band can be considerably increased by adding, for example, to silicon small amounts of elements from Group V of the Periodic Table called *donor atoms*. Dopants such as phosphorous or arsenic are commonly utilized, which are added in amounts of, for example, 0.0001%. These dopants replace some regular lattice atoms in a substitutional manner; Figure 11.12. Since phosphorous has five valence

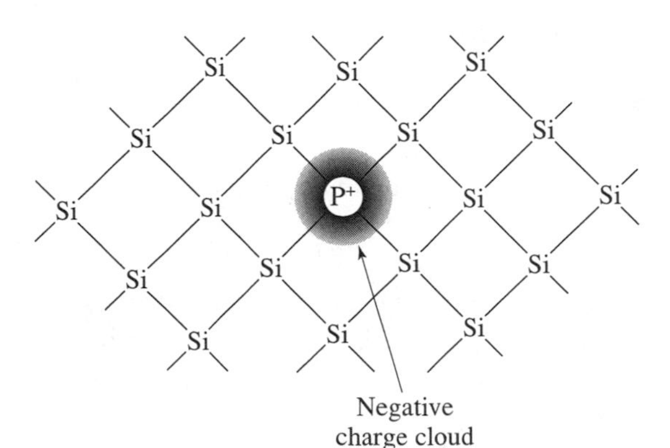

FIGURE 11.12. Two-dimensional representation of a silicon lattice in which a phosphorous atom substitutes a regular lattice atom, and thus introduces a negative charge cloud about the phosphorous atom. Each electron pair between two silicon atoms constitutes a covalent bond [see Chapter 3, particularly Figure 3.4(a)].

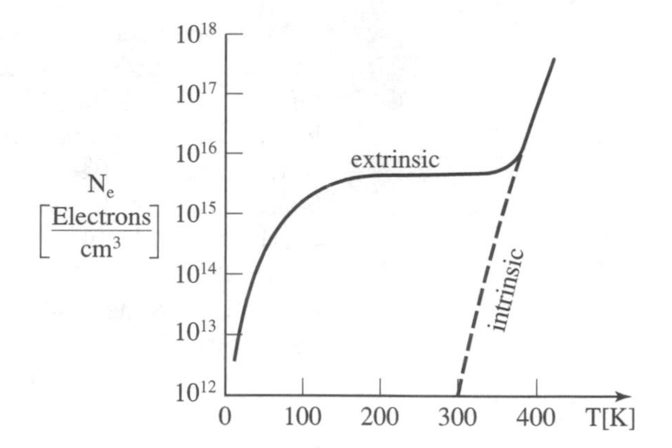

FIGURE 11.13. Schematic representation of the number of electrons per cubic centimeter in the conduction band as a function of temperature for extrinsic semiconductors, assuming low doping.

electrons, that is, one more than silicon, the extra electron, called the *donor electron*, is only loosely bound. The binding energy of phosphorous donor electrons in a silicon matrix, for example, is about 0.045 eV. Thus, the donor electrons can be disassociated from their nuclei by only a slight increase in thermal energy; see Figure 11.13. Indeed, at room temperature all donor electrons have already been excited into the conduction band.

It is common to describe this situation by introducing into the forbidden band so-called *donor levels*, which accommodate the donor electrons at 0 K; see Figure 11.14(a). The distance between the donor level and the conduction band represents the energy that is needed to transfer the extra electrons into the conduction band (e.g., 0.045 eV for P in Si). The electrons that have been excited from the donor levels into the conduction band are free and can be accelerated in an electric field as shown in Figures 11.2 and 11.4. Since the conduction mechanism in semiconductors with donor impurities is predominated by *negative* charge carriers, one calls these materials *n-type semiconductors*.

Similar considerations may be carried out with respect to im-

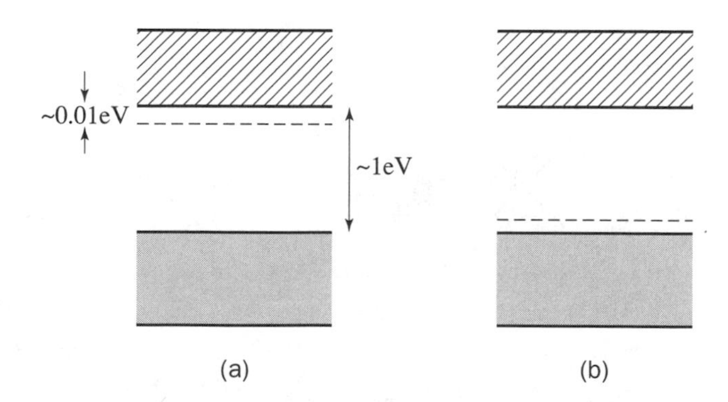

FIGURE 11.14. (a) Donor and (b) acceptor levels in extrinsic semiconductors.

purities from the third group of the Periodic Table (B, Al, Ga, In). They are deficient in one electron compared to silicon and therefore tend to accept an electron. The conduction mechanism in these semiconductors with *acceptor impurities* is thus predominated by *positive* charge carriers (holes) which are introduced from the *acceptor levels* [Figure 11.14(b)] into the valence band. They are therefore called *p-type* semiconductors. In other words, the conduction in p-type semiconductors under the influence of an external electric field occurs in the valence band and is predominated by holes.

The conductivity for n-type semiconductors is

$$\sigma = N_{de}\, e\, \mu_e, \tag{11.16}$$

where N_{de} is the number of donor electrons and μ_e is the mobility of the (donor) electrons. A similar equation holds for hole conduction.

Near room temperature, only the majority carriers need to be considered. For example, at room temperature, all donor electrons in an n-type semiconductor have been excited from the donor levels into the conduction band. At higher temperatures, however, intrinsic effects may considerably contribute to the conduction, as seen in Figure 11.13. For certain applications (high temperatures, space, military) this needs to be prevented, and semiconductor materials with wider band gaps, such as compound semiconductors (see below) are utilized.

Compound Semiconductors

Compounds made of Group III and Group V elements, such as gallium arsenide, have similar semiconducting properties as the Group IV materials silicon or germanium. GaAs is of some technical interest because of its above-mentioned wider band gap, and because of its larger electron mobility, which aids in high-speed applications; see Appendix II. Further, the ionization energies of donor and acceptor impurities in GaAs are one order of magnitude smaller than in silicon, which ensures complete electron (and hole) transfer from the donor (acceptor) levels into the conduction (valence) bands, even at relatively low temperatures. However, GaAs is about ten times more expensive than Si and its heat conduction is smaller. Other compound semiconductors include II–VI combinations such as ZnO, ZnS, ZnSe, or CdTe, and IV–VI materials such as PbS, PbSe, or PbTe. Silicon carbide, a IV–IV compound, has a band gap of 3 eV and can thus be used up to 700°C before intrinsic effects set in. The most important application of compound semiconductors is, however, for opto-electronic purposes (e.g. for light-emitting diodes, and lasers). We shall return to this in Chapter 13.

Semiconductor Devices

The evolution of solid-state microelectronic technology started in 1947 with the invention of the point contact transistor by Bardeen, Brattain, and Shockley at Bell Laboratories. (Until then, vacuum tubes were used, the principle of which was invented in 1906 by Lee deForest. Likewise, silicon rectifiers have been in use since the beginning of the 20th century.) The development went via the germanium junction transistor (Shockley, 1950), the silicon transistor (Shockley, 1954), the first integrated circuit (Kilby, Texas Instruments, 1959), the planar transistor (Noyce and Fairchild, 1962), to the ultra-large–scale integration (ULSI) of today with several million transistors on one chip.

A **rectifier** or **diode** is used to "convert" an alternating current (as used in household circuits) into a direct current, which is needed for many electronic devices and for automobile batteries. A rectifier consists of a junction of p- and n-type semiconductors [manufactured by ion implanting P and B dopants into different parts of a silicon substrate; Figure 11.15(a)]. As explained above, electrons are the predominant charge carriers in n-type semiconductors, whereas holes are the majority carriers in p-type semiconductors. If a negative voltage is applied to the p-side (and consequently a positive charge to the n-side), then the resulting electrostatic attraction forces cause the charge carriers to separate, that is, they migrate to their respective ends of the device. This causes an area which is free of carriers near the junction between n and p. It is called the *barrier region* or the *space charge region* [Figure 11.15(a–c)]. Once a barrier region of appropriate width has been formed, no current flows in the circuit, as depicted in Figures 11.15(b), (d), and (e). This operational mode is called *reverse bias*. On the other hand, a positive charge to the p- and a negative charge to the n-side forces electrons and holes to drift towards the center, where they mutually recombine. This constitutes a current, I, in the circuit which increases (exponentially) with rising applied voltage, as depicted in Figures 11.15 (c), (d), and (e) (*forward bias*). This current is

$$I = I_s \left[\exp\left(\frac{eV}{k_B T} \right) - 1 \right], \tag{11.17}$$

where V is the applied voltage on the device and I_s is a very small current (microamps!) which still flows during reverse bias. It is called the *saturation current* [see Figure 11.15(d)].

A **solar cell** (or **photovoltaic device**) consists of a p–n junction. If light of sufficiently large energy ($E = \nu \cdot \mathrm{h}$) falls on this device, electrons are lifted from the valence band into the conduction band. The electrons in the barrier region drift quickly to the n-side, whereas the concomitantly created holes in the va-

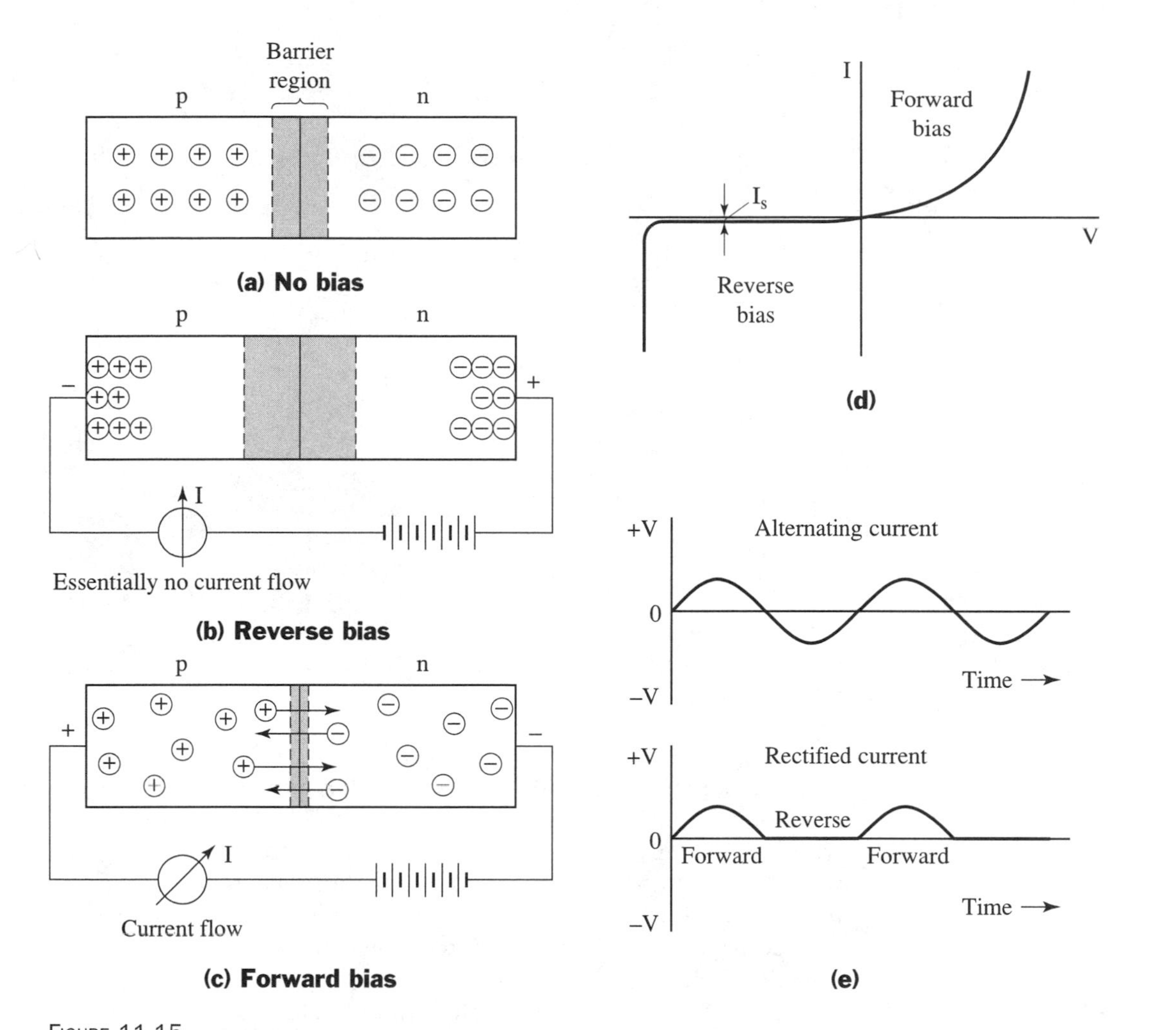

FIGURE 11.15. Schematic representation of (a) an unbiased p–n junction, (b) a p–n junction in reverse bias, (c) a p–n junction in forward bias, (d) current voltage characteristics of a p–n rectifier, and (e) voltage versus time curves.

lence band migrate into the p-region. This provides a potential difference (voltage) between the n and p terminals. (A more complete description makes use of the appropriate electron band diagram.)

The **bipolar junction transistor** may be considered to be an n–p diode back-to-back with a p–n diode, as depicted in Figure 11.16. The three terminals are called *emitter* (E), *base* (B), and *collector* (C). The transistor is used for amplification of a signal (music, voice) and as an effective switching device for computers. The "diode", consisting of emitter and base, is forward-biased, thus allowing an injection of electrons from the emitter into the base. In order to effectively reduce a possible electron–hole recombination in the base area, this region needs to

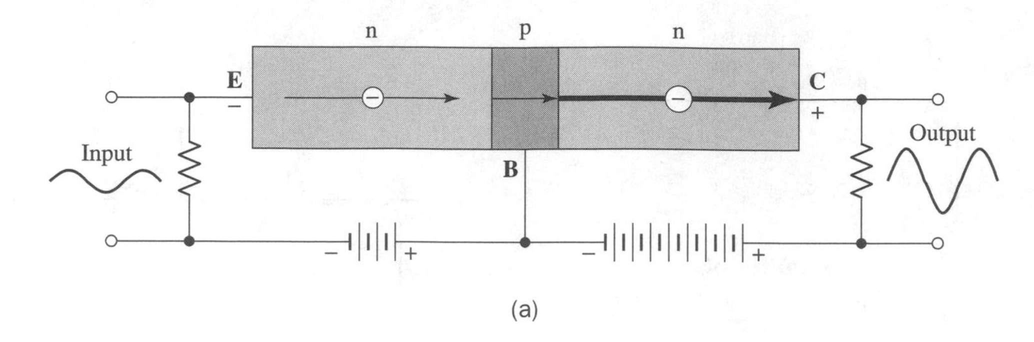

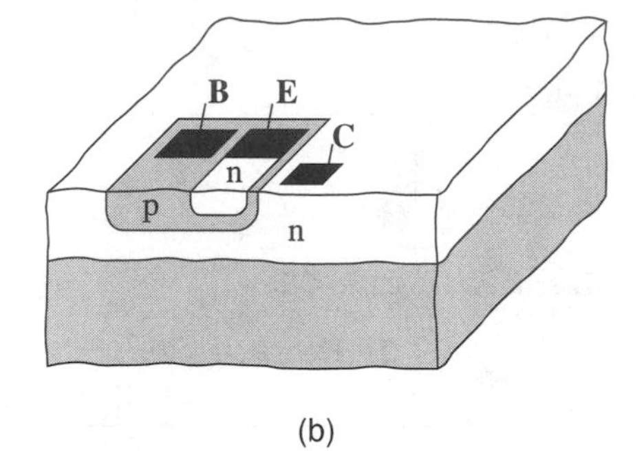

FIGURE 11.16. (a) Biasing of an n–p–n bipolar transistor. (b) Schematic representation of an n–p–n bipolar transistor. The dark areas are the contact pads.

be extremely thin and only lightly doped. As a consequence, most electrons diffuse through the base and reach the collector area, which is strongly reverse-biased with respect to the base. In other words, the strong positive charge on the collector terminal attracts the electrons and accelerates them towards the collector. This constitutes an amplification of the signal which was injected into the emitter/base section.

The electron flow from emitter to collector can be controlled by the bias voltage on the base: A *large* positive (forward-) bias on the base increases the electron injection into the p-area. In contrast, a *small*, but still positive base voltage results in a smaller electron injection from the emitter into the base area. As a consequence, the strong collector signal mimics the waveform of the input signal between emitter and base.

One may consider the amplification from a more quantitative point of view. The forward-biased emitter–base diode is made to have a small resistivity (approximately 10^{-3} Ω cm), whereas the reverse-biased base–collector diode has a much larger resistivity (about 10 Ω cm). Since the current flowing through the device is practically identical in both parts, the power ($P = I^2R$) is larger in the collector circuit. This results in a power gain.

In another application, a transistor is used as an electronic

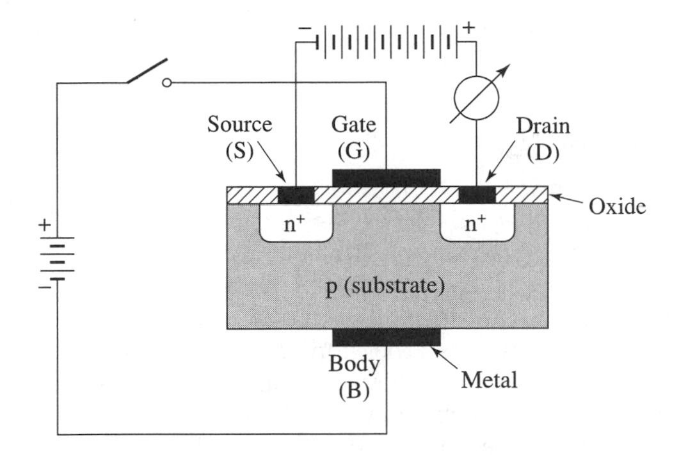

FIGURE 11.17. Normally-off–type n-channel MOSFET. The dark areas symbolize the (aluminum) metallizations. The "oxide" layer consists of SiO_2 or Si_3N_4. This layer is about 10 nm thick. Quite often, the B and S terminals are interconnected.

switch. The electron flow from emitter to collector can be essentially terminated by applying only a small or even a zero emitter/base voltage which causes little or no electron injection into the base area. Conversely, a large emitter/base voltage causes a large emitter to collector current. Again, the size of the emitter/base voltage determines the magnitude of the collector current.

In a p–n–p transistor, the majority carriers are holes. Its function and features are similar to the above-discussed n–p–n transistor.

The **metal-oxide–semiconductor field effect transistor (MOSFET)** consists of highly doped (n^+) *source* and *drain* regions (as emitter and collector are called for this device); Figure 11.17. They are laid down (ion implanted) on a p-substrate, called the body. A thin oxide layer, on which a metal film has been deposited, electrically insulates the *gate* metal from the body. Now, consider a positive voltage on the drain with respect to the source. In essence, there is no electron flow from source to drain as long as the gate voltage is zero. (The device is therefore called a *normally-off MOSFET*). If, however, a large enough positive voltage is applied to the gate (with respect to the body), most of the holes immediately below the gate oxide are repelled, that is, they are driven into the p-substrate, thus removing possible recombination sites. Concomitantly, negative charge carriers are attracted into the layer below the gate oxide (called the *channel*). This provides a path (or a bridge) for the electrons between source and drain. Thus, in this case, a current flows between source and drain.

The MOSFET dominates the integrated circuit industry at present. It is utilized in memories, microcomputers, logic circuits, amplifiers, analog switches, and operational amplifiers. One of the advantages is that no current flows between gate and body.

Thus, only little power is expended for its operation, which is vital to prevent overheating of the device.

11.5 • Conduction in Polymers

Materials that are electrical (and thermal) insulators are of great technical importance and are, therefore, used in large quantities in the electronics industry, for example, as handles for a variety of tools, as coatings of wires, and for casings of electrical equipment. Most polymeric materials are insulating and have been used for this purpose for decades. It came, therefore, as a surprise when it was discovered that some polymers and organic substances may have electrical properties which resemble those of conventional semiconductors, metals, or even superconductors. We shall focus our attention mainly on these materials. This does not imply that conducting polymers are of technical importance at this time. Indeed, they are not yet. This is due to the fact that many presently known conducting polymers seem to be unstable at or above room temperature. In addition, some dopants used to impart a greater conductivity are highly toxic, and doping often makes the polymers brittle. These problems have been partially overcome, however, in recent years. Historically, *transoidal polyacetylene* (see Figure 11.18) has been used as a conducting polymer. It represents what is called a *conjugated organic polymer*, that is, it has alternating single and double bonds between the carbon atoms. It is obtained as a silvery, flexible, and lightweight film which has a conductivity comparable to that of silicon. Its conductivity increases with increasing temperature, similarly as in semiconductors (Figure 11.13). The conductivity of trans-polyacetylene can be made to increase by up to seven orders of magnitude by doping it with arsenic pentafluoride, iodine, or bromine, which yields a p-type semiconductor, see Figure 11.19. Thus, σ approaches the lower end of the conductivity of metals, as shown in Figure 11.20. Among other (albeit non-toxic) dopants are n-dodecyl sulfonate (soap). However, the stability of this material is very poor; it deteriorates in hours or days. This very drawback, which it shares with many other conducting polymers, nevertheless can be profitably utilized

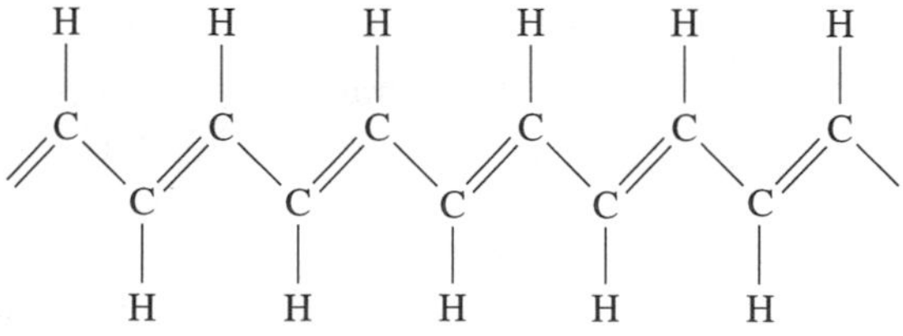

FIGURE 11.18. Transoidal isomer of polyacetylene (*trans*-$(CH)_x$). For details of the molecular structure of polymers, see Chapter 3.3, in particular Figure 3.16.

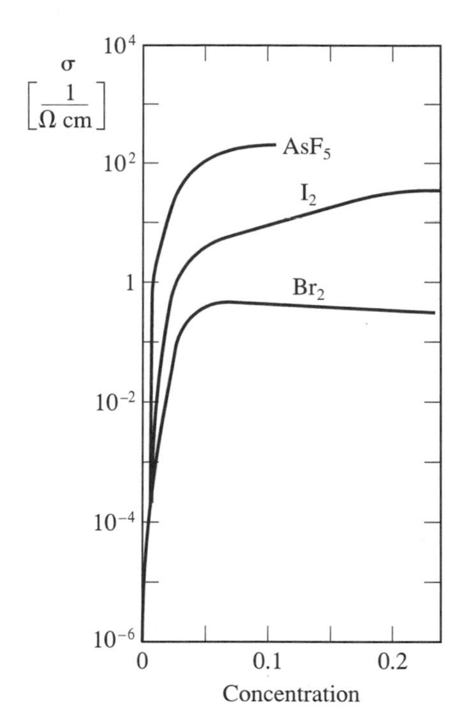

FIGURE 11.19. Conductivity change of polyacetylene as a result of doping. In polymers the "doping" concentration varies between one tenth of a percent up to 20–40%.

in special devices such as remote gas sensors, biosensors, and other remotely readable indicators which detect changes in humidity, radiation dosage, mechanical abuse, or chemical release.

Other conducting polymers include polypyrrole and polyaniline. The latter has a reasonably good conductivity and a high environmental stability. It has been used for electronic devices such as field-effect transistors, electrochromic displays, as well as for rechargeable batteries. The complex consisting of *poly*-(2-vinylpyridine) and iodine is used despite its relatively low conductivity (10^{-3} 1/Ω cm) for cathodes in lithium/iodine batteries for implantable pacemakers. These batteries have lifetimes of about 10 years and are relatively lightweight.

Arsenic pentafluoride–doped graphite has a conductivity which is about that of copper; see Figure 11.20. However, the high conductivity of graphite occurs only in the hexagonal plane of graphite (Figure 3.8). The conductivity is four orders of magnitude smaller perpendicular to this layer. Finally, and surprisingly, *poly*-sulfur nitride $(SN)_x$, which consists of alternating sulfur and nitrogen atoms and which has a room temperature conductivity of about $10^3\ \Omega^{-1}\ \text{cm}^{-1}$ along the chain direction, becomes superconducting close to 0 K!

In order to better understand the electronic properties of polymers by means of the electron theory and the band structure

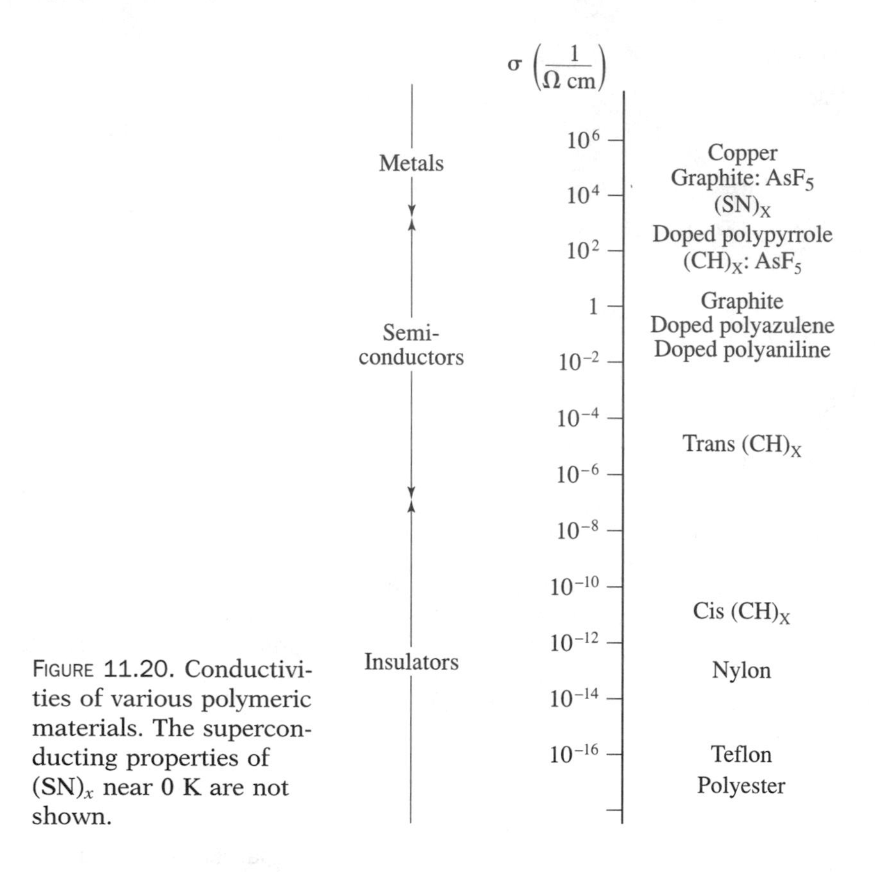

FIGURE 11.20. Conductivities of various polymeric materials. The superconducting properties of $(SN)_x$ near 0 K are not shown.

concept, one needs to know the degree of order or the degree of periodicity of the atoms because only ordered and strongly interacting atoms or molecules lead, as we know, to distinct and wide electron bands (Figure 11.5). Now, it has been observed that the degree of order in polymers depends on the regularity of the molecular structure. Certain heat treatments may influence some structural parameters. For example, if a simple polymer is slowly cooled below its melting point, one might observe that some macromolecules align parallel to each other. Actually, slow cooling yields, for certain polymers, a highly crystalline structure. The individual chains are, however, separated by regions of supercooled liquid, that is, by amorphous material (Figure 3.17).

We need to pursue this somewhat further. Figure 11.21 shows two simplified band structures for *trans*-$(CH)_x$ assuming different distances between the carbon atoms. In Figure 11.21(a), all carbon bond lengths are taken to be equal. The resulting band structure is found to be characteristic for a metal, that is, the highest band is *partially* filled by electrons.

Where are the free electrons in the conduction band of *trans*-$(CH)_x$ coming from? We realize that one of the electrons in the double bond of a conjugated polymer can be considered to be only loosely bound to the neighboring carbon atoms. Thus, this electron can be easily disassociated from its carbon atom by a relatively small energy, which may be provided by thermal energy. The delocalized electrons behave like free electrons and may be accelerated as usual in an electric field.

In reality, however, a uniform bond length between the carbon atoms does not exist in polyacetylene. Instead, the distances between the carbon atoms alternate because of the alternating single and double bonds. Band structure calculations for this case show, interestingly enough, a gap between the individual energy bands. The resulting band structure is typical for a semiconductor (or an insulator)! The width of the band gap near the Fermi level depends mainly on the degree of alternating bond lengths [Figure 11.21(b)].

It has been shown that the band structure in Figure 11.21(b) best represents the experimental observations. Specifically, one finds a band gap of about 1.5 eV between the valence band and the conduction band. This explains the semiconducting characteristics of *trans*-$(CH)_x$. In order to improve the conductivity of $(CH)_x$, one would attempt to decrease the differences of the carbon–carbon bond lengths, thus eventually approaching the uniform bond length as shown in Figure 11.21(a). This has indeed been accomplished by synthesizing $(CH)_x$ via a *non-conjugated* precursor polymer which is subsequently heat treated. Conductivities as high as 10^2 $1/(\Omega\ \text{cm})$ have been obtained this way.

It should be noted in closing that the interpretation of conducting polymers is still in flux and future research needs to clarify certain points.

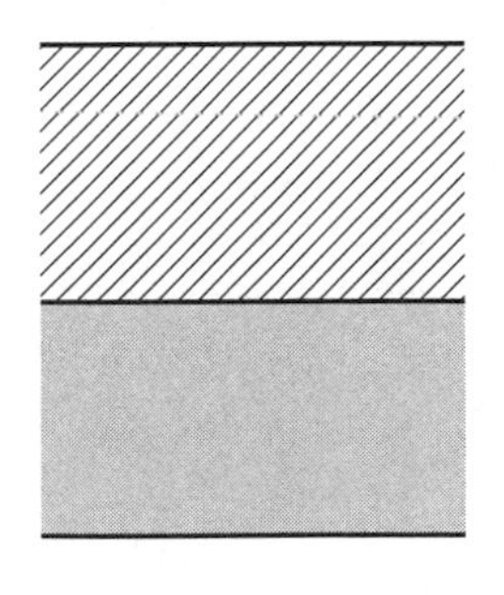

(a)

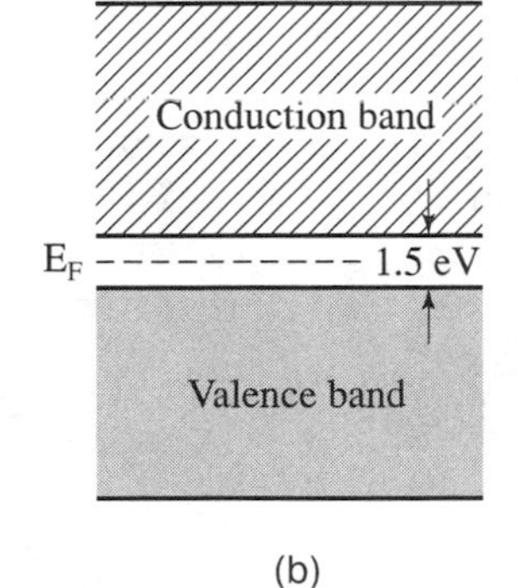

(b)

FIGURE 11.21. Schematic band structures for *trans*-$(CH)_x$ for different carbon–carbon bond lengths: (a) same length (1.39 Å) between individual carbon atoms (no band gap); (b) alternating bond lengths C = C, 1.36 Å; C − C, 1.43 Å, which yields a band gap as known for semiconductors.

11.6 • Ionic Conductors

Electrical conduction in ionically bonded materials, such as the alkali-halides, is extremely small. The reason for this is that the atoms in these chemical compounds strive to assume the noble gas configuration for maximal stability and thus transfer electrons between each other to form positively charged cations and negatively charged anions. The binding forces between the ions are electrostatic in nature, that is, they are very strong, as explained in Section 3.2. Essentially no free electrons are therefore formed. As a consequence, the room temperature conductivity in ionic crystals is about 22 orders of magnitude smaller than that for typical metallic conductors (Figure 11.1). The wide band gap in insulators allows only extremely few electrons to become excited from the valence into the conduction band [Figure 11.6(d)].

The main contribution to the electrical conduction in ionic crystals (as little as it may be) is, however, due to a mechanism which we have not yet discussed, namely, ionic conduction. Ionic conduction is caused by the movement of some negatively (or positively) charged ions which *hop* from lattice site to lattice site under the influence of an electric field; see Figure 11.22(b). (This type of conduction is similar to that which is known to occur in aqueous electrolytes; see Chapter 9.) The ionic conductivity:

$$\sigma_{\text{ion}} = N_{\text{ion}}\, e\, \mu_{\text{ion}} \tag{11.18}$$

is, as outlined before [Eq. (11.14)], the product of three quantities. In the present case, N_{ion} is the number of ions per unit volume which can change their position under the influence of an electric field whereas μ_{ion} is the mobility of these ions.

In order for ions to move through a crystalline solid, they must have sufficient energy to pass over an *energy barrier* (Figure 11.22). Further, an equivalent lattice site next to a given ion must be empty in order for an ion to be able to change its position. Thus, N_{ion} in Eq. (11.18) depends on the vacancy concentration in the crystal (i.e., on the number of *Schottky defects*; see Section 6.1.1). In short, the theory of ionic conduction contains essential elements of diffusion theory, as introduced in Chapter 6.

Diffusion theory links the mobility of the ions, which is contained in Eq. (11.18) with the diffusion coefficient D, through the Einstein relation:

$$\mu_{\text{ion}} = \frac{De}{k_B T}. \tag{11.19}$$

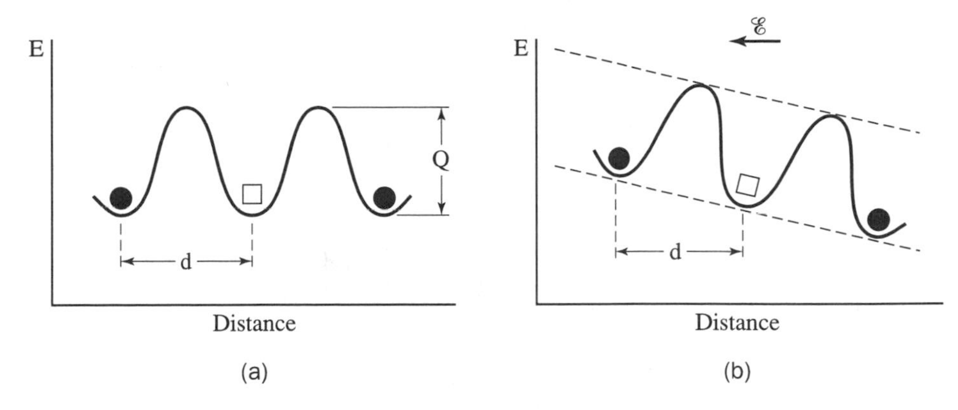

FIGURE 11.22. Schematic representation of a potential barrier which an ion (●) has to overcome to exchange its site with a vacancy (□): (a) without an external electric field; (b) with an external electric field. d = distance between two equivalent adjacent lattice sites; Q = activation energy. See Chapter 6.

The diffusion coefficient varies with temperature by an Arrhenius equation, (6.8):

$$D = D_0 \exp\left(-\frac{Q}{k_B T}\right), \tag{11.20}$$

where Q is the activation energy for the process under consideration (Figure 11.22), and D_0 is a pre-exponential factor which depends on the vibrational frequency of the atoms and some structural parameters. Combining Eqs. (11.18) through (11.20) yields:

$$\sigma_{\text{ion}} = \frac{N_{\text{ion}}\, e^2\, D_0}{k_B T} \exp\left(-\frac{Q}{k_B T}\right). \tag{11.21}$$

Equation (11.21) is shortened by combining the pre-exponential constants into σ_0:

$$\sigma_{\text{ion}} = \sigma_0 \exp\left(-\frac{Q}{k_B T}\right). \tag{11.22}$$

In summary, the ionic conduction increases exponentially with increasing temperature (as semiconductors do; Figure 11.13). Further, σ_{ion} depends on a few other parameters, such as the number of ions that can change their position, the vacancy concentration, as well as on an activation energy.

11.7 • Thermoelectric Phenomena

Assume two different types of materials (e.g., a copper and an iron wire) which are connected at their ends to form a loop, as shown in Figure 11.23. One of the junctions is brought to a higher temperature than the other. Then a potential difference, ΔV, be-

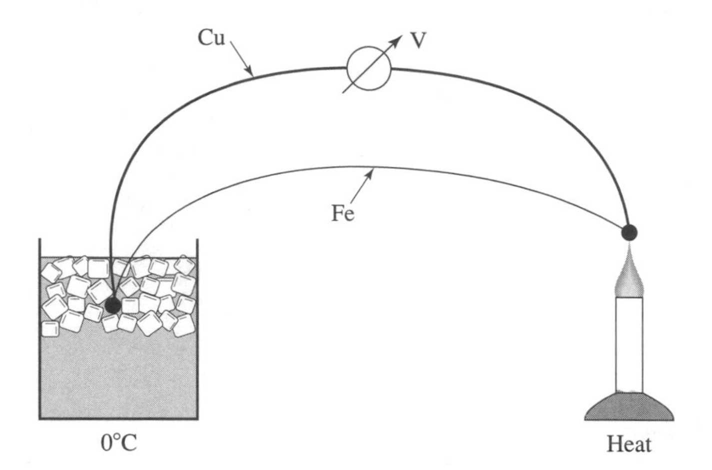

FIGURE 11.23. Schematic representation of two thermocouples made of copper and iron which are brought in contact with each other (Seebeck effect).

tween these two **thermocouples** is observed which is essentially proportional to the temperature difference, ΔT, where:

$$\frac{\Delta V}{\Delta T} = S \tag{11.23}$$

is called the *thermoelectric power*, or the Seebeck coefficient (after its inventor, T.J. Seebeck, a German physicist who discovered, in 1821, that a thermoelectric circuit like the one just described, deflected a close-by compass needle). A thermoelectric power of several microvolts per degree is commonly observed. As an example, the frequently used copper/constantan (Cu–45% Ni) combination yields about 43 μV/K. It has a useful range between -180 and $+400$°C. For higher temperatures, thermocouples of chromel (90%Ni–10%Cr) and alumel (95%Ni–2%Mn-2%Al) or platinum/Pt–13%Rh (up to 1700°C) are available. Some semiconductors have Seebeck coefficients which reach in the millivolt per degree range, that is, they are one or two orders of magnitude higher than for metals and alloys. Among them are bismuth telluride (Bi_2Te_3), lead telluride (PbTe), or silicon–30% germanium alloys.

Thermocouples made of metal wires are utilized as rigid, inexpensive, and fast probes for measuring temperatures even at otherwise not easily accessible places. *Thermoelectric power generators* (utilizing the above-mentioned semiconductors) are used particularly in remote locations of the earth (Siberia, Alaska, etc.). They contain, for example, a ring of thermoelements, arranged over the glass chimney of a kerosene lamp which is concomitantly used for lighting. The temperature difference of 300°C thus achieved yields electric power of a few watts or sometimes

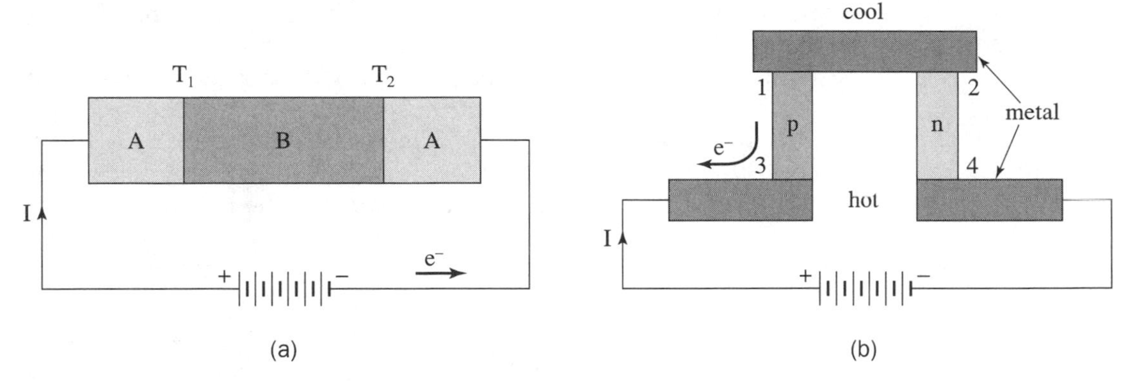

FIGURE 11.24. Thermoelectric refrigeration devices which make use of the Peltier effect. (a) Principle arrangement. (b) Efficient device utilizing p- and n-type semiconductors (see Section 11.4) in conjunction with metals.

more, which can be used for radios or communication purposes. Other fuel-burning devices provide, for example, an emf for cathodic protection of pipe lines (see Chapter 9). Heat produced by the decay of radioisotopes or by small nuclear reactors yields thermoelectric power for scientific instruments on the moon (e.g., to record moon quakes) and for relaying the information back to earth. In solar thermoelectric generators sunlight is concentrated by concave mirrors on thermocouples. Most of the above-described devices have an efficiency between 5 and 10%.

A reversion of the Seebeck effect is the **Peltier effect**: A direct electric current that flows through junctions made of different materials causes one junction to be cooled and the other to heat up (depending on the direction of the current); see Figure 11.24(a). Lead telluride or bismuth telluride in combination with metals are frequently used. One particularly effective device for which temperature differences up to 70°C have been achieved is shown in Figure 11.24(b). It utilizes n- and p-type semiconductors (see Section 11.4) in conjunction with metals. Cooling occurs on those junctions that are connected to the upper metal plate (1 and 2), whereas heat develops on the lower junctions 3 and 4. The heat on the lower plate is removed by water or air cooling. The above-quoted temperature drop can even be enhanced by cascading several devices, that is, by joining multiple *thermoelectric refrigerators* for which each stage acts as the heat sink for the next.

The thermoelectric effects can be explained by applying elements of electron theory as described in the previous sections: When two different types of conducting materials are brought into contact, electrons are transferred from the material with higher Fermi energy (E_F) "down" into the material having a lower E_F until both Fermi energies are equal. As a consequence, the material that had the smaller E_F assumes a negative charge with respect to the other. This results in the above-mentioned *contact potential* between the materials. The contact potential is temperature-dependent. Specifically, when a material is heated, a substantial amount of electrons is excited across the Fermi energy to higher energy levels. These extra electrons drift to the cold junction, which becomes negatively charged compared to the hot junction. The equivalent is true for the Peltier effect: The electrons having a larger energy (that is, those having a higher E_F) are caused by the current to transfer their extra energy into the material having a lower E_F, which in turn heats up. Concomitantly, the material having a higher E_F is caused to lose energy and thus becomes colder.

11.8 • Dielectric Properties

Dielectric materials, that is, insulators, possess a number of important electrical properties which make them useful in the electronics industry. This needs some explanation.

When a voltage is momentarily applied to two parallel metal plates which are separated by a distance, L, as shown in Figure 11.25, then the resulting electric charge essentially remains on these plates even after the voltage has been removed (at least as long as the air is dry). This ability to store an electric charge is called the *capacitance*, C, which is defined to be the charge, q, per applied voltage, V, that is:

$$C = \frac{q}{V}, \tag{11.24}$$

where C is given in coulombs per volt or farad (SI units, see Appendix II). Understandably, the capacitance is higher, the larger

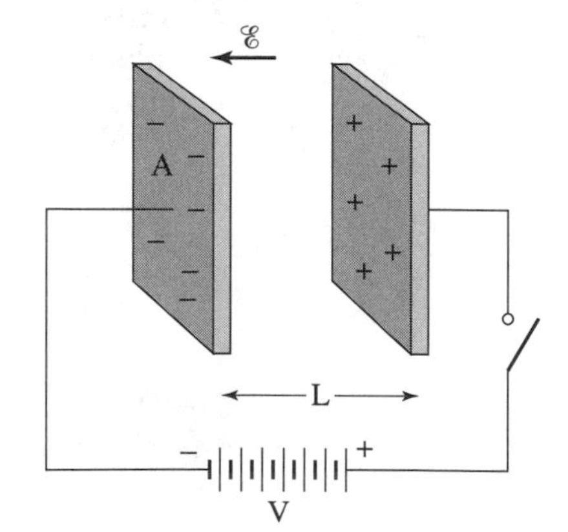

FIGURE 11.25. Two metal plates, separated by a distance, L, can store electric energy after having been charged momentarily by a battery.

the area, A, of the plates and the smaller the distance, L, between them. Further, the capacitance depends on the material that may have been inserted between the plates. The experimental observations lead to:

$$C = \epsilon \, \epsilon_0 \, \frac{A}{L} \tag{11.25}$$

where

$$\epsilon = \frac{C}{C_{\text{vac}}} \tag{11.26}$$

determines the magnitude of the added storage capability. It is called the (unitless) *dielectric constant* (or occasionally the *relative permittivity*, ϵ_r). ϵ_0 is a universal constant having the value of 8.85×10^{-12} farad per meter (F/m) or As/Vm and is known by the name *permittivity of empty space* (or of vacuum). Some values for the dielectric constant are given in Table 11.2. The dielectric constant of empty space is set to be 1 whereas ϵ of air and many other gases is nearly 1.

We now need to explain why the capacitance increases when a piece of a dielectric material is inserted between two conductors [see Eq. (11.25)]. For this, one has to realize that, under the influence of an external electric field, the negatively charged electron cloud of an atom becomes displaced with respect to its pos-

TABLE 11.2 DC dielectric constants of some materials

Barium titanate	4000	Ferroelectric
Water	81.1	
Acetone	20	
Silicon	11.8	
GaAs	10.9	
Marble	8.5	
Soda–lime–glass	6.9	
Porcelain	6.0	
Epoxy	4.0	
Fused silica	4.0	Dielectric
Nylon 6,6	4.0	
PVC	3.5	
Ice	3.0	
Amber	2.8	
Polyethylene	2.3	
Paraffin	2.0	
Air	1.000576	

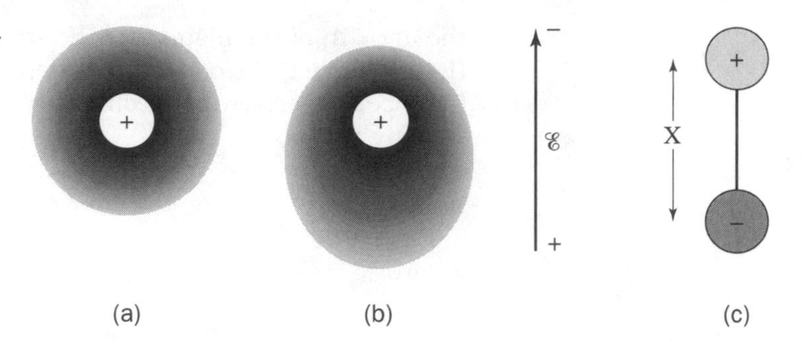

FIGURE 11.26. An atom is represented by a positively charged core and a surrounding, negatively charged electron cloud (a) in equilibrium and (b) in an external electric field. (c) Schematic representation of an electric dipole as, for example, created by separation of the negative and positive charges by an electric field, as seen in (b).

itively charged core; see Figure 11.26(b). As a result, a dipole is created which has an electric **dipole moment**:

$$p = q \cdot X, \tag{11.27}$$

where X is the separation between the positive and the negative charge as depicted in Figure 11.26(c). (The dipole moment is generally a vector pointing from the negative to the positive charge.) The process of dipole formation (or alignment of already existing dipoles) under the influence of an external electric field that has an electric field strength, $\mathscr{E}$, is called **polarization**. Dipole formation of all involved atoms within a dielectric material causes a charge redistribution so that the surface which is nearest to the positive capacitor plate is negatively charged (and vice versa), see Figure 11.27(a). As a consequence, electric field lines within a dielectric are created which are opposite in direction to the external field lines. Effectively, the electric field lines within a dielectric material are weakened due to polarization, as depicted in Figure 11.27(b). In other words, the electric field strength (11.6)

$$\mathscr{E} = \frac{\mathrm{V}}{\mathrm{L}} = \frac{\mathscr{E}_{\mathrm{vac}}}{\epsilon} \tag{11.28}$$

is reduced by inserting a dielectric between two capacitor plates.

Within a dielectric material the electric field strength, $\mathscr{E}$, is replaced by the **dielectric displacement**, D (also called the *surface charge density*), that is:

$$D = \epsilon\, \epsilon_0\, \mathscr{E} = \frac{q}{A}. \tag{11.29}$$

The dielectric displacement is the superposition of two terms:

$$D = \epsilon_0\, \mathscr{E} + P, \tag{11.30}$$

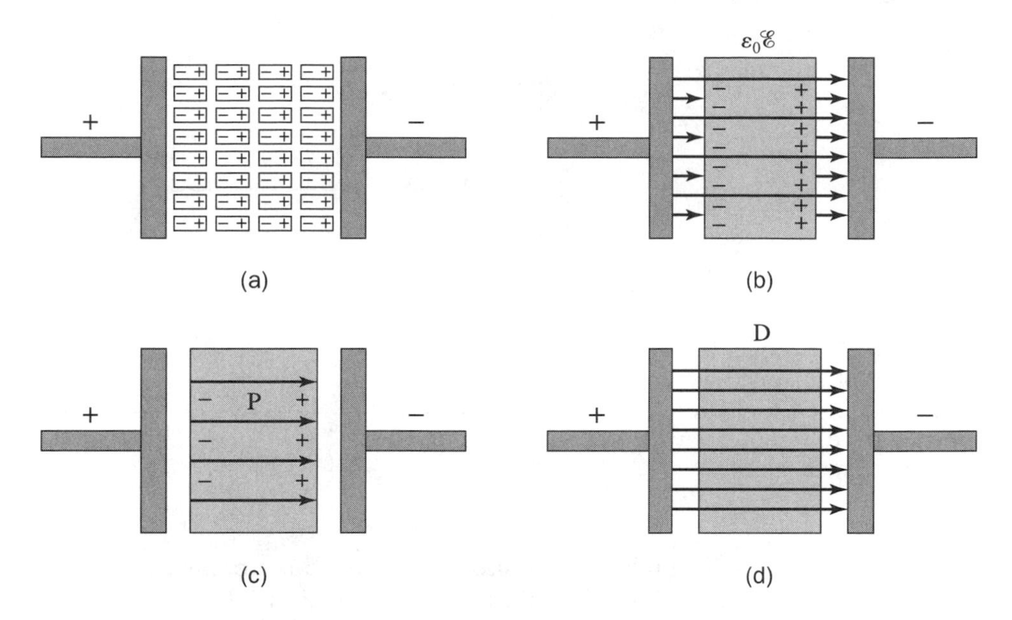

FIGURE 11.27. Schematic representation of two capacitor plates between which a dielectric material is inserted. (a) Induction of electric dipoles of opposite charge. (b) Weakening of the electric field *within* the dielectric material [Eq. (11.28)]. (c) The direction of the polarization vector is from the negative induced charge to the positive induced charge. (d) The dielectric displacement, D, within the dielectric material is the sum of $\epsilon_0\mathscr{E}$ and P [Eq. (11.30)].

where P is called the **dielectric polarization**, that is, the induced electric dipole moment per unit volume [Figures 11.27(c and d)]. The units for D and P are C m^{-2}; see Eq. (11.29). (D, $\mathscr{E}$, and P are generally vectors.) In other words, the polarization is responsible for the increase in charge density (q/A) above that for vacuum.

The mechanism just described is known by the name **electronic polarization**. It occurs in all dielectric materials that are subjected to an electric field. In ionic materials, such as the alkali halides, an additional process may occur which is called **ionic polarization**. In short, cations and anions are somewhat displaced from their equilibrium positions under the influence of an external field and thus give rise to a net dipole moment. Finally, many materials already possess permanent dipoles which can be *aligned* in an external electric field. Among them are water, oils, organic liquids, waxes, amorphous polymers, polyvinylchloride, and certain ceramics such as barium titanate ($BaTiO_3$). This mechanism is termed *orientation polarization* or

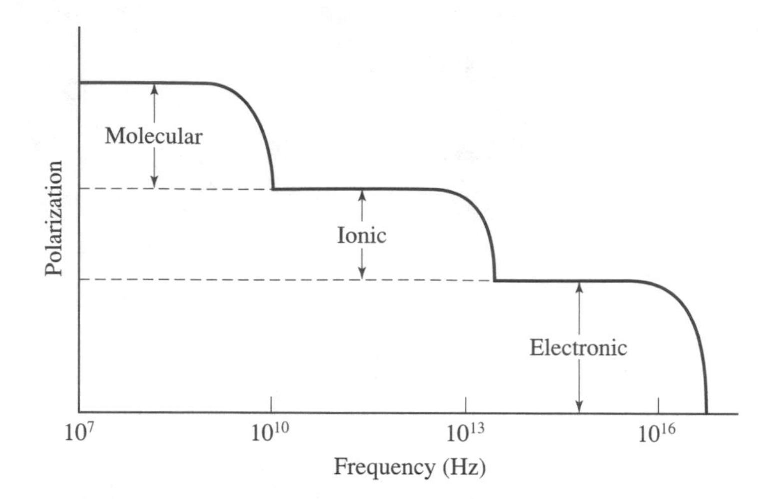

FIGURE 11.28. Schematic representation of the polarization as a function of excitation frequency for different polarization mechanisms.

molecular polarization. All three polarization processes are additive if applicable; see Figure 11.28.

Most capacitors are used in electric circuits involving alternating currents. This requires the dipoles to reorient quickly under a rapidly changing electric field. Not all polarization mechanisms respond equally quick to an alternating electric field. For example, many molecules are relatively sluggish in reorientation. Thus, molecular polarization breaks down already at relatively low frequencies; see Figure 11.28. In contrast, electronic polarization responds quite rapidly to an alternating electric field even at frequencies up to 10^{16} Hz.

At certain frequencies a substantial amount of the excitation energy is absorbed and transferred into heat. This process is called *dielectric loss*. It is imperative to know the frequency for dielectric losses for a given material so that the device is not operated in this range.

11.9 • Ferroelectricity and Piezoelectricity

Ferroelectric materials, such as barium titanate, exhibit spontaneous polarization without the presence of an external electric field. Their dielectric constants are orders of magnitude larger than those of dielectrics (see Table 11.2). Thus, they are quite suitable for the manufacturing of small-sized, highly efficient capacitors. Most of all, however, ferroelectric materials retain their state of polarization even after an external electric field has been removed. Specifically, if a ferroelectric is exposed to a strong electric field, $\mathscr{E}$, its permanent dipoles become increasingly aligned

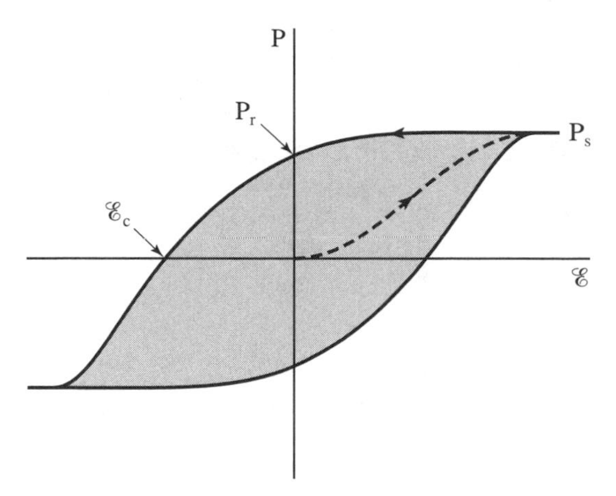

FIGURE 11.29. Schematic representation of a hysteresis loop for a *ferroelectric* material in an electric field. Compare to Figure 12.7.

with the external field direction until eventually all dipoles are parallel to $\mathscr{E}$ and saturation of the polarization, P_s, has been achieved, as depicted in Figure 11.29. Once the external field has been withdrawn, a **remanent polarization**, P_r, remains which can only be removed by inverting the electric field until a coercive field, $\mathscr{E}_c$, has been reached (Figure 11.29). By further increasing the reverse electric field, parallel orientation of the dipoles in the opposite direction is achieved. Finally, when reversing the field once more, a complete hysteresis loop is obtained, as depicted in Figure 11.29. All taken, ferroelectricity is the electric analogue to ferromagnetism, which will be discussed in Chapter 12. Therefore, ferroelectrics can be utilized for memory devices in computers, etc. The area within a hysteresis loop is proportional to the energy per unit volume that is dissipated once a full field cycle has been completed.

A critical temperature, called the *Curie temperature*, exists, above which the ferroelectric effects are destroyed and the material becomes dielectric. Typical Curie temperatures range from −200°C for strontium titanate to at least 640°C for $NaNbO_3$.

The question remains to be answered, why do certain materials such as $BaTiO_3$ possess spontaneous polarization? This can be explained by knowing that in the tetragonal crystal structure of $BaTiO_3$, the negatively charged oxygen ions and the positively charged Ti^{4+} ion are slightly displaced from their symmetrical positions, as depicted in Figure 11.30. This results in a permanent ionic dipole moment along the *c*-axis within the unit cell. A large number of such dipoles line up in clusters (also called *domains*). In the virgin state, the polarization directions of the individual domains are, however, randomly oriented, so that the

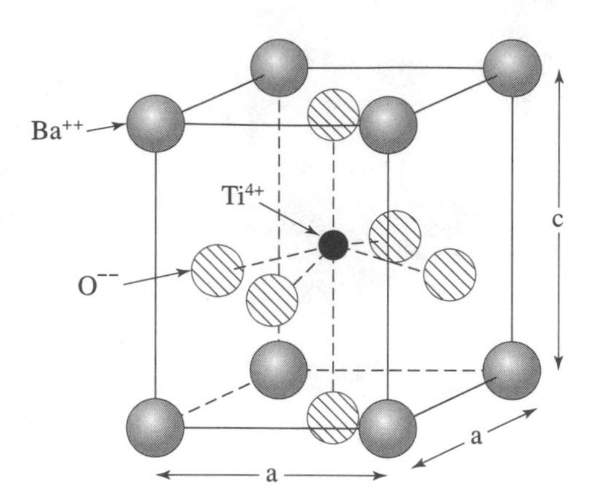

FIGURE 11.30. Tetragonal crystal structure of barium titanate at room temperature. Note the upward displacement of the Ti^{4+} ion in the center compared to the downward displacement of all surrounding O^{2-} ions. $a = 0.398$ nm; $c = 0.403$ nm.

material has no net polarization (Figure 11.31). An external field eventually orients the dipoles of the favorably oriented domains parallel to $\mathscr{E}$. Specifically, those domains in which the dipoles are already nearly parallel to $\mathscr{E}$ grow at the expense of unfavorably oriented domains.

By heating $BaTiO_3$ above its Curie temperature (120°C), the tetragonal unit cell transforms to a cubic cell whereby the ions now assume symmetric positions. Thus, no spontaneous alignment of dipoles remains and $BaTiO_3$ becomes dielectric.

If pressure is applied to a ferroelectric material, such as $BaTiO_3$, a change in the just-mentioned polarization may occur which results in a small voltage across the sample. Specifically, the slight change in dimensions causes a variation in bond lengths between cations and anions. This effect is called **piezoelectricity**.[3] It is found in a number of materials such as quartz (however, much weaker than in $BaTiO_3$), in ZnO, and in complicated ceramic compounds such as $PbZrTiO_6$. Piezoelectricity is utilized in devices that are designed to convert mechanical strain into electricity. Those devices are called *transducers*. Applications include strain gages, microphones, sonar detectors, and phonograph pickups, to mention a few.

The inverse mechanism, in which an electric field produces a change in dimensions in a ferroelectric material, is called **electrostriction**. An earphone utilizes such a device. Probably the most important application, however, is the quartz crystal resonator which is used in electronic devices as a *frequency selective element*. Specifically, a periodic strain is exerted to a quartz crystal by an alternating electric field which excites this crystal

---

[3] *Piezo* (latin) = pressure.

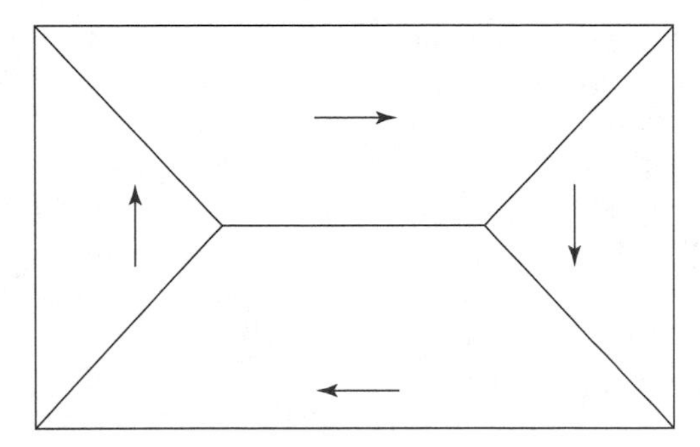

FIGURE 11.31. Schematic representation of spontaneous alignments of electric dipoles within a domain and random alignment of the dipole moments of several domains in a ferroelectric material such as $BaTiO_3$. Compare to Figure 12.8.

to vibrations. These vibrations are monitored in turn by piezoelectricity. If the applied frequency coincides with the natural resonance frequency of the molecules, then amplification occurs. In this way, very distinct frequencies are produced which are utilized for clocks or radio frequency signals.

Problems

11.1. A current of 2 A flows through a slab of tin, 1 mm thick and 2 mm wide, whose conductivity is 10^6 1/Ωm. Calculate the voltage drop between two potential probes that are 10 cm apart along the slab.

11.2. Calculate the number of free electrons per cubic centimeter for sodium from resistance data ($\rho = 4.2 \times 10^{-8}$ Ωm; relaxation time 3.1×10^{-14}s).

11.3. Calculate the population density of electrons for a metal. (Take $\sigma = 5 \times 10^5$ 1/Ωcm; $v_F = 10^8$ cm/s, and $\tau = 3 \times 10^{-14}$s.)

11.4. Calculate the number of free electrons per cm^3 for gold using its density (19.3 g/cm^3) and its atomic mass (196.967 g/mol). (*Hint:* Look in Chapter 3.)

11.5. Calculate the number of electrons in the conduction band for silicon at $T = 300$ K.

11.6. At what (hypothetical) temperature would all 10^{22} cm^{-3} valence electrons be excited to the conduction band in a semiconductor with $E_g = 1$ eV? (*Hint:* Use a programmable calculator.)

11.7. In the figure below, σ is plotted as a function of the reciprocal temperature for an intrinsic semiconductor. Calculate the gap energy. (Hint: Combine (11.12) and (11.15) and take the ln from the resulting equation assuming $N_e \equiv N_h \cdot$ Why?).)

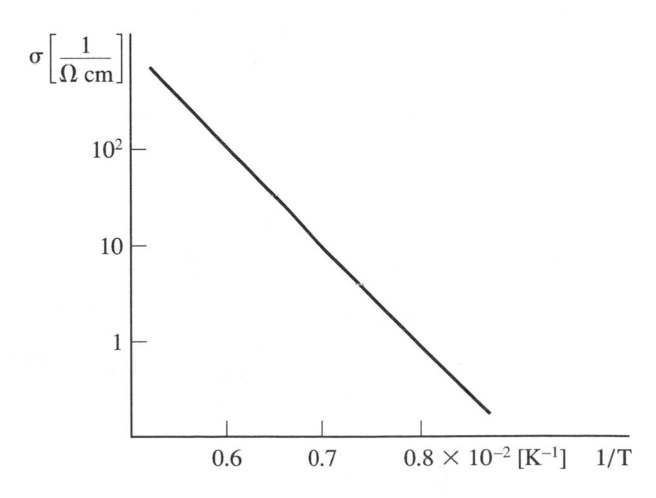

11.8. Calculate the conductivity at room temperature for germanium containing 5×10^{16} arsenic atoms per cubic centimeter. (*Hint:* Use the mobility of the electrons in the host material.)

11.9. What current flows through a p–n diode to which a voltage of 0.3V is applied and a saturation current of 1 μA is assumed at 300 K?

11.10. Calculate the mobility of electrons in a metal when the time between two electron/atom collisions is 2.5×10^{-12}s.

11.11. Show that $\mathcal{E} = \mathcal{E}_{vac}/\epsilon$ [Eq. (11.28)] by combining Eqs. (11.6), (11.24), and (11.26) and their equivalents for vacuum.

11.12. Show that the dielectric polarization is $P = (\epsilon - 1)\ \epsilon_0\mathcal{E}$. What values do P and D have for vacuum?

11.13. Show that $\epsilon\epsilon_0\mathcal{E} = q/A$ [Eq. (11.29)] by combining some pertinent equations.

Suggestions for Further Study

A.R. Blythe, *Electrical Properties of Polymers*, Cambridge University Press, Cambridge (1979).

R.H. Bube, *Electronic Properties of Crystalline Solids*, 3rd Edition, Academic, Cambridge, MA (1992).

S.K. Ghandi, *The Theory and Practice of Microelectronics*, Wiley, New York (1968).

L.L. Hench and J.K. West, *Principles of Electronic Ceramics*, Wiley, New York (1990).

R.E. Hummel, *Electronic Properties of Materials*, 3rd Edition, Springer-Verlag, New York, (2001).

J. Mort and G. Pfister, Editors, *Electronic Properties of Polymers*, Wiley, New York (1982).

M.A. Omar, *Elementary Solid State Physics*, Addison-Wesley, Reading, MA (1978).

B.G. Streetman, *Solid State Electronic Devices*, 2nd Edition, Prentice-Hall, Englewood Cliffs, NJ (1980).

S.M. Sze, *Physics of Semiconductor Devices*, 2nd Edition, Wiley, New York (1981).

12

Magnetic Properties of Materials

12.1 • Fundamentals

Modern technology would be unthinkable without magnetic materials and magnetic phenomena. Magnetic tapes or disks (for computers, video recorders, etc.) motors, generators, telephones, transformers, permanent magnets, electromagnets, loudspeakers, and magnetic strips on credit cards are only a few examples of their applications. To a certain degree, magnetism and electric phenomena can be considered to be siblings since many common mechanisms exist such as dipoles, attraction, repulsion, spontaneous or forced alignment of dipoles, field lines, field strengths, etc. Thus, the governing equations often have the same form. Actually, electrical and magnetic phenomena are linked by the famous Maxwell equations, which were mentioned already in Chapter 10.

At least five different kinds of magnetic materials exist. They have been termed para-, dia-, ferro-, ferri-, and antiferromagnetics. A qualitative as well as a quantitative distinction between these types can be achieved in a relatively simple way by utilizing a method proposed by Faraday. The magnetic material to be investigated is suspended from one of the arms of a sensitive balance and is allowed to reach into an inhomogeneous magnetic field (Figure 12.1). Diamagnetic materials are expelled from this field, whereas para-, ferro-, antiferro-, and ferrimagnetics are attracted in different degrees. It has been found empirically that the apparent loss or gain in mass, that is, the force, F, on the sample exerted by the magnetic field, is:

$$F = V \chi \mu_0 H \frac{dH}{dx}, \tag{12.1}$$

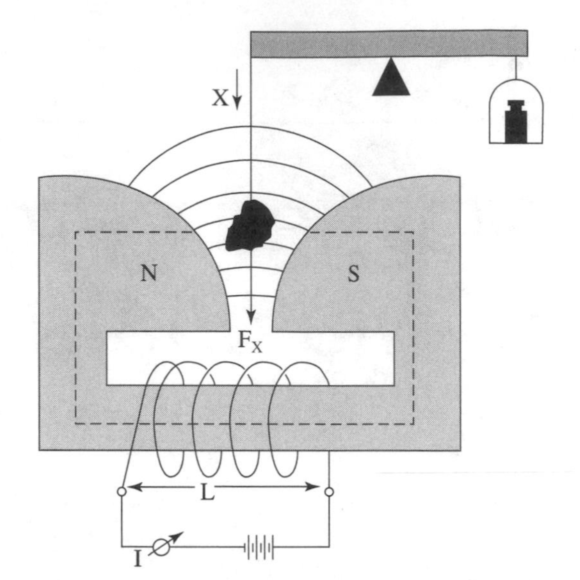

FIGURE 12.1. Measurement of the magnetic susceptibility in an inhomogeneous magnetic field. The magnetic field lines (dashed) follow the iron core.

where V is the volume of the sample, μ_0 is a universal constant called the permeability of free space (1.257×10^{-6} H/m or Vs/Am), and χ is the **susceptibility**, which expresses how responsive a material is to an applied magnetic field. Characteristic values for χ are given in Table 12.1. The term dH/dx in Eq. (12.1) is the change of the **magnetic field strength** H in the x-direction. The field strength H of an electromagnet (consisting of helical windings of a long, insulated wire as seen in the lower portion of Figure 12.1) is proportional to the current, I, which flows through this coil, and on the number, n, of the windings (called *turns*) that have been used to make the coil. Further, the magnetic field strength is inversely proportional to the length, L, of the *solenoid*. Thus, the magnetic field strength is expressed by:

$$H = \frac{In}{L}. \tag{12.2}$$

The field strength is measured (in SI units) in "Amp-turns per meter" or shortly, in A/m.

The magnetic field can be enhanced by inserting, say, iron, into a solenoid, as shown in Figure 12.1. The parameter which expresses the amount of enhancement of the magnetic field is called the **permeability** μ. The magnetic field strength within a material is known by the names **magnetic induction**[1] (or *magnetic*

[1]Calling B "magnetic induction" is common practice but should be discouraged because it may be confused with electromagnetic induction, as shown in Figure 10.3.

TABLE 12.1. Magnetic constants of some materials at room temperature

Material	χ (SI) unitless	χ (cgs) unitless	μ unitless	Type of magnetism
Bi	-165×10^{-6}	-13.13×10^{-6}	0.99983	
Ge	-71.1×10^{-6}	-5.66×10^{-6}	0.99993	
Au	-34.4×10^{-6}	-2.74×10^{-6}	0.99996	Diamagnetic
Ag	-25.3×10^{-6}	-2.016×10^{-6}	0.99997	
Be	-23.2×10^{-6}	-1.85×10^{-6}	0.99998	
Cu	-9.7×10^{-6}	-0.77×10^{-6}	0.99999	
Superconductors[a]	-1.0	$\sim -8 \times 10^{-2}$	0	
β-Sn	$+2.4 \times 10^{-6}$	$+0.19 \times 10^{-6}$	1	
Al	$+20.7 \times 10^{-6}$	$+1.65 \times 10^{-6}$	1.00002	Paramagnetic
W	$+77.7 \times 10^{-6}$	$+6.18 \times 10^{-6}$	1.00008	
Pt	$+264.4 \times 10^{-6}$	$+21.04 \times 10^{-6}$	1.00026	
Low carbon steel	Approximately the same as μ because of $\chi = \mu - 1$.		5×10^3	
Fe–3%Si (grain-oriented)			4×10^4	Ferromagnetic
Ni–Fe–Mo (supermalloy)			10^6	

[a] See Sections 11.3 and 12.2.1

Note: The table lists the unitless susceptibility, χ, in SI and cgs units. (The difference is a factor of 4π, see Appendix II.) Other sources may provide mass, atomic, molar, volume, or gram equivalent susceptibilities in cgs or mks units.

Source: Landolt-Börnstein, *Zahlenwerte der Physik*, Vol. 11/9, 6th Edition, Springer-Verlag, Berlin (1962).

flux density) and is denoted by B. Magnetic field strength and magnetic induction are related by the equation:

$$B = \mu \, \mu_0 H. \tag{12.3}$$

The SI unit for B is the tesla (T) and that of μ_0 is henries per meter (H/m or Vs/Am); see Appendix II. The permeability (sometimes called relative permeability, μ_r) in Eq. (12.3) is unitless and is listed in Table 12.1 for some materials. The relationship between the susceptibility and the permeability is

$$\mu = 1 + \chi. \tag{12.4}$$

For empty space and, for all practical purposes, also for air, one defines $\chi = 0$ and thus $\mu = 1$ [See Eq. (12.4)]. The susceptibility is small and negative for diamagnetic materials. As a consequence, μ is slightly less than 1 (see Table 12.1). For para- and antiferromagnetic materials, χ is again small, but positive. Thus, μ is slightly larger than 1. Finally, χ and μ are large and positive for ferro- and ferrimagnetic materials. The magnetic constants are temperature-dependent, except for diamagnetic materials, as we will see later. Further, the susceptibility for ferromagnetic materials depends on the field strength, H.

The magnetic field parameters at a given point in space are, as explained above, the magnetic field strength H and the magnetic induction B. In free (empty) space, B and $\mu_0 H$ are identical, as seen in Eq. (12.3). Inside a magnetic material the induction B consists of the free-space component ($\mu_0 H$) plus a contribution to the magnetic field ($\mu_0 M$) which is due to the presence of matter [Figure 12.2(a)], that is,

$$B = \mu_0 H + \mu_0 M, \tag{12.5}$$

where M is called the *magnetization* of the material. Combining Eqs. (12.3) through (12.5) yields:

$$M = \chi H. \tag{12.6}$$

H, **B**, and **M** are actually vectors. Specifically, outside a material, **H** (and **B**) point from the north to the south pole. Inside of a ferro- or paramagnetic material, **B** and **M** point from the south

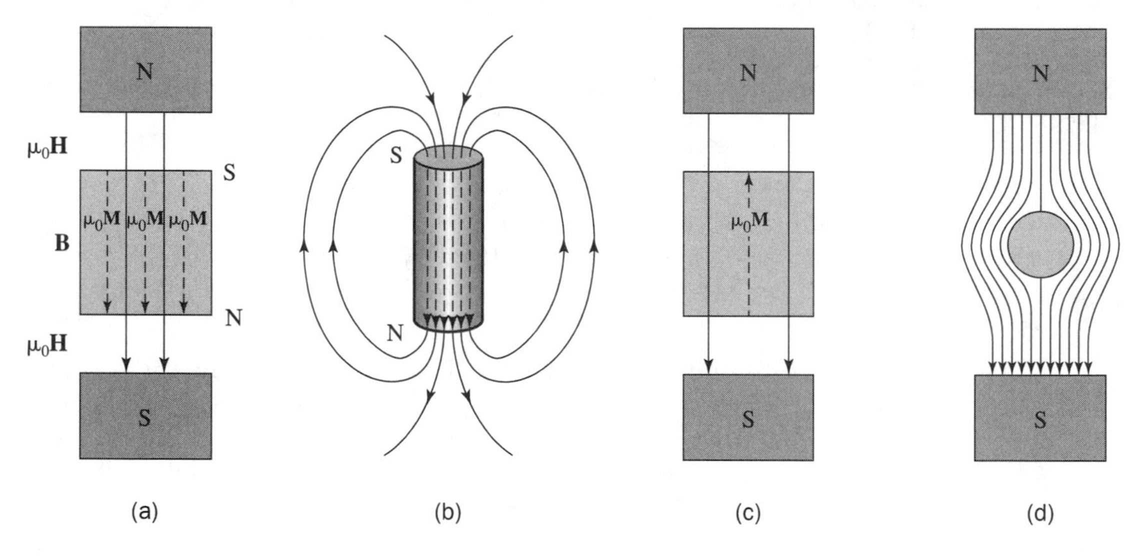

FIGURE 12.2. Schematic representation of magnetic field lines in and around different types of materials. (a) *Para- or ferromagnetics*. The magnetic induction **(B)** inside the material consists of the free-space component (μ_0**H**) plus a contribution by the material (μ_0**M**); see Eq. (12.5). (b) The magnetic field lines outside a material point from the north to the south poles, whereas inside of para- or ferromagnetics, **B** and μ_0**M** point from south to north in order to maintain continuity. (c) In *diamagnetics*, the response of the material counteracts (weakens) the external magnetic field. (d) In a thin surface layer of a *superconductor*, a supercurrent is created (below its transition temperature) which causes a magnetic field that opposes the external field. As a consequence, the magnetic flux lines are expelled from the interior of the material. Compare to Figure 11.27.

to the north; see Figures 12.2(a) and (b). However, we will mostly utilize their moduli in the following sections and thus use lightface italic letters.

B was called above to be the magnetic flux density in a material, that is, the magnetic flux per unit area. The **magnetic flux** ϕ is then defined as the product of B and area A, that is, by

$$\phi = B\,A. \tag{12.7}$$

Finally, we need to define the **magnetic moment** μ_m (also a vector) through the following equation:

$$M = \frac{\mu_{\mathrm{m}}}{V}, \tag{12.8}$$

which means that the magnetization is the magnetic moment per unit volume.

A short note on units should be added. This book uses SI units throughout. However, the scientific literature on magnetism (particularly in the United States) is still widely written in electromagnetic cgs (emu) units. The magnetic field strength in cgs units is measured in Oersted and the magnetic induction in Gauss. Conversion factors from SI into cgs units and for rewriting Eqs. (12.1)–(12.8) in cgs units are given in Appendix II.

12.2 • Magnetic Phenomena and Their Interpretation

We stated in the last section that different types of magnetism exist which are characterized by the magnitude and the sign of the susceptibility (see Table 12.1). Since various materials respond so differently in a magnetic field, we suspect that several fundamentally different mechanisms must be responsible for the magnetic properties. We shall now attempt to unfold the multiplicity of the magnetic behavior of materials by describing some pertinent experimental findings and giving some brief interpretations.

12.2.1 Diamagnetism

Ampère postulated more than one hundred years ago that so-called *molecular currents* are responsible for the magnetism in solids. He compared these molecular currents to an electric current in a loop-shaped piece of wire which is known to cause a magnetic moment. Today, we replace Ampère's molecular currents by *orbiting valence electrons*.

To understand diamagnetism, a second aspect needs to be considered. As explained in Chapter 10 a current is induced in a wire loop whenever a bar magnet is moved toward (or from) this loop.

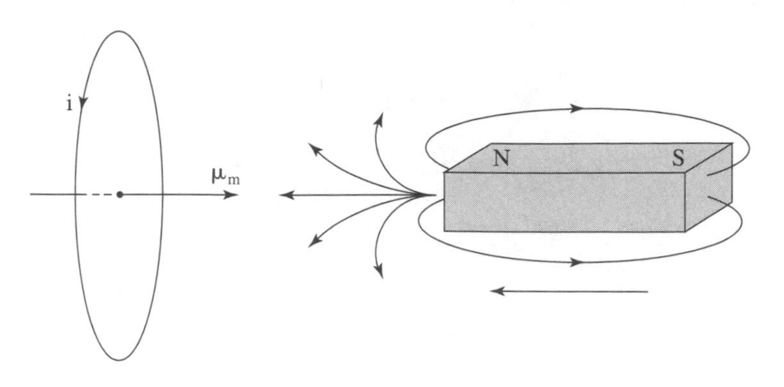

FIGURE 12.3. Induction of a current in a loop-shaped piece of wire by moving a bar magnet toward the wire loop. The current in the loop causes a magnetic field that is directed opposite to the magnetic field of the bar magnet (Lenz law).

The current thus induced causes, in turn, a magnetic moment that is opposite to the one of the bar magnet (Figure 12.3). (This has to be so in order for mechanical work to be expended in producing the current, i.e., to conserve energy; otherwise, a perpetual motion would be created!) Diamagnetism may then be explained by postulating that the external magnetic field induces a change in the magnitude of the atomic currents, i.e., *the external field accelerates or decelerates the orbiting electrons*, so that their magnetic moment is in the opposite direction to the external magnetic field. In other words, the responses of the orbiting electrons counteract the external field [Figure 12.2(c)].

Superconductors have extraordinary diamagnetic properties. They completely expel the magnetic flux lines from their interior when in the superconducting state (*Meissner effect*). In other words, a superconductor behaves in a magnetic field as if B would be zero inside the material [Figure 12.2(d)]. Thus, with Eq. (12.5) one obtains:

$$H = -M, \tag{12.9}$$

which means that the magnetization is equal and opposite to the external magnetic field strength. The result is a perfect diamagnet. The susceptibility,

$$\chi = \frac{M}{H}, \tag{12.6}$$

in superconductors is therefore -1 compared to about -10^{-5} in the normal state (see Table 12.1). This strong diamagnetism can be used for frictionless bearings, that is, for support of loads by a repelling magnetic force. The often-demonstrated levitation effect in which a magnet hovers above a superconducting material also can be explained by these strong diamagnetic properties of superconductors.

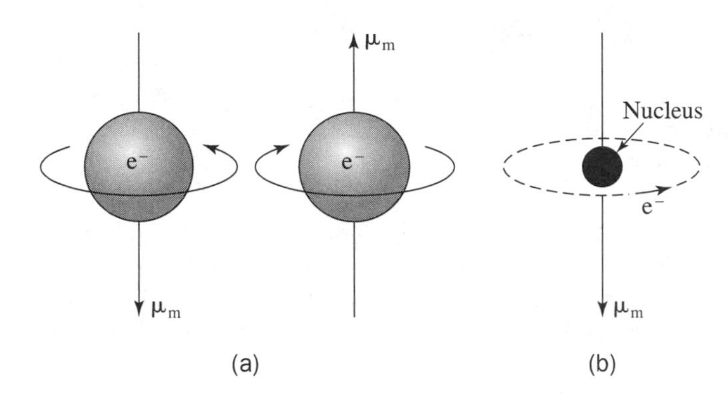

FIGURE 12.4. (a) Schematic representation of electrons which spin around their own axis. A (para)magnetic moment $\boldsymbol{\mu}_\mathbf{m}$ results; its direction depends on the mode of rotation. Only two spin directions are shown (called "spin up" and "spin down"). (b) An orbiting electron is the source for *electron-orbit paramagnetism*.

12.2.2 Paramagnetism

Paramagnetism in *solids* is attributed to a large extent to a magnetic moment that results from electrons which spin around their own axis; see Figure 12.4(a). The spin magnetic moments are generally randomly oriented so that no net magnetic moment results. An external magnetic field tries to turn the unfavorably oriented spin moments in the direction of the external field, but thermal agitation counteracts the alignment. Thus, *spin paramagnetism* is slightly temperature-dependent. It is generally weak and is observed in some metals and in salts of the transition elements.

Free atoms (dilute gases) as well as rare earth elements and their salts and oxides possess an additional source of paramagnetism. It stems from the magnetic moment of the *orbiting electrons*; see Figure 12.4(b). Without an external magnetic field, these magnetic moments are, again, randomly oriented and thus mutually cancel one another. As a result, the net magnetization is zero. However, when an external field is applied, the individual magnetic vectors tend to turn into the field direction which may be counteracted by thermal agitation. Thus, *electron-orbit paramagnetism* is also temperature-dependent. Specifically, paramagnetics often (not always!) obey the experimentally found *Curie–Weiss law*:

$$\chi = \frac{C}{T - \theta} \tag{12.10}$$

where C and θ are constants (given in Kelvin), and C is called the Curie Constant. The Curie–Weiss law is observed to be valid for rare earth elements and salts of the transition elements, for example, the carbonates, chlorides, and sulfates of Fe, Co, Cr, Mn.

From the above-said it becomes clear that in paramagnetic materials the magnetic moments of the electrons eventually point in the direction of the external field, that is, the magnetic moments enhance the external field [see Figure 12.2(a)]. On the other hand, diamagnetism counteracts an external field [see Figure 12.2(c)].

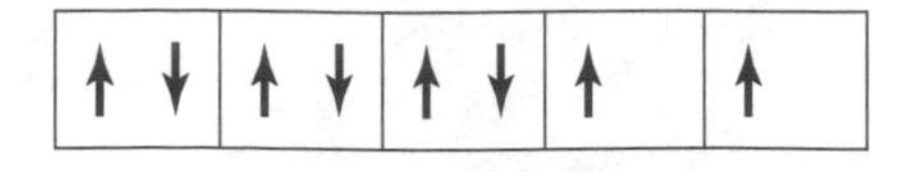

FIGURE 12.5. Schematic representation of the spin alignment in a d-band which is partially filled with eight electrons (Hund's rule). See also Appendix I.

Thus, para- and diamagnetism oppose each other. Solids that have both orbital as well as spin paramagnetism are consequently paramagnetic (since the sum of both paramagnetic compounds is commonly larger than the diamagnetism). Rare earth metals are an example of this.

In many other solids, however, the electron orbits are essentially coupled to the lattice. This prevents the orbital magnetic moments from turning into the field direction. Thus, electron-orbit paramagnetism does not play a role, and only spin paramagnetism remains. The possible presence of a net spin-paramagnetic moment depends, however, on whether or not the magnetic moments of the individual spins cancel each other. Specifically, if a solid has *completely filled electron bands*, then a quantum mechanical rule, called the **Pauli principle**, requires the same number of electrons with spins up and with spins down [Figure 12.4(a)]. The Pauli principle stipulates that each electron state can be filled only with two electrons having opposite spins, see Appendix I. The case of completely filled bands thus results in a cancellation of the spin moments and no net paramagnetism is expected. Materials in which this occurs are therefore diamagnetic (no orbital and no spin paramagnetic moments). Examples of filled bands are intrinsic semiconductors, insulators, and ionic crystals such as NaCl.

In materials that have partially filled bands, the electron spins are arranged according to **Hund's rule** in such a manner that the total spin moment is maximized. For example, for an atom with eight valence d-electrons, five of the spins may point up and three spins point down, which results in a net number of two spins up; Figure 12.5. The atom then has two units of (para-) magnetism or, as it is said, two Bohr magnetons per atom. The **Bohr magneton** is the smallest unit (or quantum) of the magnetic moment and has the value:

$$\mu_B = \frac{eh}{4\pi m} = 9.274 \times 10^{-24} \left(\frac{J}{T}\right) \equiv (A \cdot m^2). \qquad (12.11)$$

(The symbols have the usual meanings as listed in Appendix II.)

12.2.3 Ferromagnetism

Figure 12.6 depicts a ring-shaped solenoid consisting of a newly cast piece of iron and two separate coils which are wound around the iron ring. If the magnetic field strength in the solenoid is temporally increased (by increasing the current in the primary wind-

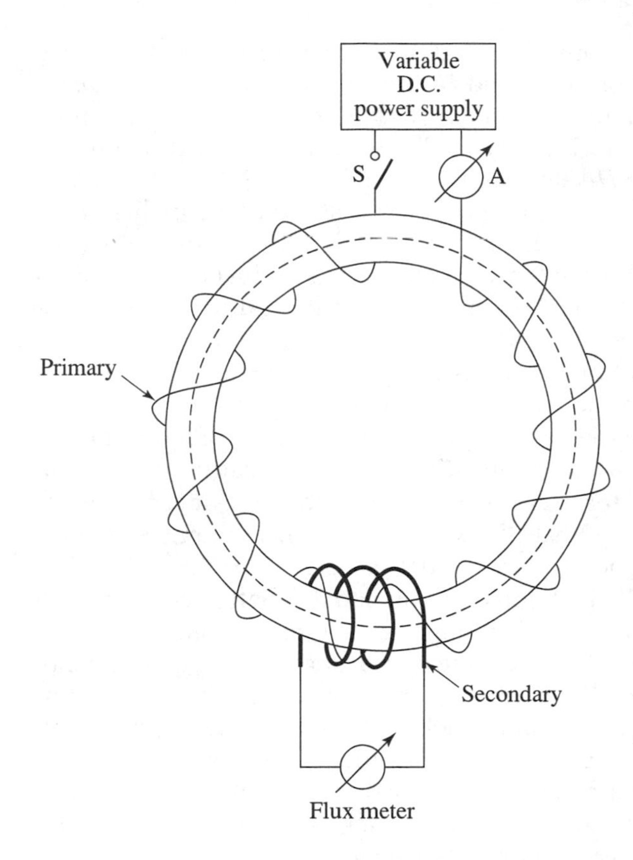

FIGURE 12.6. A ring-shaped solenoid with primary and secondary windings. The magnetic flux lines are indicated by a dashed circle. Note that a current can flow in the secondary circuit only if the current (and therefore the magnetic flux) in the primary winding changes with time. An on–off switch in the primary circuit may serve this purpose. A flux meter is an ampmeter without retracting springs.

ing), then the magnetization (measured in the secondary winding with a flux meter) rises slowly at first and then more rapidly, as shown in Figure 12.7 (dashed line). Finally, M levels off and reaches a constant value, called the **saturation magnetization**, M_s. When H is reduced to zero, the magnetization retains a positive value, called the *remanent magnetization*, or **remanence**, M_r. It is this retained magnetization which is utilized in permanent magnets. The remanent magnetization can be removed by reversing the magnetic

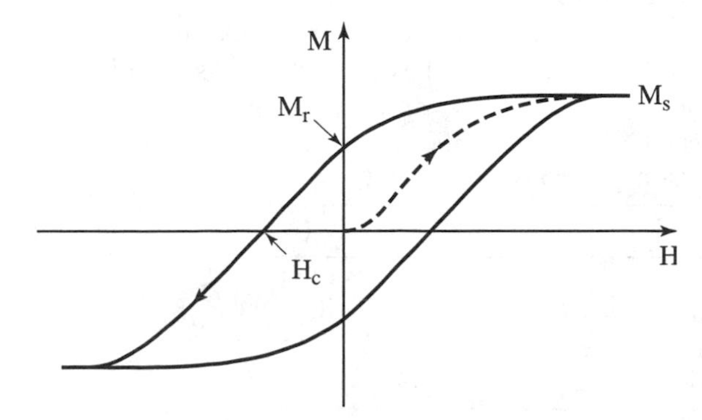

FIGURE 12.7. Schematic representation of a hysteresis loop of a ferromagnetic material. The dashed curve is for a newly cast piece of iron (called virgin iron). Compare to Figure 11.29.

field strength to a value H_c, called the **coercive field**. Solids having a large combination of M_r and H_c are called *hard magnetic materials* (in contrast to *soft* magnetic materials, for which the area inside the loop of Figure 12.7 is very small). A complete cycle through positive and negative H-values as shown in Figure 12.7 is called a **hysteresis loop**. It should be noted that a second type of hysteresis curve is often used in which B (instead of M) is plotted versus H. No saturation value for B can be observed; see Eq. (12.3). Removal of the residual induction requires a field that is called *coercivity*, but the terms coercive field and coercivity are often used interchangeably. The area within a hysteresis loop (B times H or M times $\mu_0 H$) is proportional to the energy per unit volume, which is dissipated once a full field cycle has been completed; see also Section 11.9.

The saturation magnetization is temperature-dependent. Above the **Curie temperature**, T_c, ferromagnetics become paramagnetic. Table 12.2 lists Curie temperatures for some elements.

In ferromagnetic materials, such as iron, cobalt, and nickel, the spins of unfilled d-bands spontaneously align parallel to each other below T_c, that is, they align within small domains (1–100 μm in size) without the presence of an external magnetic field; Figure 12.8(a). The individual domains are magnetized to saturation. The spin direction in each domain is, however, different, so that the individual magnetic moments for virgin ferromagnetic materials as a whole cancel each other and the net magnetization is zero. An external magnetic field causes those domains whose spins are parallel or nearly parallel to the external field to grow at the expense of the unfavorably aligned domains; Figure 12.8(b). When the entire crystal finally contains only *one* single domain, having spins aligned parallel to the external field direction then the material is said to have reached *technical saturation magnetization*, M_s [Figure 12.8(c)]. An increase in temperature progressively destroys the spontaneous alignment, thus reducing the saturation magnetization, Figure 12.8(d).

We have not yet answered the question of whether or not the flip from one spin direction into the other occurs in one step, that is, between two adjacent atoms or over an extended range of atoms instead. Indeed, a gradual rotation over several hun-

TABLE 12.2. Curie temperature, T_c, for some ferromagnetic materials

Metal	T_c (K)
Fe	1043
Co	1404
Ni	631
Gd	289

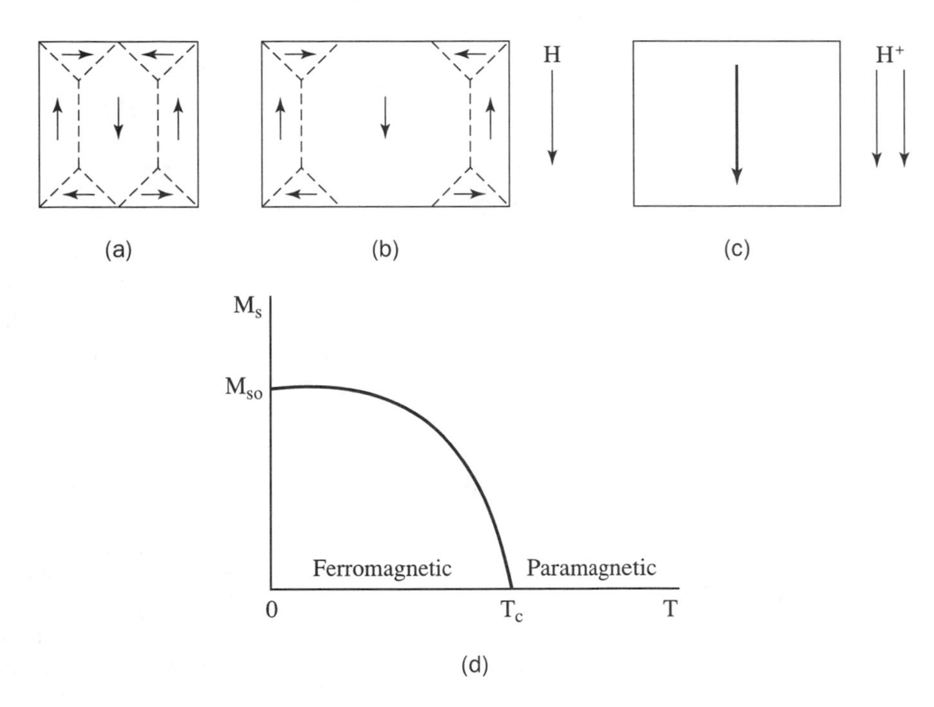

FIGURE 12.8. (a) Schematic representation of individual magnetic domains whose spins are spontaneously aligned in one direction. (b) and (c) Same as above but after an external magnetic field of increasing strength has been applied. Compare to Figure 11.31. (d) Temperature-dependence of the saturation magnetization of ferromagnetic materials.

dred atomic distances is energetically most favorable. The region between individual domains in which the spins rotate from one direction into the next is called a **domain wall** or a *Bloch wall*.

12.2.4 Antiferromagnetism

Antiferromagnetic materials exhibit, just as ferromagnetics, a spontaneous alignment of spin moments below a critical temperature (called the **Néel temperature**). However, the responsible neighboring atoms in antiferromagnetics are aligned in an antiparallel fashion (Figure 12.9). Actually, one may consider an antiferromagnetic crystal to be divided into two interpenetrating sublattices, A and B, each of which has a spontaneous alignment of spins. Figure 12.9 depicts the spin alignments for two manganese compounds. (Only the spins of the manganese ions contribute to the antiferromagnetic behavior.) Figure 12.9(a) implies that the ions in a given {110} plane possess parallel spin alignment, whereas ions in the adjacent plane have antiparallel spins with respect to the first plane. Thus, the magnetic moments of the solid cancel each other and the material as a whole has no net magnetic moment.

Most antiferromagnetics are found among ionic compounds such

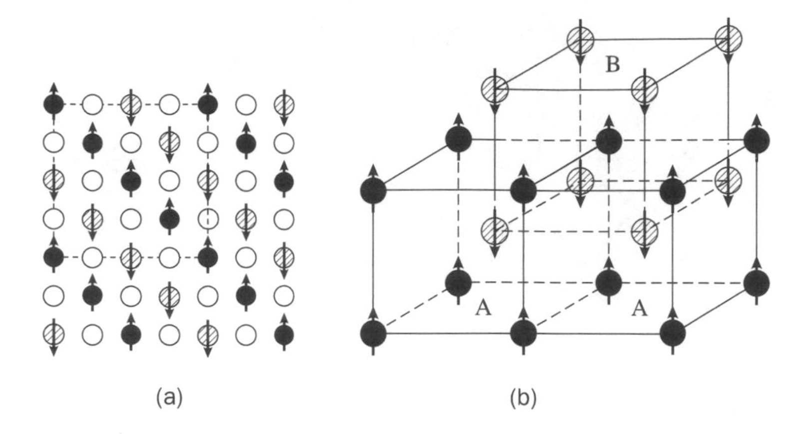

FIGURE 12.9. Schematic representation of spin alignments for antiferromagnetics at 0 K. (a) Display of a (100) plane of MnO. The gray (spin down) and black (spin up) circles represent the Mn ions. The oxygen ions (open circles) do not contribute to the antiferromagnetic behavior. MnO has an NaCl structure. (b) Three-dimensional representation of the spin alignment of manganese ions in MnF_2. (The fluorine ions are not shown.) This figure demonstrates the interpenetration of two manganese sublattices A and B having antiparallel aligned moments.

as MnO, MnF_2, FeO, NiO, and CoO. They are generally insulators or semiconductors. Additionally, α-manganese and chromium are antiferromagnetic. Antiferromagnets are a crucial component of giant magnetoresistance heads in hard disk drives. Their constant θ in the Curie–Weiss law (12.10) is negative.

12.2.5 Ferrimagnetism

Ferrimagnetic materials such as $NiO \cdot Fe_2O_3$ or $FeO \cdot Fe_2O_3$ are of great technical importance. They exhibit a spontaneous magnetic moment (Figure 12.8) and hysteresis (Figure 12.7) below a Curie temperature, just as iron, cobalt, and nickel do. In other words, ferrimagnetic materials possess, similar to ferromagnetics, small domains in which the electron spins are spontaneously aligned in parallel. The main difference to ferromagnetics is, however, that ferrimagnetics are ceramic materials (oxides); they are therefore poor electrical conductors. Poor conduction is often desired for high-frequency applications (e.g., to prevent eddy currents in cores of coils; see Section 12.3).

To explain the spontaneous magnetization in ferrimagnetics, Néel proposed that two sublattices, say A and B, should exist in these materials (just as in antiferromagnetics), each of which contains ions whose spins are aligned parallel to each other. Again, the spins of the ions on the A sites are antiparallel to the spins of the ions on the B sites. The crucial point is that each of the

A Sites	B Sites	
Fe^{3+}	Fe^{3+}	Ni^{2+}
↑	↓	↓↓

FIGURE 12.10. Distribution of spins upon A and B sites for $NiO \cdot Fe_2O_3$. The iron ions are equally distributed among the A and B sites. The nickel ions are only situated on the B sites.

sublattices contains a different amount of magnetic ions. This causes some of the magnetic moments to remain uncancelled. As a consequence, a net magnetic moment results. Ferrimagnetic materials can thus be described as *imperfect antiferromagnetics*.

We will now discuss as an example nickel ferrite, $NiO \cdot Fe_2O_3$. The Fe^{3+} ions are equally distributed between the A and B sites (Figure 12.10), and since ions on the A and B sites exhibit spontaneous magnetization in opposite directions, we expect overall cancellation of spins for the iron ions.

In contrast to this, all nickel ions are accommodated on the B sites only (Figure 12.10). Two electrons have been stripped in the Ni^{2+} ion (i.e., the two 4*s*-electrons; see Appendix I), so that the eight *d*-electrons per atom remain. They are arranged according to Hund's rule (Figure 12.5) to yield two net magnetic moments. In other words, nickel ferrite has two uncancelled spins and therefore two Bohr magnetons per formula unit. This is essentially observed by experiment.

The crystallography of ferrites is rather complex. It suffices to know that there are two types of lattice sites which are available to be occupied by the metal ions. (As before, oxygen ions do not contribute to the magnetic moments.) The unit cell of cubic ferrites contains a total of 56 ions. Some of the metal ions are situated inside a tetrahedron formed by the oxygen ions. These are the above-mentioned A sites. Other metal ions are arranged in the center of an octahedron formed by the oxygen ions and are said to be on the B sites. This distribution is called an *inverse spinel structure*.

12.3 • Applications

12.3.1 Electrical Steels (Soft Magnetic Materials)

Electrical steel is used to multiply the magnetic flux in the core of electromagnets. These materials are therefore widely incorporated in many electrical machines which are in daily use. Among their applications are cores of transformers, electromotors, generators, and electromagnets.

In order to make these devices as energy efficient and economical as possible, one needs to find magnetic materials that have the highest possible permeability (at the lowest possible price). Furthermore, magnetic core materials should be capable of being easily magnetized or demagnetized. In other words, the area within the hysteresis loop (or the coercive force, H_c) should be as small as possible; see Figure 12.7. We remember that materials whose hysteresis loops are narrow are called *soft* magnetic materials.

Electrical steels are classified by some of their properties, for example, by the amount of their core losses (see below), by their composition, by their permeability, and whether or not they are grain-oriented. We shall discuss these different properties momentarily.

The energy losses encountered in electromotors (efficiency between 50 and 90%) or in transformers (efficiency 95–99.5%) are estimated in the United States to be as high as 3×10^{10} KWh per year, which is equivalent to the energy consumption by about 3 million households. This wastes about 2×10^9 dollars per year. If, by means of improved design of the magnetic cores, the energy loss would be reduced by only 5%, one could save about 10^8 dollars per year and several electric power stations. Thus, there is a clear incentive for improving the properties of magnetic materials.

The **core loss** is the energy that is dissipated in the form of heat within the core of electromagnetic devices when the core is subjected to an alternating magnetic field. Several types of losses are known, among which the *eddy current loss* and the *hysteresis loss* contribute most. Typical core losses are between 0.3 and 3 watts per kilogram of core material (Table 12.3).

Let us first discuss **eddy current losses**. Consider a transformer whose primary and secondary coils are wound around the legs of a rectangular iron yoke [Figure 12.11(a)]. An alternating electric current in the primary coil causes an alternating magnetic flux, ϕ, in the core, which in turn induces in the secondary coil an alternating electromotive force, V_e, proportional to $d\phi/dt$ [see Eq. (12.7)]:

$$V_e \propto \frac{d\phi}{dt} = -A\,\frac{dB}{dt}. \tag{12.12}$$

Concurrently, an alternating emf is induced within the core itself, as shown in Figure 12.11(a). This emf gives rise to the eddy current I_e. The eddy current is larger, the larger the permeability, μ [because $B = \mu\,\mu_0 H$; Eq. (12.3)]. Further, I_e increases the larger the conductivity, σ, of the core material, the higher the applied frequency, and the larger the cross-sectional area, A, of the core. [A is perpendicular to the magnetic flux ϕ; see Figure 12.11(a).]

In order to decrease the eddy current, several remedies are possible. First, the core can be made of an insulator in order to decrease σ. Ferrites are thus effective but also expensive materials to build magnetic cores (see Section 12.2). They are indeed used for high-frequency applications. The most widely applied method to reduce eddy currents is the utilization of cores made out of

TABLE 12.3. Properties of some soft magnetic materials

Name	Composition (mass %)	Permeability μ_{max} (unitless)	Coercivity H_c (Oe)	Coercivity H_c (A/m)	Saturation induction[a] B_s (kG)	Saturation induction[a] B_s (T)	Resistivity $\rho(\mu\Omega \cdot \text{cm})$	Core loss at 15 kG and 60 Hz (W/kg)
Low carbon steel	Fe–0.05% C	5×10^3	1.0	80	21.5	2.1	10	2.8
Nonoriented silicon iron	Fe–3% Si, 0.005% C, 0.15% Mn	7×10^3	0.5	40	19.7	2	60	0.9
Grain-oriented silicon iron	Fe–3% Si, 0.003% C, 0.07% Mn	4×10^4	0.1	8	20	2	47	0.3
78 Permalloy	Ni–22% Fe	10^5	0.05	4	10.8	1.1	16	≈2
Mumetal	77% Ni; 16% Fe, 5% Cu, 2% Cr	10^5	0.05	4	6.5	0.6	62	
Supermalloy	79% Ni; 16% Fe, 5% Mo	10^6	0.002	0.1	7.9	0.8	60	
Supermendur	49% Fe, 49% Co, 2% V	6×10^4	0.2	16	24	2.4	27	
Metglas # 2605 annealed	$Fe_{80}B_{20}$	3×10^5	0.04	3.2	15	1.5	≈200	0.3

[a] Above B_s the magnetization is constant and $\frac{1}{\mu_0}\frac{dB}{dH}$ is unity.

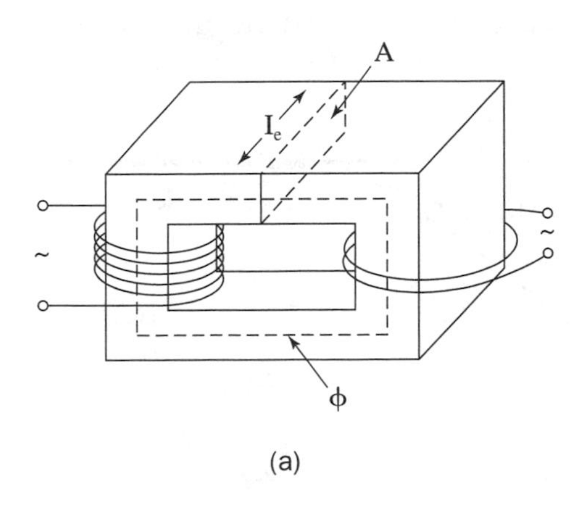

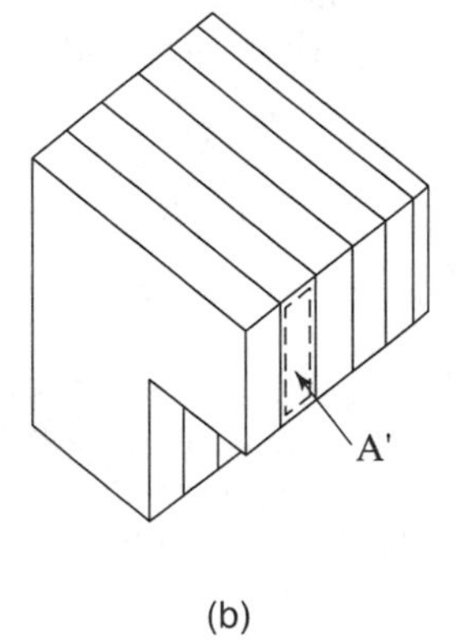

FIGURE 12.11. (a) Solid transformer core with eddy current I_e in a cross-sectional area A. Note the magnetic flux lines ϕ. (b) Cross section of a laminated transformer core. The area A' is smaller than area A in (a).

thin sheets that are electrically insulated from each other [Figure 12.11(b)]. In this way, the cross-sectional area A is split into several smaller areas, A', which in turn decreases V_e [Eq. (12.12)]. Despite the lamination, a residual eddy current loss still exists which is caused by current losses within the individual laminations, and by interlaminar losses that may arise if the laminations are not sufficiently insulated from each other. These losses are, however, less than 1% of the total energy transferred.

Hysteresis losses are encountered when the magnetic core is subjected to a complete hysteresis cycle (Figure 12.7). The electrical energy thus dissipated into heat is proportional to the area enclosed by a B/H loop. Proper materials selection and rolling of the materials with subsequent heat treatment greatly reduces the area of a hysteresis loop (see below).

Grain Orientation

The permeability of electrical steel can be substantially increased and the hysteresis losses can be decreased by making use of favorable grain orientations in the material. This needs some explanation. The magnetic properties of crystalline ferromagnetic materials depend on the crystallographic direction in which an external field is applied, an effect which is called *magnetic anisotropy*. Let us use iron as an example. Figure 12.12 shows magnetization curves of single crystals for three crystallographic directions. We observe that, if the external field is applied in the ⟨100⟩ direction, saturation is achieved with the smallest possible field strength. The ⟨100⟩ direction is thus called the *easy direction*.

A second piece of information also needs to be considered. Metal sheets, which have been manufactured by rolling and heating, often possess a *texture*, that is, they have a preferred orientation of the grains. It just happens that in α-iron and α-iron alloys the ⟨100⟩ direction is parallel to the rolling direction. This property is exploited when utilizing electrical steel.

Grain-oriented electrical steel is produced by initially hot-rolling the alloy followed by two stages of cold reduction with

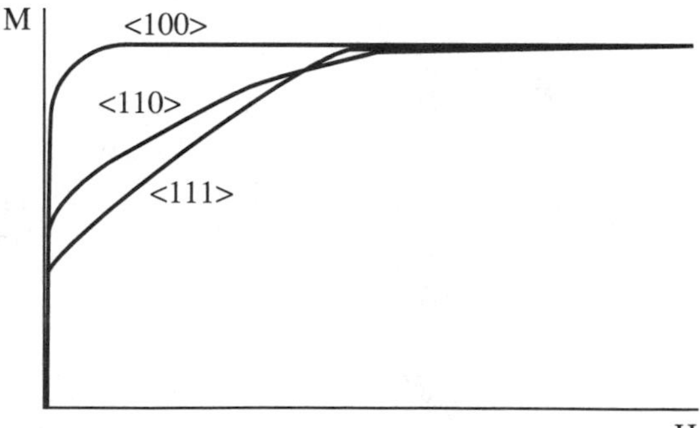

FIGURE 12.12. Schematic magnetization curves for rod-shaped *iron single crystals* of different orientations (virgin curves). The magnetic field was applied in three different crystallographic directions. (Compare with Figure 12.7, which refers to polycrystalline materials.) For direction indices, see Figure 3.13.

intervening anneals. During the rolling, the grains are elongated and their orientation is altered as described above. Finally, the sheets are recrystallized, whereby some crystals grow in size at the expense of others (occupying the entire sheet thickness). This process is called *secondary recrystallization*.

In summary, the magnetic properties of grain-oriented steels are best in the direction parallel to the direction of rolling. Electrical machines having core material of grain-oriented steel need less iron and are therefore smaller. The price increase due to the more elaborate fabrication procedure is often compensated for by the savings in material. For details, see Table 12.3.

Composition of Core Materials

The least expensive core material is commercial low carbon steel (0.05% C). It possesses a relatively small permeability and has about ten times higher core losses than grain-oriented silicon iron (Table 12.3). Low carbon steel is used where low cost is more important than the efficient operation of a device. Iron–silicon alloys containing between 1.4 and 3.5% Si and very little carbon have a higher permeability and a lower conductivity than low carbon steel (see Table 12.3). Grain-oriented silicon "steel" is the favored commercial product for highly efficient–high flux multiplying core applications such as transformer sheets. The highest permeability is achieved for certain multicomponent nickel-based alloys such as Permalloy, Supermalloy, or Mumetal (Table 12.3). The latter can be rolled into thin sheets and is used to shield electronic equipment from stray magnetic fields. Amorphous ferromagnets (metallic glasses) consisting of iron, nickel, or cobalt with boron, silicon, or phosphorous have, when properly annealed below the crystalline temperature, a considerably higher permeability and a lower coercivity than commonly used grain-oriented silicon. Further, the electrical resistivity of amorphous alloys is generally larger than their crystalline counterparts, which results in smaller eddy-current losses. However, amorphous ferromagnetics possess a somewhat lower saturation induction (which sharply decreases even further at elevated temperatures), and their core losses increase rapidly at higher flux densities (e.g., above 1.4 T). Thus, their application is limited to magnetic sensors, magnetorestrictive transducers, and transformers.

12.3.2 Permanent Magnets (Hard Magnetic Materials)

Permanent magnets are devices that retain their magnetic field indefinitely. They are characterized by a large remanence B_r (or M_r), a relatively large coercivity H_c, and a large area within the hysteresis loop. They are called *hard magnetic materials* (see Section 12.2). Another parameter which is used to characterize hard magnetic materials is the *maximum energy product*, $(BH)_{\text{max}}$, which is related to the maximum area within a hysteresis loop.

TABLE 12.4. Properties of materials used for permanent magnets. [See also Figure 19.1(a)]

Material	Composition (mass %)	Remanence B_r		Coercivity H_c		Maximum energy product $(BH)_{max}$ per volume	
		(kG)	(T)	(Oe)	(A/m)	(MGOe)	(kJ/m^3)
Steel	Fe–1% C	9	0.9	51	4×10^3	0.2	1.6
36 Co steel	36 Co, 3.75 W, 5.75 Cr, 0.8 C	9.6	0.96	228	1.8×10^4	0.93	7.4
Alnico 2	12 Al, 26 Ni, 3 Cu, 63 Fe	7	0.7	650	5.2×10^4	1.7	13
Alnico 5	8 Al, 15 Ni, 24 Co, 3 Cu, 50 Fe	12	1.2	720	5.7×10^4	5.0	40
Alnico 5 DG	same as above	13.1	1.3	700	5.6×10^4	6.5	52
Ba-ferrite (Ceramic 5)	$BaO \cdot 6\,Fe_2O_3$	3.95	0.4	2,400	1.9×10^5	3.5	28
PtCo	77 Pt, 24 Co	6.45	0.6	4,300	3.4×10^5	9.5	76
Remalloy	12 Co, 17 Mo, 71 Fe	10	1	230	1.8×10^4	1.1	8.7
Vicalloy 2	13 V, 52 Co, 35 Fe	10	1	450	3.6×10^4	3.0	24
Cobalt–samarium	Co_5Sm	9	0.9	8,700	6.9×10^5	20	159
Iron–neodymium–boron	$Fe_{14}Nd_2B_1$	13	1.3	14,000	1.1×10^6	40	318

The values of B_r, H_c, and $(BH)_{max}$ for some materials which are used as permanent magnets are listed in Table 12.4.

Today, many permanent magnets are made of **Alnico alloys**, which contain various amounts of aluminum, nickel, cobalt, and iron, along with some minor constituents such as copper and titanium (Table 12.4). Their properties are improved by heat treatments (homogenization at 1250°C, fast cooling, and tempering at 600°C; Alnico 2). Further improvement is accomplished by cooling the alloys in a magnetic field (Alnico 5). The best properties are achieved when the grains are made to have a preferred orientation. This is obtained by cooling the bottom of the crucible after melting, thus forming long columnar grains with a preferred ⟨100⟩ axis in the direction of heat flow. A magnetic field parallel to the ⟨100⟩ axis yields Alnico 5-DG (directional grain).

The superior properties of heat-treated Alnico stems from the fact that, during cooling and tempering of these alloys, rod-shaped iron and cobalt-rich α-precipitates are formed which are parallel to the ⟨100⟩ directions (*shape anisotropy*). These strongly magnetic precipitates are single-domain particles and are imbedded in a weakly magnetic nickel and aluminum matrix (α). Alnico alloys possess, just as iron, a <100> easy direction (see Fig. 12.12) and have also a cubic crystal structure. Alnico alloys are mechanically hard and brittle and can, therefore, only be shaped by casting or by pressing and sintering of metal powders.

The newest hard magnetic materials contain rare-earth elements[1] and are made mainly of iron with neodymium and boron; see Table 12.4. They were discovered in 1983 and possess a superior coercivity and, thus, a much larger $(BH)_{max}$. Their disadvantages are a relatively low Curie temperature and poor corrosion resistance. **Neomagnets** allow motors, speakers, and magnetic resonance imaging devices (MRI; for medical applications) to do the same job for less magnetic material. For example, a Magnetic Resonance Imaging system that would require 21 tons of ferrites and a system weight of 70 tons uses only 2.6 tons of neomagnets and a system weight of 24 tons. Without rare-earth magnets, motors of the same power rating would be at least five times heavier and would be less efficient than those using electromagnets.

Ceramic ferrite magnets, such as barium or strontium ferrite ($BaO \cdot 6Fe_2O_3$ or $SrO \cdot 6Fe_2O_3$), are brittle and relatively inexpensive. They crystallize in the form of plates with the hexagonal *c*-axis (which is the easy axis) perpendicular to the plates. Some preferred orientation is observed because the flat plates arrange parallel to each other during pressing and sintering. Ferrite pow-

---

[1]"Rare earth elements" are not really *rare*. They are only difficult to separate. However, samarium is rare and expensive.

der is often imbedded in plastic materials, which yields flexible magnets. They are used, for example, in the gaskets of refrigerator doors. Ferrites account for about 90% of the permanent-magnet market mainly because of their price advantage.

High carbon steel magnets with or without cobalt, tungsten, or chromium are only of historic interest. Their properties are inferior to other magnets. It is believed that the permanent magnetization of quenched steel stems from the martensite-induced internal stress which impedes the domain walls from moving through the crystal.

12.3.3 Magnetic Recording

Magnetic recording tapes, disks, drums, and magnetic strips on credit cards consist of small, needlelike magnetic oxide particles about 0.1×0.5 μm^2 in size which are imbedded in a nonmagnetic binder. The particles are too small to sustain a domain wall. They therefore consist of a single magnetic domain which is magnetized to saturation along the major axis. The elongated particles are aligned by a field during manufacturing so that their long axes are parallel with the length of the tape or the track. The most popular magnetic material has been ferrimagnetic γ-Fe_2O_3. Its coercivity is 20–28 kA/m (250–350 Oe). More recently, ferromagnetic chromium dioxide has been used, having a coercivity between 40–80 kA/m (500–1000 Oe) and a particle size of 0.05 by 0.4 μm^2. High coercivity and high remanence prevent self-demagnetization and accidental erasure; they provide strong signals and permit thinner coatings. A high H_c also allows tape duplication by "contact printing." However, CrO_2 has a relatively low Curie temperature (128°C compared to 600°C for γ-Fe_2O_3). Thus, chromium dioxide tapes which are exposed to excessive heat (glove compartment!) may lose their stored information.

Thin magnetic films consisting of Co–Ni–P or Co–Cr–Ta are frequently used in hard-disk devices. They are laid down on an aluminum substrate and are covered by a 40-nm-thick carbon layer for lubrication and corrosion resistance. The coercivities range between 60–120 kA/m (750–1500 Oe). Thin film magnetic memories can be easily fabricated (vapor deposition, sputtering, or electroplating), they can be switched rapidly, and they have a small unit size. Thin-film recording media are not used for tapes, however, because of their rapid wear.

The hysteresis loop of materials used for magnetically storing information generally has a nearly square shape as shown in Figure 12.13. If a sufficiently high magnetic field has been applied, then the material becomes magnetically saturated. An opposite directed magnetic field turns the spins in the reverse direction. These two spin orientations constitute the necessary values (0 and 1) for the binary system on which all computers operate.

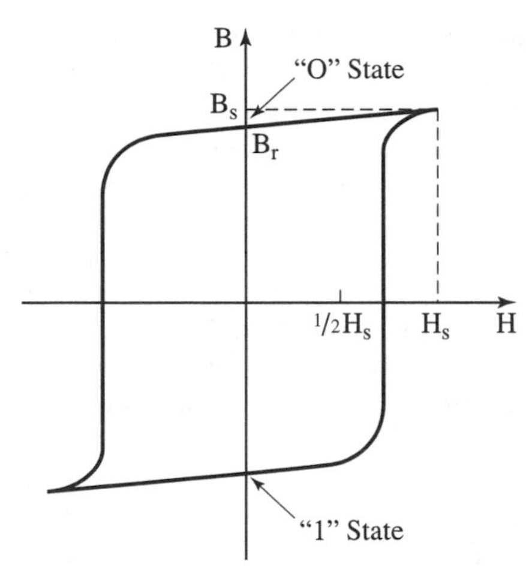

FIGURE 12.13. Square-shaped hysteresis loop for magnetic storage devices. A stray magnetic field whose value is only, say, 1/2 H_s does not switch the spins in a given magnetic domain in the opposite direction.

The recording head of a tape machine consists of a laminated electromagnet made of permalloy or soft ferrite (Table 12.3) which has an air gap about 0.3 μm wide (Figure 12.14). The tape is passed along this electromagnet whose fringing field redirects the spin moments of the particles in a certain pattern proportional to the current which is applied to the recording head. This leaves a permanent record of the signal. In the playback mode, the moving tape induces an alternating emf in the coil of the same head. The emf is amplified, filtered, and fed to a loudspeaker.

Magnetic disks (for random access) or tapes (sequential access, used mainly for music or video recordings) are the choices for long-term, large-scale information storage, particularly since no electric energy is needed to retain the information. This is in contrast to many (not all!) semiconductor storage devices which need a constant energy supply to hold the information. A third storage mechanism, called optical storage, such as in compact-disk

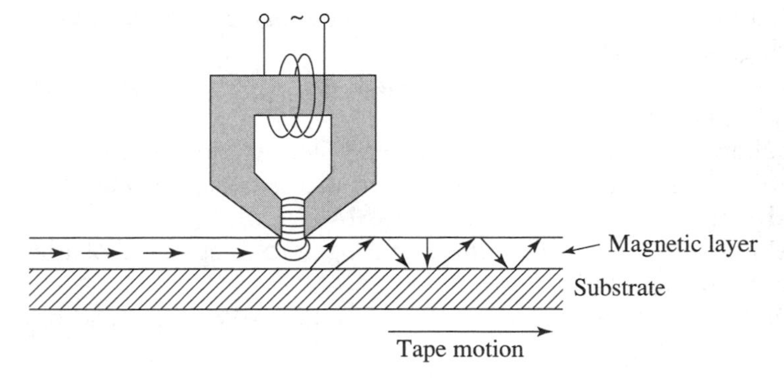

FIGURE 12.14. Schematic arrangement of a recording (playback) head and a magnetic tape (recording mode). The gap width is exaggerated. The plastic substrate is about 25 μm thick.

(CD) players (see Figure 13.15), also does not need a constant energy supply to keep the information. It should be noted in closing that magnetic tapes make contact with the recording (and playback) head and are therefore subject to wear.

Problems

12.1. Calculate the field strength of an electromagnet that is 10 cm long, has 1000 turns, and is passed by a current of 2 A. Give your answer in SI as well as in cgs units.

12.2. Calculate the Curie constant for a paramagnetic material that was inserted at room temperature (300 K) in a magnetic field of 1.2×10^4 A/m and whose magnetic induction was measured to be 0.05 T. Assume that $\theta = 0$.

12.3. A piece of material whose volume is 3 cm^3 was inserted into an inhomogeneous magnetic field ($dH/dx = 9 \times 10^4$ A/m^2) and was found to be pulled into the magnet with a force of 20 μN. What is the permeability when the magnetic field strength of the magnet is 5×10^4 A/m? Is the material dia-, para-, or ferromagnetic?

12.4. Calculate the value of one Bohr magneton from fundamental constants.

12.5. Calculate the number of Bohr magnetons for iron and cobalt ferrite from their electron configuration as done in the text. Give the chemical formula for these ferrites.

12.6. You are given two identical rectangular iron rods. One of the rods is a permanent magnet, the other is a plain piece of iron made from a soft magnetic material. The rods are now placed on a wooden table. Using only the two rods and nothing else, you are asked to determine which is which. Can this be done?

12.7. A piece of platinum is inserted into a magnetic field that has a magnetic field strength of 10^3 [Oe]. Calculate the magnetic induction in the material. ($\chi_{Pt} = 264.4 \times 10^{-6}$ mks units.)

Suggestions for Further Study

B.D. Cullity, *Introduction to Magnetic Materials*, Addison-Wesley, Reading, MA (1972).

R.E. Hummel, *Electronic Properties of Materials*, 3rd ed., Springer-Verlag, New York (2001).

D. Jiles, *Magnetism and Magnetic Materials*, Chapman and Hall, London (1991).

E. Kneller, *Ferromagnetismus*, Springer-Verlag, Berlin (1962).

J.D. Livingston, *Driving Force, The Natural Magic of Magnets*, Harvard University Press, Cambridge, MA (1996).

J.C. Mallionson, *The Foundation of Magnetic Recording*, Academic, San Diego (1987).

F.W. Sears, *Electricity and Magnetism*, Addison-Wesley, Reading, MA (1953).

13

Optical Properties of Materials

13.1 • Interaction of Light with Matter

The most apparent properties of metals, their luster and their color, have been known to mankind since materials were known. Because of these properties, metals were already used in antiquity for mirrors and jewelry. The color was utilized 4000 years ago by the ancient Chinese as a guide to determine the composition of the melt of copper alloys: the hue of a preliminary cast indicated whether the melt, from which bells or mirrors were to be made, already had the right tin content.

The German poet Goethe was probably the first one who explicitly spelled out 200 years ago in his *Treatise on Color* that *color* is not an absolute property of matter (such as the resistivity), but requires a living being for its perception and description. Goethe realized that the perceived color of a region in the visual field depends not only on the properties of light coming from that region, but also on the light coming from the rest of the visual field. Applying Goethe's findings, it was possible to explain qualitatively the color of, say, gold in simple terms. Goethe wrote: "If the color *blue* is removed from the spectrum, then blue, violet, and green are missing and red and yellow remain." Thin gold films are bluish-green when viewed in transmission. These colors are missing in reflection. Consequently, gold appears reddish-yellow. On the other hand, Newton stated quite correctly in his "Opticks" that light rays are not colored. The nature of color remained, however, unclear.

This chapter treats the optical properties from a completely different point of view. Measurable quantities such as the index of refraction or the reflectivity and their spectral variations are used to characterize materials. In doing so, the term "color" will almost completely disappear from our vocabulary. Instead, it will be postulated that the interactions of light with the electrons of a material are responsible for the optical properties.

At the beginning of the 20th century, the study of the interactions of light with matter (black-body radiation, etc.) laid the foundations for quantum theory. Today, optical methods are among the most important tools for elucidating the electron structure of matter. Most recently, a number of optical devices such as lasers, photodetectors, waveguides, etc., have gained considerable technological importance. They are used in telecommunication, fiber optics, CD players, laser printers, medical diagnostics, night viewing, solar applications, optical computing, and for optoelectronic purposes. Traditional utilizations of optical materials for windows, antireflection coatings, lenses, mirrors, etc., should be likewise mentioned.

We perceive light intuitively as a wave (specifically, an electromagnetic wave) that travels in undulations from a given source to a point of observation. The color of the light is related to its wavelength. Many crucial experiments, such as diffraction, interference, and dispersion, clearly confirm the wavelike nature of light. Nevertheless, at least since the discovery of the photoelectric effect in 1887 by Hertz, and its interpretation in 1905 by Einstein, do we know that light also has a particle nature. (The photoelectric effect describes the emission of electrons from a metallic surface after it has been illuminated by light of appropriately high energy, e.g., by blue light.) Interestingly enough, Newton, about 300 years ago, was a strong proponent of the particle concept of light. His original ideas, however, were in need of some refinement, which was eventually provided in 1901 by quantum theory. We know today (based on Planck's famous hypothesis) that a certain minimal energy of light, that is, at least one **light quantum**, called a **photon**, with the energy:

$$E = \nu h = \omega \hbar \tag{13.1}$$

needs to impinge on a metal in order that a negatively charged electron may overcome its binding energy to its positively charged nucleus, and can escape into free space. (This is true regardless of the *intensity* of the light.) In Eq. (13.1), h is the Planck constant whose numerical value is given in Appendix II and ν is the frequency of light given as the number of vibrations (cycles) per second or hertz (Hz). Frequently, the reduced Planck constant:

$$\hbar = \frac{h}{2\pi} \tag{13.2}$$

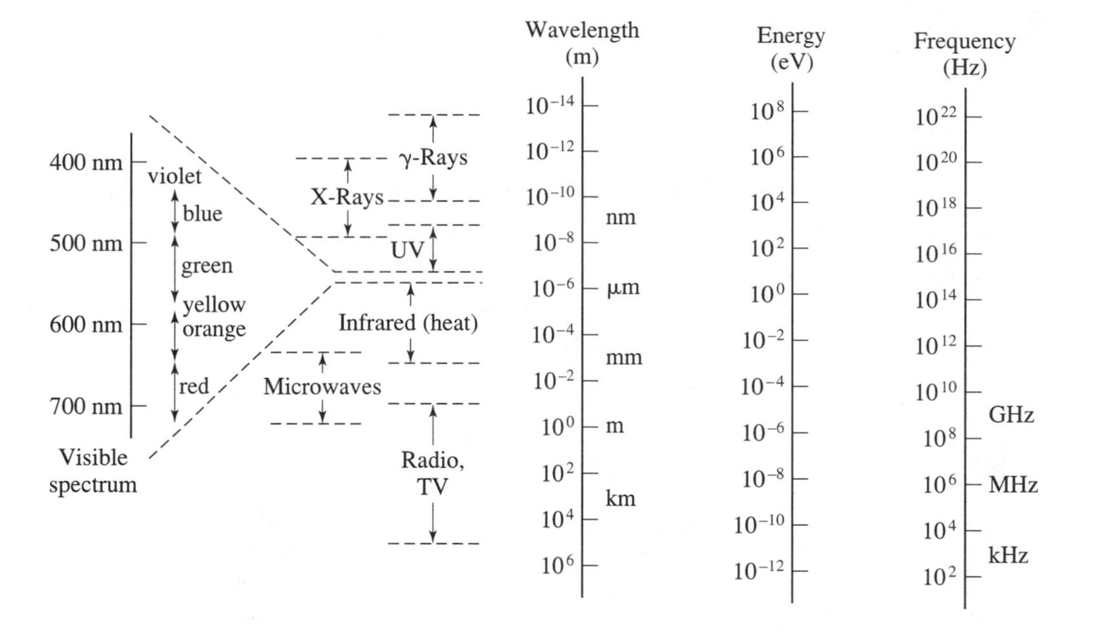

FIGURE 13.1. The spectrum of electromagnetic radiation. Note the small segment of this spectrum that is visible to human eyes.

is utilized in conjunction with the angular frequency, $\omega = 2\pi\nu$. In short, the *wave-particle duality of light* (or, more generally, of electromagnetic radiation) had been firmly established at about 1924. The *speed of light*, c, and the frequency are connected by the equation:

$$c = \nu\lambda, \tag{13.3}$$

where λ is the *wavelength* of the light.

Light comprises only an extremely small segment of the entire electromagnetic spectrum, which ranges from radio waves via microwaves, infrared, visible, ultraviolet, X-rays, to γ rays, as depicted in Figure 13.1. Many of the considerations which will be advanced in this chapter are therefore also valid for other wavelength ranges, i.e., for radio waves or X-rays.

13.2 • The Optical Constants

When light passes from an optically "thin" medium (e.g., vacuum, air) into an optically dense medium one observes that in the dense medium, the angle of refraction β (i.e., the angle between the refracted light beam and a line perpendicular to the surface) is smaller than the angle of incidence, α. This well-known phenomenon is used for the definition of the refractive power of a material and is called the *Snell law*:

$$\frac{\sin \alpha}{\sin \beta} = \frac{n_{\text{med}}}{n_{\text{vac}}} = n. \tag{13.4}$$

Commonly, the **index of refraction** for vacuum n_{vac} is arbitrarily set to be unity. The refraction is caused by the different velocities, c, of the light in the two media:

$$\frac{\sin \alpha}{\sin \beta} = \frac{c_{\text{vac}}}{c_{\text{med}}}. \tag{13.5}$$

Thus, if light passes from vacuum into a medium, we find:

$$n = \frac{c_{\text{vac}}}{c_{\text{med}}} = \frac{c}{\text{v}}, \tag{13.6}$$

where $\text{v} = c_{\text{med}}$ is the velocity of light in the material. The magnitude of the refractive index depends on the wavelength of the incident light. This property is called **dispersion**. In metals, the index of refraction varies also with the angle of incidence. This is particularly true when n is small.

The index of refraction is generally a complex number, designated as $\hat{n}$, which is comprised of a real and an imaginary part n_1 and n_2, respectively, i.e.,

$$\hat{n} = n_1 - i\, n_2. \tag{13.7}$$

In the literature, the imaginary part of $\hat{n}$ is often denoted by k. Equation (13.7) is then written as:

$$\hat{n} = n - i\, k. \tag{13.8}$$

We will call n_2 or k the **damping constant**. (In some books, n_2 and k are named *absorption constant*, *attenuation index*, or *extinction coefficient*. We will not follow this practice because of its potential to be misleading.) The square of the (complex) index of refraction is equal to the (complex) dielectric constant (Section 11.8):

$$\hat{n}^2 = \hat{\epsilon} = \epsilon_1 - i\, \epsilon_2, \tag{13.9}$$

which yields, with Eq. (13.8),

$$\hat{n}^2 = n^2 - k^2 - 2nki = \epsilon_1 - i\, \epsilon_2. \tag{13.10}$$

Equating individually the real and imaginary parts in Eq. (13.10) yields:

$$\epsilon_1 = n^2 - k^2 \tag{13.11}$$

and

$$\epsilon_2 = 2nk. \tag{13.12}$$

ϵ_1 is called **polarization** whereas ϵ_2 is known by the name **absorption**. Values for n and k for some materials are given in Table 13.1. For insulators, k is nearly zero, which yields for dielectrics $\epsilon_1 \approx n^2$ and $\epsilon_2 \to 0$.

When electromagnetic radiation (e.g., light) passes from vacuum (or air) into an optically denser material, then the amplitude of the wave decreases exponentially with increasing damping constant k and for increasing distance, z, from the surface, as shown in Figure 13.2. Specifically, the intensity, I, of the light (that is, the square of the electric field strength, $\mathscr{E}$) obeys the following equation (which can be derived from the Maxwell equations):

$$I = \mathscr{E}^2 = I_0 \exp\left(-\frac{4\pi\nu k}{c} z\right). \tag{13.13}$$

TABLE 13.1. Optical constants for some materials (λ = 600 nm)

	n	k	R %[b]
Metals			
Copper	0.14	3.35	95.6
Silver	0.05	4.09	98.9
Gold	0.21	3.24	92.9
Aluminum	0.97	6.0	90.3
Ceramics[c]			
Silica glass (Vycor)	1.46	[a]	3.50
Soda-lime glass	1.51	[a]	4.13
Dense flint glass	1.75	[a]	7.44
Quartz	1.55	[a]	4.65
Al_2O_3	1.76	[a]	7.58
Polymers			
Polyethylene	1.51	[a]	4.13
Polystyrene	1.60	[a]	5.32
Polytetrafluoroethylene	1.35	[a]	2.22
Semiconductors			
Silicon	3.94	0.025	35.42
GaAs	3.91	0.228	35.26

[a]The damping constant for dielectrics is about 10^{-7}; see Table 13.2.
[b]The reflection is considered to have occurred on one reflecting surface only.
[c]See also Table 15.1.

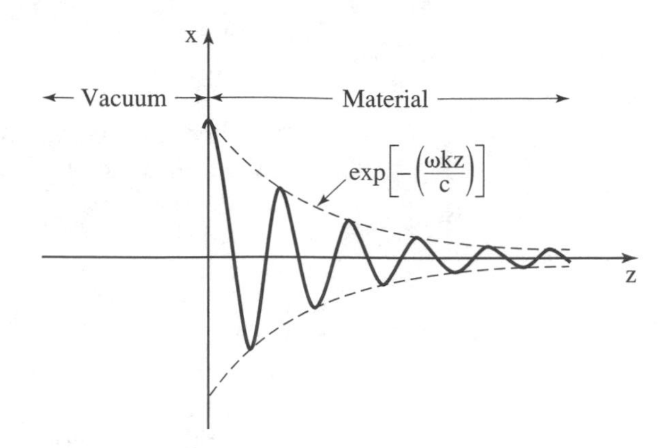

FIGURE 13.2. Exponential decrease of the amplitude of electromagnetic radiation in optically dense materials such as metals.

We define a **characteristic penetration depth**, W, as that distance at which the intensity of the light wave, which travels through a material, has decreased to $1/e$ or 37% of its original value, i.e., when:

$$\frac{I}{I_0} = \frac{1}{e} = e^{-1}. \tag{13.14}$$

This definition yields, in conjunction with Eq. (13.13),

$$z = W = \frac{c}{4\pi\nu k} = \frac{\lambda}{4\pi k}. \tag{13.15}$$

Table 13.2 presents experimental values for k and W for some materials obtained by using sodium vapor light ($\lambda = 589.3$ nm).

The inverse of W is sometimes called the (exponential) *attenuation* or the **absorbance**, α, which is, by making use of Eq. (13.15) and (13.12),

$$\alpha = \frac{4\pi k}{\lambda} = \frac{4\pi\nu k}{c} = \frac{2\pi\epsilon_2}{\lambda n}. \tag{13.16}$$

It is measured, for example, in cm^{-1}. The **energy loss** per unit length (given, for example, in decibels, dB, per centimeter) is obtained by multiplying the absorbance, α, with 4.34, see Problem 13.7. (1 dB = 10 log I/I_0).

TABLE 13.2. Characteristic penetration depth, W, and damping constant, k, for some materials ($\lambda = 589.3$ nm)

Material	Water	Flint glass	Graphite	Gold
W (cm)	32	29	6×10^{-6}	1.5×10^{-6}
k	1.4×10^{-7}	1.5×10^{-7}	0.8	3.2

The ratio between the reflected intensity I_R and the incoming intensity I_0 of the light is the **reflectivity**:

$$R = \frac{I_R}{I_0}. \tag{13.17}$$

Quite similarly, one defines the ratio between the transmitted intensity, I_T, and the impinging light intensity as the **transmissivity**:

$$T = \frac{I_T}{I_0}. \tag{13.18}$$

The reflectivity is connected with n and k (assuming normal incidence) through:

$$R = \frac{(n-1)^2 + k^2}{(n+1)^2 + k^2} \tag{13.19}$$

(*Beer equation*). The reflectivity is a unitless material constant and is often given in percent of the incoming light (see Table 13.1). R is, like the index of refraction, a function of the wavelength of the light. For insulators ($k \approx 0$) one finds that R depends solely on the index of refraction:

$$R = \frac{(n-1)^2}{(n+1)^2}. \tag{13.20}$$

Metals are characterized by a large reflectivity. This stems from the fact that light penetrates metals only a short distance, as shown in Figure 13.2 and Table 13.2. Thus, only a small part of the impinging energy is converted into heat. The major part of the energy is reflected (in some cases as much as 99%, see Table 13.1). In contrast to this, visible light penetrates into glass (and many other dielectrics) much farther than into metals, that is, approximately seven orders of magnitude more; see Table 13.2. As a consequence, very little light is reflected by glass. Nevertheless, a piece of glass about 1 or 2 m thick eventually dissipates a substantial part of the impinging light into heat. (In practical applications, one does not observe this large reduction in light intensity because windows are, as a rule, only a few millimeters thick.) It should be noted that window panes, lenses, etc., reflect the light on the front as well as on the back side.

An energy conservation law requires that the intensity of the light impinging on a material, I_0, must be equal to the reflected intensity, I_R, plus the transmitted intensity, I_T, plus that intensity which has been extinct, I_E, for example, transferred into heat, that is,

$$I_0 = I_R + I_T + I_E. \tag{13.21}$$

Dividing Eq. (13.21) by I_0 and making use of Eq. (13.17) and (13.18) yields:

$$R + T + E = 1. \tag{13.22}$$

(It has been assumed for these considerations that the light which has been scattered inside the material may be transmitted through the sides and is therefore contained in I_T and I_E.)

The reflection losses encountered in optical instruments such as lenses can be significantly reduced by coating the surfaces with a thin layer of a dielectric material such as magnesium fluoride. This results in the well-known blue hue on lenses for cameras.

Metals are generally *opaque* in the visible spectral region because of their comparatively high damping constant and thus high reflectivity. Still, very thin metal films (up to about 50 nm thickness) may allow some light to be transmitted. Dielectric materials, on the other hand, are often *transparent*. Occasionally, however, some opacifiers are inherently or artificially added to dielectrics which cause the light to be internally deflected by multiple scattering. Finally, if the diffuse scattering is not very severe, dielectrics might appear *translucent*, that is, objects viewed through them are vaguely seen, but not clearly distinguishable. Scattering of light may occur, for example, due to residual porosity in ceramic materials, or on grain boundaries (which have a small variation in refractive index compared to the matrix), or on finely dispersed particles, or on boundaries between crystalline and amorphous regions in polymers, to mention only a few mechanisms.

There exists an important equation which relates the reflectivity of light at low frequencies (infrared spectral region) with the direct-current conductivity, σ:

$$R = 1 - 4\sqrt{\pi \epsilon_0 \frac{\nu}{\sigma}}. \tag{13.23}$$

This relation, which was experimentally found at the end of the 19th century by *Hagen and Rubens*, states that materials having a large electrical conductivity (such as metals) also possess essentially a large reflectivity (and vice versa).

13.3 • Absorption of Light

If light impinges on a material, it is either re-emitted in one form or another (reflection, transmission) or its energy is extinct, for example, transformed into heat. In any of these cases, some interaction between light and matter will take place, as was explained in the preceding section. One of the major mechanisms by which this interaction occurs is called *absorption* of light.

The classical description of absorption and reemission of light was developed at the turn of the 20th century by P. Drude, a German physicist. His concepts were described in Chapter 11.1 when we discussed electrical conduction in metals. As explained there, Drude postulated that some electrons in a metal (essentially the valence electrons) can be considered to be free, that is, they can be separated from their respective nuclei. He further assumed that the free electrons within the crystal can be accelerated by an external electric field. This preliminary Drude model was refined by considering that the moving electrons on their path collide with certain metal atoms in a nonideal lattice. If an alternating electric field (as through interaction with light) is envolved, then the free electrons are thought to perform oscillating motions. These vibrations are restrained by the above-mentioned interactions of the electrons with the atoms of a nonideal lattice. Thus, a *friction force* is introduced which takes this interaction into consideration. The calculation of the frequency dependence of the optical constants is accomplished by using the classical equations for vibrations whereby the interactions of electrons with atoms are taken into account by a damping term which is assumed to be proportional to the velocity of the electrons. The Newtonian-type equation (Force = mass times acceleration) is essentially identical to that of Eq. (11.9) except that the direct-current excitation force $e\mathscr{E}$ is now replaced by a periodic (i.e., sinusoidal) excitation force:

$$F = e\,\mathscr{E}_0 \sin(2\pi\nu t), \tag{13.24}$$

where ν is the frequency of the light, t is the time, and $\mathscr{E}_0$ is the maximal field strength of the light wave. In short, the equation describing the motion of free electrons which are excited to perform forced, periodic vibrations under the influence of light can be written as:

$$m\frac{d\text{v}}{dt} + \gamma\,\text{v} = e\,\mathscr{E}_0 \sin(2\pi\nu t), \tag{13.25}$$

where γ is the damping strength which takes the damping of the electron motion into account.

The solution of this equation, which shall not be attempted here, yields the frequency dependence (or *dispersion*) of the optical constants.

The free electron theory describes, to a certain degree, the dispersion of the optical constants quite well. This is schematically shown in Figure 13.3, in which the spectral dependence of the reflectivity is plotted for a specific case. The Hagen–Rubens relation (13.23) reproduces the experimental findings only up to about $\nu = 10^{13}\ \text{s}^{-1}$. In contrast to this, the Drude theory correctly reproduces the frequency dependence of R even in the visible spectrum. Proceeding to yet higher frequencies, however, the experimentally found reflectivity eventually rises and then decreases again. Such an **absorption band** cannot be explained by the free electron theory. For its interpretation, a different concept needs to be applied; see below. By making use of the Drude theory, one can obtain the number of free electrons per unit volume, N_f, from optical measurements:

$$N_f = \frac{(1 - n^2 + k^2)\ \nu^2 m\ 4\pi^2 \epsilon_0}{e^2} \tag{13.26}$$

where m is the mass of the electrons, e their charge, and ϵ_0 is the permittivity of empty space; see Appendix II and Section 11.8. The number of free electrons is a parameter which is of great interest because it is contained in several nonoptical equations (Hall effect, electromigration, superconductivity, etc.).

The Drude theory also provides the **plasma frequency** which is that frequency at which all electrons perform collective, fluid-like oscillations:

$$\nu_1 = \sqrt{\frac{e^2 N_f}{4\pi^2 \epsilon_0 m}}. \tag{13.27}$$

A selection of plasma frequencies is given in Table 13.3 and a value for ν_1 is shown in Figure 13.3.

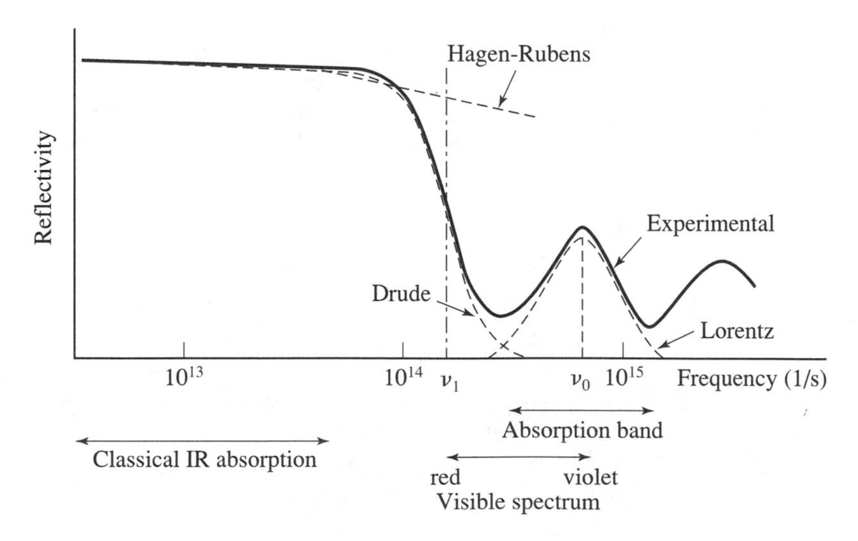

FIGURE 13.3. Schematic frequency dependence of the reflectivity of metals, experimentally (solid line) and calculated according to three models. The spectral dependence of the reflectivity is often quite similar to that of the absorption, ϵ_2. The plasma frequency is marked by ν_1.

TABLE 13.3. Experimentally found plasma frequencies for some materials

	Cs	K	Na	Au	Ag	Cu
$\nu_1 \times 10^{14}$ (1/s)	6.81	9.52	14.3	21.3	22.2	22.5

Lorentz postulated in contrast to Drude that the electrons should be considered to be bound to their nuclei and that an external electric field displaces the negatively charged electron cloud against the positively charged atomic nucleus as shown in Figure 11.26. In other words, he represented each atom as an electric dipole. If one shines light onto a solid, that is, if one applies an alternating electric field to the atoms, then the dipoles are thought to perform forced vibrations. Thus, a dipole is considered to behave similarly as a mass which is suspended on a spring (see Figure 13.4), i.e., the equations for a harmonic oscillator may be applied. The displacement of the positive and the negative charge is counteracted by a restoring force $\kappa \cdot x$ which is proportional to the displacement, x. Then the vibration equation becomes:

$$m \frac{d\mathrm{v}}{dt} + \gamma' \mathrm{v} + \kappa x = e\,\mathscr{E}_0 \sin(2\pi\nu t) \qquad (13.28)$$

The factor κ is the *spring constant* which determines the binding strength between the atom and an electron. Each vibrating dipole (such as an antenna) loses energy by radiation. Thus, $\gamma'\mathrm{v}$ represents the damping of the oscillator by radiation (γ' = damping constant).

An oscillator is known to absorb a maximal amount of energy when excited near its resonance frequency,

$$\nu_0 = \frac{1}{2\pi}\sqrt{\frac{\kappa}{m}}. \qquad (13.29)$$

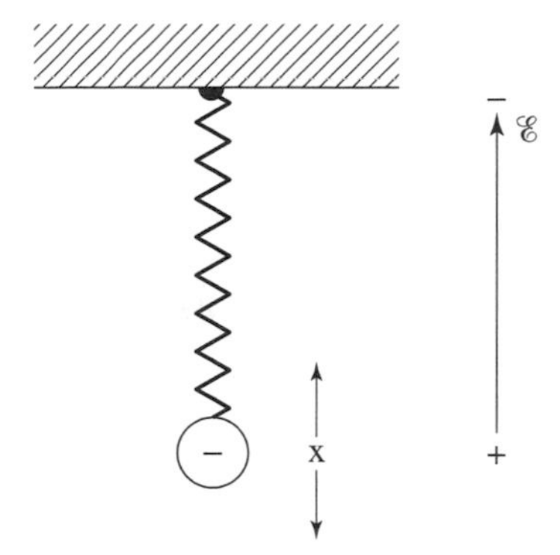

FIGURE 13.4. Quasi-elastic bound electron in an external electric field (harmonic oscillator).

(See Figure 13.3.) Very often, the electrons oscillate with more than one frequency ν_0. In this case, more than one absorption band is observed as depicted in Figure 13.3.

Thirty or forty years ago, many scientists considered the electrons in metals to behave at low frequencies as if they were free and at higher frequencies as if they were bound. In other words, electrons in a metal under the influence of light were described to behave as a series of classical free electrons and a series of classical harmonic oscillators. Insulators, on the other hand, were described by harmonic oscillators only.

Quantum mechanics provides a deeper understanding of the absorption of light. To explain this, one needs to make use of the electron band diagrams introduced in Section 11.1.

When light, having sufficiently large energy, impinges on a solid, the electrons in the crystal are thought to absorb the energy of the photons and in turn are excited into higher energy states, provided that unoccupied higher energy states are available. This is shown in Figure 13.5(a) for the specific case of a semiconductor or an insulator in which essentially a completely filled valence band and an empty conduction band are encountered. In order that electrons may be excited into a higher energy state, the light (photon) has to have an energy, $h\nu$, which is equal to or larger than the band gap. This process is called *interband transition* since an electron transfer from one band to a higher band is involved. The smallest possible energy at which photons can be absorbed to yield interband transitions is called the *threshold energy* for photon absorption. The threshold energy is understandably larger, the larger the band gap energy. Thus, the onset for interband absorption in insulators occurs at relatively high energies. For example, the gap energy for fused silica "glass" (SiO_2) is characteristically about 6.2 eV. Thus, silica does

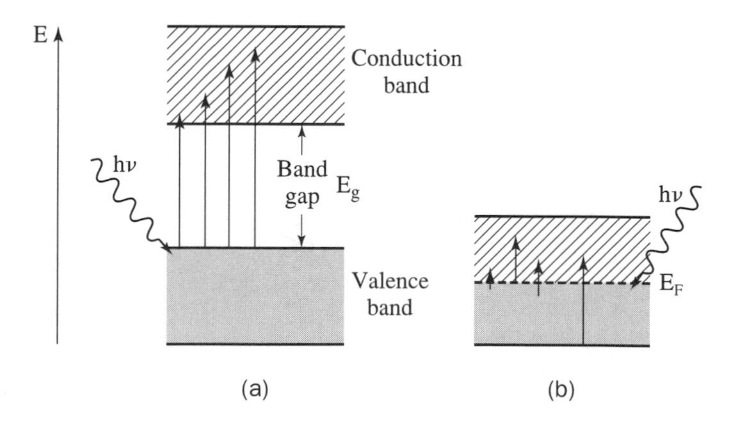

FIGURE 13.5. Schematic representation of the absorption of light and excitation of electrons into a higher energy state by (a) *interband* transitions, as encountered for insulators and semiconductors (as well as in metals, which is not shown), and (b) *intraband* transitions, as observed in metals only.

not absorb light in the visible region by the process of interband transitions (see Figure 13.1) and is therefore essentially transparent in this spectral range. However, silica eventually absorbs light in the ultraviolet, that is, at energies larger than 6.2 eV (or at wavelengths smaller than 200 nm).

Semiconductors, such as silicon, on the other hand, having gap energies in the neighborhood of 1 eV, start to absorb light already in the near IR region and are therefore opaque in the visible spectral range. They transmit, however, far-infrared wavelengths (equivalent to, say, 1 eV and smaller). *Interband* transitions are equivalent to the behavior of *bound* electrons in classical physics.

Metals, finally, have partially filled electron bands (Figure 11.6). Evidently, interband transitions into higher bands also take place in this case. Interestingly enough, a second absorption mechanism, involving transitions *within* the same band, additionally occurs. This process is called *intra*band transition; see Figure 13.5(b). No threshold energy for *intra*band transitions exists, that is, even very small photon energies can be absorbed. This causes a lift of electrons into higher energy states already at small energies. Specifically, an electron residing just below the Fermi energy can be excited by a photon to an energy state barely above E_F and larger energies. In other words, metals absorb light across the entire spectral region and are therefore opaque in the visible, as well as, in the IR or UV ranges. *Intra*band transitions are equivalent to the behavior of *free* electrons in classical physics.

It should be mentioned in passing that quantum mechanics provides a further refinement to the above-stated facts. During the absorption of light by *intraband* transitions, an additional mechanism may take place. It involves lattice vibration quanta, called *phonons*. The same is also true for many (but not all) *interband* transitions, particularly in semiconductors. Whenever this occurs, a phonon is said to have been exchanged with the lattice, which causes the solid to receive thermal energy. Electron transitions that involve phonons are called *indirect transitions*. For its understanding, a more complete band structure needs to be considered, which is beyond the scope of this book.

So far we have discussed only the absorption of photons by *electrons*. However, in the infrared region another absorption mechanism may become effective. It involves the light-induced **vibrations of lattice atoms**, that is, the excitation of **phonons** by photons. The individual atoms are thought to be excited by light of an appropriate frequency to perform oscillations about their point of rest. Now, the individual atoms are surely not vi-

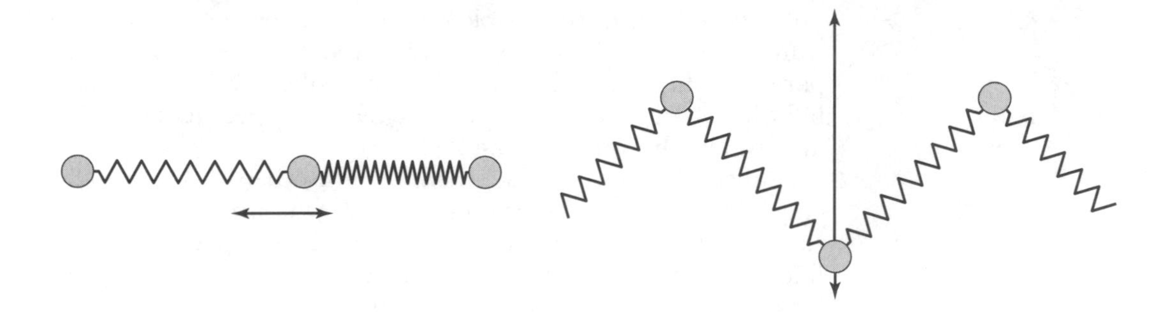

FIGURE 13.6. One-dimensional representations of possible vibration modes of atoms which have been excited by IR electromagnetic radiation (heat).

brating independently; instead they interact with their neighbors to a large degree. For simplicity, one usually models the atoms to be interconnected by elastic springs; Figure 13.6. Thus, the interaction of light with the lattice atoms can be mathematically represented in quite a similar manner as above when we calculated the classical electron theory for bound electrons. In other words, an equation similar to (13.28) can be used whereby κ now represents the binding strength between *atoms*. The damping of the oscillations is thought to be caused by interactions of the phonons with lattice imperfections, or with external surfaces of the crystal, or with other phonons. The oscillations possess a resonance frequency, ν_0, which depends on the mass of the atoms and on their degree of mutual interaction. The absorption of light is particularly strong at the resonance frequency. Thus, an absorption peak is observed at or near ν_0. Phonon absorption causes the lattice to heat up.

We consider as an example fused silica (quartz), which is essentially transparent for wavelengths between 0.2 μm (in the UV) and 4.28 μm (in the IR). Fused quartz has, however, two pronounced absorption peaks, one near 1.38 μm (similarly as seen in Figure 13.7) and the other near 2.8 μm. (Window glass has a comparable absorption spectrum with the exception that the UV cutoff wavelength is already near 0.38 μm.) In recently developed sol-gel silica "glasses," the absorption peaks near 1.38 and 2.8 μm are virtually suppressed, which causes this material to be transparent from 0.16 to 4 μm. The absorption spectrum for the commercially important borosilicate/phosphosilicate glass, used for optical fibers, is shown in Figure 13.7. The just-mentioned absorption peak near 1.38 μm is discernable.

Ordinary window glass, when used for telecommunication purposes, that is, when drawn into a fiber, has a transmission loss of 1000 dB/km [for a definition of dB, see Eq. (13.16)]. High-quality optical glass fibers have a loss of 100 dB/km. In 1970, Corning Glass developed a glass fiber that brought the losses

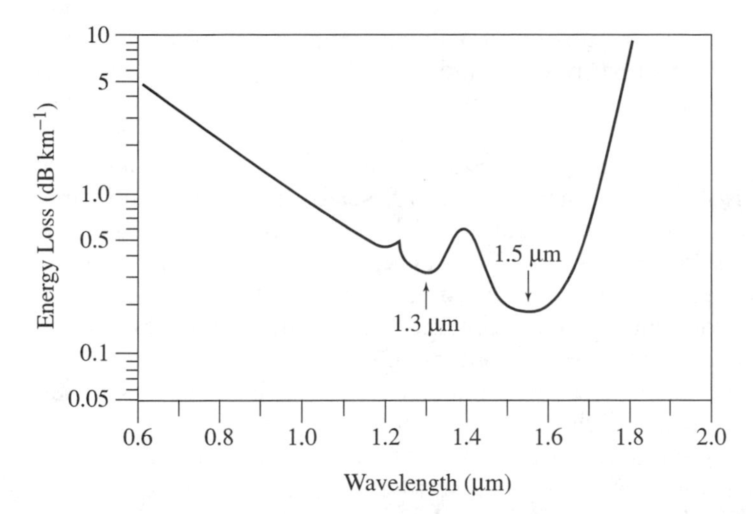

FIGURE 13.7. Energy loss spectrum of highly purified glass for fiber-optic applications which features a phosphosilicate core surrounded by a borosilicate cladding. Two communication channels near 1.3 μm and 1.5 μm are indicated.

down to 20 dB/km. Finally, in 1982, the loss was again reduced to 0.02 dB/km, near where it stands today.

In summary, the absorption of light (or generally of electromagnetic radiation) occurs by several mechanisms which cause either the *atoms* to oscillate about their equilibrium positions (far-IR spectral region) or cause *electrons* to be excited into higher energy states within the same band (*intraband* transitions) or between two or more bands (*interband* transitions). Additionally, other absorption processes have been observed (e.g., involving *excitons*, *solitons*, etc.) which shall not be covered here.

13.4 • Emission of Light

An electron, once excited, must eventually revert back into a lower, empty energy state. This occurs, as a rule, spontaneously within a fraction of a second and is accompanied by the emission of a photon and/or the dissipation of heat, that is, *phonons*. The emission of light due to reversion of electrons from a higher energy state is called **luminescence**. If the electron transition occurs within nanoseconds or faster, the process is called **fluorescence**. In some materials, the emission takes place after microseconds or milliseconds. This slower process is referred to as **phosphorescence**. A third process, called afterglow, which is even slower (seconds), occurs when excited electrons have been temporarily trapped, for example, in impurity states from which they eventually return after some time into the valence band. Commonly used phosphorescing materials consist, for example, of metal sulfides (such as ZnS), tungstates, oxides, and many organic substances.

Photoluminescence is observed when photons impinge on a material which in turn re-emits light of a lower energy. *Electroluminescing* materials emit light as a consequence of an applied voltage or electric field. *Thermoluminescence* is experienced when heating a substance, such as wax in a candle. *Cathodoluminescence*, finally, is the term which is used to describe light emission from a substance that has been showered by electrons of higher energy. All of these effects have commercial applications. For example, the insides of picture tubes of television sets are coated with a cathodoluminescing material, basically ZnS, which emits light when hit by electrons generated by a hot filament (cathode ray). In fluorescent lamps, the inside of a glass tube is covered with tungstates or silicates, which emit light as a consequence of bombardment with ultraviolet light that has been generated by a mercury glow discharge. The image generated in electron microscopes is made visible by a screen which consists of a *phosphor* such as ZnS. The same is true when X-rays or γ-rays need to be made visible.

One distinguishes between *spontaneous light emission*, which occurs in candles, incandescent light bulbs, etc., and *stimulated emission*, which is the mechanism by which **lasers** operate. Spontaneous emission possesses none of the characteristic properties of laser light: The radiation is emitted through a wide angular region in space, the light is phase-*in*coherent (see below) and it is often polychromatic (more than one wavelength). Spontaneous emission occurs particularly when materials are substantially heated (*thermal emission*). Heating excites the valence electrons to higher energies from which they spontaneously revert under release of photons. The larger the temperature, the higher the energy of the photons and the shorter their wavelength. For example, heating to about 700°C yields a dark red color whereas heating near 1600°C results in orange hues. At still higher temperatures, the emitted light appears to be white since large portions of the visible spectrum are emitted.

The situation changes considerably when stimulated emission is induced. Consider two energy levels E_1 and E_2, and let us assume for a moment that the higher energy level, E_2, contains more electrons than the lower level, E_1, that is, let us assume a *population inversion* of electrons [Figure 13.8(a)]. We further assume that, by some means (which we shall discuss in a moment), the electrons in E_2 are made to stay there for an appreciable amount of time. Nevertheless, one electron will eventually revert to the lower state. As a consequence, a photon with energy $E_{21} = h\nu_{21}$ is emitted [Figure 13.8(b)]. This photon might stimulate a second electron to descend in step to E_1, thus causing the emission of another photon that vibrates in phase with the first one.

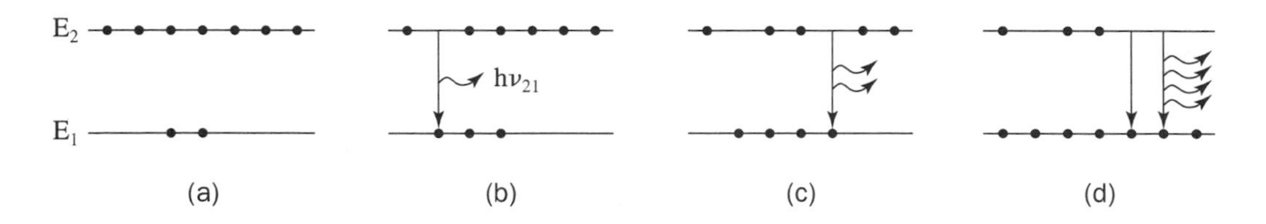

FIGURE 13.8. Schematic representation of stimulated emission between two energy levels E_2 and E_1. The dots symbolize electrons.

The two photons are consequently *phase-coherent* [Figure 13.8(c)]. They might stimulate two more electrons to descend in step [Figure 13.8(d)] and so on until an avalanche of photons is created. In short, stimulated emission of light occurs when electrons are forced by incident radiation to add more photons to an incident beam. The acronym LASER can now be understood; it stands for *light amplification by stimulated emission of radiation*.

Laser light is highly monochromatic (one color only) because it is generated by electron transitions between two narrow energy levels. (As a consequence, laser light can be focused with a lens to a spot less than 1 μm in diameter which cannot be done with "white" light because of dispersion.) Another outstanding feature of laser light is its strong collimation, i.e., the parallel emergence of light from a laser window. (The cross section of a laser beam transmitted to the moon is only 3 km in diameter!) We understand the reason for the collimation best by knowing the physical setup of a laser.

The lasing material is embodied in a long, narrow container called the *cavity*; the two faces at opposite ends of this cavity must be absolutely parallel to each other. One of the faces is silvered and acts as a perfect mirror, whereas the other face is silver-coated by a thin film only and thus transmits some of the light (Figure 13.9). The laser light is reflected back and forth by these mirrors, thus increasing the number of photons during each pass. After the laser has been started, the light is initially emitted in all possible directions (left part of Figure 13.9). However, only photons that travel strictly parallel to the cavity axis will remain in action, whereas the photons traveling at an angle will eventually be extinct in the cavity walls (center part in Figure 13.9). A fraction of the photons escape through the partially transparent mirror. They constitute the emitted beam.

We now need to explain how the electrons arrive at the higher energy level, that is, we need to discuss how they are *pumped* from E_1 into E_2. One of the methods is, of course, *optical pumping*, i.e., the absorption of light stemming from a polychromatic light source. (Xenon flashlamps for pulsed lasers, or tungsten-iodine lamps for continuously operating lasers, are often used for optical pumping. The lamp is either wrapped in helical form

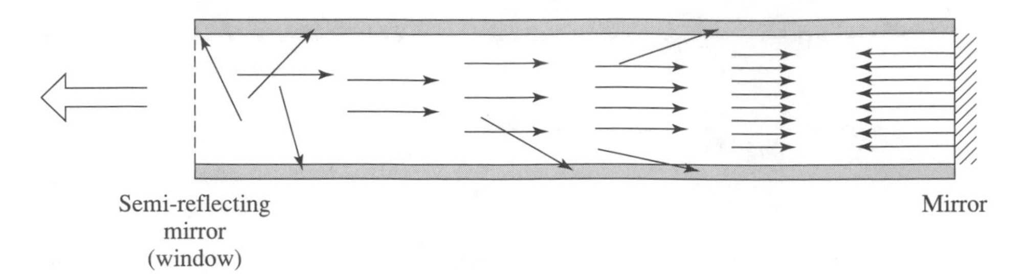

FIGURE 13.9. Schematic representation of a laser cavity (containing, for example, a mixture of helium and neon gases) and the buildup of laser oscillations. The stimulated emission eventually dominates over the spontaneous emission. The light leaves the cavity at the left side.

around the cavity, or the lamp is placed in one of the focal axes of a specularly reflecting elliptical cylinder, whereas the laser rod is placed along the second focal axis.) Other pumping methods involve electron/gas-ion collisions in an electric discharge, chemical reactions, nuclear reactions, or external electron beam injection.

The *pumping efficiency* is large if the bandwidth δE of the upper electron state is broad. In this way, an entire frequency range (rather than a single wavelength) leads to excited electrons [Figure 13.10(a)].

Next we discuss how population inversion can be achieved. For this we need to quote the Heisenberg uncertainty principle:

$$\delta E \cdot \delta t \propto h, \tag{13.30}$$

which states that the time span δt, for which an electron remains at the higher energy level E_2, is large when the bandwidth δE of E_2 is narrow. In other words, a sharp energy level (δE small, δt large) supports the population inversion; Figure 13.10(b). On the other hand, a large pumping efficiency requires a large δE [Figure 13.10(a)], which results in a small δt and a small population inversion. Thus, high pumping efficiency and large population inversion mutually exclude each other in a two-level configuration. In essence, a two-level configuration as depicted in Figure 13.10 does not yield laser action.

The three-level laser (Figure 13.11) provides improvement. There, the "pump band" E_3 is broad, which enables a good pumping efficiency. The electrons revert after about 10^{-14} s into an intermediate level, E_2, via a nonradiative, phonon-assisted process. Since E_2 is sharp, and not strongly coupled to the ground state, the electrons remain much longer, i.e., for some micro- or milliseconds on this level. This provides the required population inversion. An even larger population inversion is achieved by a four-level laser in which the energy level E_1 is rapidly emptied by electron transitions into a lower level E_0. Pumping occurs then between E_0 and E_3.

Laser materials can be selected from hundreds of substances to suit a specific purpose. Laser materials include *crystals* (such as ruby), *glasses* (such as neodymium-doped glass), *gases* (such

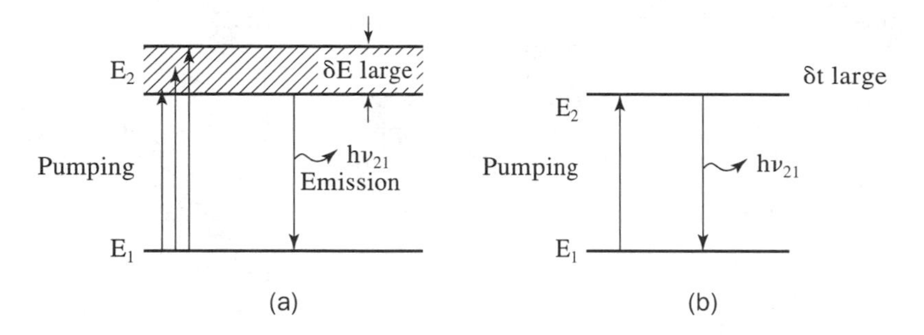

FIGURE 13.10. Examples of possible energy states in a two-level configuration. (a) δE large, i.e., large pumping efficiency but little or no population inversion. (b) Potentially large population inversion (δt large) but small pumping efficiency. (Note: Actual two-level lasers do not produce a population inversion because absorption and emission compensate each other.)

as helium, argon, xenon), *metal vapors* (such as cadmium, zinc, or mercury), *molecules* (such as carbon dioxide), or *liquids* (solvents which contain organic dye molecules). Table 13.4 lists the properties of some widely used lasers. Lasers can be operated in a *continuous mode (CW)* or with a higher power output, in the *pulsed mode*.

Semiconductor lasers are widely used for laser printers, CD players, distance measurements (land surveying), and particularly for optical telecommunication purposes. The "cavity" for a semiconductor laser consists of a combination of heavily doped (10^{18} cm^{-3}) n- and p-type semiconductors such as GaAs. The p–n junction is forward-biased (Figure 11.15). This causes electrons to be excited from the valence band into the conduction band if the applied voltage is equal to or larger than the gap energy. In this case, one observes a population inversion of electrons in the junction region between the p- and n-layers. Two opposite end faces of this *p–n* junction are made parallel and are polished or

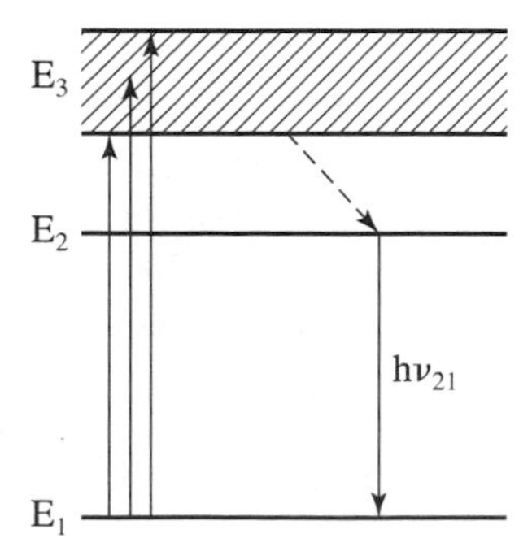

FIGURE 13.11. Three-level laser. The nonradiative, phonon-assisted decay is marked by a dashed line. Lasing occurs between levels E_2 and E_1. High pumping efficiency to E_3. High population inversion at E_2.

TABLE 13.4. Properties of some common laser materials

Type of laser	Wavelength(s) (nm)	Beam divergence (mrad)	Peak power output (W)	Comments
Ruby (Cr^{3+}-doped Al_2O_3)	694.3	10 5 0.5	CW:[a] ~5 pulsed (1–3 ms): 10^6–10^8 *Q*-switched[c] (10 ns): 10^9	Optically pumped three-level laser. Lasing occurs between Cr^{3+} levels. Low efficiency (0.1%). Historic device.
Neodymium (Nd^{3+}-doped glass or YAG)[b]	1,064	3–8	CW: 10^3 pulsed (0.1–1 μs): $\sim 10^4$	Optically pumped four-level laser. High efficiency (2%).
HeNe	632.8 (1150; 3390)	1	10^{-3}–10^{-2}	Most widely used.
CO_2	10,600; 9,600	2	CW: 10–1.5×10^4 pulsed (10^2–10^3 ns): 10^5	High efficiency (20%). Lasing occurs between vibrational levels
Semiconductor				Small size, direct conversion of electrical energy into optical energy. 10–55% efficiency. See Figure 13.12.
GaAs	~870	250	Homojunction, pulsed: (10^2 ns) 10–30	
GaAlAs	~850	500	Heterojunction, CW: 1–4×10^{-1}	
Dye (organic dyes in solvents)	350–1000	3 10	CW: $\sim 10^{-1}$ pulsed (6 ns) $\sim 10^5$	Lasing occurs between vibrational sublevels of molecules. Tunable.

[a] CW: Continuous wave.
[b] Yttrium aluminum garnet ($Y_3Al_2O_{15}$).
[c] Mirror in Fig. 13.9 is turned sideways during pumping. After some time it is returned to vertical position.

cleaved along crystal planes. The other faces are left untreated to suppress lasing in unwanted directions (Figure 13.12). A reflective coating of the window is usually not necessary since the reflectivity of the semiconductor is already 35% (Table 13.1). Semiconductor lasers are small and can be quite efficient. The wavelength of a binary GaAs laser is about 0.87 μm. This is, however, not the most advantageous wavelength for telecommunica-

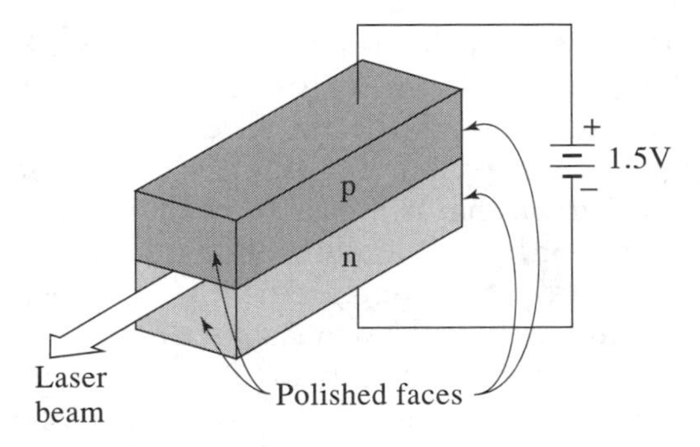

FIGURE 13.12. Schematic setup for a semiconductor laser (Homojunction laser).

tion purposes because some glasses attenuate light of this wavelength appreciably. In general, the optical absorption in glass is quite wavelength-dependent, as shown, for example, in Figure 13.7, where a minimum in absorption at 1.3 μm is seen. Fortunately, the bandgap energy, i.e., the wavelength at which a laser emits light, can be adjusted to a certain degree by utilizing ternary or quaternary compound semiconductors. Among them, $In_{1-x}Ga_xAs_yP_{1-y}$ plays a considerable role for telecommunication purposes, because the useful emission wavelengths of these compounds can be varied between 0.886 and 1.55 μm (which corresponds to gap energies from 1.4 to 0.8 eV). In other words, the above-mentioned desirable wavelength of 1.3 μm can be conveniently obtained by utilizing a properly designed indium–gallium–arsenide–phosphide laser.

Light-emitting diodes (*LEDs*) are of great technical importance as inexpensive, rugged, small, and efficient light sources for display purposes. The LED consists, like the semiconductor laser, of a forward-biased *p–n* junction. The above-mentioned special facing procedures are, however, omitted during the manufacturing process. Thus, the LED does not operate in the lasing mode. The emitted light is therefore neither phase-coherent nor collimated. It is, of course, desirable that the light emission occurs in the visible spectrum. Certain III/V compound semiconductors, such as $Ga_xAs_{1-x}P$, GaP, and $Ga_xAl_{1-x}As$, fulfill this requirement. The light is emitted in the red, green, yellow, or orange part of the spectrum. Blue color is achieved when using Ga-N-8%In, silicon carbide or zinc selenide. The radiation may leave the device through a window that has been etched through the metallic contact (*surface emitter*). For efficiencies of LEDs, see Section 18.6.

The emission of electromagnetic radiation of higher energies than that characteristic for UV light is called **X-rays**. (Still higher energetic radiation are γ-rays). X-rays were discovered in 1895 by Wilhelm Conrad Röntgen, a German scientist. In 1901, he received the first Nobel Prize in physics for this discovery. The wavelength of X-rays is in the order of 10^{-10} m; see Figure 13.1. For its production, a beam of electrons emitted from a hot filament is accelerated in a high electric field towards a metallic (or other) electrode. On impact, the energy of the electrons is lost either by *white X-radiation*, that is, in the form of a continuous spectrum (within limits), or by essentially monochromatic X-rays (called characteristic X-rays) that are specific for the target material. The white X-rays are emitted as a consequence of the deceleration of the electrons in the electric field of a series of atoms, whereas each interaction with an atom may lead to photons of different energies. The maximal energy that can be emitted this way (assuming only *one* interaction with an atom) is proportional to the acceleration voltage, V, and the charge of the electron, e, that is:

$$E_{\text{max}} = eV = h\nu = \frac{hc}{\lambda} \tag{13.31}$$

[see Eqs. (13.1) and (13.3)]. From this equation the minimum wavelength, λ (in nm), can be calculated using the values of the constants as listed in Appendix II and inserting V in volts, that is:

$$\lambda = \frac{1240}{V}. \tag{13.32}$$

Figure 13.13 depicts the voltage dependence of several white X-ray spectra. The cutoff wavelengths, as calculated by Eq. (13.32), are clearly detected. White X-radiation is mostly used for medical and industrial applications such as dentistry, bone fracture detection, chest X-rays, and so on. Different densities of the transmitted materials yield variations in the blackening of the exposed photographic film which has been placed behind the specimen.

The wavelength of *characteristic X-rays* depends on the mater-

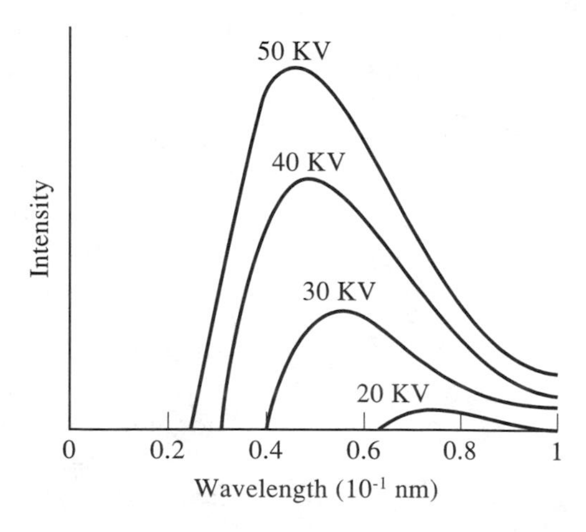

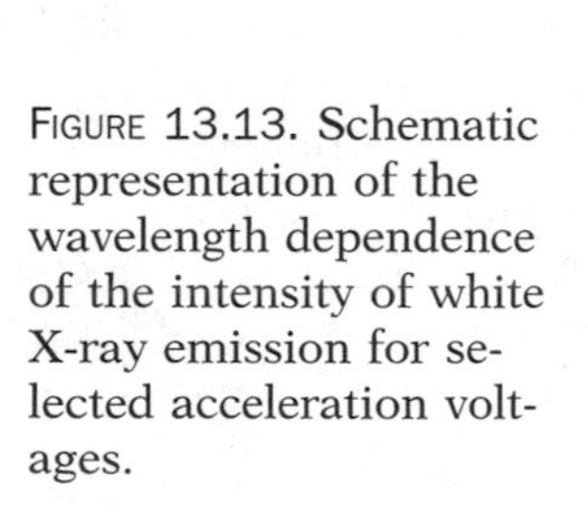

FIGURE 13.13. Schematic representation of the wavelength dependence of the intensity of white X-ray emission for selected acceleration voltages.

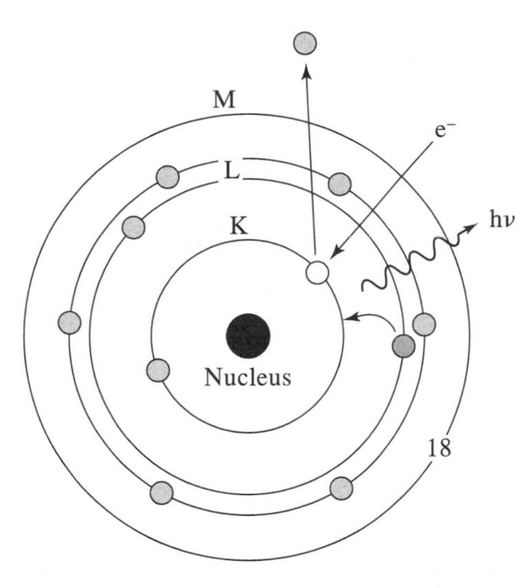

FIGURE 13.14. Schematic representation of the emission of characteristic X-radiation by exciting a *K*-electron and refilling the vacancy thus created by an *L*-electron.

ial on which the accelerated electrons impinge. Let us assume that the impinging electrons possess a high enough energy to excite inner electrons, for example, electrons from the *K*-shell, to leave the atom. As a consequence, an *L* electron may immediately revert into the thus created vacancy while emitting a photon having a narrow and characteristic wavelength. This mechanism is said to produce K_α *X-rays*; see Figure 13.14. Alternately and/or simultaneously, an electron from the *M* shell may revert to the *K* shell. This is termed to produce K_β-radiation.

For the case of copper, the respective wavelengths are 0.1542 nm and 0.1392 nm. (As a second example, aluminum yields K_α and K_β radiations having characteristic wavelengths of 0.8337 nm and 0.7981 nm.) Characteristic (monochromatic) X-radiation is frequently used in materials science, for example, for investigating the crystal structure of materials. For this, only one of the possible wavelengths is used by eliminating the others utilizing appropriate filters, made, for example, of nickel foils which strongly absorb the K_β-radiation of copper while the stronger K_α-radiation is only weakly absorbed. The characteristic X-radiation is superimposed on the often weaker, white X-ray spectrum.

13.5 • Optical Storage Devices

Optical techniques have been used for thousands of years to store information. Examples are ancient papyrus scrolls or stone carvings. The book you are presently reading likewise belongs in this category. It is of the *random access* type, because a particular page can be viewed immediately without first exposing all of the

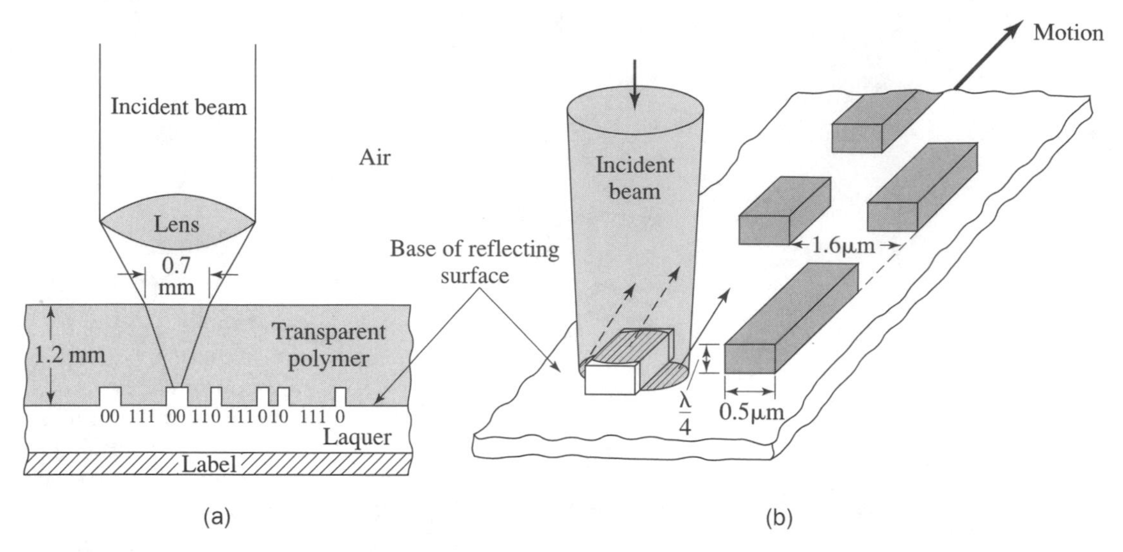

FIGURE 13.15. Schematic of a compact disk optical storage device. Readout mode. (Not drawn to scale.) The reflected beams in part (b) are drawn under an angle for clarity. The base and bump areas covered by the probing light have to be of equal size in order that destructive interference can occur [See the hatched areas covered by the incident beam in part (b)].

previous pages. Other examples of optical storage devices are the conventional photographic movie film or the microfilms used in libraries. The latter are *sequential* storage media because all of the previous material has to be scanned before the information of interest can be accessed. They are also called *read-only memories* (ROM) because the information content *cannot* be changed by the user. All examples given so far are *analog* storage devices.

Another form of storage utilizes the optical disk which has recently gained widespread popularity. Here, the information is generally stored in *digital* form. The most common application, the compact disk (CD), is a *random-access, read-only memory* device. However, *write once, read many* (WORM), rewritable CDs, and erasable magneto-optic disks are also available. The main advantage of optical techniques is that the readout involves a non-contact process (in contrast to magnetic tapes or mechanical systems). Thus, no wear is encountered.

Let us now discuss the optical compact disk. Here, the information is stored below a transparent, polymeric medium in the form of bumps, as shown in Figure 13.15. The height of these bumps is one-quarter of a wavelength ($\lambda/4$) of the probing light. Thus, the light which is reflected from the base of these bumps travels half a wavelength farther than the light reflected from the bumps. If a bump is encountered, the combined light reflected from the bump and the base is extinguished by destructive interference. No *light* may be interpreted as a *zero* in binary code, whereas *full intensity* of the reflected beam would then constitute a *one*. For audio purposes, the initial analog signal is sampled at a frequency of 44.1 kHz (about twice the audible frequency) to digitize the information into a series of ones and zeros (similarly as known from computers). Quantization of the signal into 16-digit binary numbers gives a scale of 2^{16} or 65,536 different values. This information is transferred to a disk (see below) in the form of bumps and absences of bumps. For readout from the disk, the probing light is pulsed with the same frequency so that it is synchronized with the digitized storage content.

The spiral path on the useful area of the 120-mm diameter CD is 5.7 km long and contains 22,188 tracks which are 1.6 μm apart. (As a comparison, 30 tracks can be accommodated on a human hair.) The spot diameter of the readout beam near the bumps is about 1.2 μm. The information density on a CD is 800 kbits/mm^2, i.e., a standard CD can hold about 7×10^9 kbits. This provides a playback time of about one hour. A disk of the same diameter can be digitally encoded with 600–800 megabytes of computer data, which is equivalent to three times the text of a standard 24-volume encyclopedia. The digital video disk (DVD) can store 4.7 gigabytes of information, which allows videos on CD-ROM.

Problems

13.1. The intensity of Na light passing through a gold film was measured to be about 15% of the incoming light. What is the thickness of the gold film? ($\lambda = 589$ nm; $k = 3.2$)

13.2. Calculate the reflectivity of silver at $\lambda = 0.6$ μm and compare it with the reflectivity of flint glass ($n = 1.59$).

13.3. Calculate the characteristic penetration depth in aluminum for Na light ($\lambda = 589$ nm, $k = 6$).

13.4. The transmissivity of a piece of glass of thickness $d = 1$ cm was measured at $\lambda = 589$ nm to be 89%. What would the transmissivity of this glass be if the thickness were reduced to 0.5 cm?

13.5. The plasma frequency, ν_1, can be calculated for the alkali metals by assuming *one* free electron per atom, i.e., by substituting N_f by the number of atoms per unit volume (atomic density N_a). Calculate ν_1 for potassium and lithium.

13.6. Calculate the number of free electrons per cubic centimeter and per atom for silver from its optical constants at 600 nm ($n = 0.05$; $k = 4.09$).

13.7. Show that the energy loss in an optical device, expressed in decibels per centimeter, is indeed equal to 4.3 α as stated below Eq. (13.16).

Suggestions for Further Study

M. Born and E. Wolf, *Principles of Optics*, 3rd Edition, Pergamon, Oxford (1965).

O.S. Heavens, *Optical Properties of Thin Solid Films*, Academic, New York (1955).

R.E. Hummel, *Optische Eigenschaften von Metallen und Legierungen*, Springer-Verlag, Berlin (1971).

R.E. Hummel, *Electronic Properties of Materials*, 3rd ed., Springer-Verlag, New York (2001).

T.S. Moss, *Optical Properties of Semiconductors*, Butterworth, London (1959).

A.H. Simmons and K.S. Potter, *Optical Materials*, Academic Press, San Diego (2000).

F. Wooten, *Optical Properties of Solids*, Academic, New York (1972).

14

Thermal Properties of Materials

14.1 • Fundamentals

The thermal properties of materials are important whenever heating and cooling devices are designed. Thermally induced expansion of materials has to be taken into account in the construction industry as well as in the design of precision instruments. Heat conduction plays a large role in thermal insulation, for example, in homes, industry, and spacecraft. Some materials such as copper and silver conduct heat very well; other materials, like wood or rubber, are poor heat conductors. Good electrical conductors are generally also good heat conductors. This was discovered in 1853 by Wiedemann and Franz, who found that the ratio between heat conductivity and electrical conductivity (divided by temperature) is essentially constant for all metals.

The **thermal conductivity** of materials varies only within five orders of magnitude (Figure 14.1). This is in sharp contrast to the variation in electrical conductivity, which spans about twenty-five orders of magnitude (Figure 11.1). The thermal conductivity of metals and alloys can be readily interpreted by making use of the electron theory, elements of which were explained in previous chapters of this book. The electron theory postulates that *free* electrons perform random motions with high velocity over a large number of atomic distances. In the hot part of a metal bar they pick up energy by interactions with the vibrating lattice atoms. This thermal energy is eventually transmitted to the cold end of the bar.

In electric insulators, in which no *free* electrons exist, the conduction of thermal energy must occur by a different mechanism. This new mechanism was found by Einstein at the beginning of

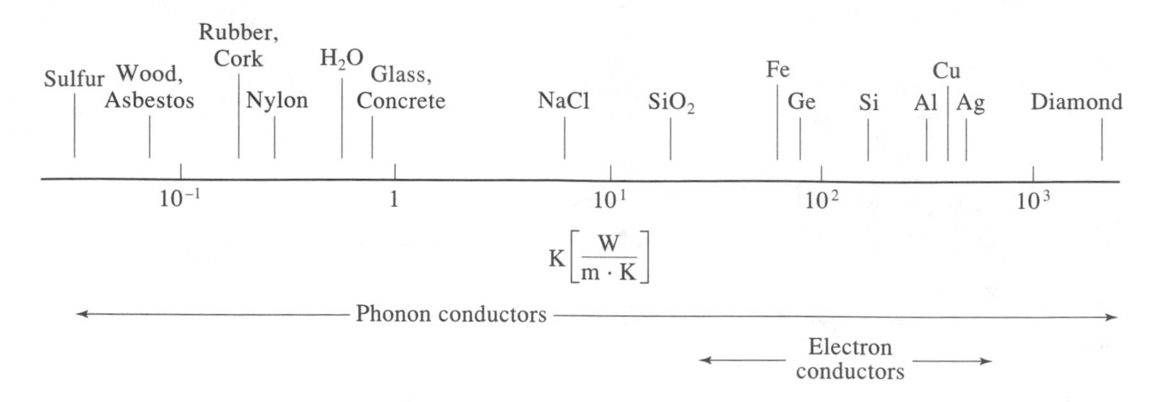

FIGURE 14.1. Room-temperature thermal conductivities for some materials. See also Table 14.3.

the 20th century. He postulated the existence of *phonons* or lattice vibration quanta, which are thought to be created in large numbers in the hot part of a solid and partially eliminated in the cold part. The transfer of heat in dielectric solids is thus linked to a flow of phonons from hot to cold. Figure 14.1 indicates that in a transition region both electrons as well as phonons may contribute, in various degrees, to thermal conduction. Actually, phonon-induced thermal conduction occurs even in metals, but its contribution is negligible compared to that of electrons.

Other thermal properties are the *specific heat capacity*, and a related property, the *molar heat capacity*. Their importance can best be appreciated by the following experimental observations: Two substances with the same mass but different values for the specific heat capacity require different amounts of thermal energy to reach the same temperature. Water, for example, which has a relatively high specific heat capacity, needs more thermal energy to reach a given temperature than, say, copper or lead of the same mass. Specifically, it takes 4.18 J[1] to raise 1 g of water by 1 K. But the same heat raises the temperature of 1 g of copper by about 11 K. In short, water has a larger heat capacity than copper. (The large heat capacity of water is, incidentally, the reason for the balanced climate in coastal regions and the heating of North European countries by the warm water of the Gulf Stream.) We need to define the various versions of heat capacities for clarification.

The **heat capacity, C′**, is the amount of heat, dQ, that needs to be transferred to a substance in order to raise its temperature by a certain temperature interval. The unit for the heat capacity is J/K.

---

[1]The unit of energy is the joule; see Appendix II. Obsolete units are the calorie (1 cal = 4.18 J), or the British thermal unit (BTU) which is the heat required to raise the temperature of one pound of water by one degree fahrenheit (1 BTU = 1055 J).

The heat capacity is not defined uniquely, that is, one needs to specify the conditions under which the heat is added to the system. Even though several choices for the heat capacities are possible, one is generally interested in only two: the *heat capacity at constant volume* C'_v and the *heat capacity at constant pressure* C'_p. The former is the most useful quantity because C'_v is obtained immediately from the energy, E, of the system. The heat capacity at constant volume is defined as:

$$C'_v = \left(\frac{\partial E}{\partial T}\right)_v. \tag{14.1}$$

On the other hand, it is much easier to measure the heat capacity of a solid at constant pressure than at constant volume. Fortunately, the difference between C'_p and C'_v for *solids* vanishes at low temperatures and is only about 5% at room temperature.

The **specific heat capacity** is the heat capacity *per unit mass*:

$$c = \frac{C'}{m} \tag{14.2}$$

where m is the mass of the system. Again, one can define it for constant volume or constant pressure. It is a material constant and is temperature-dependent. Characteristic values for c_p are given in Table 14.1. The unit of the specific heat capacity is J/g · K. We note from Table 14.1 that the c_p values for solids are considerably smaller than the specific heat capacity of water. Combining Eqs. (14.1) and (14.2) yields:

$$\Delta E = \Delta T m\, c_v, \tag{14.3}$$

which expresses that the thermal energy (or heat) which is trans-

TABLE 14.1. Experimental thermal parameters of various substances at room temperature and ambient pressure

Substance	Specific heat capacity, (c_p) $\left(\frac{J}{g \cdot K}\right)$	Molar (atomic) mass $\left(\frac{g}{mol}\right)$	Molar heat capacity (C_p) $\left(\frac{J}{mol \cdot K}\right)$	Molar heat capacity (C_v) $\left(\frac{J}{mol \cdot K}\right)$
Al	0.897	27.0	24.25	23.01
Fe	0.449	55.8	25.15	24.68
Ni	0.456	58.7	26.8	24.68
Cu	0.385	63.5	24.48	23.43
Pb	0.129	207.2	26.85	24.68
Ag	0.235	107.9	25.36	24.27
C (graphite)	0.904	12.0	10.9	9.20
Water	4.184	18.0	75.3	

ferred to a system equals the product of mass, increase in temperature, and specific heat capacity.

A further useful material constant is the heat capacity *per mole*. It compares materials that contain the same number of molecules or atoms. The **molar heat capacity** is obtained by multiplying the specific heat capacity c_v (or c_p) by the molar mass, M (see Table 14.1):

$$C_v = \frac{C'_v}{n} = c_v \cdot M, \tag{14.4}$$

where n is the amount of substance in mol.

We see from Table 14.1 that the room-temperature molar heat capacity at constant volume is approximately 25 J/mol · K for most solids. This was experimentally discovered in 1819 by Dulong and Petit. The experimental molar heat capacities for some materials are depicted in Figure 14.2 as a function of temperature. We notice that some materials, such as carbon, reach the Dulong–Petit value only at high temperatures. Some other materials such as lead reach 25 J/mol · K at relatively low temperatures.

All heat capacities are zero at $T = 0$ K. The C_v values near $T = 0$ K climb in proportion to T^3 and reach 96% of their final value at a temperature Θ_D, which is defined to be the Debye temperature. Θ_D is an approximate dividing point between a high-temperature region, where classical models can be used for the interpretation of C_v, and a low-temperature region, where quantum theory needs to be applied. Selected Debye temperatures are listed in Table 14.2.

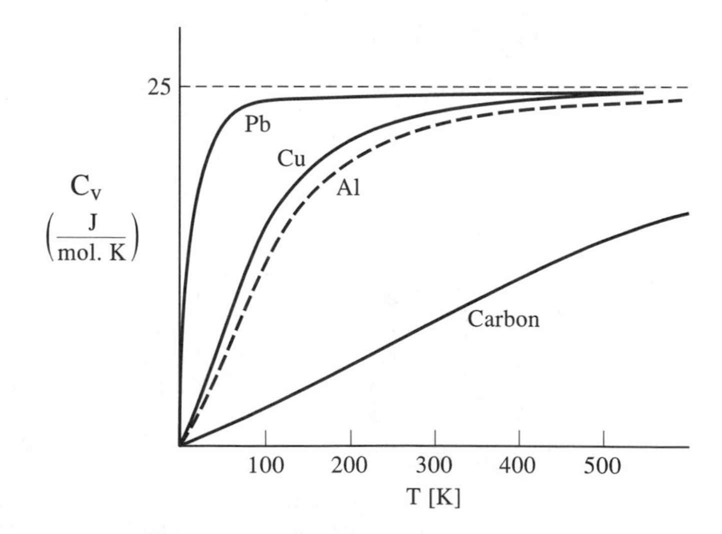

FIGURE 14.2. Temperature dependence of the molar heat capacity C_v for some materials.

TABLE 14.2. Debye temperatures of some materials

Substance	Θ_D (K)
Pb	95
Au	170
Ag	230
W	270
Cu	340
Fe	360
Al	375
Si	650
C	1850
GaAs	204
InP	162

14.2 • Interpretation of the Heat Capacity by Various Models

The classical (atomistic) theory for the interpretation of the heat capacity postulates that each atom in a crystal is bound to its site by a harmonic force similar to a spring. A given atom is thought to be capable of absorbing thermal energy, and in doing so it starts to vibrate about its point of rest. The amplitude of the oscillation is restricted by electrostatic repulsion forces of the nearest neighbors. The extent of this thermal vibration is therefore not more than 5 or 10% of the interatomic spacing, depending on the temperature. In short, we compare an atom with a sphere which is held at its site by two springs [Figure 14.3(a)].

The thermal energy that a harmonic oscillator of this kind can absorb is proportional to the absolute temperature of the environment. The proportionality factor has been found to be the Boltzmann constant k_B (see Appendix II). The average energy of the oscillator is then:

$$E = k_B T. \tag{14.5}$$

Now, solids are three-dimensional. Thus, a given atom in a cubic crystal also responds to the harmonic forces of lattice atoms in the other two directions. In other words, it is postulated that each atom in a cubic crystal represents three oscillators [Figure 14.3(b)], each of which absorbs the thermal energy $k_B T$. Therefore, the *average energy per atom* is:

$$E = 3k_B T. \tag{14.6}$$

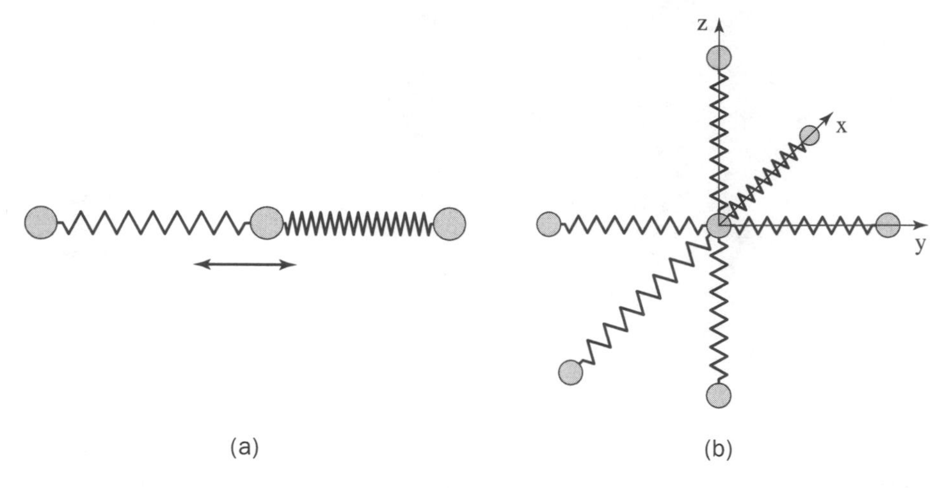

FIGURE 14.3. (a) A one-dimensional harmonic oscillator and (b) a three-dimensional harmonic oscillator.

We consider now nN_0 atoms, where N_0 is the Avogadro number; see Appendix II. Then the *total internal energy* of these atoms is:

$$E = 3nN_0k_BT. \tag{14.7}$$

Finally, the molar heat capacity is given by combining Eqs. (14.1), (14.4), and (14.7), which yields:

$$C_v = 3N_0k_B. \tag{14.8}$$

Inserting the numerical values for N_0 and k_B (see Appendix II) into Eq. (14.8) yields

$$C_v = 24.95 \text{ J/mol} \cdot K,$$

i.e., about 25 J/mol $\cdot$ K which is quite in agreement with the experimental findings at high temperatures (Figure 14.2).

It is satisfying to see that a simple model involving three harmonic oscillators per atom can readily explain the experimentally observed heat capacity. However, one shortcoming is immediately evident: the calculated molar heat capacity turned out to be temperature-independent, according to Eq. (14.8), and also independent of the material. This discrepancy with the observed behavior (Figure 14.2) was puzzling to scientists at the turn of the 20th century and had to await quantum theory to be explained properly. Einstein postulated in 1907 that the energies of the above-mentioned classical oscillators should be quantized, i.e., he postulated that only certain vibrational modes should be allowed, quite in analogy to the allowed energy states of electrons. These lattice vibration quanta were called *phonons*.

The term **phonon** stresses an analogy with *electrons* or *photons*. As we know from Chapter 13, photons are quanta of electromagnetic radiation, i.e., photons describe (in the appropriate frequency range) classical light. Phonons, on the other hand, are quanta of the ionic displacement field which (in the appropriate frequency range) represent classical sound.

The word *phonon* conveys the *particle* nature of an oscillator. Moreover, Einstein also postulated a particle-wave duality. This suggests *phonon waves* which propagate through the crystal with the speed of sound. Phonon waves are *not* electromagnetic waves; they are *elastic waves*, vibrating in a longitudinal and/or in a transversal mode.

The quantum theoretical treatment of the heat capacity, as developed by Einstein and improved by Debye is too involved for the present book. The result of the Einstein theory may be given here, nevertheless:

$$C_v = 3N_0 k_B \left(\frac{\hbar\omega}{k_B T}\right)^2 \frac{\exp\left(\frac{\hbar\omega}{k_B T}\right)}{\left(\exp\left(\frac{\hbar\omega}{k_B T}\right) - 1\right)^2}. \tag{14.9}$$

We discuss C_v for two special temperature regions. For large temperatures the approximation $e^x \simeq 1 + x$ can be applied, which yields $C_v \cong 3N_0 k_B$ (see Problem 14.6) in agreement with (14.8), i.e., we obtain the classical Dulong–Petit value. For $T \to 0$, C_v approaches zero, again in agreement with experimental observations. Thus, the temperature dependence of C_v is now in qualitative accord with the experimental findings. One minor discrepancy, however, has to be noted: At very small temperatures the experimental C_v decreases by T^3, as stated above. The Einstein theory predicts, instead, an exponential decrease. The Debye theory which we shall not discuss here alleviates this discrepancy by postulating that the individual oscillators interact with each other.

At very high and very low temperatures, the phonon theory does not yield a complete description of the observed behavior. The reason for this is that at these temperatures the **free electrons** (if present) provide a noticeable contribution to C_v. Again, quantum mechanical considerations as developed in Section 11.1 need to be applied. The electron contribution yields (for monovalent metals) the following expression:

$$C_v^{el} = \frac{\pi^2 N_0 k_B^2 T}{2 E_F}, \tag{14.10}$$

where E_F is the Fermi energy (see Section 11.1). We notice a linear relationship between heat capacity and temperature. Figure 14.4 summarizes the experimentally observed C_v-values as well

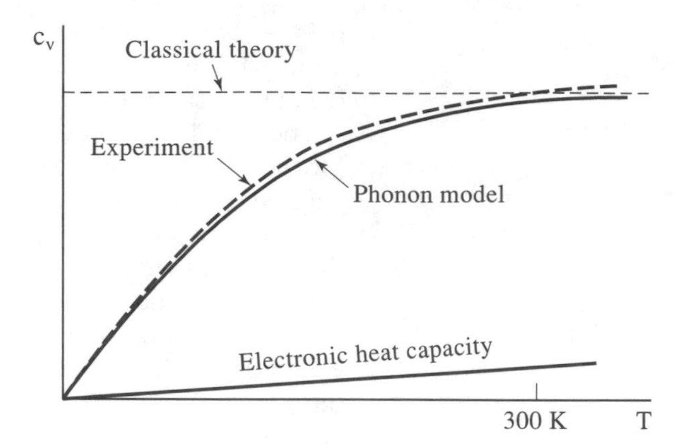

FIGURE 14.4. Schematic representation of the temperature dependence of the molar heat capacity, experimental and according to three models.

as the contributions of electron theory, phonon theory, and classical considerations as outlined above.

14.3 • Thermal Conduction

Heat conduction (or thermal conduction) is the transfer of thermal energy from a hot body to a cold body when both bodies are brought into contact. For best visualization we consider a bar of a material of length x whose ends are held at different temperatures (Figure 14.5). The heat that flows through a cross section of the bar divided by time and area (i.e., the *heat flux*, J_Q) is pro-

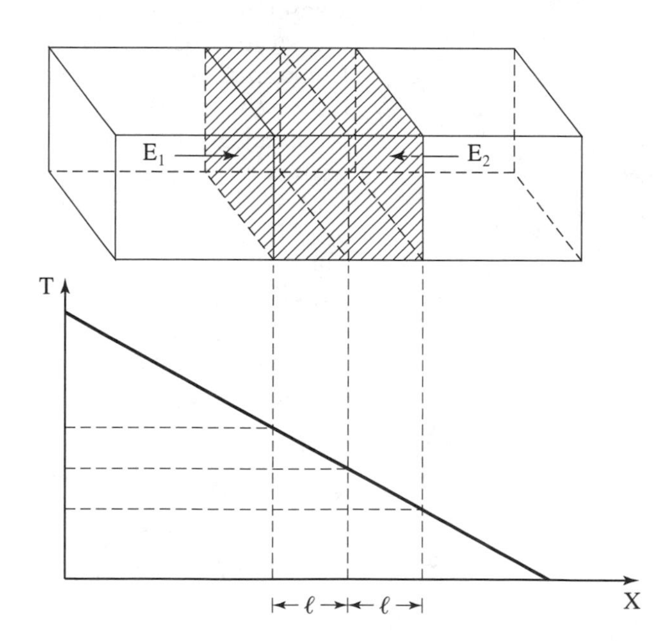

FIGURE 14.5. Schematic representation of a bar of a material whose ends are at different temperatures.

portional to the temperature gradient dT/dx. The proportionality constant is called the *thermal conductivity* K (or λ). We thus write:

$$J_Q = -K \frac{dT}{dx}. \tag{14.11}$$

The negative sign indicates that the heat flows from the hot to the cold end (*Fourier Law*, 1822). Possible units for the heat conductivity are J/(m·s·K) or W/(m · K). The heat flux J_Q is measured in J/(m^2 · s). Table 14.3 gives some characteristic values for K. The thermal conductivity decreases slightly with increasing temperature. For example, K for copper or Al_2O_3 decreases by about 20% within a temperature span of 1000°C. In the same temperature region, K for iron decreases by 10%.

For the interpretation of thermal conduction we postulate that the heat transfer in solids may be provided by *free electrons* as

TABLE 14.3. Thermal conductivities at room temperature[a]

Substance	$K\left(\frac{W}{m \cdot K}\right) \equiv \left(\frac{J}{m \cdot s \cdot K}\right)$
Diamond Type IIa	2.3×10^3
SiC	4.9×10^2
Silver	4.29×10^2
Copper	4.01×10^2
Aluminum	2.37×10^2
Silicon	1.48×10^2
Brass (leaded)	1.2×10^2
Iron	8.02×10^1
GaAs	5×10^1
Ni-Silver[b]	2.3×10^1
Al_2O_3 (sintered)	3.5×10^1
SiO_2 (fused silica)	1.4
Concrete	9.3×10^{-1}
Soda-lime glass	9.5×10^{-1}
Water	6.3×10^{-1}
Polyethylene	3.8×10^{-1}
Teflon	2.25×10^{-1}
Snow (0°C)	1.6×10^{-1}
Wood (oak)	1.6×10^{-1}
Sulfur	2.0×10^{-2}
Cork	3×10^{-2}
Glass wool	5×10^{-3}
Air	2.3×10^{-4}

[a] See also Figure 14.1. Source: *Handbook of Chemistry and Physics*, CRC Press, Boca Raton, FL (1994).
[b] 62% Cu, 15% Ni, 22% Zn.

well as by *phonons*. We understand immediately that in insulators, which do not contain any free electrons, the heat must be conducted exclusively by phonons. In metals and alloys, on the other hand, the heat conduction is dominated by electrons because of the large number of free electrons which they contain. Thus, the phonon contribution is usually neglected in this case.

One particular point should be clarified. Electrons in metals travel in equal numbers from hot to cold and from cold to hot in order that the charge neutrality is maintained. This is indicated in Figure 14.5 by two arrows marked E_1 and E_2. Now, the electrons in the hot part of a metal possess and transfer a high energy. In contrast to this, the electrons in the cold end possess and transfer a lower energy. The heat transferred from hot to cold is thus proportional to the difference in the energies of the electrons.

The situation is quite different in phonon conductors. We know from Section 14.1 that the number of phonons is larger at the hot end than at the cold end. Thermal equilibrium thus involves, in this case, a net transfer of phonons from the hot into the cold part of a material.

Returning to the portion of thermal conduction caused by electrons, we consider a volume at the center of the bar depicted in Figure 14.5 whose faces have the size of a unit area and whose length is $2l$, where l is the mean free path between two consecutive collisions between an electron and lattice atoms. A simple energy balance, taking in account the electrons of energy E_1 that travel from left to right and electrons having a lower energy, E_2, drifting from right to left, yields for the **classical equation** for the **heat conductivity of metals and alloys**:

$$K = \frac{N_{\mathrm{v}} \mathrm{v}\, k_B l}{2}. \tag{14.12}$$

Equation (14.12) thus reveals that the heat conductivity is larger the more electrons, N_{v}, per unit volume are involved, the larger their velocity, v, and the larger the mean free path, l, between two consecutive electron–atom collisions. The connection between thermal conductivity and the *heat capacity per volume* is:

$$K = \tfrac{1}{3}\, C_{\mathrm{v}}^{el}\, \mathrm{v}\, l. \tag{14.13}$$

All three variables contained in Eq. (14.13) are temperature-dependent, but while C_{v}^{el} increases with temperature (Figure 14.4), l and, to a small degree, also v are decreasing. Thus, K should change very little with temperature, which is indeed experimentally observed. As mentioned above, the thermal conductivity decreases about 10^{-5} W/(m · K) per degree. K also changes at the melting point and when a change in atomic packing occurs.

The question arises as to what velocity the electrons (that participate in the heat conduction process) have? Further, do all of the electrons participate in the heat conduction? We have raised similar questions in Section 11.1. We know from there that only those electrons which have an energy close to the Fermi energy, E_F, are able to participate in the conduction process. Thus, the velocity in Eqs. (14.12) and (14.13) is essentially the Fermi velocity, v_F, which can be calculated with:

$$E_F = \frac{1}{2} m v_F^2 \tag{14.14}$$

if the Fermi energy is known (see Appendix II).

Second, the number of participating electrons contained in Eq. (14.12) is proportional to the population density at the Fermi energy, $N(E_F)$, that is, in first approximation, by the number of free electrons N_f per volume. Combining (14.10) (per volume) with (14.13) yields the **quantum mechanical** expression for the **heat conductivity**:

$$K = \frac{\pi^2 N_f k_B^2 T v_F l_F}{6E_F} \tag{14.15}$$

where l_F is now the mean free path of the electrons near the Fermi energy E_F. Both the classical equation (14.12) and the quantum mechanical relation (14.15) contain similar variables and constants, whereas quantum mechanics deepens our understanding.

Heat conduction in dielectric materials occurs as already explained by a flow of phonons. The hot end possesses more phonons than the cold end, causing a drift of phonons down a concentration gradient.

The thermal conductivity can be calculated similarly as above, which leads to the same equation as (14.13):

$$K^{ph} = \frac{1}{3} C_v^{ph} v^{ph} l^{ph}. \tag{14.16}$$

In the present case C_v^{ph} is the (lattice) heat capacity per volume of the phonons, v^{ph} is the phonon velocity, and l^{ph} is the phonon mean free path. A typical value for v^{ph} is about 5×10^5 cm/s (sound velocity) with v^{ph} being relatively temperature-independent. In contrast, the mean free path varies over several orders of magnitude, that is, from about 10 nm at room temperature to 10^4 nm near 20 K. The drifting phonons interact on their path with lattice imperfections, with external boundaries, and with other phonons. These interactions constitute a thermal resistivity which is quite analogous to the electrical resistivity. Thus, we may treat the thermal resistance just as we did in Chapter 11, i.e., in terms of interactions between particles (here phonons) and matter, or in terms of the scattering of phonon waves on lattice imperfections.

14.4 • Thermal Expansion

The length L of a rod increases with increasing temperature. Experiments have shown that in a relatively wide temperature range, the linear expansion ΔL is proportional to the increase in temperature ΔT. The proportionality constant is called the *coefficient of linear expansion* α_L. The observations can be summarized in:

$$\frac{\Delta L}{L} = \alpha_L \, \Delta T. \tag{14.17}$$

Experimentally observed values for α_L are given in Table 14.4.

The expansion coefficient has been found to be proportional to the molar heat capacity C_V, i.e., the temperature dependence of α_L is similar to the temperature dependence of C_V. As a consequence, the temperature dependence of α_L for *dielectric materials* follows closely the $C_V = f(T)$ relationship shown in Figure 14.4. Specifically, α_L approaches a constant value for $T > \Theta_D$ and vanishes as T^3 for $T \to 0$. The thermal expansion coefficient for *metals*, on the other hand, decreases at very small temperatures in proportion to T, and depends, in other temperature regions,

TABLE 14.4. Linear expansion coefficients α_L for some solids measured at room temperature. (See also Table 15.1 for glasses.)

Substance	$\alpha_L \times 10^{-5}$ [K^{-1}]
Hard rubber	8.00
Lead	2.73
Aluminum	2.39
Brass	1.80
Copper	1.67
Iron	1.23
Glass (ordinary)	0.90
Glass (pyrex)	0.32
NaCl	0.16
Invar (Fe–36% Ni)	0.07
Quartz	0.05

on the sum of the heat capacities of phonons and electrons (Figure 14.4).

We turn now to a discussion of a possible mechanism that may explain thermal expansion from an atomistic point of view. We postulate, as in the previous sections, that the lattice atoms absorb thermal energy by vibrating about their equilibrium position. In doing so, a given atom responds with increasing temperature and vibrational amplitude to the repulsive forces of the neighboring atoms. Let us consider for a moment two adjacent atoms only, and let us inspect their potential energy as a function of internuclear separation (Figure 14.6). We understand that as two atoms move closer to each other, strong repulsive forces are experienced between them. As a consequence, the potential energy ($U(r)$) curve rises steeply with decreasing r. On the other hand, we know that two somewhat separated atoms also attract each other in some degree. This results in a slight decrease in $U(r)$ with decreasing r.

Now, at low temperatures, a given atom may rest in its equilibrium position r_0, i.e., at the minimum of potential energy. If, however, the temperature is raised, the amplitude of the vibrating atom also increases. Since the amplitudes of the vibrating atom are symmetric about a median position, and since the potential curve is not symmetric, a given atom moves farther apart from its neighbor, that is, the average position of an atom moves to a larger r, say r_T, as shown in Figure 14.6. In other words, the thermal expansion is a direct consequence of the asymmetry of the potential energy curve. The same arguments hold true if all of the atoms in a solid are considered.

A few substances are known to behave differently from that described above. They contract during a temperature increase.

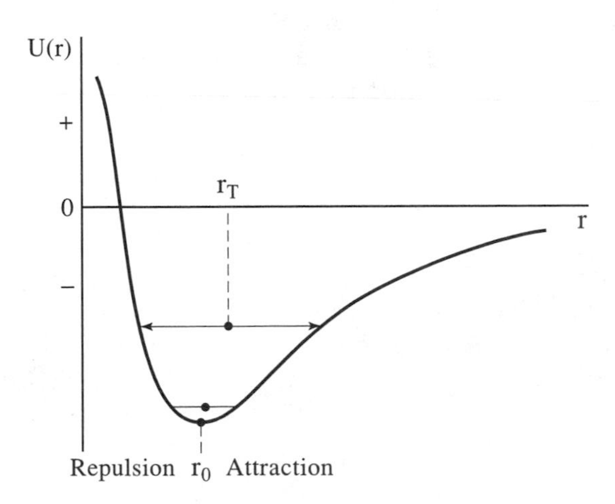

FIGURE 14.6. Schematic representation of the potential energy, $U(r)$, for two adjacent atoms as a function of internuclear separation, r.

This happens, however, only within a narrow temperature region. For its explanation, we need to realize that longitudinal as well as transverse vibrational modes may be excited by thermal energy. The lattice is expected to contract if transverse modes predominate. Interestingly enough, only one known liquid substance, namely water, behaves in a limited temperature range in this manner. Specifically, water has its largest density at 4°C. (Furthermore, the density of ice is smaller than the density of water at the freezing point.) As a consequence, water of 4°C sinks to the bottom of a lake during winter, while ice stays on top. This prevents the freezing of a lake at the bottom and thus enables aquatic life to survive during the winter. This exceptional behavior of water suggests that the laws of physics do not just "happen" but rather, they were created by a superior being.

Problems

14.1. Calculate the average difference in % between C_p and C_v for the metals listed in Table 14.1. Compare this result with what is said in the text.

14.2. A block of copper whose mass is 100 g is quenched directly from an annealing furnace into a 200-g glass container that holds 500 g of water. What is the temperature of the furnace if the water temperature rises from 0° to 15°C? ($c_{glass} = 0.5$ J/g · K)

14.3. Calculate the rate of heat loss in a 5-mm-thick window glass when the exterior temperature is 0°C and the room temperature is 20°C. Compare your result with the heat loss in an aluminum and a wood frame of 10-mm thickness. How can you decrease the heat loss through the window?

14.4. Calculate the mean free path of electrons in a metal such as silver at room temperature from heat capacity and heat conduction measurements. Take $E_F = 5$ eV, $K = 4.29 \times 10^2$ J/s·m·K and $C_v^{el} = 1\%$ of the lattice heat capacity. (*Hint:* Remember that heat capacity is per volume!)

14.5. Calculate the gap that has to be left between 10-m-long railroad tracks when they are installed at 0°C and if no compression is allowed at 40°C.

14.6. Confirm that Equation (14.9) reduces for large temperatures to the Dulong–Petit value.

Suggestions for Further Reading

R.E. Hummel, *Electronic Properties of Materials*, 3rd ed., Springer-Verlag, New York (2001).

C. Kittel and H. Kroemer, *Thermal Physics*, 2nd Edition, W.H. Freeman, San Francisco, (1980).

J.M. Ziman, *Electrons and Phonons*, Oxford University Press, Oxford (1960).

PART III
MATERIALS AND THE WORLD

15

No Ceramics Age?

15.1 • Ceramics and Civilization

The utilization of ceramic materials by man is probably as old as human civilization itself. Stone, obsidian, clay, quartz, and mineral ores are as much a part of the history of mankind as the products which have been made of them. Among these products are tools, earthenware, stoneware, porcelain, as well as bricks, refractories, body paints, insulators, abrasives, and eventually modern "high-tech ceramics" used, for example, in electronic equipment or jet engines. Actually, fired or baked ceramic objects are probably the oldest existing samples of handicraft which have come to us from ancient times. They are often the only archaeologic clues that witness former civilizations and habitats. Moreover, there are scholars who believe that life took its origin from ceramics. And some ancient mythologies relate that man was created from clay. It might be of interest in this context that the Hebrew word for soil, dirt, clay, or earth is "Adamá." Taking all of these components into consideration, it might be well justified to ask why historians did not specifically designate a ceramics age. The answer is quite simple: stone, copper, bronze, and iron can be associated with reasonably well-defined time periods that have a beginning and frequently also an end during which these materials were predominantly utilized for the creation of tools, weapons, and objects of art. In contrast to this, ceramic materials have been used continuously by man with essentially unbroken vigor commencing from many millennia ago until the present time. Thus, ceramics can be compared to a basso obstinato in a piece of music in which other instruments play the melody. Still, some identification of cultural stages through pottery is common among historians and archaeologists, who distinguish a "pre-pottery" era from a pottery period. The latter one is classed by color, shape, hardness, and, notably, by decoration.

Beginning from early times, **clay** was of particular interest to man for a number of reasons. First, clay is abundantly found in many parts of the world, albeit in different compositions, quantities, and qualities. Second, clay is pliable if it has the right consistency (e.g., water content). Third, the shape of an object made from clay is retained as long as it is not exposed to water for an extended period of time. Fourth, clay becomes hard when dried, for example, in the sun or near fires. A fifth and most important property of clay was eventually discovered in ancient times: clay permanently hardens to a virtually indestructible but brittle and porous material when heated above about 500°C. At this temperature, an irreversible chemical reaction begins to take place which precludes the substance from returning to its original ductile state and makes it water-resistant. Moreover, the fired product is much less susceptible to environmental interactions than many metals and alloys such as iron. In primitive pottery-making the objects were placed in a shallow pit in the ground and the fire (using wood or dung) was built over them.

Some early artifacts made of fired clay are at least 9,000 years old and consist mostly of pottery or building materials, such as bricks. Moreover, some figurines have been found (such as in French caves) portraying animals or human bodies. The oldest samples of baked clay include more than 10,000 fragments of statuettes which were found in 1920 near Dolní Věstonice, Moravia, in the Czech Republic. They portray wolves, horses, foxes, birds, cats, bears, or women with exaggerated female attributes (Figure 15.1). One of these prehistoric female figurines remained almost undamaged. It has been named the "Venus of Věstonice" and is believed to have been a fertility charm. Scholars date the statuette, which stands about 10 cm tall, as far back as 23,000 B.C. Much speculation has evolved about these fragments, which, incidentally, contain some mammoth bone ash in the clay. Specifically, one group of anthropologists proposes that the figurines served some divinational purpose and were designed to shatter in the fire (by wetting the clay) so that priests or shamans could foretell future events from the fragments. In any event, this practice apparently did not lead to the invention of pottery.

A still earlier example for prehistoric pyrotechnology, dating back to about 50,000 B.C., consists of ground-up iron oxide powders which have been fired to yield various colors. Pigments of this kind, together with a lead oxide binder, have probably been used for millennia, for body painting and other decorations.

Unfired clay as building material was likewise rather common. In some cases, the mud was inserted between and around wooden

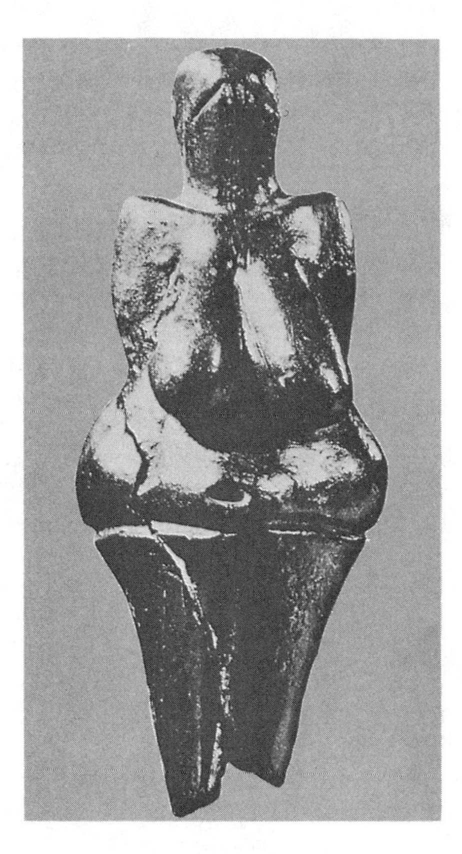

FIGURE 15.1. Baked clay figurine called the "Venus of Věstonice" found in 1920 in the Czech Republic. Approximate age: 23,000 B.C.

structures. In other instances, pressed clay bricks formed in baskets were used (called *adobe*) which were sometimes fortified with straw. (This constitutes one of the first examples of the production of a *composite material*, that is, a technique to strengthen clay by fibers.) Additionally, asphalt from natural oil wells was sometimes used as mortar. However, buildings of this type were highly vulnerable to the weather. They easily crumbled and thus needed constant renewal (except for cliff-dwellings; see Plate 15.11). As a consequence, layers of settlements were often built on top of previous ones, creating a mound of occupation debris (called a *tell* or a *tall*). This is characteristic for ruins found in Mesopotamia. Mass production of bricks was not performed in this region until the sixth millennium B.C., probably because of a shortage of fuel. Later, however, that is, in the sixth century B.C., the buildings of Babylon utilized fired and glazed bricks (see Plate 15.5).

As outlined above, pottery-making is one of the oldest forms of crafts. The resulting vessels are quite useful for storing property and dry food, for cooking, and for transporting water. Still, pottery-making is seldom found among nomadic tribes in particular as long as natural materials, such as gourds, skins, and large leaves, can be found or when baskets can be woven for the above-mentioned purposes. Moreover, and most importantly, potters must live within the reach of their raw materials and their kilns. Finally, ceramic pots are heavy and might break during travel. It is therefore not surprising that the art of pottery did not commence or was not practiced at all locations of the world at the same time and with equal sophistication, even though the raw materials were certainly available. Instead, pottery-making has been mainly exercised in areas in which agriculture has been firmly established, that is, where the population was reasonably settled. It is interesting to know in this context that pottery appears in the Americas approximately 5,000 years later than in the "old world" and that glazes, as well as the potter's wheel, and frequently also the kiln, were not known there in pre-Columbian times.

The discovery that fired clay objects are water-resistant and sturdy eventually led to a systematic development of kilns with permanent walls and open tops. Certain modifications on the outlay of the kilns and the type of fuel used eventually allowed increasingly higher temperatures. This has been discussed already in Chapter 1 in the context of the interrelationship between pottery-making and copper-smelting.

15.2 • Types of Pottery

Pottery is broadly divided into *vitrified ware* and *unvitrified ware*, a distinction that is based on whether or not clay composition and firing temperature cause the clay to melt or fuse into a glassy (vitreous) substance. **Earthenware** is made from "earthenware clay" fired at relatively low temperatures, that is, between 800 and 1200°C, depending on the raw material; see Plate 15.1. It is porous when not subsequently glazed (see below) and is relatively coarse and often red or buff-colored, even black after firing. Bricks and other construction materials, such as tiles, as well as terra cotta vessels are the major products in this category. Earthenware was probably the earliest kind of ceramics that was made, dating back to about 7,000 or possibly 8,000 B.C. Specimens of this age were found, for example, in Catal Hüyük in Anatolia (today's Turkey).

Because of its porosity, the water which is stored in vessels made of unglazed earthenware percolates eventually through the walls and evaporates on the free surface, thus cooling the water in the container. For the same reason, terra cotta containers cannot be used for storing milk or wine. Examples of more recent unglazed earthenware include Chinese teapots, first made during the Ming dynasty, as well as red stoneware made in Meißen (Germany) or by Wedgwood in England at the beginning of the eighteenth century A.D.

In order to hermetically seal the pores of goods made of earthenware, an additional processing step called **glazing** was introduced around or probably even before 3000 B.C. by the Egyptians. It involved the coating of the fired objects with a water suspension of finely ground quartz sand mixed with sodium salts (carbonate, bicarbonate, sulfate, chloride) or plant ash. This was followed by a second firing, during which the glassy particles fused into an amorphous layer. The second firing is often at a lower temperature, being just sufficient to fuse the glaze. Incidentally, the Egyptians also glazed beads and bowls made of steatite, a soft stone which could be easily shaped, drilled, and abraded. Another technique to produce beads involved hand-molding a mixture of crushed quartz with sodium salts and malachite (Plate 1.3), which was then fired.

Two other types of glazes which have been applied to earthenware are likewise several millennia old. One of them is a transparent *lead glaze*. Lead reduces the melting or fusion point of the glaze mixture, which allows the temperature of the second firing to be even lower. Lead glaze was invented in China during the Han dynasty (206 B.C.–A.D. 200) and was subsequently widely used by many civilizations. However, some lead from the glaze on tableware may be leached by the food. It is believed that this poisoned a large number of Roman nobility and thus contributed (together with the lead from water pipes) to the fall of the Roman empire. Lead glazing for tableware is outlawed today by many countries unless *fritted glazes* are utilized which convert lead into a nontoxic form.

An alternative technique involves an opaque, white *tin glaze* which hides possible color blemishes, for example, from iron impurities. Tin glazing was probably first discovered by the Assyrians who lived in Mesopotamia (today's Northern Iraq) during the second millennium B.C. It was utilized for decorating bricks but fell eventually into disuse possibly because of the sudden interruption of the tin supply, as explained already in Chapter 7. In the ninth century A.D., tin glazing was reinvented and again extensively utilized in Mesopotamia. From there it spread to Italy via the Spanish island of *Maiolica* (Majorca), after which the product was later named (Plate 15.7). French earthenware,

Faïence, and Dutch earthenware, *Delft*, are likewise tin-glazed ceramic products whose names are derived from the Italian city of Faenza and the city of Delft in Holland. Faïence is generally distinguished by its elaborate decorations which flourished particularly in the sixteenth and seventeenth centuries (Plate 15.9). (Some people apply the word Faïence to a much broader range of ceramic products dating back to the Egyptians.) It should be noted that unglazed ceramic products are called *bisqueware*.

Stoneware is fired at temperatures around 1200 to 1300°C, which causes at least partial vitrification of certain clays stemming typically from sedimentary deposits that are low in iron content. Stoneware is hard and opaque and sometimes translucent. Its color varies from black via red, brown, and grey to white. Fine white stoneware was made in China as early as 1400 B.C. (Shang dynasty) and was glazed with feldspar. Korea and Japan followed at about 50 B.C. and the thirteenth century A.D., respectively. The first European stoneware was produced in Germany after *Johann Friedrich Böttger*, an alchemist looking for gold (Chapter 17), together with E.W. von Tschirnhaus, rediscovered red stoneware in 1707 (Plate 15.8). Josia Wedgwood, an Englishman, followed somewhat later with black stoneware called basalte, and with white stoneware, colored by metal oxides, called jasper. Stoneware may remain unglazed or may receive lead or salt glazes. The latter one (first used in the Cologne region in Germany) involves NaCl which is tossed into the kiln when it has reached its highest temperature, allowing sodium and silica from the clay to form sodium silicate. This yields a pitted appearance like an orange peel. Alternatively, the objects are dipped into a salt solution before firing. Salt glazes give off poisonous chlorine gas during firing and are thus environmentally objectionable. Large-scale salt glazing was therefore discontinued some years ago.

The climax of the art of pottery was reached when the Chinese invented **porcelain**, a white, thin, and translucent ceramic that possesses a metal-like ringing sound when tapped. It is believed that Marco Polo, when seeing it in China (about 1295), named it *porcellana* (shell) because of its translucency. In its initial form, porcelain was produced during the T'ang dynasty (A.D. 618–907) but was steadily improved to the presently known configuration starting with the Yüan dynasty (A.D. 1279–1368). Many western and Islamic countries tried in vain to duplicate (or vaguely imitate) this ultimate form of tableware until eventually, in 1707–1708, the above-mentioned J.F. Böttger succeeded, which laid the ground for the Meißen porcelain manufacture in Saxony (Germany) in 1710.

The secret of porcelain was found in a combination of raw materials, namely, in pure, white, *kaolin clay* (see Section 15.4) which was mixed with quartz and feldspatic rock. Pure kaolin, having a melting point of 1260°C, is, however, too difficult to shape due to its poor ductility. Further, its high refractory property does not allow it to be fired to a hard and dense body at acceptable temperatures. The combination with other ingredients, such as "ball clay," increases the plasticity of kaolin and reduces its firing temperature. In addition, alumina and silica serve as glazing ingredients. Only the careful balance between the ingredients produces porcelain that is white, dense, completely vitrified when fired above 1260°C, and translucent when thin. Interestingly enough, kaolin was used in Böttger's time to powder wigs, and it is said that this inspired him to experiment with kaolin.

Böttger probably utilized for his exploratory experiments the solar furnace developed by von Tschirnhaus, who reported on it in 1699. In this device the sunlight was focused with a large, 1-m-diameter lens which allowed it to reach at least 1436°C, that is, the melting temperature of a sand–lime mixture. Such a high temperature could not be achieved at that time in Europe with conventional means. Later endeavors by Böttger, however, to build a horizontal high-temperature kiln allowed mass production of porcelain. During the firing, the feldspar vitrifies while the clay ensures that the vessel maintains its shape. In other words, the body and glaze of most hard porcelains can be fired in *one* operation since the fusion temperature of both components is roughly the same. This one-step firing process is, however, not always performed, in particular if colored decorations need to be added.

Early Western attempts to imitate porcelain included **milk glass** (a mixture of glass and tin oxide), **soft porcelain** (a mixture of clay and ground glass) manufactured particularly in Italy and Egypt, and an English version of soft porcelain in which bone ash (from cattle) was added to ground glass and clay. Following this practice, the British also added in later years some bone ash to the true, hard porcelain which renders an ivory-white color. **Bone china** is somewhat easier to manufacture and better resists chipping. A different type of translucent ware was made in Persia during the seventeenth century and was called Gombroon. In Italy, under the patronage of the Grand Duke Francesco I de Medici, a hard, white, translucent ware was produced between 1575 and 1587 which, however, because of its high content in alkali and alkali earth (total 13%) and alumina (9.5%), liquefied rapidly when the temperature was raised. Probably all specimens of this production sustained some distortion during firing which led eventually to the abandonment of this technique. Likewise, French soft-paste porcelain (with high lime and low clay content) were difficult to form and fire.

For the interested reader some recipes for porcelain and glazes are listed below.[1] The quantities are given in mass percent.

White porcelain (for casting bodies)

Kaolin	46%
Silica	34.2%
Potassium feldspar	19.8%
Add Sodium Carbonate	0.4%
Firing temperature	1285–1325°C
Shrinkage	11.5%

White/light-gray porcelain (for throwing bodies)

Kaolin	40%
Silica	25%
Potassium feldspar	25%
Ball clay (Kentucky #4)	7%
Bentonite (volcanic ash clay)	3%
Firing temperature	1260–1325°C
Shrinkage	15%

Clear semi-mat glaze

Feldspar	58%
Silica	12.5%
Whiting (natural $CaCo_3$)	12.5%
Kaolin	11%
Zinc oxide	6%
Add: C.M.C.[2]	1 tsp.
Firing temperature:	1170–1250°C

15.3 • Shaping and Decoration of Pottery

One of the earliest methods for *shaping* clay included pressing the clay into a basket which was eventually consumed by the fire. Other techniques utilized paddling clay over the exterior of a mold pot to form the base, whereby the upper portion was formed with a series of coils laid layer upon layer. Also, hand-modelling was frequently practiced. The potter's wheel is, in contrast, a rel-

[1]Adapted from J. Chappell, *Clay and Glazes* (see Suggestions for Further Study).
[2]Carboxymethyl cellulose (acts as thickener and binder).

atively late invention. It appeared in the Near East around 3500 B.C. and in China between 2600 and 1700 B.C. Some areas of the world, such as the Western Hemisphere, never used the potters wheel until contact with European settlers was made.

A number of *decorations* have been applied even to the earliest pottery. Among them are impressing or stamping the clay before firing with fingernails, pointed sticks, or ropes (Japanese Jomon ware of the second and first millennium B.C.) or rolling a cylinder with a design over the clay body, thus producing relief ornaments (Etruscans, first millennium B.C.). Washing or painting the pottery with semi-liquid clay, called *slip* (with or without coloring metal oxides), has been quite popular over many millennia; see Plate 15.1. White slip, when covered with a transparent glaze, looks quite similar to tin glazing. Ancient Egyptians, for example, painted animals and scenic motifs with slip on red potteries. Metal oxides have often been added to glazes or slip for color. Specifically, tin oxide provides a white color, cobalt oxide and cupric oxide yield various bluish hues, and cuprous oxide, a series of greens. The colors obtained from ferric iron vary from pale yellow via orange-red to black. Manganese gives colors ranging from bright red to purple and antimony yields yellow. To prevent intermingling of the different hues, patterns are outlined with clay threads, thus exercising a cloisonné technique. The colors can be applied either *under* the glaze or *over* the glaze. When the decoration is painted on a white tin glaze, a third firing, utilizing, for example, a transparent lead glaze, needs to be applied.

Many more decoration techniques are (and have been) used including luster decoration (invented by early Islamic potters involving a colloidal suspension of gold, silver, or platinum to the glazed object, requiring an additional, gentle firing). Early pottery, dating back as far as 6500 B.C., was polished or burnished after firing by rubbing with a soft, smooth stone (Turkey, 6500 B.C., Incas A.D. 500, North American Indians A.D. 1000 Plate 15.10) or varnished (Fiji islands). The wealth of art work evolving from pottery over 9000 years is just overwhelming and cannot be done justice in a few paragraphs as presented here. The interested reader is referred to the art books listed at the end of this chapter.

It might be of interest to know how the **age** of ancient pottery can be determined. Certainly, the common *carbon fourteen method* which requires organic material and which measures the radioactive decay of C-14, cannot be applied for inorganic materials such as clay. However, in some cases pottery has been added to human burial sites which allows an estimate of the age of ceramics by knowing the age of the bones. In other cases where such a comparison is not possible, the *thermoluminescence* technique is utilized. This method makes use of the fact that many clays and soils contain minute amounts of radioactive elements such as uranium, thorium, or potassium. Their emitted α-, β-, or γ-radiations excite under certain circumstances some electrons of the clay into higher energy states where they might be trapped in impurity states (see Chapter 13). If thermal energy is supplied to this substance, the electrons may be forced to leave their metastable positions and revert to a lower energy state by concomitantly emitting light. In other words, thermoluminescence unlocks the stored energy that has been radiation-induced over time by slightly heating the object under investigation. For this it is essential that the clay has been fired at some time in order that "the time clock is reset to zero." The age is then assigned to an object by measuring the amount of emitted light and by knowing its and the surrounding soil's radioactive content as well as by taking into consideration how susceptible the material is to radiation damage. Another dating technique that is, however, still in its developmental state involves electron spin resonance.

At this point the reader probably wants to know about the chemistry and physical properties of ceramics. Specifically, why is clay pliable and what is the composition of clay? This will be explained in the next section.

15.4 • The Science Behind Pottery

The principal purpose of this section is to provide some understanding of why clay is such a remarkable material which is ductile when wet and hard after firing. First of all, "clay" is not just *one* substance but a whole family of minerals whose common characteristic is that they have a *sheetlike* crystal structure, as we shall see momentarily, which allows the platelets (that are <1 μm in diameter) to slide easily past one another even when only little force is applied. Further, clays are so-called *hydrous aluminum or magnesium silicates*, which are distinguished by the property that they lose physically adsorbed or structural water when heated. Clays have formed as a result of marine sediments or from hydrothermal activities during all ages. Clays vary in composition and additional constituents depending on the environment in which they formed and the hydrological or climatic conditions. They are found in mudstones, shales, and soils almost everywhere on the earth.

As was just indicated, clays are composed of silica (SiO_2), alumina (Al_2O_3), possibly magnesia (MgO), and water, along with impurities of iron, alkalies, or alkaline earths. They are sometimes permeated with materials such as feldspar [$(K, Na)_2O \cdot Al_2O_3 \cdot 6SiO_2$] or mica [$KAl_3Si_3O_{10}(OH)_2$]. In other words, the clay minerals belong to a larger family of *silicates* which are common among an extended number of ceramic materials. It is therefore necessary to digress for a moment from our theme and promulgate a few concepts on the physics and crystallography of silicates.

Silica, or silicon dioxide, is composed of the two most abundant elements of the earth's crust, which consists of 59 mass% SiO_2. Silica is the major component (at least 95%) of rocks, soils, clays, and sands. Silica exists in several allotropic forms upon raising the temperature, having slightly different crystal structures. Specifically, the room temperature *α-quartz* transforms at 573°C into *β-quartz* by a displacive transformation (similar to that found in martensite; see Section 8.3) involving a rapid but slight distortion of the crystal lattice. A further transformation (of the *reconstructive* type) takes place at 870°C from β-quartz to *β-tridymite*, during which the bonds between atoms are broken and a new crystal structure is formed by nucleation and growth. A third allotropic transformation occurs at 1470°C from β-tridymite into *β-cristobalite*, which is again of the reconstructive type. The melting point of pure SiO_2 is finally reached at 1723°C but can be reduced by additional constituents as shown in Figure 15.5.

It is common to represent silicates as a series of $(SiO_4)^{4-}$ tetrahedra,[3] which means that four oxygen atoms tetrahedrally surround one silicon atom; Figure 15.2. This basic tetrahedral unit of silicates is fourfold negatively charged. The bonding between the silicon and the oxygen atoms is mostly covalent. Thus, each bond is strong and directional (see Section 3.2 and Figure 3.4). The melting temperature of silica is therefore high (1723°C). Bulk silica can be represented by a three-dimensional network of the just-discussed tetrahedral units whereby each corner oxygen atom is shared by an adjacent tetrahedron (Figure 15.2). The silica tetrahedra can combine, for example, to chains [Figure 15.3 (a)] or to rings [Figure 15.3 (b)]. To satisfy the charge balance, each oxygen ion (or group of oxygen ions) can combine with, say, metal ions. For example, two Mg^{2+} ions may combine with one $(SiO_4)^{4-}$ tetrahedral unit, thus forming Mg_2SiO_4, called foresterite. Compounds of this type are termed *orthosilicates* (or *olivines*).

Of particular interest in the present context are the clays. In this case, the silicon combines with oxygen to yield $(Si_2O_5)^{2-}$; see Figure 15.3 (c), which forms a sheet-type structure. Specifically, in kaolinite the silicate sheets are ionic-covalently bound

---

[3]*Tetraetros* (Greek) = four-faced.

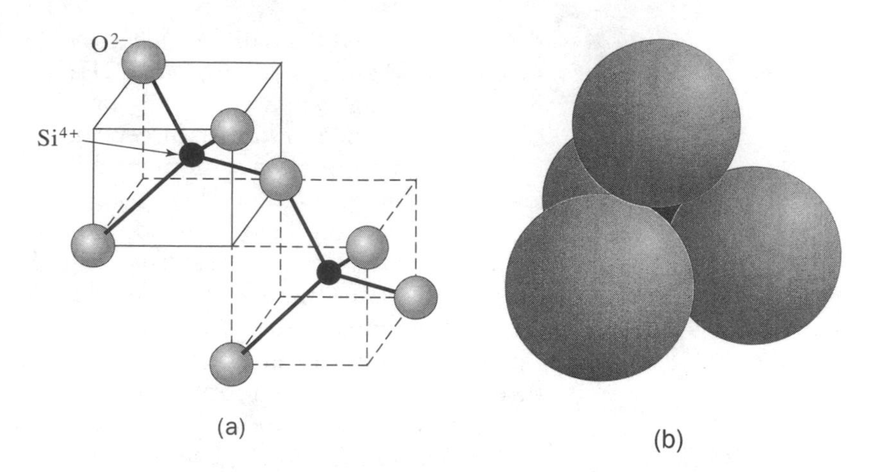

FIGURE 15.2. Two schematic representations of an $(SiO_4)^{4-}$ tetrahedron. (a) Spacial arrangement of the oxygen atoms with respect to a silicon atom. (b) The atoms touch each other when assuming a hard-sphere-model. Note: The ionic radii are not drawn to scale. The silicon atom (black) is barely visible in the center between the four oxygen atoms.

to $Al_2(OH)_4{}^{2+}$-layers to yield $Al_2(Si_2O_5)(OH)_4$; see Figure 15.4. Now, it is important to know that adjacent $Al_2(Si_2O_5)(OH)_4$ sheets are quite weakly bound to one another involving van der Waals forces (Section 3.2). It is this weak van der Waals force that allows for the easy gliding of the individual sheets or platelets past each other, rendering the above-mentioned ductility (plasticity) of clay, particularly when water is present between the sheets.

Interestingly enough, the silicate sheet structure is not restricted to clays. It is also found in other minerals, such as mica ($KAl_3Si_3O_{10}(OH)_2$). Moreover, *graphite*, one of the polymorphic forms of carbon, is likewise composed of layers whereby the carbon atoms assume the corners of a hexagon. Each atom is bonded to its three coplanar neighbors by strong covalent bonds. The fourth bond to the next layer is, however, of the van der Waals type. For this reason, graphite sheets slide easily past each other and can therefore be used as a low-temperature lubricant. (For lubrications at higher temperatures, another substance with a hexagonal-layered structure is used, namely, *boron nitride*, which is also called *white graphite.*)

The crystallography of clay minerals is certainly one of the most complex among inorganic materials. Kaolinite is only one group (but the most common) of these minerals which are classified into allophanes, halloysites, smectides, vermiculites, etc., to mention just a few. They all have different crystal structures and compositions. Moreover, clay minerals are able to adsorb on the outside of their structural unit various impurity elements

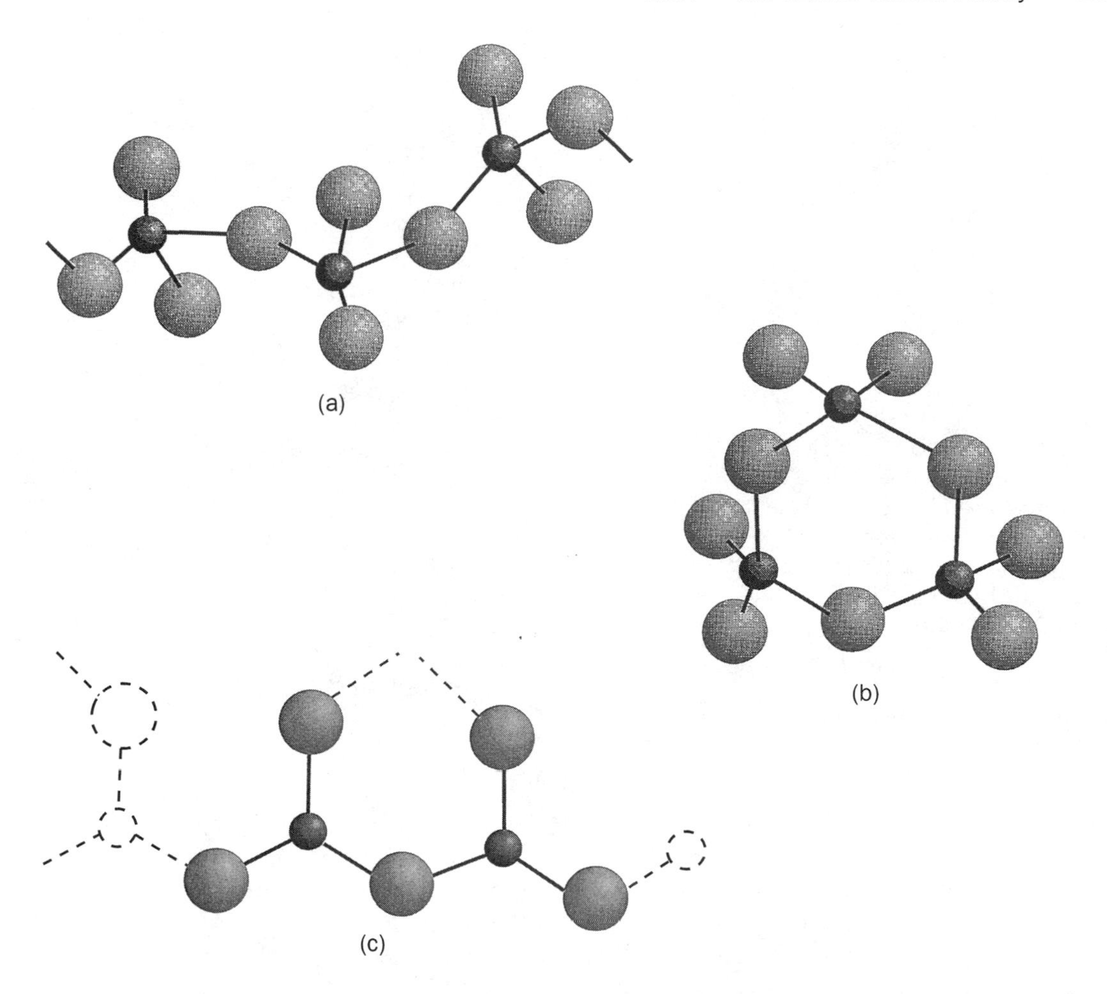

FIGURE 15.3. Schematic representation of (a) a silica chain $(SiO_3)_n^{2n-}$, (b) a silica ring $(Si_3O_9)^{6-}$, and (c) a silica sheet $(Si_2O_5)^{2-}$. [Note: The fourth Si bond in (c) points into the paper plane.] See also Figure 15.4.

such as calcium, iron, or sodium, which can be easily mutually exchanged. As an example, for a characteristic composition of kaolin, the one found in North-Central Florida may serve:

47%	SiO_2
37.9%	Al_2O_3
0.45%	Fe_2O_3
0.18%	TiO_2
0.08%	CaO
0.3%	MgO
0.2%	K_2O
0.24%	Na_2O
13.5%	Loss, i.e., mainly H_2O.

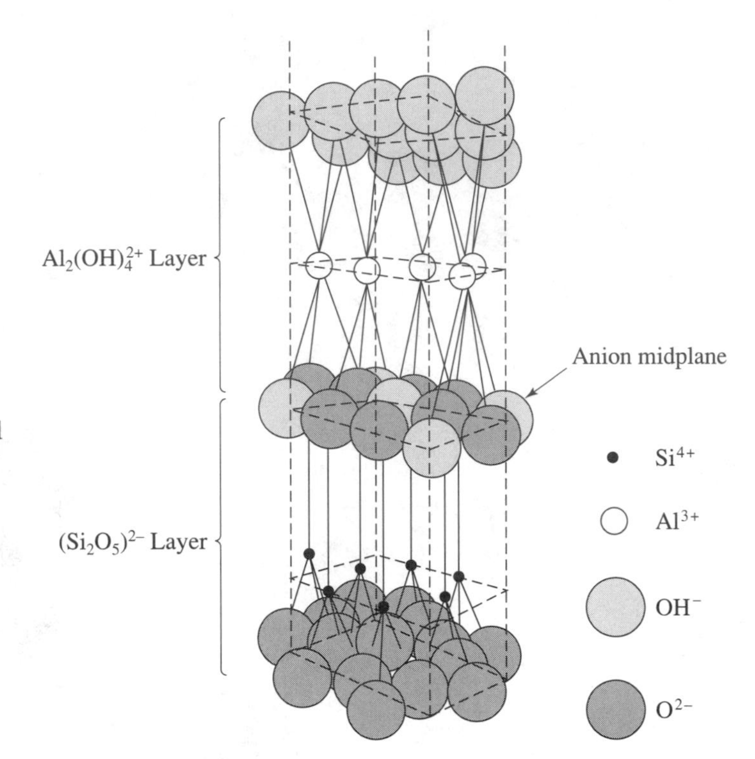

FIGURE 15.4. Schematic exploded view of the layered structure of kaolinite clay. The center plane contains O^{2-} ions from the $(Si_2O_5)^{2-}$ layer and $(OH)^-$ ions from the $Al_2(OH)_4{}^{2+}$ layer. (From W.E. Hauth, *American Ceramic Society Bulletin*, 30 (1951), p. 140. Reprinted by permission from the American Ceramics Society.)

We turn now to the important question involving the drying and firing of clays. Clay minerals hold water in a variety of ways. If water is contained in pores, it may be removed by drying at ambient conditions. On the other hand, physically adsorbed water is driven off by heating at temperatures of about 100–200°C. However, it is readily readsorbed by exposing dried clay to water. Finally, and most importantly, hydroxyl (OH^-) ions are removed by heating clay at temperatures above about 400°C, depending on the mineral at hand. This causes a change in the lattice structure. In the case of kaolinite, the *dehydroxylation* of $Al_2(Si_2O_5)(OH)_4$ forms $Al_2O_3 \cdot 2SiO_2 + 2H_2O$. (Other sources maintain that a "defect spinel structure" is probably formed.) This crystal structure transforms at still higher temperatures into mullite and cristobalite (see Fig. 15.5). Eventually, fusion (vitrification) takes place. This concept needs some further explanation.

Figure 15.5 depicts the high-temperature portion of the relevant SiO_2–Al_2O_3 phase diagram in which an intermediate phase, namely the just-mentioned mullite ($3Al_2O_3 \cdot 2SiO_2$), is evident. Mullite probably forms by a peritectic reaction. The melting point of SiO_2 (cristobalite) decreases dramatically by additions of only

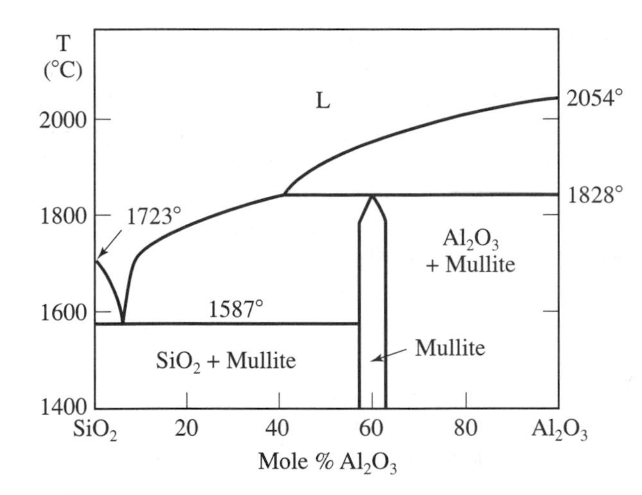

FIGURE 15.5. High-temperature portion of the SiO_2–Al_2O_3 phase diagram. A non-stoichiometric intermediate compound ($3Al_2O_3 \cdot 2SiO_2$) called mullite is formed around 60 mole% (72 mass%) Al_2O_3.

a few percent Al_2O_3. It is also seen that, by further increasing the Al_2O_3 content, the fired clay becomes more heat-resistant, that is, it gains better refractory qualities particularly if at least 60 mole% Al_2O_3 is present. The crystal structure of cristobalite is shown in Figure 15.6, which displays the tetrahedral symmetry of the oxygen around the silicon atoms. It should also be mentioned that considerable shrinkage occurs during drying (2–9%) and during firing (6–20%).

The use of clay is not restricted to pottery. Indeed, large quantities of kaolinite are used in the paper industry to fill and coat paper (see Section 16.2). Only 5% of the kaolinite produced in the United States is actually used in the fine ceramic industry.

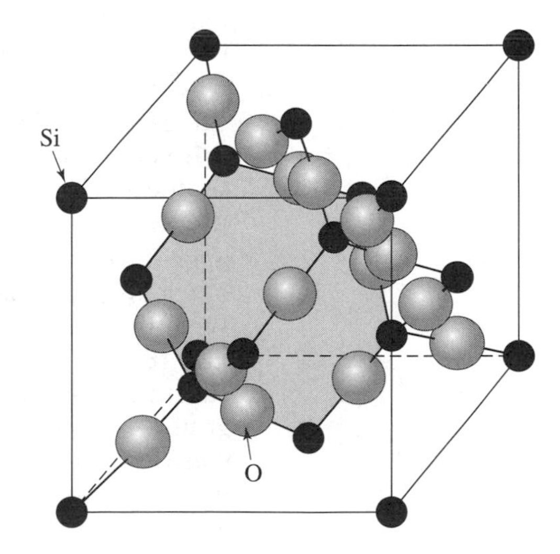

FIGURE 15.6. The crystal structure of cristobalite. The tetrahedral symmetry for some of the Si atoms can be clearly seen (compare to Figure 15.2). Note that in some other cases the third and fourth oxygen atoms are contained in the adjacent unit cell and are therefore not depicted. A (non-coplanar) hexagon with silicon atoms on the corners and oxygen atoms on the edges has been shaded. A series of hexagons of this type is shown in Figure 15.7(a) in a two-dimensional representation.

Kaolinite and bentonite (a clay derived from volcanic ash having an extremely fine grain) are utilized as bonding agents in foundry sands for casting metals. Certain clays such as attapulsite are used in the oil-drilling industry for their wall-building capabilities. And naturally, clays are used to manufacture refractory (fire brick) materials which are capable of withstanding high stresses at high temperatures. Further uses are for high-tension insulators, wall and floor tiles, plumbing fixtures, and so on.

15.5 • History of Glass-Making

The history of glass-making and glass utilization is relatively young compared to that of, say, copper smelting or pottery. The reason for this may be found, to a large extent, in the high melting temperature of silica (sand), from which glass is made and which is above 1700°C; see Figure 15.5. As we know today, however, adding soda and limestone to silica causes the melting point of SiO_2 to decrease substantially (see Figure 15.8). Concomitantly, the properties of glass are favorably changed, yielding a viscous, pliable material which can be formed in a temperature range of about 700–1000°C. These facts, however, were probably not fully recognized until late into the first millennium B.C. The oldest objects of man-made glass date back to about 2500 B.C. and consist of beads which were found in Egypt. Their green, sometimes blue color allowed duplication of semiprecious stones, such as turquoise or lapis lazuli, which were scarce and highly priced at that time. Blue and green were colors of water, fertility, and good luck.

A green glass rod that was unearthed in Eshnunma (Babylonia) may be even about 100 years older. And a piece of blue glass from Eridu was probably made around 2200 B.C. Before that time, glass was known to man only in its native form, that is, as volcanic glass called obsidian; see Plate 1.2.

It is quite interesting to note that the first civilization which introduced glazing of earthenware (namely ancient Egypt) is also the country where the oldest glass beads have been found. There appears to be a certain interrelationship between one and the other, in particular when taking into consideration that the first Egyptian glazes consisted essentially of glass as outlined in Section 15.2. It is therefore quite conceivable to assume that the glassy substance which was used for glazing (composed probably of ground-up obsidian or fine, pure quartz sand) interacted with the constituents of the pottery and thus caused a reduction in melting temperature and a fusing together of the SiO_2 particles at the available temperatures. It is possible that pyrotechnic experimentations with a silica–clay mixture eventually led to the creation of the above-mentioned glass beads. This may be true in particular since the first glazings were done about 500 years earlier (about 3000 B.C.) than the creation of the glass beads (about 2500 B.C.), if the datings are correct. One might therefore postulate that two major technologies, namely, smelting of metals (Chapter 1) and glass-making, both originated from the art of pottery.

The creation of utilitarian objects made out of glass took an amazingly long time to develop. At first man did not make use of the ductile property of heated glass. This may be understandable by knowing that pottery already fulfilled the respective needs, and is much easier to handle. The earliest glass vessels that have been found so far were chiseled or ground out of solid glass blocks. They are of Mesopotamian origin and are believed to have been sculptured around 2000 B.C. A technological breakthrough came at the middle of the 16th century B.C., when the *core-forming technique* was invented in northern Mesopotamia (Plate 15.2). A removable core probably consisting of clay, mud, sand, and an organic binder was built around a metal rod into the shape of the interior of the vessel. The core was then covered with hot glass by dipping or by trailing threads of glass over the core as it was rotated.

After repeatedly heating and rolling, the metal rod was removed and the core was scraped out, leaving a rough, pitted interior surface. The vessels (mostly used for cosmetics, perfumes, or scented oils in sanctuaries or to anoint the dead) often had pointed bottoms or disc bases. Around the rim there occasionally would be a cane of spirally twisted threads of different colors. The flasks essentially resembled in shape those of the pottery of that time. Not long afterwards polychrome vessels made of *mosaic glass* were introduced in Mesopotamia. It involved colored rods or canes made from rods of different colors that were fused together by heating. These rods or canes were cut into pieces and laid side by side on the inner surface of a mold. A second mold was inserted above them, and the combination was heated with the purpose of fusing the glass together. A spirally wound trail of contrasting colors was finally applied to form a rim that was fused on by renewed heating. The last manufacturing step involved a smoothing of the surface by grinding and polishing (Plate 15.3).

For decoration, coiled threads of different colored glass were alternately pulled back and forth to form zig-zag patterns. Because of their flowerlike design, these creations were called *millefiori*, which means "thousand flowers." Other techniques developed in Alexandria involved pressing hot glass into molds and melting powdered glass in a clay form. Moreover, the millefiori and molding techniques were combined, thus yielding bowls of exquisite beauty and variety. Coloring was achieved by small additions of metal oxides to the melt as described in Section 15.2.

As a rule, sand contained some iron which rendered the glass a bluish-green hue. Almost colorless glass could be achieved by utilizing "silver sand," which is free from iron, or by employing some *decolorants* such as antimony or manganese. Copper-ore additions yielded ruby red, turquoise, dark green, or opaque dark red colors depending on the firing temperature. Glass was a luxury material produced for an aristocratic, royal, or priestly market. Most of the pieces have been excavated from temples, palaces, or tombs rather than from private homes.

Another method utilized molten glass that was poured in successive layers into clay or sand forms, thus creating a vessel of proper sturdiness. In essence, this technique was borrowed from metal casting, which was widely practiced at that time. The vessels, generally of blue glass, were decorated with molten droplets of colored glass which were pressed into the still ductile surface.

Egypt was not the only glass-producing country. Indeed, glass-casting centers during the first and second millennium B.C. were in Assyria, Greece, Italy, Sicily, and probably even further to the west. Interestingly enough, China had no part in this development, being occupied with practicing fine porcelain-making. The above-described techniques, which had their origin in glass casting and which started in Mesopotamia and Egypt after 1500 B.C., were utilized for about 1000 years until eventually glass-blowing was invented in Babylonia around 200 B.C., which revolutionized the entire trade.

Glass-blowing utilized a *blow tube*, made of iron, about one and one half meters long. Bronze would not have served this purpose because of its comparatively lower solidus temperature (~850°C versus 1538°C for Fe). Thus glass-blowing had to await the iron age for its realization.

For creating a vessel by this new technique, a blob of soft glass, approximately 700–1000°C hot, depending on the composition, is picked up with the knob-end of the blowing iron and is at first twirled on a polished block of iron (called a marver) into a suitable shape. Subsequently, air is blown through the tube with the mouth, which inflates the blob into a globe. A shape can be given by making use of a mold or without constraints. If desired, a second piece of glass, held by a solid iron rod (called the pontil), can be again shaped by twirling and rolling a blob on the marver. It is then welded on the still hot globe to facilitate a stem or a foot. The finished objects are eventually separated from the blow tube by shears or a sharp tap on the iron. Subsequent slow cooling and/or annealing for many hours reduces the *thermal stress* in the glass and thus reduces the chance for *fracture*, called, in the present context, *thermal shock*.

This revolutionary new technique of glass processing was

quickly adapted by the Romans (and other empires) who eventually became masters in this trade. Certain modifications to the methods were made with time, but the principles of glass-making stayed essentially the same for many centuries.

Glass making centers were founded all over the Roman Empire, as far North as in Cologne (Germany) and York (England). Where the respective glass blowers obtained their ingredients for glass-making has been a matter of speculation. Today, one assumes that the Egyptians and Palestinians produced raw glass by combining soda (Na_2CO_3), which was mined in *Wadi Natrun* east of the pyramids, with fine quartz sand that was found near the Belus River in Palestine and which contained the necessary calcium. These ingredients were melted and then quenched in water yielding lumps of raw glass that resembled rock candy. It was exported through the nearby harbor of Alexandria and could be remelted at 800°C at faraway places.

The high craftsmanship of glass-making known from the time of the Roman empire eventually declined in Europe after about A.D. 200 but was finally revived by the Venetians in the 13th century through trading contacts with the Byzantium (eastern part of the Roman empire). The Venetians redeveloped the former skills and became so successful that they had to move their entire operations to the island of Murano for fire protection of the city of Venice and to prevent their artisans from migrating to other places and thus spreading their skills. From contemporary descriptions of the work place it could have well been that about 100 glass-blowers labored at the same time in these extended factories.

Despite the fact that considerable accomplishments have been achieved over the centuries in creating fine drinking and eating ware from glass, the task of producing flat panels for windows remained a problem for a long time. Attempts by the Romans to pour molten glass on flat surfaces was as unsuccessful as the 19th century practice of rapidly spinning a flattened glass balloon until the centrifugal forces drew the soft glass into a large disk. Both methods were plagued by uneven products, poor transparency, and required polishing at almost prohibitive costs. By the end of the nineteenth century, however, a glass cylinder blown by compressed air was slit lengthwise and was allowed to flatten under its own weight after reheating. Today flat glass (plate glass) is mostly produced by casting molten glass on a bed of liquid tin (confined in a closed chamber to prevent oxidation of the metal). Bottles and containers are blown automatically using compressed air and split molds made of hollow iron blocks for reusability. Additionally, pressing glass into molds is quite common.

Different glass compositions have been developed over the past 200 years, especially for the emerging optics industry. The pioneering work by an ideal team consisting of Carl Zeiss, a microscope maker, Ernst Abbé, a university professor, and Otto Schott, a glass specialist, all from Jena (Germany), gave glass optics a sound scientific basis. The humanitarian deeds by Ernst Abbé in surrendering eventually his shares of the Zeiss factory into a foundation that benefited workers and their families should not be forgotten in this context. The optical industry needs glasses of various refractive powers which, when combined into lens systems, alleviate certain imaging problems such as the chromatic and the spheric aberrations. Flint glass containing up to 80% lead oxide has a stronger refractive power than the more common crown glass; see Table 15.1. Flint glass is therefore also used as "crystal glass," which stronger divides the light into its individual color components and thus effects some sparkling in light. For most applications, however, the soda–lime–silica glass is the product of choice because of its price effectiveness, its moderate softening temperature, its chemical stability, and its reasonable hardness. The raw materials are plentiful, consisting of quartz sand (SiO_2), lime (CaO from limestone, that is, calcium carbonate, $CaCO_3$), and soda, Na_2O (from soda ash, i.e., sodium carbonate, Na_2CO_3). The latter has been obtained from burning hardwood which was used as fuel.

Glass is utilized for a large number of applications beyond windows, optics, and tableware. Examples are flexible *optical fibers* for telecommunication devices, each strand having the thickness of a human hair. Glass fibers are increasingly used because transmission of information by light is considerably faster than by electrons in wires. Moreover, 32,000 times more information can be transmitted. Further uses of glass include *light bulbs*, *glass wool* (for thermal insulation), *photochromic glass* (which darkens under the influence of light), *foam glass* (a mixture of ground glass and carbon which swells when heated to an agglomerate of thin-walled bubbles), nuclear waste immobilization material, biological glasses that bond to human tissue, dental restoration, and more. The U.S. glass industry sales in 1995 approached 20 billion dollars annually.

Pure silica is, contrary to soda–lime glass, transparent for ultraviolet light and as such is desirable for spectroscopic instruments. However, its large viscosity even at very high temperatures does not allow molding by common means. As a substitute, 99.5% pure silica glass, called *fused quartz*, is satisfactory for most applications; see Section 15.6.

A word on glass **recycling** might be in order. Reutilization of waste glass (called *cullet*) does not only relieve the environment from debris, but also makes economic sense since molten glass having a lower melting point than its constituents acts as a solvent for new raw material and thus reduces the melting temperature of the raw material and consequently the energy costs for melting. Today, between 25 and 50% of ingots in glass factories consist of recycled glass; see also Section 18.4.

Glass has been used for millennia by artists who used this material to create lasting objects of beauty. This has not changed in modern times. New and exciting forms are created every day. Artists are still using the individual blowing technique and involve the blow pipe similarly as 2000 years ago.

15.6 • Scientific Aspects of Glass-Making

Glasses are defined to be *noncrystalline* (amorphous) solids. Members of this group are the noncrystalline silicates (which may contain additional oxides such as Na_2O and CaO). They are of particular interest in the present context. Other examples of amorphous solids are the *metallic glasses* such as Cu–Zr or $Fe_{32}Ni_{36}Cr_{14}P_{12}B_6$ (termed Metglas[4]). A number of compounds can be obtained in both modifications, i.e., crystalline or amorphous. The amorphous state of a solid is achieved by more or less rapid cooling from the liquid to prevent crystallization. In the case of metallic glasses, this cooling needs to be extremely fast and is then called rapid solidification or quenching. In the case of silicate glasses, however, the tendency towards crystallization is more sluggish. Nevertheless, even silicate glass will transform eventually into the equilibrium, crystalline, modification. At room temperature, however, this may take in excess of 100 years.

"Amorphous" in the present context does not mean a completely random arrangement of atoms, as in gases. Such a complete randomness is seldom found, even in liquids. Indeed, the relative positions of nearest neighbors surrounding a given atom in an amorphous solid are almost identical to the positions in *crystalline* solids because of the ever present binding forces between the atoms or ions; see, for example, Figures 15.7(a) and (b). In short, the atomic order in amorphous materials is restricted to the nearest neighbors. Amorphous materials therefore exhibit only *short-range order*.

Silica in its various allotropic, crystalline modifications has a relative open crystal structure as seen in Figure 15.6 for cristobalite and again in Figure 15.7(a) in a two-dimensional and rather schematic representation. The open structure allows impurity atoms in large quantities to be incorporated into the network of

[4]Metglas 2826A is a trademark of Allied Chemical

Si^{4+} and O^{2-} ions. In other words, the impurity atoms do *not* substitute silicon or oxygen atoms, that is, they do not occupy regular lattice sites; instead, they are squeezed in between the atoms; Figure 15.7(c). This may have beneficial effects as in the case of sodium or other monovalent ions, which substantially de-

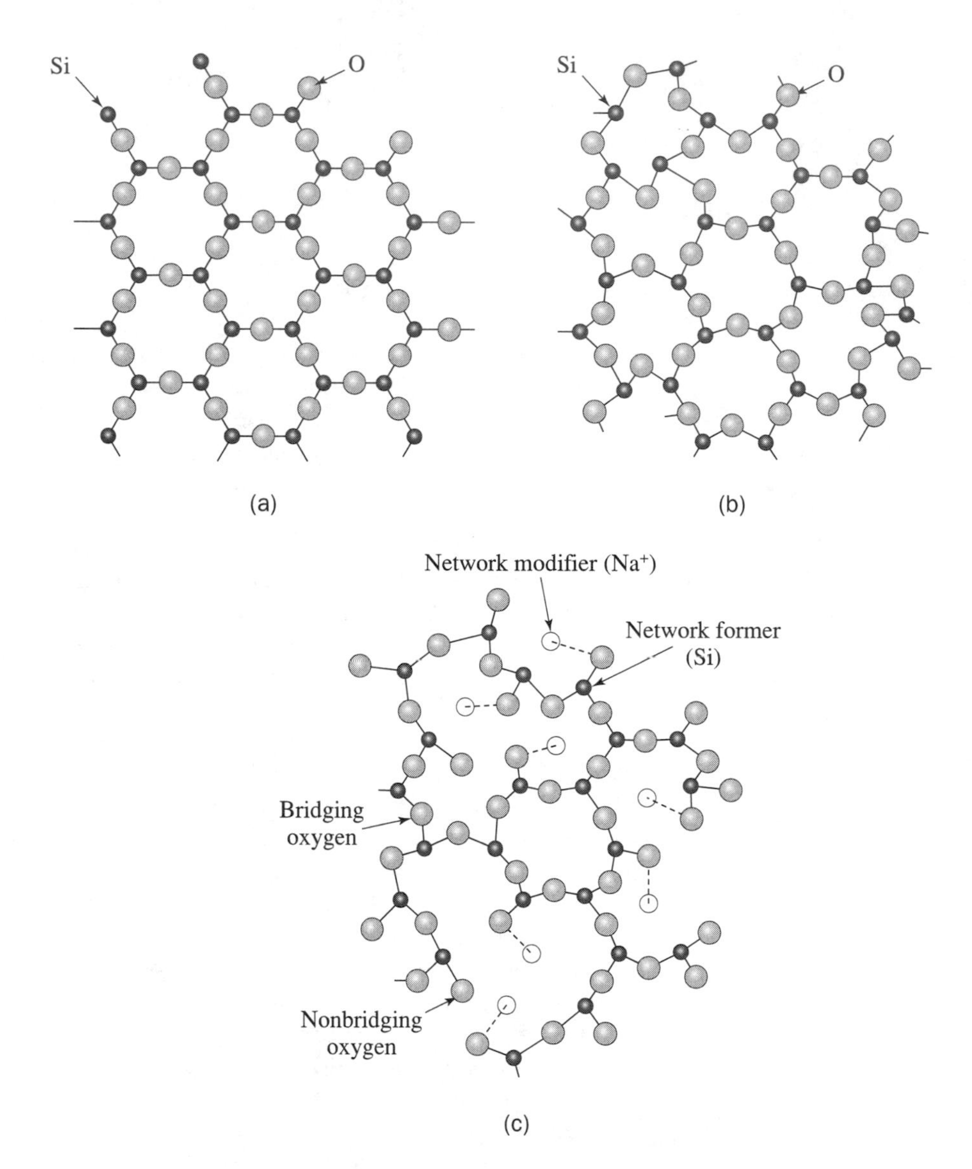

FIGURE 15.7. Schematic, two-dimensional representation of (a) crystalline silica, (b) amorphous silica, and (c) soda–silicate glass. Note that the fourth bond of Si points *into* the paper plane and is therefore not seen. Compare to Figure 15.6 and the shaded hexagon shown there.

crease the liquidus temperature of silica from 1723 to below 800°C (at the eutectic composition); see Figure 15.8. Thus, the working temperature (and the cost of glass production) is reduced. Low valence cations of this type are called *network modifiers*. Soda (Na_2O) in this context is termed a *flux*. Glasses that contain only monovalent modifiers are quite prone to corrosion caused, for example, by water. Thus, CaO, MgO, Al_2O_3, and/or B_2O_3 are frequently added to increase the stability or otherwise favorably modify the physical properties of silica glass; see below and Table 15.1. Further constituents may be Na_2SO_4 or KNO_3, which are added to remove gas inclusions from the melt. Briefly, they cause small gas bubbles to combine with larger oxygen bubbles which eventually escape.

It should be noted in passing that oxides of Pb, Al, Be, and Ti (called *intermediates*) cannot form a glass by themselves because of their lower metal–oxygen bond strength compared to the *glass formers* or *network formers* (such as Si, Ge, P, As, and others). The same is true for oxides of the *modifiers* which we mentioned already above such as Na, Ca, Cs, K, etc., and which have an even lower metal–oxide bonding strength. Actually, the modifiers may cause a glass to crystallize over a long period of time because they expand and thus disrupt the network of atoms in glasses, see Figure 15.7(c).

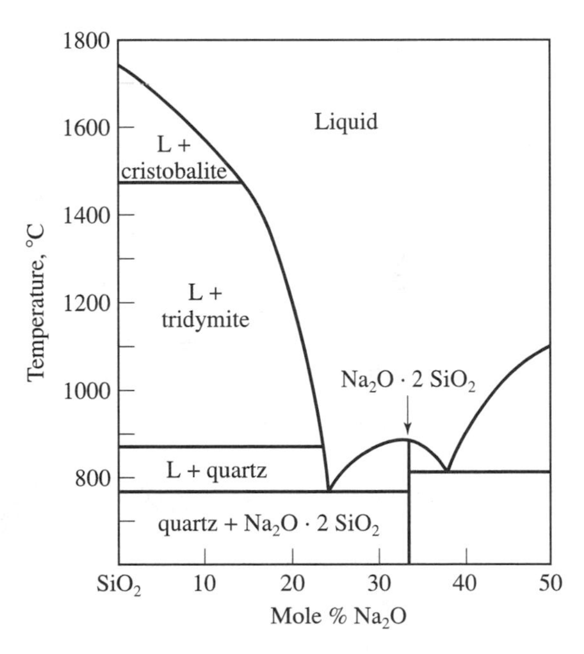

FIGURE 15.8. Binary SiO_2–Na_2O phase diagram up to 50 mole% Na_2O.

TABLE 15.1. Physical properties and typical compositions of some silica glasses

Name	Composition (mass %)	Softening point (°C)	Thermal expansion coefficient $(°C^{-1}) \times 10^{-7}$	Index of refraction (at Na–D line, i.e., 589.3 nm)	Dielectric constant (at 10^6 Hz and 20°C)	Density (g/cm^3)	Elastic modulus (GPa)
Fused silica	>99.5%SiO_2	1,580	5.5	1.459	3.8	2.2	72
Vycor	96% SiO_2 4% B_2O_3	1,530	5.5	1.458	3.8–3.9	2.18	
Borosilicate glass (Pyrex 7740)[a]	81% SiO_2 13% B_2O_3 0–10% ZnO 4% Na_2O 2% Al_2O_3	820	33	1.474	4.6	2.23	63
Soda-lime glass (standard window glass) (0081)	73% SiO_2 17% Na_2O 5% CaO 4% MgO 1% Al_2O_3	695	93	1.512	7.3	2.47	69
Alumino silicate glass (1720)	62% SiO_2 17% Al_2O_3 8% CaO 7% MgO 5% B_2O_3 1% Na_2O	915	42	1.530	7.2	2.52	88

Fiberglass	55% SiO_2 16% CaO 15% Al_2O_3 10% B_2O_3 4% MgO	845	60	1.552	6.6	2.605	74
Optical flint glass	54% SiO_2 37% PbO 8% K_2O 1% Na_20	630	89	1.560	6.7	3.05	59
Optical dense flint (8395)	65% PbO 32% SiO_2 2% K_2O 1% Na_2O	547	87	1.751		4.73	
Optical borosilicate crown (8370)	69.5% SiO_2 11.5% B_2O_3 9% Na_2O 7% K_2O 3% BaO	720	80	1.517		2.53	
Optical dense barium crown (8410)	41% BaO 36% SiO_2 10% B_2O_3 9% ZnO 4% Al_2O_3	759	70	1.611		3.56	

[a] Numbers in parentheses refer to "Corning Glass Works" designations.

Figure 15.7 indicates that O^{2-} ions are, as a rule, shared by two Si^{4+} ions and are then called *bridging oxygens*. If, however, the regular network structure is broken as, for example, by inserting mono- or divalent cations (such as Na^+ or Ca^{2+}), direct oxygen links between the Si ions are not possible. The involved oxygen ions are then termed nonbridging oxygens; see Figure 15.7(c). They have important ramifications in the optical properties of amorphous structures.

Another important aspect needs to be discussed at this point. Glasses do not solidify at an exact temperature (the melting temperature, T_m,) as is known for crystalline materials, but instead in a rather broad *temperature range*. Specifically, upon cooling, a glass becomes increasingly viscous (like honey). Concomitantly, the specific volume, V_s, of the glass (that is, the volume per unit mass), decreases continuously upon cooling whereas for crystalline solids a sudden drop of V_s at the melting temperature is observed; Figure 15.9. The contraction of glasses during cooling is a combination of two effects. The first one occurs, as in most crystalline substances, by reducing the interatomic distances. The second contraction mechanism is due to a rearrangement of atoms. As the glass cools, the atomic rearrangement becomes slower until a temperature is eventually reached at which the viscosity is so high that any further structural change is nearly impossible. Below this temperature, T_g, the volume of the glass contracts at a fixed rate that is determined by the present structure. The intermediate range between T_g and T_m is called the *glass transition* or *glass transformation range*. T_g is then defined to be that temperature at which, during cooling, the V_s versus T curve

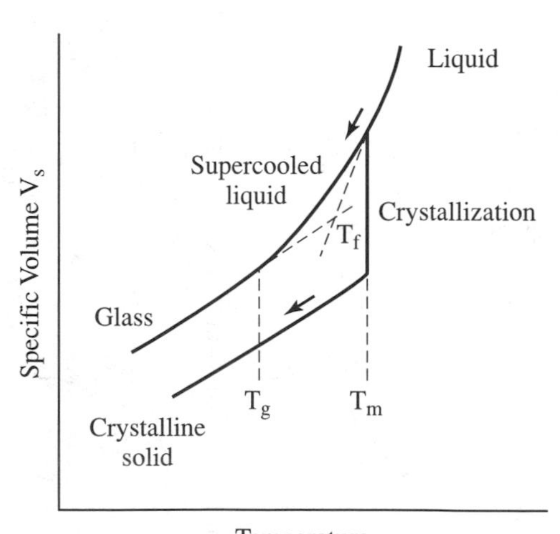

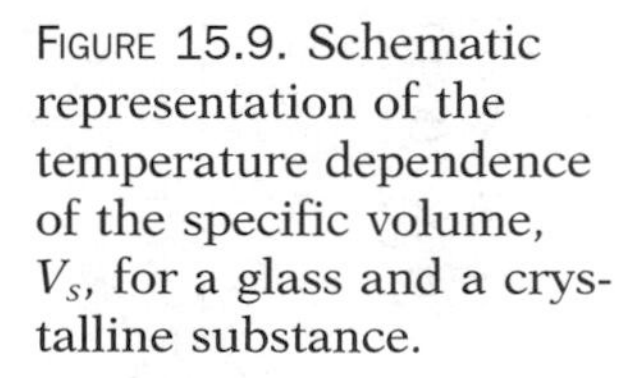
FIGURE 15.9. Schematic representation of the temperature dependence of the specific volume, V_s, for a glass and a crystalline substance.

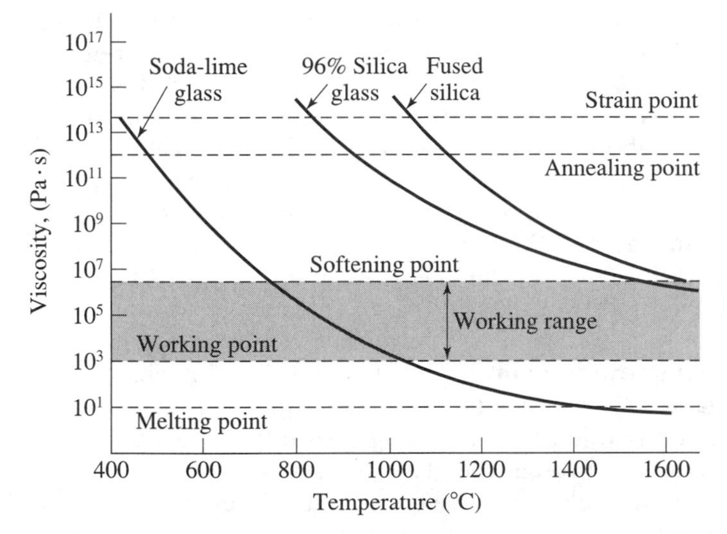

FIGURE 15.10. Schematic representation of a viscosity versus temperature diagram for three glass-forming materials.

reaches an essentially constant slope, as is schematically depicted in Figure 15.9. Above T_g, the material is defined to be a *supercooled liquid*, or eventually a liquid. Below T_g, it is a solid (i.e., a glass).

For practical reasons, the *"melting point"* of glasses is defined to be that temperature at which the viscosity is 10 Pa·s; see Figure 15.10. The intersection of the two straight portions of the V_s versus T curve for glasses is called the *fictive temperature*, T_f (see Figure 15.9). Because of the high viscosity of most glasses and the consequential low mobility of the atoms, any crystallization (called deglassing or *devitrification*) is very sluggish. Nevertheless, extremely slow cooling rates or prolonged heating at high temperatures eventually causes devitrification.

In order that effective glass-blowing can be accomplished, the viscosity of the glass has to be just right. Specifically, the glass may not be too soft to prevent dripping or not too stiff so that deformation is essentially impractical. To assess this, a viscosity–temperature diagram for a given glass is established, from which such properties as the working range, can be taken; see Figure 15.10. The process of glass blowing is performed between the *softening point* (4×10^6 Pa·s), and the *working point* (10^3 Pa·s). At the *annealing point* (10^{12} Pa·s), the diffusion is fast enough to facilitate thermal stress removal within a reasonable amount of time (e.g., 15 minutes). At temperatures below the *strain point* (3×10^{13} Pa·s), fracture of the glass takes place before plastic deformation sets in.

Glass is generally perceived to be a very rigid material. Nevertheless, glass behaves at room temperature just as an elastic solid (Figure 2.5) and can be bent as long as the breaking strength is not exceeded. This is particularly exploited in optical fibers. Glass also possesses a remarkable tensile strength (if free of surface flaws), being about five times as large as that for steel (see Table 2.1), that is, glass can support a mass of 70,000 kg/mm^2. However, the tensile strength may be reduced by surface defects, as is known from glass "cutting," which involves scratching a glass with a diamond tip. Once a scar has been inflicted and the glass is stressed, the rigidity of the glass causes a concentration of the entire tensile force on the few interatomic bonds at the base of a crack and thus exerts an immense stress at this point which causes a propagation of the crack into the material. Since glass does not possess grain boundaries which might stop this propagation, the crack progresses through the glass and leads to fracture.

The strength of glass may be enhanced, however, by introducing compressive stresses at its surface. This is accomplished by annealing the glass above the glass transition temperature (but below the softening point) followed by a rapid quench to room temperature, which causes the surface layer to cool and contract first while the center is still plastic. Eventually the center solidifies, too, but it is restrained from contraction by the already rigid surface. This causes a compressive stress on the surface and a tensile stress in the center. As a consequence, it is much harder to generate a surface crack. The resulting product is called *tempered glass* and is particularly used for eye wear, large doors, and windows for automobiles. Tempered glass shatters on impact into small round-edged pieces which lessen injuries during automobile accidents. An alternate method to accomplish surface hardening of glass (especially for spectacles) is to immerse it for 12–24 hours into a molten potassium nitrate (KNO_3) salt bath, which causes an exchange of some of the sodium ions by larger potassium ions. As a consequence, the surface is under compressive stress.

Another physical parameter that is important for many technical applications describes how much glass will expand when the temperature is raised (see Section 14.4). As a matter of fact, glass is a material that has quite a small linear expansion coefficient, which is about 1/5 of that for crystalline silica; Figure 15.11. (Incidentally, the expansion coefficient of crystalline silica changes abruptly at the temperature at which the allotropic transformation between α-quartz to β-quartz takes place.) Commercially important is *Pyrex,*[5] a borosilicate glass whose thermal

---

[5]Trade name.

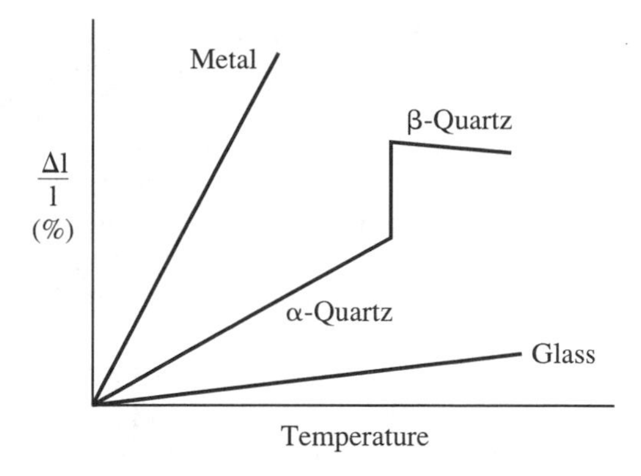

FIGURE 15.11. Comparison of the linear expansion $\Delta l/l$ of glass, crystalline silica, and a typical metal as a function of temperature (See also Table 15.1 and Section 14.4).

expansion coefficient is only 1/3 of that for common soda–lime glass; see Table 15.1.

Finally, a few words about **glass-ceramics**. This expression is probably a misnomer because a glass is, by definition, a noncrystalline material; see above. What is meant, however, is a glass that has been transformed by proper heat treatment or other means into a polycrystalline material. For example, if a piece of glass is heated at a relatively high temperature for an extended length of time, *devitrification* takes place, which renders the material translucent. This is generally not desirable. However, in glass ceramics, devitrification is purposely induced by adding a nucleation agent such as titanium dioxide to the glass, which produces a very fine-grained material after heat treatment. Glass-ceramics possess increased mechanical strength and considerable resistance towards thermal shock (see above). Glass ceramics also possess a smaller thermal expansion and higher thermal conductivity than glass of the same composition. Hammers, knives, and scissors have been made of glass-ceramics, but their principal application is in ovenware, tableware, stovetops, and substrates for printed circuit boards. The most important glass ceramic materials are based on LiO_2, Al_2O_3, and SiO_2.

15.7 • Cement, Concrete, and Plaster

A special class of ceramic materials includes **hydraulic cement**.[6] This compound, when mixed with sand and gravel, is called *concrete*. It constitutes the most widely utilized construction material in the industrialized nations. In essence, concrete is a composite material in which the matrix and dispersed phases are both ceramic components.

[6]*Caementum* (Latin) = stone chippings.

It is known that the ancient Romans and Greeks, already 2000 years ago, mastered the art of producing mortar that hardens under water. Their cement consisted of a mixture of powdered limerock (calcium oxide) and volcanic ash (mostly silica, alumina, and iron oxide). This blend forms a hard body in the presence of water, undergoing a process called *hydration*, during which the water is rigidly attached to the minerals, thus producing a gel. Contemporary hydraulic cement, for example, *Portland cement* (named after a natural stone in England which it resembles), essentially has a similar composition. Further ingredients include about 5% gypsum (see below) to retard the setting time, or kaolin for whitening purposes. Modern hydraulic cement was developed between 1756 and 1850 in England by Smeaton, Aspdin, and Johnson. The processing involves the burning of various raw materials such as clays, shales, slates, blast furnace slag, fly ash, etc., in rotary kilns at 1300–1500°C. This process is called *calcination*. Specifically, tricalcium silicate ($3CaO \cdot SiO_2$), dicalcium silicate ($2CaO \cdot SiO_2$), tricalcium aluminate ($3CaO \cdot Al_2O_3$), and tetracalcium aluminoferrite ($4CaO \cdot Al_2O_3Fe_2O_3$) are the major ingredients of Portland cement. Typical compositions are 60–67% lime, 19–25% silica, 3–8% alumina, 0.3–6% iron oxide, 1–3% sulfur trioxide (from gypsum), plus magnesia and alkalides. During hydration, the cement reacts with water according to the following equations:

$$3CaO \cdot Al_2O_3 + 6H_2O \rightarrow Ca_3Al_2(OH)_{12} + \text{heat}$$

or

$$3CaO \cdot SiO_2 + (X + 1)H_2O \rightarrow Ca_2SiO_4 \cdot XH_2O + Ca(OH)_2 + \text{heat},$$

where X is variable. Concrete made of these components possesses a fast setting time but low strength. To improve the strength at the expense of setting time, $2CaO{\cdot}SiO_2$ is used. Typical curing times are several days to one month. Concrete may continuously harden for years. The water/cement ratio has an influence on the strength of the hydrated cement. Specifically, a smaller water content increases the compressive strength. Some cements, such as $3CaO \cdot Al_2O_3$, are quite vulnerable to seawater or sulfates. Moreover, the heat development during hydration may be as high as 40°C above the ambient temperature. This may eventually lead to cracking during cooling, particularly if the involved masses are substantial, such as in dams. The construction industry distinguishes between five types of Portland cement whose setting time, strength, heat generation, and corrosion resistance varies with composition and proportions of the raw materials. Recently, "High-Performance Concrete" has been developed which is said to be denser, 20% stronger in compression, more water-resistant, and less susceptible to cracks. It is made from 74% cement, 20% fly ash (from power plant emissions), and 6% microsilica (which makes the concrete less porous since it fills microscopic voids). Details can be found in specialized handbooks.

Plaster is made by heating naturally occurring gypsum ($CaSO_4 \cdot 2H_2O$), which drives off some or all of the water yielding, for example, *calcium sulfate hemihydrate* ($CaSO_4 \cdot 1/2H_2O$), also called *plaster of paris*. When mixed with water, this substance sets within a few minutes and requires a retarding agent (a protein called keratin) when slower setting times are wanted. On the other hand, anhydrous calcium sulfate plaster sets at a slow rate and needs another sulfate salt to accelerate the curing. The hardening process of plaster of paris occurs by a cementation reaction that involves the creation of interlocking solid crystals. The reaction is as follows:

$$CaSO_4 \cdot {}^1/_2H_2O + {}^3/_2\ H_2O \rightarrow CaSO_4 \cdot 2H_2O.$$

15.8 • High-Tech Ceramics

The accounts given in the above sections have by no means covered the entire field of ceramic materials, in particular not the fast developing field of **advanced**, i.e., manmade **ceramics.** Classical ceramic materials, which we discussed so far, are mainly made from naturally occurring substances such as clays, sands, or kaolins. In contrast to this, advanced ceramics are manufactured from highly purified oxides, nitrides, carbides, and/or borides. Powders of these substances need to possess well-defined particle sizes, particle shapes, compositions, and distributions of individual particle sizes. For producing useful objects from these powders, they are generally compacted by pressing and sintering at high temperatures. The resulting properties of high-tech ceramics are critically dependent on the microstructure (more than for many metallic alloys) so that considerable skill and research efforts need to be expended to produce these new materials.

One generally distinguishes between two broad groups of ceramics. **Functional ceramics** are used for their electrical, magnetic, optical, thermal, or chemical properties and occasionally even for their aesthetic appearance. They were discussed in previous chapters of this book. On the other hand, **structural ceramics** are utilized for their load-bearing capacities. A few characteristic examples on the latter category will be presented on the following pages.

One family of materials, called **silicon nitride ceramics**, has attracted considerable attention in the past 30 years (55,000 publications) even though, at present, Si_3N_4 serves only 1% of the total market of advanced (electronic) ceramic materials and about 5% of the structural ceramics sector. Their useful properties range from good thermal shock resistance, to high temperature stability, high hardness (12–20 GPa), high strength (800–2000 MPa, depending on defects), light weight, good fracture toughness (3–12 $MPa.m^{1/2}$) corrosion resistance, and wear resistance. They are used for cutting tools, cooking plates with integrated heaters, engine components (injection links, check bells, brake pads, fuel pump rollers), ball bearings, metal-forming tools, gas turbines, and diffusion barriers for microelectronics. However, a full ceramic engine is not commercially available. The main flaw of Si_3N_4 at present is its high cost of production and its metastability in air and combustion gas environments, just as known for all other nonoxide ceramics. The majority of Si_3N_4 ceramic products are made from powders that need to be well characterized and pure. In 1998, 350 tons of Si_3N_4 powders have been produced. These powders are mainly manufactured by direct nitridation of silicon, or by diimide synthesis (involving liquid ammonia, $SiCl_4$, and an organic solvent). The production of Si_3N_4 parts applies techniques known from powder metallurgy, such as pressing, shaping, densification by sintering, and surface finishing. The slurry for the manufacturing process contains a number of additives, such as dispersants (e.g., fatty acids, amines), binders (e.g., wax polyacrylates), plasticisers (e.g., stearic acids), and solvents (e.g., water, organics) that evaporate during the presintering treatment. For complete densification various compounds are added resulting in a substance that is a multicomponent mixture, or an alloy containing Si_3N_4. Three crystalline modifications are known, called α,β, and γ, which vary in their lattice parameter, density, oxygen content, and other physical properties so that Si_3N_4 ceramics can be tailored to specific applications.

Another material, zirconium dioxide (ZrO_2), or **zirconia** is a hard ceramic whose melting point is 2,700°C. It is widely used as an abrasive, as refractory material, as opacifier, and for additions to acid- and alkali-resistant glasses or ceramic parts in fuel cells and other electrical systems. Its inherent brittleness can be alleviated by adding yttrium, calcium, or magnesium, yielding **transformation-toughened zirconia** (TTZ). Indeed, TTZ is so tough that it can be hit with a hammer without breaking, and a hammer can be made of it. The mechanism, which causes TTZ to be shock resistant, involves the action of the above-mentioned nanometer-sized, needle-shaped particles imbedded in ZrO_2, which serve as obstacles for crack propagation. [It is this lack of obstacles for crack movement that makes common ceramics to shatter on impact. Specifically, at the tip of a given crack the applied stress is magnified (called a stress raiser), which causes the cracks to proceed without hindrance through the crystal.] In TTZ however, the above-mentioned crystallites are stimulated by the stress on the tip to transform into a different phase (i.e., micro constituent) having a larger lattice parameter. In other words, the small crystallites at the tip of a crack increase in size under stress causing compression on the crack and thus, prevent further crack propagation. Transformation-toughened zirconia has been found to be tougher than silicon nitride (see above) and is therefore considered for use in wear-resistant and thermally insulating components in heat engines, but also for sensors in automobile exhaust systems, scissors, knife blades (which stay sharper for longer times), faces of golf clubs, tools for wire drawing and beverage can production, and artificial hips (which make use of the chemical inertness and wear resistance of TTZ).

The crack-stopping properties of TTZ can be transferred to other ceramic materials, such as alumina to improve their toughness. For this, 15–20% of ground-up TTZ powder is added to Al_2O_3, which doubles its toughness. **Zirconia-toughened alumina** (ZTA) is used for cutting tools, particularly for steel, which cannot be cut by silicon nitride.

Alumina (Al_2O_3) that is, aluminum oxide, is found in nature as *corundum* mineral. It has a relatively high melting point (2,015°C), a high specific gravity[1] (four times larger than water), and is unsolvable in most solvents. Single-crystalline varieties of alpha alumina are rubies and sapphires whose colors result from small amounts of impurities (i.e., Cr^{+++} in case of ruby). Alumina is also commercially produced by calcinating natural hydrated oxides of aluminum (e.g., bauxite) at about 1,100°C. Alumina is used in dental restoration, spark plugs, insulators, high-temperature crucibles, furnace tubes, lab ware, and certain types of abrasive wheels. Its hardness is, however, lower than that of many other abrasives (such as diamond, boron carbide, cubic boron nitride, silicon carbide, silicon nitride), and cutting tools made of alumina are quite brittle. The above-mentioned zirconia-toughened alumina alleviates many of these shortcomings, which makes the latter applicable for cutting tools.

Activated alumina (obtained by moderately heating hydro-

---

[1] The specific gravity is the mass of a given volume of substance divided by the mass of an equal volume of water.

genated alumina) is porous and is therefore capable of absorbing water vapor and some other gaseous molecules. It is utilized for drying gases and some liquids, and as carriers for catalysts. Alumina and alumina-based composites are also used for armor but there are better and lighter ceramics available for this purpose; see below.

We already mentioned several times the usefulness of certain ceramic materials as **cutting tools**. They mostly replaced those tools manufactured from high-grade steel that were utilized in the early to mid 1900s. Better cutting speeds became available through the development of tungsten carbide, silicon carbide, tungsten carbide-cobalt (WC-Co) **cermets,** that is, ceramic/metal composites consisting essentially of tungsten carbide and 5–6% cobalt that are cemented together. These tools were in turn partially displaced in the 1960s and 1970s by alumina with or without insertions of titanium carbide, followed by silicon nitride (see above), transformation-reinforced Al_2O_3 (see above), and alumina, reinforced with silicon carbide whiskers. (Whiskers are single-crystalline microcrystals, which range in size from 100 to 600 nanometers and diameters less than one-tenth of the thickness of a hair. They are 20 times stronger than bulk ceramics. But their relatively high price and their health hazard during fabrication, similar to asbestos, limit their applicability.) The new ceramic tools last longer, and the machining times of metals are reduced by 75–90% compared to using earlier tools.

TABLE 15.2. Hardness of some ceramics and ceramic-based materials (Note: 1 kg/mm^2 = 9.81 MPa)

Material	Hardness, in kg/mm^2
Boron sub-oxides	~10,000
Single-crystal diamond	7000–9500
Polycrystalline diamond	7000–8600
Cubic boron nitride (c–BN)	3500–4750
Boron carbide (B_4C)	3200
Titanium carbide (TiC)	2800
Silicon carbide (SiC)	2300–2900
Aluminum oxide (Al_2O_3)	2000
Silicon nitride (Si_3N_4)	1200–2000
Tungsten carbide–cobalt "cermet"	1500
Zirconium dioxide (ZrO_2)	1100–1300
Vycor (SiO_2)	700
Silicon dioxide (SiO_2)	550–750
Borosilicate glass	530

Single-crystalline diamond is generally considered to be the hardest material (followed by polycrystalline diamond), as shown in Table 15.2. However, even harder materials have been found, named **ultra hard ceramics**. They consist of boron suboxides having compositions ranging from B_8O up to $B_{22}O$ that easily scratch (111) diamond faces. Other superhard ceramics are cubic boron nitride (c-BN) and boron carbonitrides ($B_xC_yN_z$). Some ceramics such as boron carbide (B_4C), silicon carbide (SiC), and titanium diborate (TiB_2) are used for **armor.**

A completely different approach than that discussed so far is the utilization of ceramics in the medical field. The resulting materials have been appropriately named **bioceramics**. They have been utilized in the past 20 years for repairs caused by periodontal diseases (jaw bones), for replacements of the small bones of the middle ear, damaged by chronic infection, and in spinal fusion and bone repair after tumor surgery, to name a few examples. The field has been pioneered by my good friend, Professor Larry Hench, at the University of Florida, who, in 1967, on the way to a conference was challenged by a military man, returning from Vietnam, to develop materials for bone healing rather than for their destruction. This suggestion fell on fertile grounds, and Professor Hench developed in turn silica-based materials containing calcium, phosphorous, and sodium, which were found to bond to living tissues and bones, and which he eventually trade-named Bioglass.®

The crucial point in successful application of any prosthesis is the formation of a stable interface and a good match of the mechanical properties between implant material and the connecting tissue to be replaced. Now, it is known that no material implanted in living tissue is completely inert. Instead, all materials (metals, ceramics, plastics) elicit a certain response from the living tissue they surround. Several responses are possible:

- If the material is toxic, the surrounding tissue dies.
- If the material is nontoxic and biologically inactive (almost inert), a fibrous tissue of variable thickness forms.
- If the material is nontoxic and biologically active (that is, bioactive), an interfacial bond forms.
- If the material is nontoxic and dissolves, the surrounding tissue replaces it.

All responses, except the first one, are utilized for specific applications. For example, *inert materials* (metals in hip replacements) may be cemented into place. This technique is called *morphological fixation.*

An increased resistance toward movement of the implant is accomplished by *biological fixation*, that is, by allowing the living bone to grow into pores of, say Al_2O_3, whereby these pores need to have a diameter larger than 100–150 mm in order to provide blood supply to the ingrown connective tissue. Al_2O_3, (α-alumina) is corrosion resistant, wear resistant, has a high strength, high fracture toughness, low friction (see above), and has also a good biocompatibility. It is therefore often used for the balls of hip joints and for dental implants. In a similar way metal implants can be made porous on the surface by coating them with porous hydroxyapetite or similar material using a plasma spray technique. (The stems in hip prostheses are still made of metal).

Resorbable biomaterials gradually degrade concomitant with the formation of natural host tissue, whereby resorption and repair rates need to be quite similar in order to maintain enough strength during healing. Examples for these materials are tricalcium phosphate ceramics or polylactic-polyglycolic acid polymers, the latter being used for sutures.

Bioactive materials are intermediates between resorbable and bioinert materials. Specifically, a bioactive ceramic elicits a biological response at the interface of the material, resulting in the formation of a bond between the tissue and the material. Some ceramic compositions bond to both soft tissue as well as to bones, others only to one of them. The first bioactive ceramic created by Hench was essentially based on 45 mass % SiO_2, 24.5% Na_2O, 24.5% CaO, and 6% P_2O_5. (Larger amounts of P_2O_5 or additions of 3% Al_2O_3 did not allow the ceramic to bond to bone.) Today, the clinically most important bioactive glass-ceramic is composed of apatite [$Ca_{10}(PO_4)_6(OH_1F_2)$] and wollastonite (CaO-SiO_2) crystals and a residual CaO-SiO_2-rich glassy matrix, termed A/W glass-ceramic and was developed by Professors Yamamuro and Kokubo in Kyoto (Japan). This material has excellent mechanical properties and bond strengths. Specifically, a bending strength of 215 (MPa), a Young's modulus of 218 (GPa), and a fracture toughness of 2 ($MPa.m^{1/2}$) is encountered.

High strength, hardness, heat resilience, structural stability at high temperatures, wear and corrosion resistance, biocompatibility, and toughness are only a few properties that make ceramics attractive. Their utilization extends, however, far beyond these mechanical properties. We have discussed already in previous chapters the use of functional ceramics for optical materials, i.e., for lenses, mirrors, windows, lasers (ruby, neodymium-doped glass), fiber optics (phosphosilicate cores surrounded by a borosilicate cladding), or luminescence (tungstates, oxides, sulfides). Some of the permanent magnets are made from ceramics (barium ferrite), and there even exists a ceramic-based soft magnetic material, called Metglas. Magnetic storage media consist mostly of oxides (CrO_2, γ-Fe_2O_3). Electronic devices could not exist without highly insulating or semiconducting materials (SiO_2, Si), and without ceramics having high dielectric constants (barium titanite, $BaTiO_3$) for capacitors and $PbZrTiO_6$ (PZT) for piezoelectric applications. Superconductors having the highest critical temperatures are based on copper-oxides. Finally, the thermal properties of ceramics are of technical importance, such as a small expansion coefficient for some ceramics used for cooking ware, precision instruments, and telescopes (Jena glass, Pyrex). One should keep in mind, however, that ceramics constantly compete with other materials, particularly metallic alloys that possess generally a definite price advantage and presently do not suffer as much from hard-to-reproduce fabrication techniques and a high sensitivity towards mechanical and thermal shock, that is, brittleness.

Some ceramists feel that the considerable contributions that ceramic materials have made to the advancement of science and particularly technology are grossly neglected by many individuals in the scientific community and elsewhere and state that modern transportation, computer technology, electronics, communication services, space exploration, illumination, fuel cells, sensors, and much more would be impossible without advanced ceramics. Agreed. It is hoped that this lack of support has been somewhat remedied in this book to the extent possible in the available space.

A final point: The word *ceramics* is derived from the Greek word *keramos*, which means "burned earth." Of course, not all ceramics involves pyrotechnology. The **sol gel technique** may serve as an example, in which a liquid colloidal solution that contains metallic ions and oxygen (the sol) is caused to precipitate into a rigid metal oxide (the gel). This product is, after drying, sintered, i.e., densified at relatively low temperatures, and yields high-strength components.

In short, recent technologies utilize relatively new avenues for producing ceramic products. Thus, it may be argued that the modern ceramics age has just about started.

Suggestions for Further Study

L. Camusso and S. Bortone, *Ceramics of the World*, Abrams, New York (1992).

J. Chappell, *The Potter's Complete Book of Clay and Glazes*, Watson–Guptill Publications, New York (1991).

R.J. Charleston, *Roman Pottery*, Faber and Faber, London (1955).

R.J. Charleston (Editor), *World Ceramics*, McGraw-Hill New York (1968).

D.E. Clark and B.K. Zoitos (Editors), *Corrosion of Glass, Ceramics, and Ceramic Superconductors*, Noyes Publications, Park Ridge, NJ (1992).

W.E. Cox, *The Book of Pottery and Porcelain*, Crown, New York (1944).

S. Ducret, *German Porcelain and Faience*, Universe, New York (1962).

G.A. Godden, *An Illustrated Encyclopedia of British Pottery and Porcelain*, Crown, New York (1966).

R.E. Grim, *Clay Mineralogy*, 2nd Edition, McGraw-Hill, New York (1968).

Handbook of Advanced Ceramics, Vol. I and II, Elsevier and Academic Press (London) 2003.

L.L. Hench and J.K. West, *Principles of Electronic Ceramics*, Wiley, New York (1990).

L.L. Hench, *Bioceramics*, J. Am. Ceram. Soc. *81*, 1705 (1998).

W.B. Honey, *European Ceramic Art from the End of the Middle Ages to About 1815*, 2nd Edition, 2 Volumes Faber and Faber, London (1949–1952).

S. Jenyns, *Japanese Porcelain*, Faber and Faber, London (1965).

W.D. Kingery, H.K. Bowen, and D.R. Uhlmann, *Introduction to Ceramics*, 2nd Edition, Wiley, New York (1976).

W.D. Kingery (Editor), *Ceramics and Civilization*, 3 Volumes, The American Ceramic Society, Inc., Westerville, OH (1986).

W.G. Lawrence, *Ceramic Science for the Potter*, Chilton, Philadelphia (1972).

B. Leach, *A Potter's Book*, Transatlantic, New York (1949).

E. Levin, *The History of American Ceramics*, Abrams, New York (1988).

F. Litchfield, *Pottery and Porcelain: A Guide to Collectors*, Black, London (1925).

G.W. McLellan and E.B. Shand, *Glass Engineering Handbook*, 3rd Edition, McGraw-Hill, New York (1984).

F.H. Norton, *Fine Ceramics, Technology and Applications*, R.E. Krieger, New York (1978).

G. Petzow, The Optimization of the Microstructure of High Technology Ceramics, *Pract. Met.*, *25* (1988), 53.

G. Petzow and M. Herrmann, *"Silicon Nitride Ceramics"* Structure and Bonding, Springer-Verlag, Berlin, Heidelberg, 102 (2002)1.

B.A. Purdy, *How To Do Archaeology the Right Way*, University of Florida Press, Gainesville, FL (1996).

D.W. Richardson, *"The Magic of Ceramics,"* The American Ceramic Society, Westerville, OH (2000).

G. Savage, *Porcelain Through the Ages*, 2nd Edition, Cassell, London (1963).

16

From Natural Fibers to Man-Made Plastics

16.1 • History and Classifications

Fibers

Natural fibers were utilized by mankind considerably earlier than metals, alloys, and ceramics. Indeed, it can be reasonably assumed that fibers were applied by humans long before recorded history. Moreover, even some animals use fibers, for example, when building nests (birds, mammals), webs (spiders), for protection during pupation (caterpillars, silkworms), or for retrieving insects out of narrow holes (chimpanzees). In short, some animals produce fibers for their needs, whereas others collect them. The history of the utilization of fibers is, however, much harder to trace than that of metals and ceramics because fibers often deteriorate through rot, mildew, and bacterial action. In other words, only a few specimens of early fibers have been found so far.

The first raw material that man turned into fabrics was probably **wool**. Scholars assume that this might have occurred as early as during the Paleolithic period, that is, during the Old Stone Age, about 2 million years ago (see Chapter 1). Fabric from wool may have been produced by *felting*, a process that yields a nonwoven mat upon the application of heat, moisture, and mechanical action to some animal fibers, as will be explained in Section 16.2. That there was trade in wool can be inferred from documents and seals dating back to 4200 B.C. which have been found in Tall al-Asmar (Iraq). Breeding and raising wool-producing sheep apparently commenced in Central Asia and spread from there to other areas of the world. This was possible by the fact that sheep adapt easily to different climates. For example, it is reported that the Phoenicians brought the ancestors of the Merino sheep from Asia

Minor to Spain several millennia ago. Today, Merino sheep are raised essentially on all of the continents. New Zealand, for example, with a population of only 3.5 million people, hosts some 50 million sheep of various breeds, whose forebearers were introduced there by British settlers about 150 years ago.

Wild sheep have long, coarse fibers (called hairs) and a softer undercoat of short and fine fibers which provides thermal insulation. The Merino sheep has been bred to eliminate the outer coat and the annual shedding, allowing instead a continuously growing fine and soft fleece which can be repeatedly shorn off.

Fibers retrieved from plants likewise played an important role in early civilizations because of their usefulness for clothing, storage, shelter (e.g., tents), and cordage. It is, however, not fully known when specific plants were first cultivated for fiber production. Nevertheless, some records indicate that **hemp** was presumably the oldest cultivated plant for this purpose. Hemp was first grown in Southeast Asia, from where it spread to China in approximately 4500 B.C. **Flax** was probably cultivated in Egypt before 3400 B.C., at which time the art of spinning and weaving linen was already well developed. Woven flax and wool fabrics were found at the sites of the "Swiss lake dwellers" dating back to the seventh and sixth centuries B.C.

Cotton was spun in India as early as 3000 B.C., as some finds in tombs of that time indicate. A Hindu hymn written around 1400 B.C. describes the fabrication of cotton yarn and the weaving of cotton cloth. In contrast, the Egyptians seem to have started the cultivation of cotton much later, that is, at about A.D. 600–700. From there cotton spread to the Greek mainland and to the Romans.

The production of **silk**, that is, the cultivation of the larva of *Bombyx mori* (commonly called mulberry *silkworm*) is attributed to the Chinese empress, Hsi-ling Shih, who, in 2640 B.C., discovered that the silk filament from a cocoon could be unwound. (Other sources claim that Japan, at about 3,000 B.C., was the first country in which silkworms were domesticated.) The technique of silk-making (called *sericulture*) was kept a secret by the Chinese for about 3,000 years but eventually spread to Persia, Japan, and India. Legend has it that two Persian monks smuggled some silkworm eggs and seeds of the mulberry tree (on whose leaves the larva feed) out of China. This triggered a silk industry in Byzantium during the reign of emperor Justinian (A.D. 527–565) and in Arabic countries beginning with the eighth century A.D. Eventually, the art of sericulture spread in the twelfth century to Italy and thus to Europe. Silk was and still is regarded even today as a highly esteemed, luxury fabric because it is the finest of all natural fibers and its production is cumbersome, as will be described later. Chinese silk textiles manufactured during the Han dynasty (206 B.C.–A.D. 220) have been found in Egypt, in graves located in northern Mongolia, and in Chinese Turkistan.

During the years of the industrial revolution, that is, in the eighteenth and nineteenth centuries, a number of machines were invented and put into service which transferred spinning, weaving, and other fiber-processing techniques from individual homes to centralized factories with consequential economic hardships for some people and concomitant social upheavals. These machines which produced relatively inexpensive fabrics triggered, however, an increase in fiber demand and production.

In the 1880s, it was eventually learned how to dissolve cellulose (from soft wood) and to extrude the resulting substance through narrow nozzles to form **regenerated cellulose fibers** such as **artificial silk**, which was later called **viscose** or **rayon** (see Section 16.3). Actually, the first artificial silk is said to have been made in 1879 by J.W. Swan in England for filaments of light bulbs even before Edison came up with his version![1] The first rayon stockings for women were manufactured as early as 1910 in Germany. (In France, viscose was called "mother-in-law silk" because of its extremely high flammability.) From there it was only one more (but not so easy) step to create completely synthetic fibers (from coal or oil) such as *nylon*.[2] These new products challenged the monopoly of natural fibers for textile and industrial uses. Still, even today, more than one-half of the world's fibers stem from natural sources, among which cotton constitutes the most important part (Figure 16.1). We shall return to these topics in Sections 16.3 and 16.4.

Natural fibers are generally classified by their origin. The *plant* or **vegetable fibers** are mostly cellulose-based, that is, they consist of polymers derived from carbohydrates (i.e., $C_nH_{2n}O_n$) which are manufactured by the plant from water and carbon dioxide gas through photosynthesis. They include *bast fibers* from stems of plants (jute, flax, sunn, hemp, ramie), *leaf fibers* (sisal, New Zealand flax, henequen, abaca, istle), *palm-type and brush fibers* (coir, raffia, palmyra, piassava), *seed and fruit-hair fibers* (cotton, kapok), and, of course, *wood* from trees. Table 16.1 contains usage, price, and origin of some of these fibers.

Among the **animal fibers**, which are protein-based, are wool, mostly from sheep. Specialty animal fibers include *mohair* from

---

[1]Actually, the first useful light bulb was invented by the German H. Goebel who, in 1854, inserted a carbonized bamboo fiber into an evacuated glass flask exactly like Edison 25 years later.

[2]Generic name for polyamides.

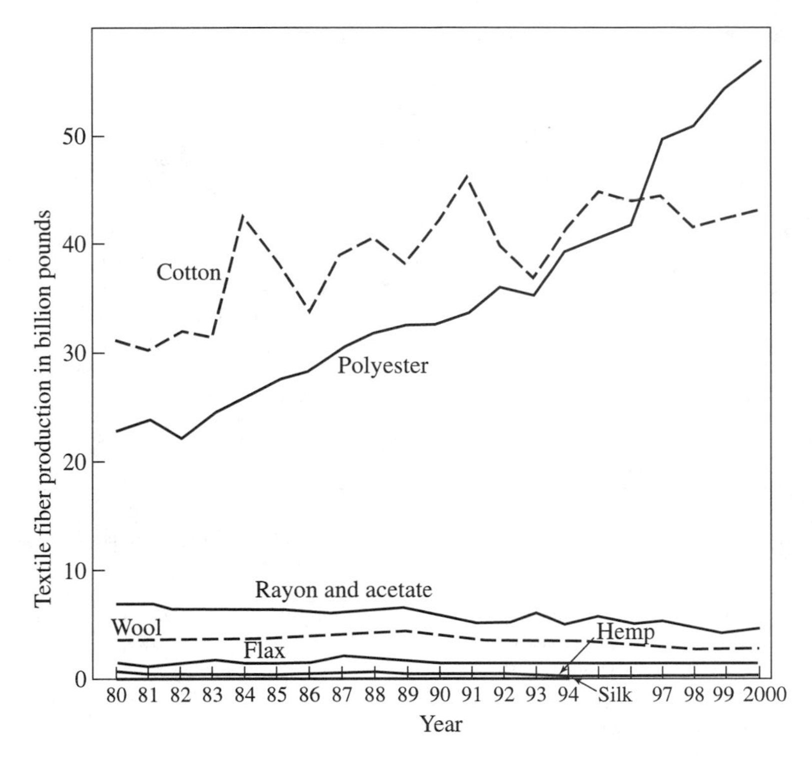

FIGURE 16.1. World textile fiber production 1980–2000. Note: Polyester means here "noncellulosic synthetic fibers." Silk $\approx 0.15 \times 10^9$ pounds. (Source: U.S. Department of Agriculture Outlook Board/Economic Research Service/Nov 2002.)

the fleece of the Angora goat (named after the ancient province of Angora, today's Ankara, in Turkey), *cashmere wool* (stemming from the fine and soft undercoat of Kashmir goats which live in the mountains of Asia), and *camel hair* (which is collected during molting). Other specialty animal fibers stem from the *llama* and the *alpaca,* which are close relatives of the camel and live predominantly in the high grasslands of the Andes in South America. Further, one uses hair from *horses, cows*, and *angora rabbits*. The highest regarded animal fiber, however, is *silk*, which is spun by a caterpillar as already mentioned above.

The third category of natural fibers is comprised of those made from **minerals**. *Asbestos*[3] is the major member of this group of about 30 crystalline magnesium silicates, of which chrysotile accounted for 95% of the world production. Asbestos is no longer utilized for general applications because of its health hazards to the lungs. The fire-resistant property of asbestos was apparently known already to the Greeks and was documented during the first century A.D. by a Roman historian. The Chinese knew about asbestos as Marco Polo reported in the thirteenth century A.D.

[3]*Asbestos* (Greek) = indestructible.

TABLE 16.1. Major sources of natural fibers, usage, and raw and retail prices

Fiber	Usage	Raw world market price (U.S. cents/pound)		Principal growing countries
		1995/96	2000	
Flax/linen raw, retted	Fine textiles, cordage, yarn	18.5	58.2	Belgium, Netherlands, Russia, France, China
Ramie farm price	Garment blend with cotton	28.3	51.8	China, Taiwan, Korea, Philippines, Brazil
Cotton farm price	Garments, paper, explosives, oil, padding	75.6	51.8	China, USA, Pakistan, India, Uzbekistan, Brazil
Wool 3" and up	Knitting yarn, tweeds, flannels, carpets, blankets, upholstery, felts	192.5	75	Australia, New Zealand, China, South Africa, Russia, Argentina

Fiber	Retail price	Fiber	Retail price ($/½ pound, Sept 2003)
Almost white cashmere	$130.00	Superwash Merino wool	$16.33
Fine baby camel	46.00	Ramie	14.10
Baby llama	40.80	Goat hair	11.88
Cultivated silk	40.80	Fine Shetland wool	11.14
Water retted flax	24.76	Moorit English wool	9.28
White alpaca	19.68	Dew retted flax	8.17
Fine mohair	17.45		

Source: Department of Commerce, U.S. Census Bureau, Foreign Trade Statistics.

Rubber

Probably the most fascinating natural material, however, is *rubber*, also known by the Maya name *caoutchouc*.[4] Knowledge of the elastic properties of rubber was brought to Europe in 1496 by Christopher Columbus, who observed inhabitants of Haiti playing with bouncing balls. Considerably later, in 1615, a Spanish explorer reported how "milk" (*latex*[5]) gathered from incisions made on specific tropical trees was brushed on cloaks, rendering them waterproof after drying, or on earthen, bottle-shaped molds to produce containers. It was not until 1735 that a French geographical expedition identified caoutchouc as the condensed sap of the *Hevea brasiliensis* tree, today called the *rubber tree*, because rubber has the capability to erase (rub off) pencil marks. Hevea trees grow only about ten degrees north or south of the equator and need heavy annual rainfalls of about 250 cm, that is, a tropical climate. The rubber tree is cultivated at present particularly in Malaysia, but also in Ceylon, Southeast Asia, and West Africa. Wild rubber is still harvested in South America (Brazil, Peru).

[4]*Caa* = wood and *o-chu* = weeping, i.e., weeping wood.
[5]*Latex* (Latin) = fluid.

Latex is only workable when freshly tapped from the rubber tree. Thus, Europeans struggled considerably to find solvents for caoutchouc to make it spreadable after it arrived in Europe in its "dried" (actually, coagulated, i.e., solid) state. Efforts utilizing ether, turpentine, or naphtha (a waste product from coal-gas plants) were only partially successful since the waterproofed items, produced from rubber, remained sticky particularly when warm, and turned to dust in hot summers. Moreover, these rubber items were odorous, perishable, and became brittle and even cracked upon the slightest use during extremely cold winters. Nevertheless, a large number of products were manufactured in the early 1800s, such as air mattresses, portable bath tubs, waterproof mailbags, boots, and, notably, "mackintoshes" (named after their Scottish inventor). The latter material consisted of a mixture of naphtha and rubber which was sandwiched between double layers of cloth. This procedure alleviated the exposure of a tacky surface which was so annoying in earlier products.

A different (nonchemical) approach was applied in the 1820s by Thomas Hancock in England. He built a machine that rapidly cuts rubber into small pieces which generated heat and thus facilitated the fusing of rubber scraps into blocks. This process is called *mastication*[6] and is still used in the rubber industry.

Riding on the rubber boom of the 1830s was Charles Goodyear of Boston (USA) who, in the cold winter of 1839, after considerable experimentation, accidentally dropped a piece of rubber coated with sulfur and lead[7] onto a hot stove. Both white lead (a common pigment) and sulfur were used before by others in this context, but it was Goodyear who recognized the transformation (curing) process that occurred during heating. The new substance did not melt (as untreated rubber would do); it was durable and retained its pliability and elasticity when cold. This technique of *vulcanization* is still used today with very little modification. However, Goodyear's discovery was made at a time when rubber had a bad reputation because many rubber products had failed in extreme weather. As a consequence, potential investors were reluctant to risk money for the support of additional experimentation. Further, Goodyear was imprisoned for debt more than once, which required him to sell even his children's school books at one point. Nevertheless, in 1842, Goodyear received a U.S. patent which became probably the most litigated one in history (about 150 suits were filed in the first 12 years). Goodyear received a gold medal for excellence at the international exhibitions in London and Paris in the 1850s, at which he displayed his entire vision about the future of rubber products, in-

---

[6]*Mastikhan* (Greek) = to grind the teeth.
[7]Other sources say zinc.

cluding "hard rubber," which he and his brother Nelson created by extending the heating and sulfurization of caoutchouc. Goodyear died in 1860 and left his widow and six children with $200,000 in debts. In contrast, John B. Dunlop, a British veterinarian, fared much better after he patented and developed (in 1888) the pneumatic rubber tire based on Goodyear's invention, which eventually made the bicycle popular and had an impact on the automobile industry several decades later. High-performance tires such as for trucks are still produced from this exceptional material.

The demand for natural caoutchouc has not decreased in this century despite fierce competition from synthetic rubber, for example, Buna, neoprene, and methyl rubber. (The latter was already produced in Germany in the 1910s.) We shall return to synthetic rubber and other synthetic materials in Section 16.3.

Other Organic Materials

There is a large number of other natural materials—not necessarily fibers—which have been used by mankind over the millennia. Among them is **cork**, which is harvested from cork oaks (*quercus suber*) by stripping their bark, boiling it, and scraping off the outer layer. (The trees need to be at least 20 years old but can be stripped again at 8–10-year intervals.) Cork was utilized as early as 400 B.C., for example, by the Romans for sandals, float anchors, and fishing nets. Bottle stoppers made of cork were introduced in the seventeenth century. Today, cork is used for heat-and-sound insulation, linoleum (by mixing cork powder with linseed oil and spreading it over burlap), gasket seals, buoys, and household goods. The cork oak is native to the Mediterranean area and is cultivated in Portugal, Spain, Italy, and India.

Sponges have been utilized by the ancient Greeks and Romans for applying paint, as mops, and as substitutes for drinking vessels. In the Middle Ages, burned sponges were used as medicine. Sponges are primitive, multicellular sea animals which attach to surfaces. They are removed by skin divers from tidal levels to depths of about 70 meters, particularly in the Eastern Mediterranean area and on the West coast of Florida.

The list of natural materials is not complete with the brief sketch given above. Indeed, it is estimated that in the Western Hemisphere alone, more than 1000 species of plants or parts of plants are utilized in one way or another to create utilitarian products. Most of them, however, are consumed locally or in such small quantities that their mention is not warranted here. Other organic materials, such as **animal skin**, **animal guts**, **horns**, **ivory** (from elephant or mammoth tusks), **straw**, **bark**, **reed**, **shell**, **amber** (fossilized tree resin), etc., likewise have been used by mankind for millennia and complement the variety of materials which are at our disposal for a more comfortable living.

16.2 • Production and Properties of Natural Fibers

Animal Fibers

Animal fibers (wool, silk, etc.) are composed mostly of *proteins*, as already mentioned in Section 16.1. (Proteins are highly complex substances which consist of long chains of alpha amino acids involving carbon, hydrogen, nitrogen, sulfur, and oxygen.) All taken, animal fibers do not contain cellulose. They are therefore more vulnerable to chemical damage and unfavorable environmental conditions than cellulose.

After extraction of the fibers as described above, they need to be spun into yarn. For this the individual fibers are arranged in parallel to overlap each other, yielding a ribbon. These ribbons are then softened with mineral oil, lubricated, and eventually drawn down to the desired sizes and twisted for securing the position of the fibers. The yarn is eventually woven into fabrics.

Wool consists mainly of the animal protein *keratin*, which is common in the outermost layers of the skin, nails, hooves, feathers, and hair. Keratin is completely insoluble in cold or hot water and is not attacked by proteolytic enzymes (i.e., enzymes that break proteins). Keratin in wool is composed of a mixture of peptides. When wool is heated in water to about 90°C, it shrinks irreversibly. This is attributed to the breakage of hydrogen bonds and other noncovalent bonds.

Wool fibers are coarser than those of cotton, linen, silk, or rayon, and range in diameter between 15 and 60 μm, depending on their lengths. Fine wool fibers are 4–7.5 cm long, whereas coarse fibers measure up to 35 cm. Unlike vegetable fibers, wool has a lower breaking point when wet. The fibers are elastic to a certain extent, that is, they return to their original length after stretching or compression and thus resist wrinkling in garments. The low density of wool results in light-weight fabrics. Wool can retain up to 18% of its weight in moisture. Still, water absorption and release are slow, which allows the wearer not to feel damp or chilled. Wool deteriorates little when properly stored and is essentially mildew-resistant. However, clothes moths and carpet beetles feed on wool fibers, and extensive exposure to sunlight may cause decomposition. Further, wool deteriorates in strong alkali solutions and chars at 300°C.

Felting shrinkage, that is, compaction, occurs when wet, hot wool is subjected to mechanical action. Thus, washing in hot water with extensive mechanical action is harmful. On the other hand, felting produces a nonwoven fabric, as already mentioned in Section 16.1. This is possible due to the fact that

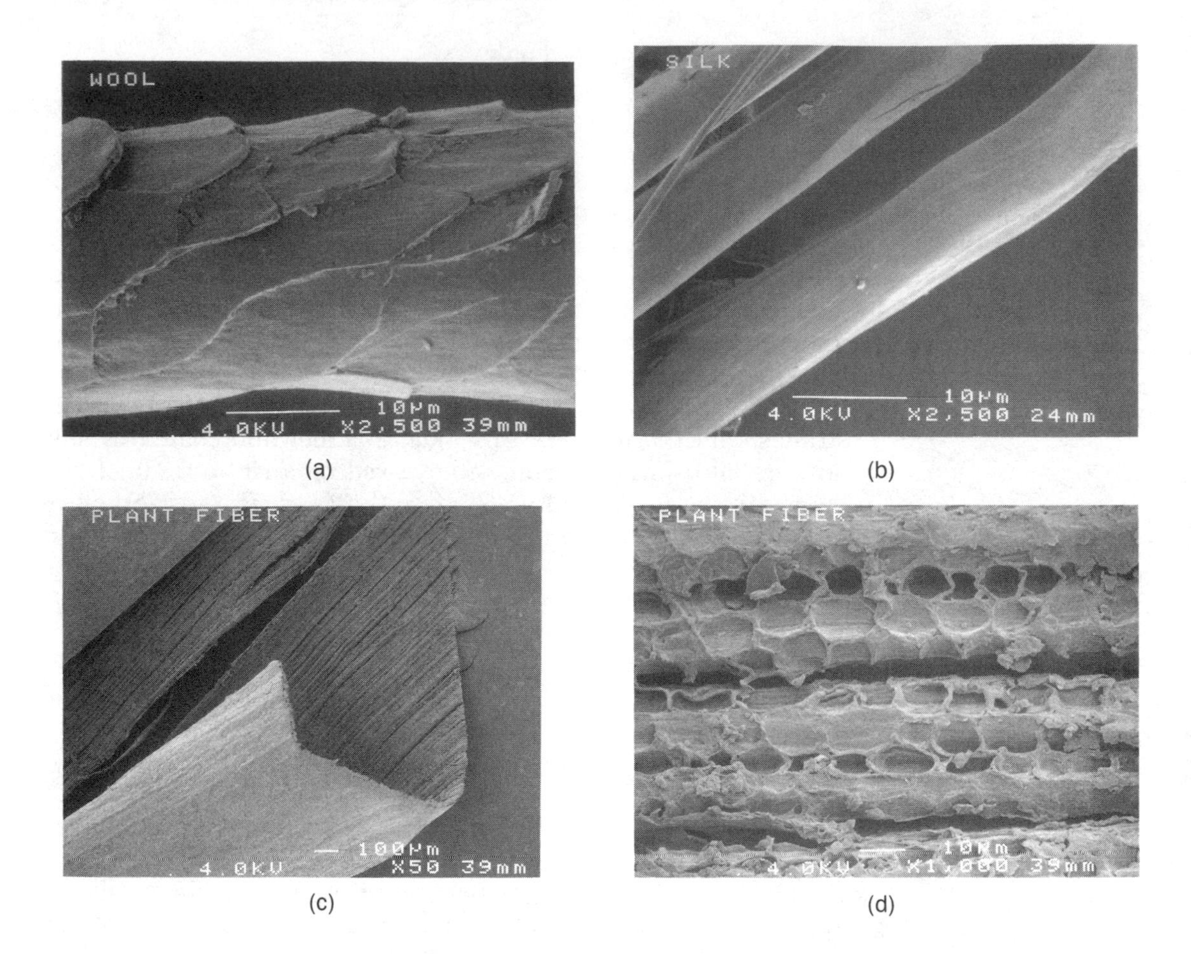

FIGURE 16.2. Scanning electron micrographs of (a) wool fiber (note the scales on the surface that overlap each other; the tips point to the free end of the hair), (b) silk fibers (note the thin synthetic fiber that has been smuggled in), (c) plant fiber at low magnification, and (d) plant fiber at high magnification. (Courtesy of R. Crockett and R.E. Hummel, MAIC, University of Florida.)

animal fibers (except silk) are covered with an outer layer of unidirectional overlapping scales, as depicted in Figure 16.2(a). Mechanical action in conjunction with heat and moisture causes the fibers to slide past each other and interlock. Felt is widely used in the hat industry and for making slippers and polishing materials.

Silk is spun by the larva of *Bombyx mori*, as was mentioned in Section 16.1. The proteins of silk contain about 80% fibroin (which makes up the filament) and about 20% sericin or silk gum (which holds the filaments together). Minor constituents are waxes, fats, salts, and ash. Silk is a continuous fiber, that is, it has no cellular structure. The life cycle of Bombyx mori includes hatching of the disk-shaped eggs in an incubator at 27°C, which requires about 10 days. The "silkworm," 3 mm long and 3 mg in mass, eventually grows into a 90-mm-long caterpillar which needs five daily feedings of chopped, young mulberry leaves. After about 6 weeks and four moltings, it stops eating, shrinks somewhat, and its head makes restless rearing movements, indicating a readiness to spin the cocoon. The silkworm is then transferred into a compartmentalized tray or is given twigs. There it spins at first a net in whose center the cocoon is spun around the silkworm. After 3 days, during which time the filament is wound in a figure-eight pattern, the completed cocoon has the shape and size of a peanut shell.

The silk substance is produced by two glands and is discharged through a spinneret, a small opening below the jaws. The spinneret is made up of several chitin plates which press and form the filament. The filament (called *bave*) actually consists of two strands (called *brins*) that are glued together and coated by silk gum (*sericin*), which is excreted by two other glands in the head of the silkworm. The liquid substance hardens immediately due to the combined action of air exposure, the stretch and pressure applied by the spinneret, and to acid that is secreted from still another gland. Under normal circumstances, the chrysalis inside the cocoon would develop into a moth within 2 weeks and would break through the top by excreting an alkaline liquid that dissolves the filament. Male and female moths would then mate within 3 days and the female would lay 400–500 eggs, after which time the moths would die. The life cycle is, however, generally interrupted after the cocoon is spun by applying hot air or boiling water (called *stoving* or *stifling*) except in limited cases when egg production is desired. The filaments of 2–7 cocoons are then unwound (called *reeling*) in staggered sequence to obtain a homogeneous thread strength; see Plate 16.1. The usable length of the continuous filament is between 600 and 900 meters. Shorter pieces are utilized for spun silk. It takes 35,000 cocoons to yield 1 kg of silk. [Note in this context the silk fibers depicted in Figure 16.2(b)].

The raw silk is usually degummed to improve luster and softness by boiling it in soap and water, which reduces its weight by as much as 30%. (Sericin is soluble in water whereas fibroin is not.) The silk is subsequently treated with metallic salt solutions (e.g., stannic chloride), called weighting, which increases the mass (and profit) by about 11% and adds density. Excessive weighting beyond 11% causes the silk to discolor and decompose. Likewise, dying adds about 10% weight. Silk fabric treated with polyurethane possesses excellent wet wrinkle recovery and dimensional stability during washing. Silk is more heat-resistant than wool (it decomposes at about 170°C); it is rarely attacked by mildew but degrades while exposed extensively to sunlight. Silk can adsorb large quantities of salts, for example during perspiration. These salts, however, eventually weaken the silk and destroy it.

Vegetable Fibers

The commercially important vegetable fibers are taken from those parts of plants that strengthen, stiffen, or otherwise support their structure. Exceptions are seed hair, as in cotton, or the seed-pod fiber kapok.

Bast fibers (see Tables 16.2 and 16.3) are taken from tall, reedlike plants whose stems they cover. The fibers are cemented together by gums and need to be separated by a decomposing process called *retting*. This can be achieved by soaking the plants in water (water retting) or by spreading them out and thus exposing them to the weather (dew retting). In either case, the gummy substance is eventually broken down by microbiological agents which allow one to separate the fibers after about one to five weeks. Subsequent vigorous mechanical action splits the fibers further apart into even finer *fibrils*, which are soft and fine, that is, hairlike. The longer ones are called *line fibers*. They are particularly suitable for processing into yarns, textiles, and cordage. The shorter pieces are called *tow*. The physical properties of bast and other fibers are listed in Table 16.2. Note the pronounced difference in fineness between bast and leaf fibers.

Leaf fibers usually extend along the entire length of a leaf to reinforce its structure and keep it rigid. The fibers are embedded in a pulpy tissue which needs to be removed by mechanical scraping, called *decortication*. Leaf fibers are generally hard and coarse compared to bast fibers (Table 16.2).

The fibrils of bast and leaf fibers consist of a large number of elongated cells whose ends are cemented together. The hollow interior of the cells is called the *lumen*. The cell walls are composed of *cellulose*, that is, of essentially linear, and complex carbohydrate macromolecules, each of which consists of hundreds or thousands of glucose units having the formula $C_6H_{10}O_5$. In other words, cellulose is a polymerized sugar. We shall return to macromolecules in Section 16.3. The fibers are mostly crystalline and are separated by small amorphous regions. Cellulose is insoluble in water and is the most abundant of all naturally occurring organic compounds. It is undigestible by humans but can be broken down by microorganisms, for example, in the stomachs of certain herbivorous animals such as cows, horses, and sheep.

The cementing constituent between the cells of woody tissue is called *lignin*, a complex macromolecule which is not based on carbohydrates. *Pectic substances*, that are capable of forming thick solutions and which consist of an associated group of polysaccharides (sugars) are also contained in plant substances al-

TABLE 16.2. Properties of selected fibers

Fiber	Fineness (den)[a]	Tenacity (g/den)	Tensile strength (MPa)	Elongation (strain) (%)	Color range
			Bast Fibers		
Jute	20	3	39	1.5	Creamy white to brown
Flax (Linen)	5	5	66	1.5	White to brown
Hemp	6	4	52	2.0	Light to grayish brown
Ramie (China grass)	5	5	67	4.0	White to grayish brown
			Leaf Fibers		
Sisal	290	4	51	3.0	Creamy white to yellowish
Henequen	370	3	39	5.0	Creamy white to reddish brown
Abaca (Manila hemp)	190	5	64	3.0	Creamy white to dark brown
Istle	360	2.5	32	5.0	White to reddish yellow
			Seed Fiber		
Cotton	2	2.5	300	8.0	Creamy to grayish white, brown, purple, blue, red
			Animal Fibers[b]		
Wool	4–20	1.5	150	40	Creamy white to brown
Silk	1.0	4	800	20–25	White
			Synthetic Fibers		
Nylon	0.5–18	3–10	350–890	15–40	White-transparent
Polyester	0.1–10	3–9	500–1100	11–40	White-transparent

[a]denier (den) is the mass in (*g*) of 9000 m of fiber. The smaller the number, the finer the yarn.
[b]See also Table 16.6 in which the mechanical properties of spider drag lines are given.

beit in small proportions. In addition, plant materials contain *extractives* (gums, fats, resins, waxes, sugars, oils, starches, alkaloids, tannins) in various amounts. Extractives are nonstructural components that are deposited in cell cavities or cell walls. They may be removed (extracted) without changing the wood structure. Further constituents of plant fibers are *minerals*, which remain as *ash* when a plant is incinerated. The compositions of selected plant fibers are presented in Table 16.3. Some commercially important types of vegetable fibers will be briefly discussed below.

Flax is a bast fiber that is extracted from *Linum usitatissimum*, an annual plant which grows in the north and south temperate

TABLE 16.3. Composition of selected plant fibers

Fiber	Cellulose (%)	Moisture (%)	Ash (%)	Lignin and pectins (%)	Extractives (%)
		Bast Fibers			
Congo jute	75.3	7.7	1.8	13.5	1.4
Flax (linen)	76.0	9.0	1.0	10.5	3.5
Hemp	77.1	8.8	0.8	9.3	4.0
Ramie (China grass)	91.0			0.6	
		Leaf Fibers			
Sisal	77.2	6.2	1.0	14.5	1.1
Henequen	77.6	4.6	1.1	13.1	3.6
Abaca (Manila hemp)	63.7	11.8	1.0	21.8	1.6
Istle	73.5	5.6	1.6	17.4	1.9
		Seed Fiber			
Cotton[a]	90.0	8.0	1.0	0.5	0.5

[a]*Note:* "Easy care" or "no-iron" fibers are treated with formaldehyde resin that emits formaldehyde fumes. They have been observed to cause tiredness, headaches, coughing, watery eyes, or respiratory problems.

zones where the soil is fertile and sandy, and the weather is cool and damp during the summer. Major flax-producing countries are Belgium (particularly the western part), The Netherlands, and Luxembourg. Water retting yields the best linen (i.e., the fabric made from flax) and takes about 8–14 days, whereas dew retting (2–5 weeks) yields lesser qualities. The subsequent *scutching* mechanically crushes the retted stems into small pieces called *shives*, which are then beaten for separation. The long fibers (30–90 cm long and 0.05–0.5 mm in diameter) are used for fine textiles. The shorter pieces (tow), which constitute one-third of the fiber yield, are used for cordage, coarse yarn, and shoe-stitching threads. Linen is characterized by high strength, low stretch, high water absorption, and pronounced swelling when wet. Flax is also grown for flax seeds, linseed oil, and flax straw for fine, strong paper.

Ramie, a vegetable fiber which was used for Chinese burial shrouds over 2000 years ago, has been relatively unknown for garments in the western part of the world. It is now often blended with cotton and resembles fine linen to coarse canvas. Ramie is produced from the stalk of a plant and is processed like linen from flax. The fiber is very fine and silk-like, naturally white, and of high luster. (See Tables 16.2 and 16.3.) Ramie is resistant to bacteria, mildew, and insect attack; it is extremely absorbent, dyes easily, increases in strength when wet, does not shrink, and improves its luster with washing. However, Ramie has a low elasticity, lacks resiliency, wrinkles easily, and is stiff and brittle. It therefore tends to break when a garment made of it is pressed too sharply.

Wood

Wood is and always has been a major material for construction, tools, paper-making, fuel, weapons, and, more recently, as a source for cellulose. Indeed, manuscripts have been found by Aristotle, Theophrastus, and other ancient writers who describe the properties of wood as it was known in those days.

Goods made of wood, when left unpainted, are aesthetically pleasing and convey a feeling of warmth. Wood is one of the few natural resources that can be renewed when forests are managed properly. Moreover, forests are a necessity for water control, oxygen production, recreation, and for providing habitat to many animal species. Approximately one-third of the land mass of the earth (27%) is presently covered with forests. However, it is estimated that the world's forests decrease by about 0.9% annually. We shall return to this subject in Section 18.3.

In relation to its weight, wood has a high strength. It is an electrical and thermal insulator (Figures 11.1 and 14.1) and has desirable acoustic properties. It can be easily shaped and finished and is inert to many chemicals. However, not all properties of wood are favorable. Many types of wood decay through interaction with water and wood-destroying organisms. Wood may burn, it is hygroscopic, and changes its size when the humidity fluctuates. Moreover, the physical properties of wood vary in different directions due to its fibrous nature (anisotropy), as shown in Table 16.4. Finally, cut lumber may have imperfections from knots, etc., which decrease its strength. In short, wood is not a homogeneous and static material as are many metals and ceramics.

TABLE 16.4. Tensile strength, compressive strength, modulus of elasticity, and densities for some wood species (12% Moisture) (see also Table 2.1)

Material	Tensile strength σ_T [MPa]		Compressive strength [MPa]		Density [g/cm^3]	Modulus of elasticity $\parallel$ to grain [GPa]
	$\parallel$ to grain	$\perp$ to grain	$\parallel$ to grain	$\perp$ to grain		
Soft Wood						
Douglas fir	78	2.7	37.6	4.2	0.45	13.5
Ponderosa pine	73	2.1	33.1	3.0	0.38	8.5
White spruce	60	2.5	38.7	4.0	0.35	9.2
Red cedar	45.5	2.2	41.5	6.3	0.3	7.7
Hard Wood						
American elm	121	4.5	38	4.7	0.46	9.2
Sugar maple	108	7.6	54	10.1	0.56	12.6
Beech	86.2	7	50.3	7	0.62	
Oaks	78	6.5	42.7	5.6	0.51–0.64	12.3

Lumber is generally classified into *softwood* and *hardwood* (even though some "hardwoods" are actually softer than some "softwoods"). Major growth areas for softwoods are located in the temperate climate zone whereas the home of hardwood is predominantly in the tropical forests of Africa and the Amazonas region.

As was discussed above, the structural components of wood consist of elongated plant cells [Figure 16.2(d)] that are made up largely of cellulose and which are held together by lignin. Wood is therefore a naturally occurring *composite material*. Two general types of cells exist: The food-storing elements are called *parenchyma*. They normally remain alive for more than one year. The cells that provide the support of the tree and serve for the conduction of the protoplasm are known by the name *prosenchyma*. They lose their function in the same year they were formed. The food-conducting tissue located at the outer part of the prosenchyma is called the *phloem*. In these specialized cells, the phloem sap that carries the products of photosynthesis streams downward from the leaves to the root. On the other hand, the water and minerals stream upward in the cells located toward the inside, called the *xylem* (see Figure 16.3). Botanists subdivide both the phloem and xylem into primary and secondary parts. The secondary phloem contributes to the formation of bark, and the secondary xylem contributes to the formation of wood. The new wood in the secondary xylem is called

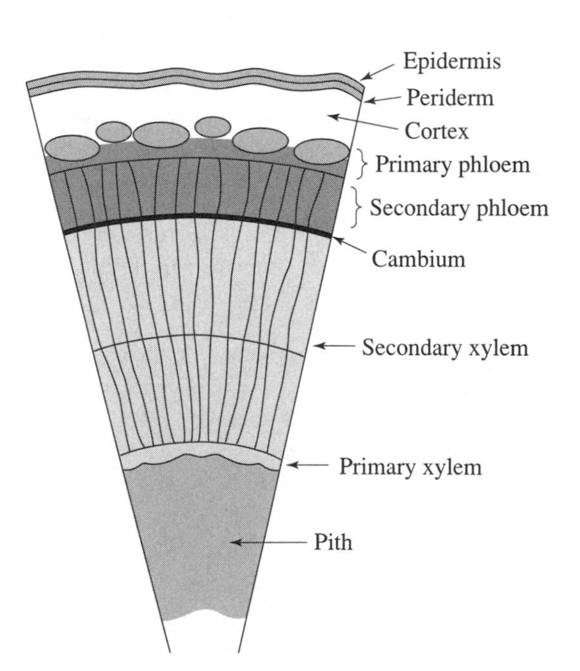

FIGURE 16.3. Schematic representation of a cross section of a young wood stem showing xylem, cambium, phloem, cortex, and epidermis.

sapwood. Eventually, over a period of years, the older sapwood dies and becomes more rigid; it is then called *heartwood*. The heartwood is the firm portion of the tree at the center and provides strength and support. It is the principle source for lumber and pulp. The xylem and phloem areas instead supply the chemicals, that is, wood distillates, latex (for rubber), and, to some extent, fuel.

The growth in diameter of a tree takes place in a single row of cells between the phloem and the xylem, which is called the *cambium*. As cell division proceeds, about 6–8 xylem cells are formed for every phloem cell. In other words, the increase in thickness of the wood is greater than that of the bark. As a consequence, the larger formation of the secondary xylem cells forces the cambium outward and the diameter of the stem increases.

The phloem in turn is surrounded by the *cortex* and the *epidermis* (Figure 16.3), which protect the underlying tissue from mechanical injury. As the plant grows, the epidermis is ruptured and one or more protection layers called *periderms* are formed.

Cell division in the cambium is larger during spring when rapid growth occurs. This leads to the formation of *spring wood* (or early wood), which is somewhat different in chemical composition, color, and physical structure than *summer wood* (or late wood). In other words, darker and lighter areas alternate. Thus, the age of a tree can be estimated by counting these *annual rings*. (In some years, for reasons not well understood, additional rings or no rings at all form. The term *growth rings* instead of annual rings is therefore preferred.) See in this context Plate 16.3.

Many cross-sectional cuts of trees show *rays*, that is, bands of tissue that radiate from the center (called *pith*) to the phloem. They are essentially parenchyma (food-storing) cells. They add to the liveliness of lumber.

Selected physical properties of some species of wood are presented in Table 16.4. It is observed there that the density of *softwood* is generally lower than that of *hardwood* because softwood contains more void spaces. The density of wood is commonly given for a moisture content of 12% (compared to green wood), which is characteristic for 65% environmental humidity. The water is held in the cavities of the cells (lumen) from where it can escape relatively effortlessly, or in the cell walls of the cellulose where it is more firmly held. The lightest wood of commercial importance (Balsa) has a density of only 0.11 g/cm^3, which is lighter than cork (0.24 g/cm^3), whereas black ironwood of South Florida has a density of 1.42 g/cm^3 and therefore does not float in water. Heartwood generally has a higher density than sapwood owing to the deposition of extraneous materials.

Table 16.4 also contains the tensile strengths of some representative types of wood. It is noted that σ_T is considerably larger when wood is stretched parallel to its fibers compared to when it is stretched in its radial direction. This can be understood when knowing that the bonds between the individual cellulose fibers are much weaker compared to those within the crystalline fibrils. Moreover, the tensile strength in *compression* parallel to the fibers is lower than that in tension since the fibers tend to buckle in compression. This anisotropic behavior is reduced in particle board and plywood, in which small wood chips of random orientations or thin layers of wood (plies) having mutual orientations of 90° are glued together.

The modulus of elasticity for wood is fairly small compared to metals and ceramics (see Table 2.1). This underscores the well-known elastic properties of wood. The moduli of elasticity in tension, compression, and bending are approximately equal, but the elastic limit is considerably lower for compression than that for tension.

When wood which is completely saturated with water is slowly and uniformly dried, the water in the lumen evaporates first and the wood does not change its dimensions. As a consequence, no change in the stiffness or strength is observed. However, further drying to less than about 30% water content causes some loss of water from the cell walls. At this point the wood shrinks due to the reduced distance between the cells. As a consequence, the density as well as the bonding between the fibers and thus the strength increase. In other words, completely dry wood is stronger than green wood (but less elastic).

Other physical properties of wood (electrical, thermal, mechanical) are given in the appropriate chapters. Wood production figures are listed in Chapter 18, particularly in Table 18.3.

Paper

It is generally believed that the art of paper-making was invented at about A.D. 105 in China by Ts'ai Lun, who utilized fibers from flax, hemp, and mulberry tree bark. His technique spread in time to central Asia, Persia, Egypt, Morocco, and eventually, during the second millennium A.D., to Europe. The word "paper" is derived from a reed plant named *Cyperus papyrus*, which was utilized by the Egyptians to make a paper-like material. In their procedure, however, the fibers were not at first separated, reworked, and then bound together into a sheet as in paper. Instead, the fibrous layers of the stem were placed side by side and crossed at right angles with another identical layer, which were subsequently dampened and pressed (see Plate 16.2). Written documents were recorded even before that time on *parchment*, that is, on processed skin of certain animals such as sheep, goats, and calves. The name is supposedly derived from the city of Pergamum (today Bergama, Turkey) where parchment is said to have originated in the second century B.C. Skins for writing materials were used still earlier, but the parchment technique involving better cleaning, stretching, and scraping allowed the utilization of both sides of the skin and thus facilitated books rather than resorting to a rolled manuscript. Particularly fine parchment, made from stillborn or newly born calves or lambs, is called *vellum*. However, this name is now broadened to include fine paper made from wood and rag pulp having a special finish (see below).

Today, wood pulp is by far the most utilized raw material for paper even though flax, cotton scrap (rags), cereal straw, hemp, jute, esparto, inorganic fibers (glass), and synthetic polymer fibers, such as nylon and polyolefins, are used for fine or specialty products. Fibers from recycled waste paper comprise a substantial portion of raw stock, particularly for low-grade paper, paper board, and wrapping paper. Waste paper utilization ranges from 48% (USA) to 3% (Finland). For details, see Chapter 18.

Paper-making commences with *pulping*, that is, with separating the fibers of wood, etc., either by mechanical means (groundwood pulp) or by chemical solutions which dissolve and remove the lignin and other wood components leaving only cellulose behind (see above). Groundwood pulp is made by involving either a grinder or by passing wood chips through a mill causing a considerable degree of fragmentation. Groundwood pulp contains *all* wood components and is thus not suitable for papers in which high whiteness and permanence are required. It is highly absorbent and is therefore used for printing and wallpaper. *Chemical wood pulp* is obtained by cooking wood chips in chemical solutions at high pressures. Among the lignin solvents that are in use are caustic soda + sodium sulfide (called kraft process), or calcium bisulfite + sulfurous acid. Chemical wood pulp is employed when high brightness, strength, and permanence are desired. An optional (and environmentally objectionable) *bleaching* process with chlorine whitens the pulp even further (called alpha or dissolving pulp).

The next step in paper-making is called *beating* or *refining*, during which the cellulose fibers that are dispersed in an aqueous slurry (slush) are mechanically squeezed and pounded. This reduces the size of the fibrils, compacts them, makes them flexible, causes them to swell, and makes them slimy. Beating also reduces the rate of water drainage and increases the ability of the fibers to bond together when dried. Beating for one hour or more eventually produces a dense paper of high tensile strength, low porosity, and high stiffness. A sheet from unbeaten pulp, instead, is light, fluffy, porous, and weak. Industry predominantly uses the *Hollander beater*, developed in 1690. It consists of an oval tank in which a heavy roll revolves against a bedplate.

In a third step, called *filling* or *loading* mineral pigments, in particular, Kaolin (Section 15.2), but also occasionally titanium dioxide or zinc sulfide are added to improve brightness, opacity, softness, smoothness, and ink receptivity. Fillers are essentially insoluble in water and have, in most cases, no affinity to fibers. Thus an agent such as alum is added to hold the filler in the future sheet. The amount of filler varies from 1 to 10% of the fiber.

Paper that contains only cellulose absorbs substantial quantities of water. This causes water-based ink to readily spread in it. If, however, the paper is impregnated with appropriate substances (called *sizing*), the wetting is prevented. Agents for sizing are starch, glue, casein, resin (from pine tree stumps), waxes, asphalt emulsions, or synthetic resins, to mention a few. Unsized papers are used for facial tissue, paper towels, and blotting paper.

Further additives include *colorants, pH controller* (acidic paper deteriorates with time), and polymers from the urea-formaldehyde family (to provide wet strength to the finished paper).

The last major steps involve *sheet forming* and *drying*. This has been done by hand for many centuries. A fine screen contained in a wooden frame is dipped into the fibrous suspension prepared as described above. The water is then allowed to drain through the screen leaving a thin mat of fibers which is removed, squeezed, and dried. The same principle is also used in modern machines.

A major concern during large-scale paper production is the waste water that may contain inert solids called slime which eventually settles on the bottoms of rivers or lakes. The waste water may also be poisonous or have a biochemical oxygen demand which is hazardous to aquatic life. Recycling of waste water employing filtration, flotation, and sedimentation should therefore be the rule.

The physical properties of papers are classified by basis weight (in grams per square meter), thickness (caliper), tensile strength, stretch, bursting strength, tearing strength, stiffness, folding endurance, moisture content, water resistance, brightness, color, opacity, gloss, and transparency. Papers that come in contact with food (packaging) are regulated by the Food and Drug Administration. Specialty papers having maximum strength, durability, color, texture, and feel are required for bank notes, high-grade stationary, paper for Bibles, cigarettes, legal documents, security certificates, and so on. They are mostly made of cotton or linen fibers that are derived from scraps obtained from the textile industry.

16.3 • Tales About Plastics

It is not quite clear why, in 1846, Professor Schönbein, a German chemist who taught at the University of Basel (Switzerland), spilled an aggressive mixture of sulfuric acid and nitric acid in his (or his wife's?) kitchen. Knowing of the potential harm that this "nitrating acid" could do to his property, he quickly mopped it up, grabbing the first thing he could lay his hands on. This was his wife's cotton apron. Fearing reprimand, he immediately rinsed the apron in running water and hung it up to dry near the stove. Soon afterwards the apron ignited and burned to dust. This was the birth of smokeless *gun cotton*. But more importantly, it was also the birth of man-made plastics, as we shall see momentarily. What happened? A transformation took place from relatively inert cellulose into reactive *cellulose nitrate*. Specifically, a compound was formed in which some of the hydroxyl groups ($-OH$) of the cellulose polymer converted into nitrate groups ($-O-NO_2$), whereby the sulfuric acid served as a catalyst.

Interestingly enough, there exist some important differences between various cellulose nitrates. In its explosive modification, cellulose nitrate is *completely* nitrated and is then called *trinitrate* of cellulose. In contrast, the lesser nitrated *dinitrate*, which contains many residual hydrogen bonds, is no longer explosive, but still highly flammable. The flammability can be reduced by further denitrification using, for example, sodium bisulfate or other substances, as we shall explain below.

Cellulose nitrate, in contrast to untreated cellulose, can be dissolved in a number of organic liquids such as in an equimolar mixture of alcohol and ether. Moreover, *nitrocellulose*, as it is also called, is pliable at elevated temperatures and thus can be molded into objects that retain their form and are hard after cooling. Last but not least, cellulose nitrate can be reshaped as often as desired by renewed heating. For this reason, nitrocellulose is called a **thermoplastic**[8] material. The applications have been manifold.

---

[8] *Thermos* (Greek) = warm; *plasticos* (Greek) = to shape, to form.

For example, artificial silk can be spun from cellulose nitrate solutions. This was accomplished in 1884 by the French Count de Chardonnet (see Section 16.2). Somewhat before that time, i.e., in 1879, J.W. Swan experimented with filaments for electric light bulbs made of artificial silk, as was already mentioned in Section 16.1. The early movie industry would not have been possible without thin cellulose nitrate layers, called *films*, on which a light-sensitive emulsion was cast. However, because of their flammability, cellulose nitrate films now have been replaced by other substances. In short, Schönbein's discovery opened the door to a completely new world, namely, that of synthetic (or semi-synthetic) polymeric materials called *plastics*.

The next story is equally seminal. In 1863, the New England firm of Phelan and Collander, which manufactured billiard balls made of ivory, became concerned about the ever increasing slaughter of elephants whose tusks provided the raw material for their products. Their demand was in direct competition with that of other companies which utilized ivory for piano keys, jewelry, ornaments, and knife handles. Indeed, about 100,000 elephants per year (and a large number of hunters) had to give their lives in the 1860s to satisfy the demand of the world's ivory industry. No wonder that the above-mentioned manufacturers were worried about a possible extinction of the elephant (and their raw material). Thus, they offered a handsome price of $10,000 in gold to the person who would be able to produce a suitable substitute. The American journeyman printer John Hyatt won this prize. One day he found a bottle tipped over which contained collodion, a solution of nitrocellulose in ether and alcohol which printers use to coat their fingertips to keep them from getting scorched by hot lead. The content was spilled out and had turned into a tough, flexible, solid layer. This gave Hyatt the idea to try it as a substitute for billiard balls. However, there was one big drawback of these billiard balls: each time they collided with force, they exploded and caused every man in a saloon to pull his gun. After some experimentation, Hyatt added some camphor (from a tree having the same name) for denitrification of nitrocellulose. This resulted in a thermoplastic material, originally called *artificial ivory* but later named *celluloid*. Billiard balls made of celluloid turned out to be somewhat too brittle, but other products such as ping-pong balls, combs, photographic films, and dental plates were quite successful. The only shortfall of celluloid was (and still is) that it inflames quite readily. (As youngsters we demonstrated this by focusing the sunlight with a magnifying glass onto old ping pong balls.)

Cellulose nitrate, celluloid, and galalith[9] (still another product which was developed in 1897 from milk casein, kaolin, and formaldehyde[10]) are made of natural materials that are chemically converted into new compounds. For their creation no new polymer is synthesized. Rather, a modification of natural polymers is performed. They are therefore classified as *semisynthetics*.

In contrast, a *fully synthetic* organic polymer was invented by the Amero-Belgian L.H. Baekeland. In 1906, he introduced a resin made of various phenols (C_6H_5OH) which he heated with formaldehyde (CH_2O) under pressure. This yielded an insoluble, hard plastic. *Bakelite*, as it was called, is inert against solvents and is a good electrical and heat insulator. Bakelite (and other phenolic resins) differ in one important aspect from the above mentioned thermoplastics. They *retain* their shape upon heating and are therefore called **thermosets**. In other words, they have a good dimensional stability. Examples are epoxies, polyesters, and phenolic resins.

Talking about classifications: The third category of plastics is called **elastomers**. Natural rubber, as well as Mr. Goodyear's vulcanized rubber (see Section 16.2), belong to this family of materials.

Polymerization

So far, we have used the word "polymer" quite casually without explaining what precisely is meant with this term. We shall now make up for this deficiency. This can best be done by recounting the work of Eduard Simon, who, in 1839, experimented with an organic liquid called *styrene*, which he distilled out of an extract from an Asian tree having the botanical name *Liquambar orientalis* and whose sap was used 3000 years ago by Egyptians for embalming bodies. In any event, Simon discovered that heating liquid styrene to 200°C did not yield a gas as observed for other substances known at that time, but instead produced a gelatinous solid. The analysis by Blyth and Hofman six years later revealed that the chemical composition of the liquid and the solid form of styrene were identical, that is, styrene and *metastyrene*[11] both had a ratio of 8 carbon atoms to 8 hydrogen atoms. This disproved an earlier hypothesis which postulated that a reaction between styrene and oxygen may have taken place. Actually, what

[9]*Gala* (Greek) = milk; *lithos* (Greek) = stone.
[10]By W. Krische and A. Spitteler of Germany.
[11]*Meta* (Greek) = after, between. In this context: The substance that comes after the heating at 200°C, i.e., the solid configuration.

happened was that several styrene molecules (called styrene *monomers*) united via covalent bonds with other styrene molecules without changing the ratio of the atoms. This process is called **polymerization.**[12] The chemical equation for this reaction is commonly written as:

$$n\,(\mathrm{CH_2 = \underset{\displaystyle C_6H_5}{\underset{|}{C}H}}) \rightarrow \sim(\mathrm{CH_2 - \underset{\displaystyle C_6H_5}{\underset{|}{C}H}})_n \sim \qquad (16.1)$$

where n is an integer and the lines between the carbon and hydrogen atoms symbolize, as usual, the covalent single (−) or double (=) bonds. [See in this context Figure 16.4(c).] The reaction equation (16.1) essentially expresses that n individual styrene molecules combine to form a polymeric chain (called here a *linear polymer*) having n members called *mers* or *repeat units*. The wavy lines to the left and right of metastyrene (or *polystyrene* as it is called today) indicate a multiple continuation of the same repeat unit until eventually the chain is terminated by an *end-group* (see below).

The question with respect to the shape of these (and some other) polymers was vigorously debated during the end of the nineteenth and the beginning of the twentieth centuries until eventually, in the early 1930s, mainly due to the extensive and persistent work of H. Staudinger,[13] the above promulgated concept involving essentially long chains (having occasionally side branches, or rings) became generally accepted. The question then arises, how long are these "supermolecules"? This information can be obtained by knowing the molar mass[14] of the polymeric chains. Extensive experimentation utilizing viscosity measurements and particularly the ultracentrifuge[15] eventually led to the conclusion that molar masses of 100,000 g/mol (as for natural rubber, nitrocellulose, starch, etc.) were quite common. Moreover, a molar mass of 6.9×10^{13} g/mol for a specific deoxyribonucleic acid has been found. In other words, these molecules must have an enormous amount of repeat units. The word *macromolecule*[16] is therefore quite an appropriate description. Macromolecules may have lengths of up to several meters (!) but di-

---

[12]*Poly* (Greek) = many; *meros* (Greek) = part, particle.
[13]H. Staudinger, German chemistry professor, taught at the Technical University of Zürich, Switzerland. He received the Nobel prize for Chemistry in 1953 for the above-mentioned work.
[14]Formerly called "molecular weight."
[15]Work done by the Swedish professor T. Svedberg during 1924–1927.
[16]*Macros* (Greek) = big.

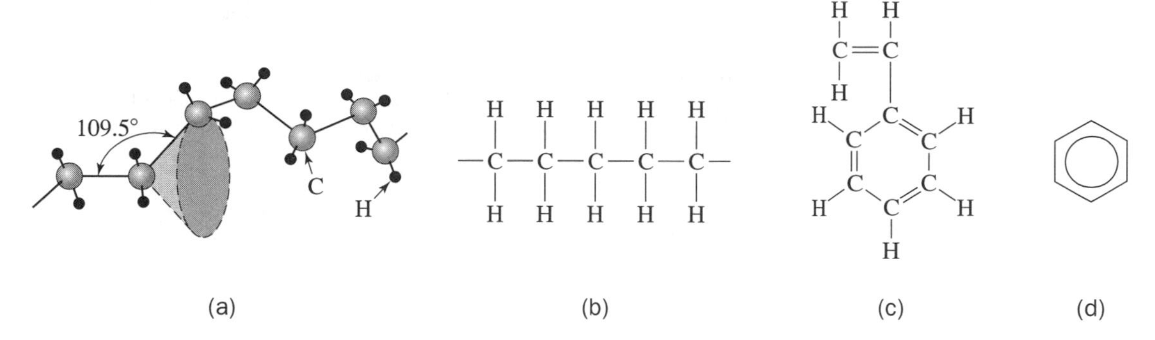

FIGURE 16.4. Schematic representation of the structure of some polymers. (a) Three-dimensional display of polyethylene. Note that each carbon atom can be positioned on a circle of a cone which has an angle of 109.5° between the bond axes. (b) Two-dimensional display of polyethylene. (c) Two-dimensional display of styrene which contains a benzene ring structure [compare to Eq. (16.1). (d) The ring is often abbreviated by a hexagon with an inscribed circle.

ameters that are merely in the nanometer range. They solidify in amorphous or crystalline structures, or in a mixture of both. Randomly arranged macromolecules are amorphous or "glass like." Crystalline macromolecules are organized by folding of the chains back and forth. Actually, chain folding was discovered at a relatively late date, namely in 1957. Before that time, a coil-shaped macromolecule was the more preferred model.

The next topic we need to discuss involves the graphic representation of polymers. Several possibilities exist. Among them, the three-dimensional model gives a vivid impression of the spacial arrangement of the atoms. We notice, for example, in Figure 16.4(a) a *backbone* of carbon atoms which twists and turns through space. Each carbon atom in the present case is additionally bonded to two hydrogen atoms. The angle between the bond axes is 109.5°; see Figure 3.4(b). The individual atoms are not drawn to have contact, as one would expect in a hard-sphere model.

More often, however, the simpler, two-dimensional representation is shown with the tacit understanding that the spacial arrangement has been neglected [Figures 16.4(b) and (c)]. In still other cases, a short-hand chain formula is used, as was presented in the reaction equation (16.1).

We turn now to the important question, dealing with how single molecules (monomers) can be induced to form polymers. Naturally, there is more than one method to do this, and the techniques in modern days have become increasingly complex.

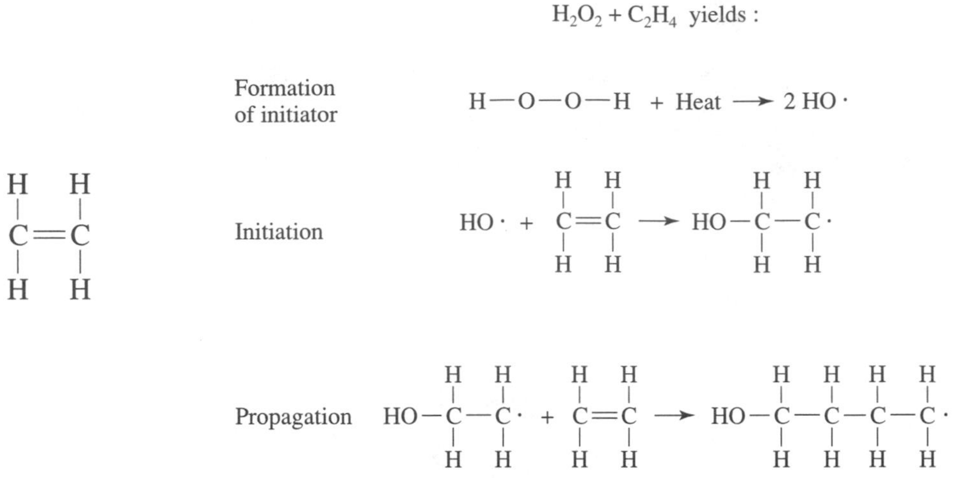

$$H_2O_2 + C_2H_4 \text{ yields:}$$

Formation of initiator $\quad H{-}O{-}O{-}H + \text{Heat} \rightarrow 2\,HO\cdot$

Initiation $\quad HO\cdot + CH_2{=}CH_2 \rightarrow HO{-}CH_2{-}CH_2\cdot$

Propagation $\quad HO{-}CH_2{-}CH_2\cdot + CH_2{=}CH_2 \rightarrow HO{-}CH_2{-}CH_2{-}CH_2{-}CH_2\cdot$

(a) $CH_2{=}CH_2$ (b)

FIGURE 16.5. (a) Ethylene monomer, (b) Reaction of an initiator for addition polymerization (here H_2O_2) to an ethylene monomer which initiates polyethylene to form.

The principles of polymerization can be described briefly, nevertheless. One of these techniques, called **addition polymerization** (or chain reaction polymerization) shall be explained first using polyethylene as an example. The ethylene monomer C_2H_4 consists of two carbon atoms and four hydrogen atoms that are covalently bonded, as shown in Figure 16.5(a). The two carbon atoms are joined by a double bond called an *unsaturated bond*. The latter term implies that, under certain conditions of heat, UV light, pressure, or catalytic action, one of the comparatively weak double bonds can be broken. This renders the molecule reactive, i.e., it may join other reactive molecules. As an example, we consider in Figure 16.5(b) the reaction of ethylene with hydrogen peroxide (H_2O_2), which is called in the present context an *initiator*. H_2O_2 splits under the influence of slight heat into two HO· groups whereby the dot symbolizes an unpaired electron, termed here a *free radical*. This HO· radical breaks the double bond in ethylene and then attaches to one of its ends, creating a free radical on the other end of ethylene, as depicted in Figure 16.5(b). In the present case, two sides of the ethylene molecule may react. The system is therefore called *bifunctional*. In other words, reacting an initiator with a monomer leaves a free radical at one end of this monomer which is then available for chain building.

Effectively, once a free radical has been formed, the addition of other repeat units to the chain (propagation) progresses with substantial speed. The reaction is driven by an energy difference between the monomer and the polymer. Specifically, the poly-

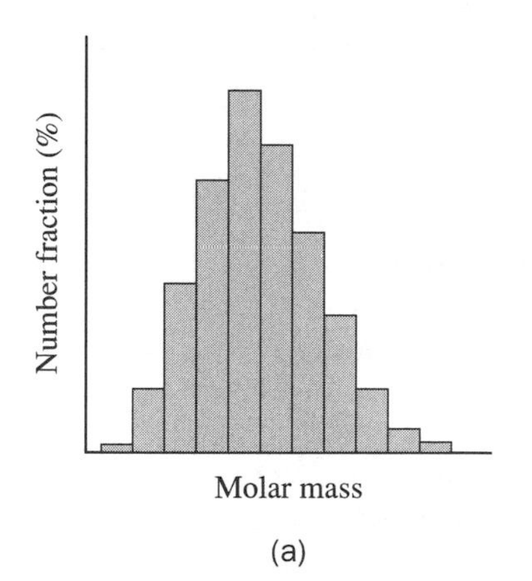

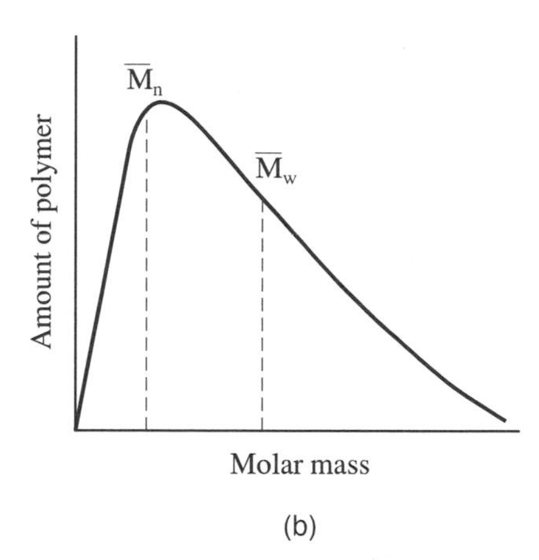

FIGURE 16.6. (a) Schematic representation of the length distribution of an assumed polymer. (b) Comparison between number-average and mass-average molar masses.

mer has a lower energy than the monomer. Thus, energy is set free (in the form of heat) during polymerization. This process eventually slows down, however, when the remaining monomers need to diffuse a comparatively long distance until they find a reaction partner. During maximal growth periods, about 10^5–10^6 mer units per second are added to a chain.

Each chain must eventually have an end. Formation of an end group may take place in several ways. First, the active ends of two chains may link together to form a larger, but nonreactive macromolecule. This, of course, stops further growth and is called termination by *combination*. Second, the active end of a chain may react with an initiator or another constituent that has a free radical, thus forming an endgroup. Third, a hydrogen atom from one chain may be removed by an active end of another chain, thus causing a double carbon bond to be formed. This mechanism (called *disproportionation*) simultaneously ends two chains, one from which the hydrogen atom was removed and one to which it was attached. Since termination is essentially a random process, a distribution of chain lengths and therefore of molar masses is encountered. In order to arrive at an average molar mass, one groups the chain lengths in a series of ranges [Figure 16.6(a)] and then determines the number fraction of chains within each group. This leads to the **number-average molar mass**:

$$\overline{M_n} = \sum n_i M_i, \tag{16.2}$$

where M_i is the mean molar mass of size range i and n_i is the fraction of the chains that have this mean molar mass. Frequently, an alternative parameter is defined called the *mass-average molar mass*, $\overline{M_w}$, which is based on the mass fraction of molecules within various size ranges. The degree of polymerization, n_n, is defined as:

$$n_n = \frac{\overline{M_n}}{\overline{m}}, \tag{16.3}$$

where $\overline{m}$ is the molar mass of the involved mer.

The length of a chain can be influenced to a certain degree by controlling the amount of initiator: A large quantity of initiator produces more and therefore shorter chains, and also terminates the chains more readily. Among the polymers that are synthesized by the addition mechanism are polyethylene, polypropylene, polystyrene, and polyvinyl chloride (PVC).

Another mechanism by which polymerization can be achieved is called **condensation** (or *step-growth reaction*). It involves more than one monomer species which react and combine. During the reaction, they release a by-product of low molar mass such as water or alcohol. This by-product is discarded. Condensation polymerization is used to make *thermosetting polymers* as well as *thermoplastic polymers* (see above). We use as an example the formation of an ester which is caused by the reaction between ethylene glycol and adipic acid:

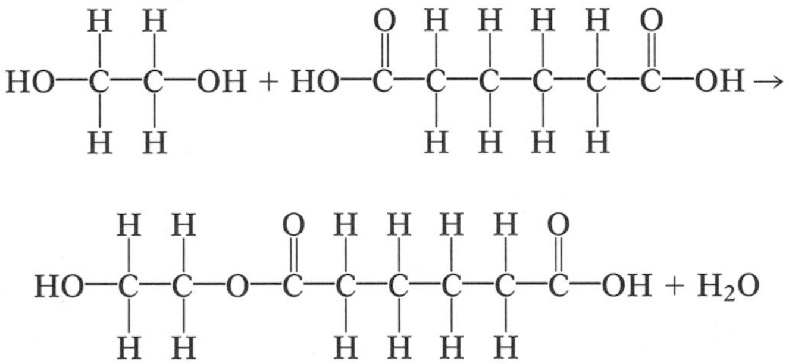

```
       H  H                O  H  H  H  H  O
       |  |                ‖  |  |  |  |  ‖
   HO—C—C—OH + HO—C—C—C—C—C—C—OH →
       |  |                   |  |  |  |
       H  H                   H  H  H  H
```

```
       H  H       O  H  H  H  H  O
       |  |       ‖  |  |  |  |  ‖
   HO—C—C—O—C—C—C—C—C—C—OH + H2O
       |  |          |  |  |  |
       H  H          H  H  H  H
```

The by-product is seen to be water. Both of the monomers in the given example are bifunctional. Thus, condensation polymerization yields in the present case a linear polymeric chain. However, if trifunctional monomers are involved, **cross-linking** yields **network polymers**. These network polymers are referred to as **thermosets** as stated earlier. Examples for condensation polymers are the nylons, phenol-formaldehyde, the polycarbonates, and polyesters.

This section gave a short overview of the early years of polymer development. The beauty of polymer science is that its history is less than 200 years old, in contrast to the history of most other materials, as was pointed out in previous chapters. Thus, it is quite possible today to follow from the literature how scientists and inventors slowly and painstakingly came to grips with the nature of synthetic and naturally occurring polymers. Quite often it took a substantial amount of time from the discovery of a polymer until its industrial production finally took place. Table 16.5 lists a few examples. The striving goes on. Present-day researchers receive cross-fertilization from other scientists who live and work all over the world (in contrast to antiquity where developments often occurred locally with very little and time-

TABLE 16.5. History of polymer discoveries, applications, and inventors

Polymer	Discovery	Production	Typical applications	Inventor
Derivatives of Natural Polymers (Semi-Synthetics)				
Vulcanized natural rubber	1839	1850	Tires, rain gear, shoes	Goodyear
Cellulose nitrate	1846	1869	Tool handles, frames for eyeglasses, films, gun cotton	Schönbein
Cellulose acetate	1865	1927	Photographic films, packaging, fibers, audio and video tapes	
Celluloid	1869	1870	Combs, films, collars, ping-pong balls	Hyatt
Nitrocellulose fiber	1884	1891	Artificial silk (rayon)	de Chardonnet
Casein/formaldehyde thermoset	1897	1904	Accessories	Krische and Spitteler
Cellophane	1908	1911	Packaging	Brandenberger
Fully Synthetic Thermoplastics				
Polyvinyl chloride	1838	1914	Shopping bags, window frames, floor tiles	Regnault
Polyvinylidene chloride	1838	1939	Packaging films	
Polystyrene	1839	1930	Small containers, foam, toys	Simon, Blyth, Hofman
Polymethyl methacrylate	1880	1928	Lamp casings, advertising signs	
Polyethylene	1932	1939	Garbage bags, milk bottles	
Fully Synthetic Thermosets				
Alkyd resins	1901	1926	Coatings	Röhm
Phenol/formaldehyde (Bakelite)	1906	1909	Electrical insulators	Baekeland
Synthetic Rubbers				
Polyisoprene	1879	1955	Tires	
Polybutadiene	1911	1929	Tires, gaskets	

Source: Adapted from H.G. Elias, *Mega Molecules*, Springer-Verlag, Berlin (1987).

delayed interaction between various regions). A gap of information exchange occurred, however, during World War II, at which time separate (and secret) developments took place in both camps. Attempts were made to bridge this gap by a group of American polymer specialists who swarmed all over Germany right after the war in order to find out what progress had been made over there. It is amazing to read a 550-page book entitled, *German Plastics Practice* which is based on "Quartermaster Reports" and which was published in 1946 under the auspices of the U.S. Department of Commerce. This document provides a detailed account of the accomplishments of that time, listing hitherto known polymer materials, know-how, production facilities, production figures, and the men and women who made it possible. Times have changed, however, and current progress is now essentially available in the open literature.

16.4 • Properties of Synthetic Polymers

The properties and the structure of polymers are substantially interrelated. This shall be explained now in some detail. Figure 16.7 schematically depicts the chain structures of the abovementioned three principal types of polymers. Specifically, **thermosetting** polymers are composed of long molecular chains that are rigidly *cross-linked* and thus form a three-dimensional network [Figure 16.7(a)]. Therefore, they belong to the group of *network polymers*. Usually 10 to 50% of the mer units are cross-linked. As a consequence, thermosetting polymers are strong to the extent of brittleness (Table 16.6). In some thermosetting polymers, such as the epoxy glues, the "hardener" irreversibly initiates the cross-linking in the resin. Thus, thermosets cannot be reshaped by heating. Instead, heating at high temperatures causes the cross-links to break and leads to the destruction of the material. Their recyclability is therefore limited; see Chapter 18.

In contrast to this, the polymeric chains in **thermoplastic** materials are not chemically interconnected and are comparable to spaghetti in a bowl [Figures 16.7(b) and 3.17(a)]. They belong to the group of *linear polymers*. Thermoplastics are normally held together by weak forces, e.g., van der Waals forces (Figure 3.7) and mechanical entanglement. At sufficiently high temperatures, they are soft and ductile. Further, they can be recycled easily by reheating and remolding them into a new product. In specific cases, some of the macromolecules are folded back and forth over small distances, thus forming platelets or lamellae about 100

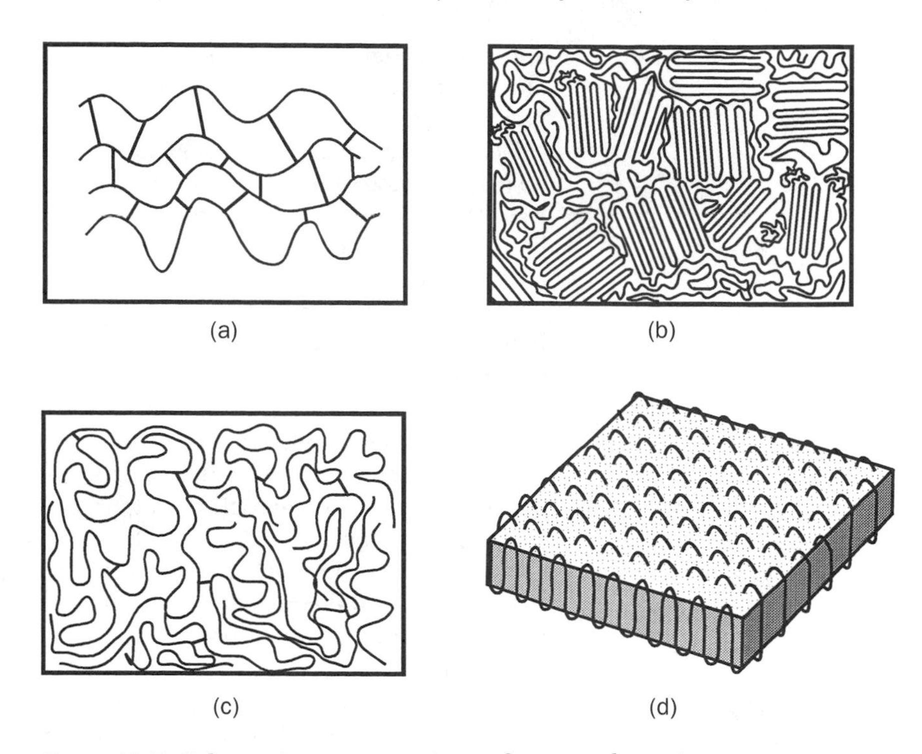

FIGURE 16.7. Schematic representations of some polymeric structures. (a) Thermosetting polymers (strong covalent bridges between individual chains exist). (b) Semicrystalline thermoplastic polymer (folded chain model). The crystallites are regularly shaped, thin platelets or lamellae which are about 100 atomic layers thick and on the order of 10 μm in diameter. The space between the crystalline areas is amorphous. (c) Elastomers. Some cross-linking takes place. (d) Three-dimensional representation of a folded chain crystalline platelet of thermoplastic polymers.

atomic layers thick and about 10 μm wide, as shown in Figures 16.7(b) and (d), which depicts the **folded chain model**. Because of the periodic arrangement of the chains, the platelets represent crystallites. Occasionally, several of these crystallites are stacked on top of each other and thus form multilayered structures. Polymers that have crystallized from dilute solutions possess such a folded chain morphology within a single crystal. When the crystallites emerge from a melt, however, several lamellae or chain-folded ribbons combine and form star-like three-dimensional structures which are called *spherulites*. The spaces between the crystalline areas are amorphous. Polymers of this type are therefore termed *semicrystalline*. Heavily crystallized polymers have an increased density, are effectively resistant to chemical attack, and retain their mechanical properties even at elevated temperatures. It should be noted, however, that not all thermoplastics have a crystalline structure.

Elastomers finally have a structure somewhat in between the just-mentioned ones; Figure 16.7(c). They are slightly, but not rigidly, cross-linked. Vulcanization at processing temperatures between 120 and 180°C (Section 16.2), for example, produces cross-links containing sulfur atoms. A low sulfur content (about 0.5%) causes the rubber to be soft and flexible (as in rubber bands) whereas sulfur contents up to 5% make the rubber stiffer and harder. This material is used to dampen vibrations of machines. Elastomers are generally amorphous. Specifically, their molecular chains are substantially coiled, twisted, and kinked. Applying a tensile force to them causes the chains to uncoil, that is, to straighten out, which results in an elongation of the material and a rather ordered state. Once the stress is released, however, the chains instantly revert back into their former condition. Interestingly enough, the temperature of natural rubber rises during stretching due to crystallization.

Mechanical Properties

A deeper understanding of the mechanical properties of polymers can be gained by inspecting their **stress–strain diagrams**. Figure 16.8, curve a, depicts this interrelationship for hard and brittle polymers as found in many thermosets. Their behavior is similar to that found in ceramics, refractory metals, diamond, etc. [Figure 2.6(a)]. The strength is high, the elastic modulus [Eq. (2.3)] is large, and only little ductile behavior is encountered (Table 16.6). This can be understood by knowing that chemically cross-linked chains are strongly tied together and thus sustain a rigid network as already mentioned above. In contrast to this, *elastomeric polymers* [Figure 16.7(c)] allow considerable deformation under little stress. They have a large nonlinear elastic re-

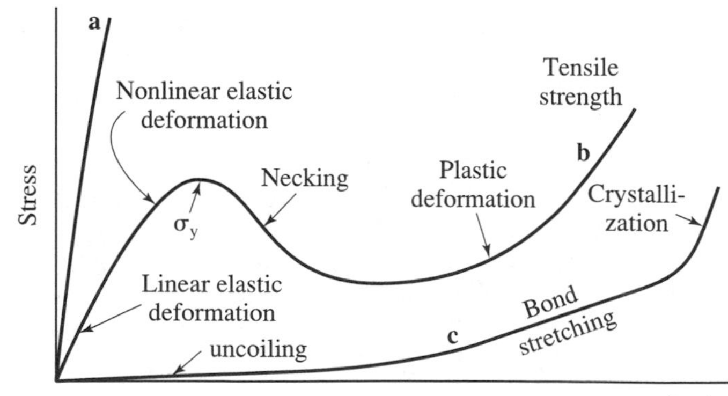

FIGURE 16.8. Mechanical behavior of three types of polymeric materials: (a) hard and brittle polymer as found in many thermosets, (b) thermoplastics, (c) elastomers. Note: The ordinate is *engineering stress*.

TABLE 16.6. Some mechanical properties of polymers at room temperature compared to metals and spider drag lines

Material	Modulus of elasticity E [GPa]	Tensile strength σ_T [MPa]	Ductility, i.e., elongation at break [%]
Thermosets *Example: Epoxy*	2.8–4.1	50–100	30–300
Thermoplastics *Example: Polyethylene*	0.17–1.1	8.3–31	100–1200
Elastomers *Example: Rubbers*	0.01–0.1	5–30	300–2000
Metals *(see Table 2.1)*	10–400	10–1500	<50
Spider drag line (single strand)	2.8–4.7	870–1420	30

gion. The elastic modulus and the tensile strength are small. This can be interpreted by considering, as outlined above, that the individual chains in elastomers initially uncoil and straighten due to the applied stress. Thus, they slide easily past each other. Once the chains have been completely straightened, further stressing leads to a stretching of the bonds and to an increase in the elastic modulus.

The stress–strain diagrams for *thermoplastic* materials are instead more involved; Figure 16.8(b). At low applied stresses the material behaves elastically, similar to most metals [Figure 2.6(a)]. The covalent bonds within the chains stretch and revert back almost instantly into their original positions upon release of a force (linear elastic deformation). At higher applied stresses, however, thermoplastics behave in a nonelastic, viscous manner. In this *nonlinear elastic range*, entire segments of the polymeric chains are distorted. Interestingly enough, this behavior is only observed if the stress is applied slowly or when the temperature is high. However, if the force is applied quickly or the temperature is low, a brittle fracture is encountered, as seen in Figure 16.8(a). In other words, the stress–strain diagrams vary with temperature and rate of deformation. In short, the nonlinear elastic behavior is linked to the viscosity of the material. [See in this context Eq. (6.16).]

Once the stress eventually exceeds the yield strength σ_y (Chapter 2), plastic deformation sets in. It occurs when the individual polymeric chains slide past each other, and when the entanglement of the chains is straightened out. (This mechanism is in sharp contrast to that encountered in metals, where dislocation movement is the prime cause for plastic deformation, as discussed in Section 3.4.) As in metals, necking eventually sets in, allowing chain sliding at a lower nominal stress until finally the chains become almost parallel to each other. At this point, van der Waals forces hold the chains together, thus causing an increase in required stress upon further straining; see Figure 16.8(b).

Table 16.6 compares the mechanical properties of synthetic polymers with those of metals and spider drag lines. The remarkable tensile strength of a spider drag line, which is comparable to that of steel, is particularly emphasized.

The mechanical properties of many polymers are quite **temperature-dependent**. This occurs already around room temperature, contrary to the much higher temperatures needed for metals and ceramics. As an example, Figure 16.9 schematically depicts stress–strain diagrams as found for *thermoplastic* polymers such as acrylics (e.g., Lucite). At relatively low temperatures (near 0°C), acrylics are brittle and comparatively strong, having a high elastic modulus and a high tensile strength. Upon moderately increasing the temperature, the elastic modulus and the tensile strength of acrylics decreases, and the material becomes ductile and soft.

Polymers undergo characteristic **transformations** at specific temperatures or temperature ranges. Similar observations are, of course, also known for ceramics, glasses, and metals. For comparison we display the temperature dependence of the specific volume, V_s (that is, the volume per unit mass). Figure 16.10 shows that *crystalline* polymers possess a sharp discontinuity in V_s at the melting temperature, T_m. (Note, however, that completely crystalline polymers are never found. Instead, as discussed above, amorphous transition regions between the crystallites are always encountered.) In contrast to this, *amorphous* polymers do not

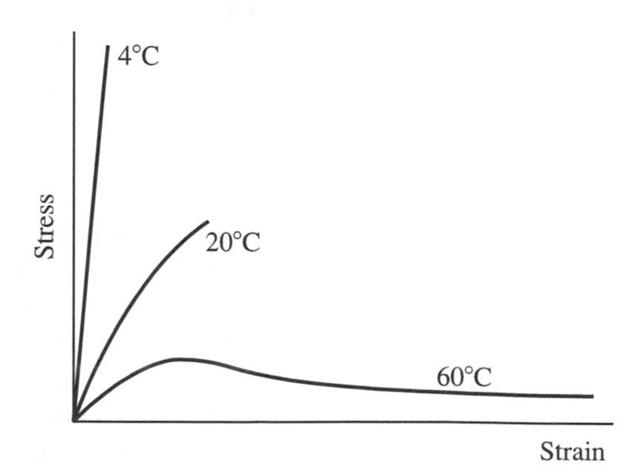

FIGURE 16.9. Schematic representation of the temperature dependence of the mechanical behavior of thermoplastic polymers such as polymethylmethacrylate (Lucite, Plexiglas).

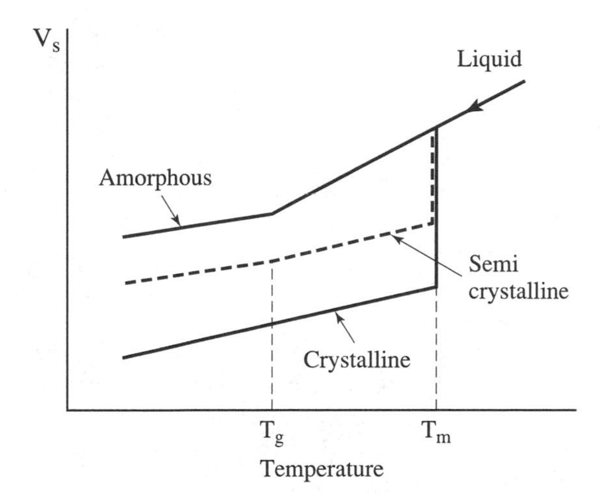

FIGURE 16.10. Schematic representation of the temperature dependence of the specific volume V_s for amorphous, semicrystalline, and crystalline polymers displaying a glass transition temperature T_g and a melting point T_m. Compare to Chapter 15, Figure 15.9.

possess a kink of V_s at T_m. Instead, when cooled from the liquid state, they pass through various stages called *viscous*, *rubbery*, *leathery*, and *glassy*. These descriptive terms vividly refer to the degree of viscosity in each state. A slight change in the slope of V_s occurs, however, at the *glass transition temperature*, T_g, below which the material is called a glassy amorphous solid.

Finally, *semicrystalline* polymers display discontinuities of V_s at T_g as well as at T_m upon cooling, thus possessing the characteristics of the crystalline as well as the amorphous phases. Generally, T_g is about two-thirds of T_m (where T is the absolute temperature), i.e., T_g/T_m is about 0.67.

Characteristic melting temperatures for polymers are between 100 and 350°C, whereas glass transition temperatures range between −150 and +300°C. Some important polymers such as polyvinyl chloride and polystyrene have a T_g above room temperature.

An additional characteristic temperature is often defined above which the polymer burns or chars. It is called the *degradation* or *decomposition temperature*, T_d.

Procedures

There exists a large number of procedures which may affect the properties of polymers. For example, the strength of polymers can be increased by **enhancing** the **molar mass**. This may be accomplished by decreasing the amount of initiators as explained above. Further, if polymers are **pre-deformed**, the strength can be improved. This procedure aligns the above-mentioned crystalline lamellaes in the direction of the tensile force, leading to a highly oriented structure.

Additives

The strength and abrasion resistance of polymers are often enhanced by adding **fillers** such as carbon black (for tires), saw dust, sand, talc, glass, or clay. The respective particles have sizes from 10 nm upward. An important additive named a *stabilizer* counteracts the deterioration of some polymers that is caused by ultraviolet light or oxidation. Specifically, UV radiation imparts enough energy to a polymer to sever some covalent bonds, thus shortening some molecular chains. As a consequence, substantial cross-linking and, thus, embrittlement may take place. **Plasticizers** are added to brittle polymers (such as polyvinylchloride) to increase the distances between chains and thus reduce T_g. This makes the polymer more pliable but also decreases its hardness and stiffness. Plasticizers consist of liquids that possess a low vapor pressure and a small molar mass. **Flame retardants**, when added to polymers, interfere with the combustion process by releasing a suffocating gas or by cooling the burning region through a chemical reaction. **Antistatic agents** are often added because polymers are generally poor electrical conductors and therefore retain static electricity. Antistatic agents can attract moisture from the air, thus increasing the surface conductivity.

The properties can also be altered by joining two or more different linear addition polymers, thus forming a so-called **copolymer**. Various types of copolymers are synthesized such as by alternating the sequence of the involved monomers either periodically or randomly or by adding side branches to the main chain (called grafting). Copolymers usually have excellent combinations of toughness, strength, and rigidity. Finally, mixing elastomers with thermoplastics yields two-phase polymers. This improves the toughness by absorbing the impact of a sudden blow. In short, the possibilities for variations seem to be virtually unlimited. However, we are presently still at a very primitive stage in controlling the synthesis of polymers. A mixture of chain lengths with usually only one or two types of mers is currently produced. In contrast, the human skin, for instance, is composed of collagen, which has units containing about ten different mers that are arranged in an exact order whereby each chain is exactly 1056 mers long.

This section emphasized essentially the structural and mechanical properties of polymers only. It goes almost without saying that some polymers also have quite interesting and commercially exploitable electrical, optical, and thermal properties. Examples are lenses for cameras, conducting polymers, superconductors, magnetic films, light-emitting polymers, and applications in the computer-chip industry. These properties have been discussed already in Part II of this book and therefore need not be repeated here.

16.5 • Composite Materials

Composites are solid substances that are produced by structurally joining two or more materials to obtain properties that can, as a rule, not be acquired by either of the original materials alone. In previous chapters, some uses and properties of selected composites were introduced. We mentioned as examples the insertion of straw fibers into clay bricks (ancient Egypt) and concrete with or without iron-rod reinforcements. We also pointed out that wood is a naturally occurring composite that consists of elongated plant cells (made largely of cellulose) that are held together by soft lignin [Figure 16.2(d)]. Man-made composite materials have recently seen a steep upswing in development and application, mainly because of their improved strength and stiffness that are caused by the strong ionic and covalent bonds of the involved fibers and their lack of flaws that, if present, would propagate cracks through the material. Specifically, many brittle materials such as ceramics, glass, porcelain, etc., and many soft materials, such as plastics, can be toughened considerably by incorporating fibers into them. Composites are often light and corrosion resistant toward oil, gasoline, or other organic fluids. If used in the automotive sector, composites are mechanically safer in cases of accidents compared to metals because of their improved energy-absorbing properties. An additional desirable feature of composites is their high thermal conductivity, which stems from the fact that they are good phonon- as well as electron conductors (see Chapter 14). Specifically, since phonons easily propagate in a stiff lattice of light atoms, particularly if they are in fibrous form (having a low level of structural defects), composites made of carbon, boron, or silicon carbide are excellent thermal conductors. Moreover, composites are frequently good insulators in cases where atoms are held together by strong ionic and covalent bonds (see Chapters 3 and 11).

On the negative side, a higher cost of production is frequently encountered. Further, at least in the case of polymer-based composites, one experiences a lower wear resistance, some degradation at temperatures above 200°C, and difficulty in recycling discarded products.

A few selected examples are given below. One of the most important types of composite materials involves polymers as a matrix (such as epoxy resins or polyester) in which fibers of glass, carbon, Al_2O_3, polyethylene, boron, beryllium, or ZrO_2 are inserted before polymerization has taken place (**fiber-reinforced composites**). In other words, the manufacturing of a composite often takes place concomitantly with the shaping of a commercial product. New types of fibers are contemplated consisting of diamond strands or carbon *nanotubes* the latter of which are molecular-scale structures like *Bucky balls*, made up of one or more cylindrical graphite basal-plane layers. They have strengths and stiffnesses in excess of those known for conventional carbon fibers. Even tiny single crystals, called *whiskers,* are utilized. The matrix does not need to be a polymer, however. For example, high-performance automotive brakes are made of a carbon matrix in which carbon fibers are inserted (called carbon/carbon).

Because of their low electrical conductivity, polymer-based composites combined with oxide fibers evade radar detection and are therefore used for stealth aircrafts. Other fibers for this purpose consist of *aramid* (a polyamide fiber similar to nylon), which is chopped into small pieces so that the size of the particles are equal to one quarter of the wavelength of the incoming radiation. This results in absorption of an endangering signal.

There is commercial interest in metallic matrices (such as aluminum) because metals are stronger, stiffer, tougher, and more heat resistant than polymers, but their exploitation is in its infancy. This is mainly due to the fact that hot metallic melts may damage specific fibers. Further, some of the advantageous properties of metals may be compromised by some fibers. Still, brake disks for sports cars are made from aluminum that is reinforced by silicon carbide particles (**particulate composites**). These Al-SiC composite brake disks have a wear resistance beyond that of pure aluminum, a smaller thermal expansion coefficient, a good thermal conduction (which prevents overheating), and are much lighter than the traditional cast-iron disks.

For obtaining extremely hard cutting tools, hard ceramic particles, such as tungsten carbide are dispersed in a metallic matrix consisting, for example, of cobalt. These composites are called "cemented carbides" or *cermets.* The purpose of the cobalt matrix is to alleviate the brittleness of the tungsten carbide and to increase the impact resistance. For manufacturing, the pressed powder of the components is heated above the melting point of Co so that, after solidification, the Co surrounds the carbide particles and thus acts essentially as a ductile binder.

In another partriculate composite, good-conducting silver is reinforced by hard tungsten, which yields electrical contacts that are wear resistant under arcing as needed in automobile distributors.

Fibers are usually packed densely parallel to one another which, in turn, causes the resulting sheets to be mechanically anisotropic; that is, the sheets are strong in its length direction and weak sideways, just as is known from wood. To alleviate this deficiency, sheets in which the fibers are lying in different directions are stacked on top of one another and then bonded, as in plywood. This technique is often used to produce, among others, sporting goods (golf clubs, tennis rackets) or wings for airplanes.

Finally, the modern automobile tire in which layers of rubber, fiber bundles (made of nylon, rayon, glass, polyaramid), and metal sheets (steel) are interleaved to withstand multiple flexing loads (**laminar composites**) should be mentioned. The occasional separation of these composites under extreme conditions and the serious consequences that are caused by these failures are known from the press.

16.6 • Advanced Fabrics

The quest for new fibrous materials, mostly for technical textile applications, continues. These new materials, which are either already on the market or are still in development, often have *smart* properties, that is, they can sense and respond to chemical or biological substances or react to physical stimuli such as blood pressure, body temperature, or blood oxygen levels. It has also been proposed that computers or other electronic systems be integrated into textiles (*"wearable mother boards"*), among them are photovoltaics, energy conversion systems (mechanical to electrical), switches, or storage elements.

In the medical field, antimicrobal, thrombonic, and superabsorbing textiles are used or under development. For the home or automobile, advanced filters that hold back pollen down to nanometer sizes and even remove odors have been developed.

Other important textiles are made from high-strength fibers (Kevlar, carbon strands, nylon, M5) for protection against impact (bulletproof vests). They possess a high stiffness in order to transfer the impact stress along the fibers. As an example, cellulose fibers possess an elastic modulus (Chapter 2) of 90–140 GPa, whereas polyacrylonitrile-based carbon has a modulus of 900–1100 GPa, mainly because of its covalent bonds between adjacent carbon planes. Somewhere in between is the so-called M5 (or PIPD) fiber whose tensile modulus is 550 GPa but whose specific strength (tenacity) is 5 GPa compared to 3.5 GPa for the above-mentioned high-strength carbon fiber.

There is also a strong interest in protective suits or overgarments that shield humans from biological contaminants or excessive heat, such as a fire. They should be lightweight and provide comfort over long time periods. This is accomplished by utilizing semipermeable adsorptive carbon liners within the garment. Concomitantly, carbon layers adsorb vapors that may pass into the clothing, thus providing protection. Other suits consist of barrier laminates that are completely impermeable, such as Tyvek-reinforced polypropylene films with a rugged outer-shell fabric such as aluminized Kevlar and polybenzobisoxazole (PBO). Recently, membranes that permit the penetration of water vapor molecules to allow transpiration and cooling, but are impermeable for larger molecules have been developed. These *"perm-selective"* membranes have been used for water purification by reverse osmosis and are made of poly(vinyl alcohol), cellulose acetate, cellulosic cotton, or poly(allylamine).

Some fabrics are self-decontaminating and kill dangerous airborne micro-organisms that attach to the protecting garment. These antimicrobal substances consist, among others, of silver or silver salts.

A particularly interesting fiber is capable of changing its color upon an electrical or biological stimulus from a bright orange to a camouflaging dark green. (This *"chameleon fiber")* may be of interest to hunters). Among these smart materials are conducting polymers made from polypyrrole or polyaniline (Section 11.5), which are coated on woven fabrics made of poly(ethylene terephthalate). Since electrochromism and electroluminescence (Section 13.4) are the pertinent driving mechanisms, a reasonable and stable electrical conduction is essential.

Suggestions for Further Study

J. Alper and G.L. Nelson, *Polymeric Materials, Chemistry for the Future*, American Chemical Society, Washington, DC (1989).

G.E. Baker and L.D. Yeager, *Wood Technology*, Howard W. Sams, Indianapolis (1974).

A.J. Barbaro, *Introduction to Composite Materials Design*, Teyler and Francis, London (1998).

J.M. DeBell, W.C. Goggin, and W.E. Gloor, *German Plastics Practice*, DeBell and Richardson, Springfield, MA (1946).

L.J. Broutman, *Mechanical Properties of Fiber Reinforced Plastics*, Composite Engineering Laminates, G.H. Dietz, ed. The M.I.T. Press (1969).

The Composites Institute, *Introduction to Composites*. CRC Press, Boca Raton (1998).

H.G. Elias, *Macromolecules*, 2nd Edition, 2 Volumes, Plenum Press, New York and London (1984).

H.G. Elias, *Mega Molecules*, Springer-Verlag, Berlin (1987).

Encyclopedia of Polymer Science and Engineering, Wiley, New York (1985).

T. Kelly and B. Clyne, *Composite Materials—Reflections on the First Half Century*, Physics Today, College Park, MD, *52*, 37 (1999).

H.F. Mark, *Giant Molecules*, Time-Life Books, New York (1966).

D.C. Martin and C. Viney, Editors, *Defects in Polymers*, MRS Bulletin, dedicated issue, September 1995, Materials Research Society, Pittsburgh, PA.

H.F. Schreuder-Gibson and M.L. Realff, Eds. *Advanced Fabrics*, MRS Bulletin *28* (2003) 558.

R.B. Seymor and C.E. Carraher, Jr., *Polymer Chemistry*, 3rd Edition, Marcel Dekker, New York (1992).

J.F. Siau, *Transport Processes in Wood*, Springer-Verlag, Berlin (1984).

A.Y. So, *The South China Silk District*, State University of New York Press, Albany, NY (1986).

17 Gold

Man's relationship to gold has always been quite different from his association to stone, copper, bronze, or iron. Because of this, a separate chapter is dedicated to this metal. If iron is considered to be the metal of strength and will, gold may be perceived to appeal, instead, to the forces of the heart. In antiquity, gold was the metal of the gods and was used to create religious vessels, utensils, and sculptures, but also ornaments and articles for adornment. In some ancient cultures, gold was not owned by individuals but belonged to the gods and their representatives on earth. Many poets have lauded gold, which they associated with the sun and its warmth and with wisdom.

Eventually, however, gold became the symbol of wealth and monetary stability. To satisfy the lust for gold and thus provide the means for a luxurious living for a few selected people and to have the means to pay soldiers, the most brutal wars have been fought and atrocities were committed. Two examples of such behavior are that of the ancient Romans and of the conquistadors who came to South and Central America. The power and wealth of empires such as those of the Sumerians, the Egyptians, the British, the Mesopotamians, or the Romans were largely based on their possession of gold and their trade with it.

Today, about 60% of the world's gold (i.e., 38 million kg) is said to be held by governments and central banks. It is supposedly used to back the paper currencies of their respective nations. U.S government sources claim that Fort Knox holds currently (May 1996) $\$54.9 \times 10^9$ worth of gold. Nevertheless, the rumor stubbornly persists that only a small portion of this gold remains presently at Fort Knox. Most of it is said to have been shipped to London in 1967/68 by the Johnson administration and was sold there in an ill-fated attempt to keep the price of gold at $35 per ounce. Despite numerous attempts by concerned citizens, who even offered to pay for an audit of the gold reserves, no such audit has been conducted since 1950 by the treasury department. Moreover, it is said that the gold in Fort Knox is mostly only 90% pure because it is derived from gold coins which went out of circulation in 1930. (Source: www.fgmr.com).

Gold is used as a means for international payment. As a result, the price of gold fluctuates with the value of the currency of a given country and due to international political events. For example, gold prices started to rise in 1973 because of the beginning of the OPEC oil embargo, and then again starting 1978 due to the increases of gold purchases of Middle Eastern investors as a consequence of upheavals in Iran (US hostage taking) and the invasion of Afghanistan by the Soviet Union a year later (see Figure 17.1). On the other hand, gold prices declined during 1981 and 1991 caused by the conflict in the Persian Gulf, the breaking up of the Soviet Union, the erosion of gold's role as a safe haven for investors, and worldwide weak economic growth. Despite of this, the world gold production has steadily increased during the past 100 years (Figure 17.2) probably due to increased industrial uses (computers) and demand for jewelry.

It may be argued that a gold-backed monetary system functions only as long as the cost of production of gold, and thus its

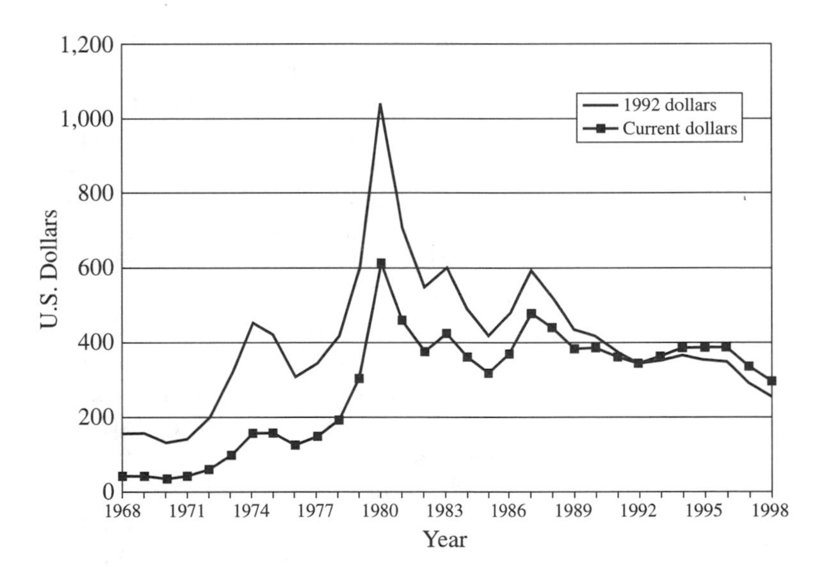

Figure 17.1. Annual average gold prices in US $ for one troy ounce of gold. (1 troy ounce = 31.1 g ; 1 "common" ounce = 28.35 g). The upper curve adjusts the current price to the unit value in constant 1992 US dollars (correction for inflation). Compiled by E.B. Amey, US Geological Survey, Minerals and Materials Analysis Section (eamey@usgs.gov).

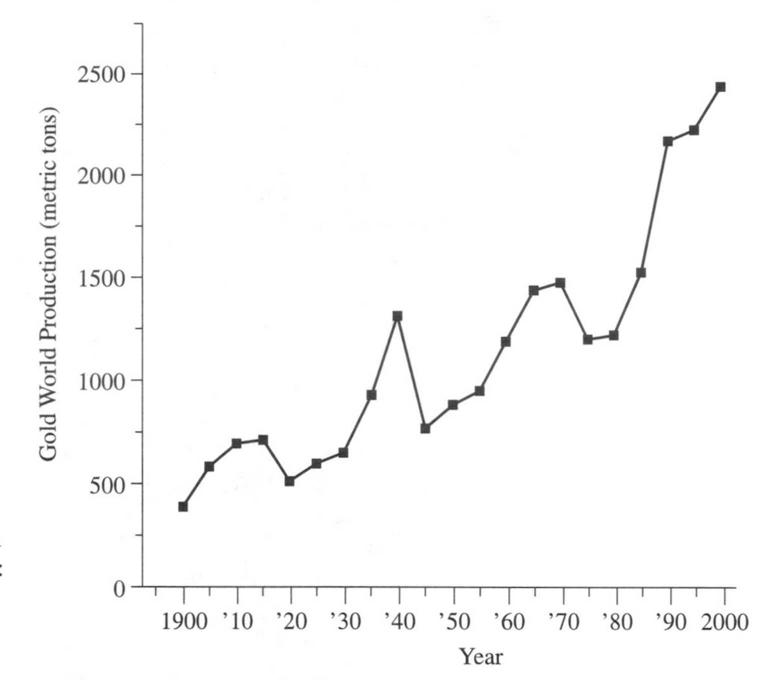

Figure 17.2. World production of gold in metric tons. *Source*: US Geological Survey, Minerals and Materials Analysis Section.

price, does not drastically drop, or as long as the world supply of gold does not suddenly and excessively increase due to hitherto unforeseen circumstances. Some economists suggest, therefore, that the gross national product of a country would be better suited for supporting a currency. In the late 1800s, many western countries adopted the **gold standard**, which pledges to exchange paper money for gold at a fixed price. (The United States went on and off the gold standard and finally abandoned it in 1971.) The world's gold production and reserves are given in Table 17.1.

Early chemistry and other sciences have greatly benefited from the quest of alchemists to produce gold from less precious elements. Many useful substances have been "accidentally" discovered through their experimentations, such as the red stoneware that was invented in 1707 by an alchemist named Johann Friedrich Böttger (who, incidentally, also reinvented, in 1707/08, white porcelain and thus laid the ground for the Meißen porcelain manufacture in Germany; see Section 15.2.).

Gold is a reasonably good conductor for electricity and heat; it is quite dense and is the most ductile, i.e., malleable, of all metals. Indeed, gold can be hammered into thin sheets, only 50 nm in thickness, called gold leaves. Specifically, one gram of gold can be spread out to cover about one square meter. The high es-

TABLE 17.1. World gold production and reserves in 2001

Country	Production [tons]	% of world production	Reserves [tons]	% of world reserves
South Africa	400	15.8	19,000	38
USA	350	13.8	5,600	11.2
Australia	290	11.5	5,000	10
China	185	7.3	1,000	2
Canada	160	6.3	1,500	3
Russia	155	6.1	3,000	6
Peru	140	5.5	200	0.4
Indonesia	120	4.7	1,800	3.6
Others (Ghana 69, Brazil 52, Colombia 22)	725	28.6	13,000	26
World total	2,530	100	50,000	100

Source: Drillbits & Tailings, September 2002.

teem for gold results, however, from its inertness to environmental influences combined with a relatively high reflectivity, which is about 93% in the yellow-red spectral region and 98% in the infrared. The latter property makes gold useful as heat shields for satellites, visors in space suits, and for thin, IR-reflecting films on windows. Because of its relatively high conductivity (71% of that found for copper) and its corrosion resistance, gold is used in large quantities by the electronics industry, for example, for conducting pads and for extremely thin wires which connect computer chips with the pins of headers. Further, gold is used in dental restoration. (Dental alloys contain, for example, 85.6% gold, 12.7% platinum, 0.2% iridium, 0.1% rhodium and traces of indium, zinc, titanium, iron, etc.) Fine particles of gold (colloidal gold) mixed with glass produces a deep red color. Gold leaves are green-blue in transmission.

Native gold, that is, gold in its pure (unoxidized) state, is widely found on all continents and in all climate zones, albeit often in finest distribution. It appears near igneous rocks, occasionally in combination with silver, selenium, tellurium, or bismuth. Gold-bearing rock formations, that is, gold accumulations in cracks of rocks, called *vein deposits* or *lode deposits*, include hydrothermal veins which are associated with quartz and pyrite (fool's gold). It is believed that superheated water dissolved gold in the earth's interior which, after rising to the surface, deposited in cracks of rocks. Another theory proposes that liquid, gold-containing rocks (magma) emerged from great depths of the earth, out of which gold eventually precipitated. Crystals as large as 3 cm in size have been found in California and even larger ones in Transylvania as depicted in Plate 17.1. Chunks up to 90 kg in weight have been supposedly encountered in Australia. Most of the time, however, quartz rocks contain gold in concentrations of one-hundredth of a percent or less. Gold has been traditionally "harvested" from river beds into which it was washed after the rocks have been eroded by wind and water. The technique of panning is commonly used. (Due to its high density, gold sinks to the bottom of a slurry of water and rocks.) Small nuggets or flakes of gold have been, and still can be, picked up in streams near surface veins or alluvial deposits. In antiquity, gold-containing mud was passed through sheep skin whose wool was subsequently burned to obtain gold dust. The Sumerians 6000 years ago already crushed gold-bearing quartz applying massive stone hammers. Eventually, "surface" gold became scarce and man had to resort to underground mining. This was done particularly since about 600 B.C. in the Mediterranean region, as well as in Asia Minor, Western and Eastern Europe, Africa, and Asia. It was made possible through advances in ventilation and later by better drainage of the ground water. In antiquity, Egypt was the land richest in gold (and poorest in silver). It derived its gold from the desert region on the Red Sea and from Nubia. (Nub in the Egyptian language means gold.) The Greeks and Romans used scores of slaves and convicts as mine workers who labored under deplorable conditions. Rome got most of its gold from Spain. Looting of the gold holdings of other countries likewise filled her treasury.

During the Middle Ages, the miners were liberated from slavery and became free agents who offered their services to the highest paying governments. Gold mining, melting, and casting pretty much paralleled similar endeavors developed for copper production, as described in Chapter 1. This can be understood by knowing that the melting point of gold (1063°C) is even 20°C lower than that of copper.

Discoveries of gold in certain regions of the world led to large migrations of people to these places. Examples are the great gold rush to California in 1849, or to New Zealand in 1861 (which doubled that country's population in six years). The city of Johannesburg in South Africa was founded as a result of a gold rush in 1886, and Canada's Yukon Territory had a similar influx during 1897/98.

Over the centuries several chemical methods have been developed to separate gold from its admixtures. Among them is the cyanide leaching process, in which gold from crushed rocks is dissolved by NaCN in the presence of oxygen. Subsequently, the impurities are separated by treating the resulting solution, for example, with metallic zinc, which causes the gold to precipitate. In another process, hot sulfuric acid or nitric acid is applied to gold alloys which dissolves silver and other constituents (but not gold) and thus separates gold from its alloying components if this is desired. In "cementation" (a process which has been in existence since about 550 B.C.), the gold–silver alloy in the form of thin leaves is heated in the presence of salt (NaCl), whereby silver chloride forms, which can be washed out. Gold is so noble that it needs a mixture of nitric and hydrochloric acids (aqua regia) to change it into a salt ($AuCl_3$).

For jewelry and many utensils and coins, pure gold would be too soft and therefore needs to be alloyed with silver, or possibly palladium or platinum, or left unpurified, as was done in antiquity; see below. Thus, goods made of gold alloys usually have a broad range of colors. (A 50% gold alloy is known by the term "12 karat" whereas pure gold is called "24 karat".)

A large number of skilled gold artisans emerged in different cultures through the millennia, as witnessed through splendid artifacts which have been found in numerous burial sites. Among them are finds near Varna (Bulgaria) on the coast of the Black Sea (see Plate 1.4) dating from 4300 B.C., artifacts from the Sumerians (4000 B.C.), and particularly from the Egyptians who skillfully covered wooden or copper articles with gold leaves (called *gilding*). But *solid* gold articles were also created by the Egyptians. One of the finest examples of this is the inner coffin belonging to the pharaoh Tutankhamen (1350 B.C.), which was made of solid gold and weighs 110 kg. A number of Irishmen during the second millennium B.C. were likewise gifted gold artisans. They traded their goods between the British islands and Central Europe. This considerably enhanced the wealth of the involved people, particularly the British. Other places of gold mining and artistic activities emerged in India, China, Greece, and Asia Minor, to mention a few. By the time of the Middle Ages (around A.D. 1500), London counted 52 goldsmith shops who had to maintain a certain standard in workmanship before they were accepted in a trades guild. A few examples of art work made of gold are depicted in Plates 17.2–17.5.

As implied above, gold coins played an important role in early trade. (To a lesser extent, coins made of silver and even bronze were likewise used.) Probably the first true coins, that is, cast disks of standard weight and value, were manufactured by the Lydians of Anatolia at about 640 B.C. using a natural alloy consisting of gold with 20 to 35% silver, which was known by the name *electrum*. The coins were embossed with the aid of a die that bore the image of a lion. However, the trade value of these coins was somewhat impaired because of the variation in the composition of electrum. Eventually, under the reign of the Egyptian King Croesus of Lydia (560–546 B.C.), purification of gold was accomplished and the quality and weight of gold coins were certified by the government. Similar minting was eventually adopted by the Greeks and finally also by the Romans. The latter used iron dies which were more resistant against wear compared to the bronze dies that were used by the Greeks and Egyptians. In contrast to this, the Chinese issued square-holed bronze coins utilizing essentially the same size and shape for almost 2,500 years until the twentieth century.

In conclusion, gold was probably known to and valued by mankind as early as during the Neolithic or even the Mesolithic time periods because of its color, luster, ductility, its inertness to environmental interactions, and its availability in the native state. Gold played a major role in mythology and folklore and allowed beautiful pieces of art to come into existence. On the other hand, gold also brought hardship, pain, and tears to many people around the world because of the desire for its possession.

Suggestions for Further Study

L. Aitchison, *A History of Metals*, Wiley, New York (1960).

R.W. Boyle, *Gold, History and the Genesis of Deposits*, Nostrand Reinhold, Inc., New York (1987).

T.A. Rickard, *Man and Metals*, Arno, New York (1974).

18

Economic and Environmental Considerations

Materials scientists often need to advise other engineers who work in different and more specialized areas as to the best suitability of certain materials for specific applications. An airplane, for example, requires light-weight and high-strength materials such as aluminum or titanium alloys, whereas rotor blades for turbine engines, which have to withstand extremely high temperatures, are better served by certain nickel alloys. However, cost, availability, safety, aesthetic appearance, and recyclability of materials likewise need substantial consideration. The latter issues shall be discussed in the present chapter.

18.1 • Price

Figure 18.1 depicts the price per unit weight of some typical industrial materials between 1900 and 2000. It is observed in this graph that the expense for aluminum decreased in the first half of this century, mostly due to more efficient production techniques but also because domestic producers held the price for aluminum at a low steady level to improve their competitive edge against copper in the electrical industry. It can be further seen that, among the materials displayed in Figure 18.1, steel is still the least expensive one if one considers the price on a weight basis. The relative price increases during the past 50 years are essentially alike for all depicted substances. Long-term changes in price are caused by increases in cost of labor, energy, trans-

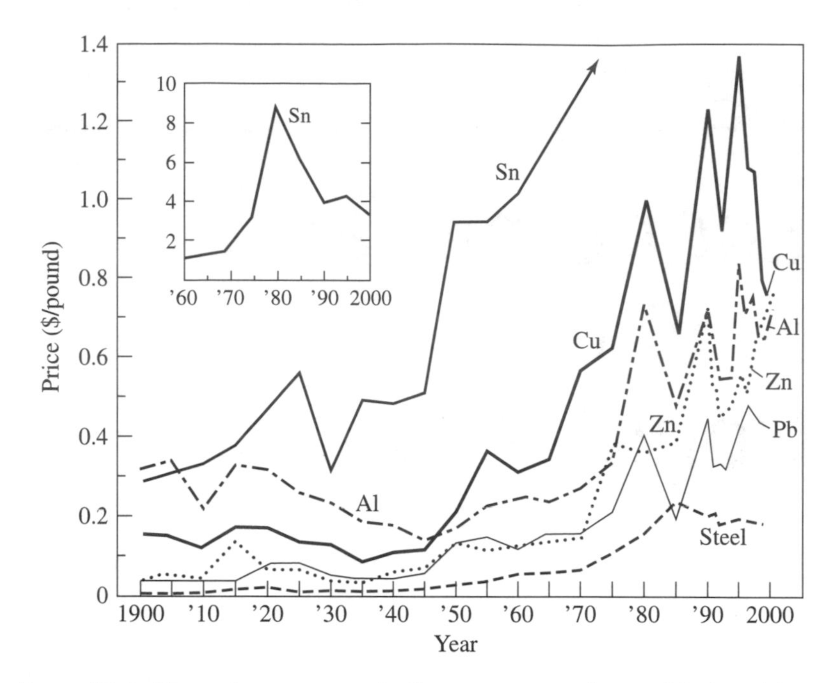

FIGURE 18.1. The price per pound of some commonly used industrial materials from 1900 to 2000 in the United States. Prices are not corrected for inflation. For plastics, see Table 18.1. [Source: U.S. Bureau of Mines, U.S. Department of the Interior.]

portation, and by the usage of leaner ores. Short-term fluctuations depend on speculators, supply and demand, and political factors such as strikes and wars. For example, the steep price increases in the late 1970s were caused by the OPEC oil embargo and by the removal of government price controls.

It is interesting in this context to compare the prices of metals with those of plastics. The prices of polymeric materials vary, however, with types and properties, and can therefore not be readily included in Figure 18.1. For this reason Table 18.1 lists the cost of plastics as published on August 4, 2003. It is noticed that the cost of steel based on weight is essentially still lower than that for plastics. This is, however, not always true if the price is based on volume.

18.2 • Production Volumes

Figure 18.2 displays the production volumes of various materials over the past 16 years. It can be learned from this graph that timber and concrete are essentially the most widely used materials in the United States. (It needs to be kept in mind that

TABLE 18.1. Average price of plastics (virgin and recycled) in dollars per pound as of 8-4-03 when bought at volumes between 2 and 5×10^6 pounds

Material	$/lb.
PVC resin (pipe grade)	0.48
Recycled PVC (clean, regrind, or flaked)	0.29
Polystyrene (high impact)	0.71
PET (PETE) packaging resin (for bottles, etc.)	0.73
ABS (high impact, for telephones, suitcases, etc.)	1.27
High-density polyethylene (HDPE) (for milk bottles)	0.54
Recycled HDPE (pellets)	0.36
LDPE (low density polyethylene) for grocery bags, wrappings	0.65
PP (polypropylene) yogurt containers, medicine bottles	0.50
Recycled PET (PETE) (clear, post-consumer, pellets)	0.61

Source: Plastics News, August 2003

the consumption of concrete is almost one order of magnitude larger than that of cement due to the addition of gravel and sand. This raises the output figures for concrete above the levels of timber and steel.) Also interesting to observe is the steady increase of polymer production over the past 16 years. It should be noted that, on a volume basis, light-weight aluminum and

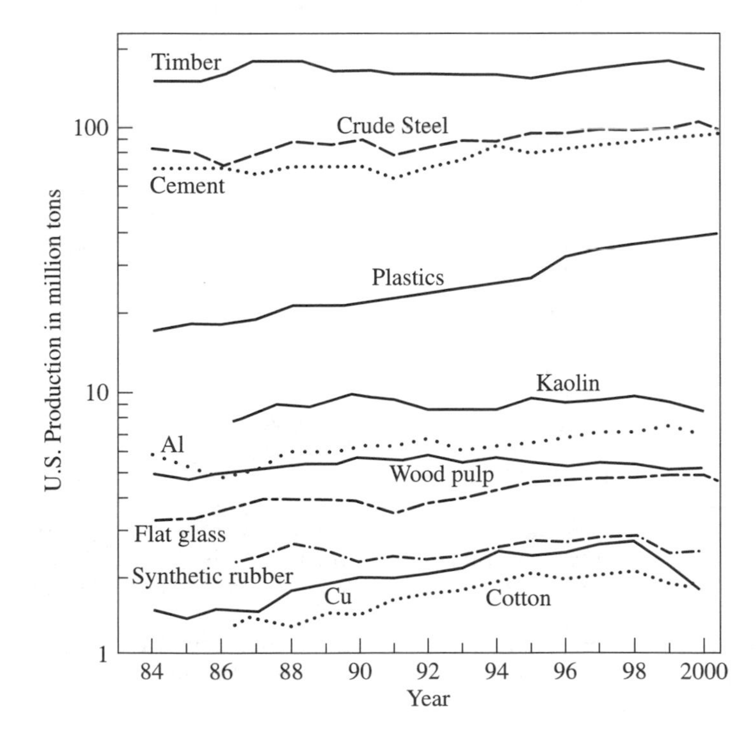

FIGURE 18.2. U.S. annual production figures on a weight basis for various materials from 1984 until 2000. [Source: *Industrial Commodities Statistics Yearbook* (2002), United Nations and US Census Bureau (Flatglass).]

plastics are even more utilized than Figure 18.2 might suggest because the presented data are based on weight rather than on volume. Figure 7.2 depicts the *world* steel production.

18.3 • World Reserves

Next, the future availability and the remaining world's supply of raw materials need to be taken into consideration when designing an industrial product. Table 18.2 lists some data concerning current world productions and known world reserves for some minerals. Table 18.2 also reveals the number of years these supplies are projected to last if usage proceeds at the present rate and no new sources are discovered. As probably expected, iron, oil, and coal top the list by far with respect to world production. Some forecasters predict an exponential growth of usage for some materials. This would deplete the resources much faster than the above-assumed constant level of consumption. A specific time interval, t_D, may be defined, during which future consumption is predicted to have doubled, that is, $t_D = 69/r$, where

TABLE 18.2. Annual world production and estimated world reserves of materials (data in 10^6 metric tons except for crude oil, which is given in 10^9 barrels (bbl), whereas 1 bbl of oil is 0.159 m^3 or 0.18 tons). The production data are for 2000 and the world reserves from 2001/02.

Materials	World production	World reserve base	Years' supply
Iron ore	620	160,000[a]	258
Copper	13.0	650	50
Aluminum	30.1	6,600[b]	219
Lead	6.0	130	22
Zinc	8.8	450	51
Tin	0.3	11	37
Nickel	0.9	140	155
Silicon	4.5	Essentially unlimited	
Potash	25.0	17,000	680
Phosphate	41.4	37,000	894
Crude oil	27.9	1,047.5	38
Coal	4,343	984,453	227

[a]World iron ore reserve base: 330,000 × 10^6 tons. Two tons of iron ore yield approximately one ton of iron. Note: Scrap iron is not included.
[b]Bauxite world reserve: 33,000 × 10^6 tons; Bauxite yields 1 ton of Al from 4 to 6 tons of ore.
Sources: USBM, Mineral Commodity Summaries 2002, "World Oil," August 2002, for crude oil, and "1999 Survey of Energy Resources," World Energy Council for coal.

r is the current rise in consumption per year in % (assuming exponential growth).[1] For example, if copper use increases by 8% per year, the consumption would double in 9 years. However, an increase in price and recycling may eventually slow down a too-rapid rise in consumption so that an exponential growth rate may not be encountered.

Table 18.2 does not contain **timber** (a renewable material), which currently grows worldwide on about 3.4×10^9 hectares (27% of the land area) and whose 1995 world harvest was 1.4×10^9 m^3. If used responsibly, and if substantial amounts of pulp are recycled (see below), the same amount of wood which has been harvested can probably be regrown. This would preserve the forests on all continents. However, between 1980 and 1990, the world lost an average of 9.95×10^6 hectares of net forest area annually, i.e., roughly the size of South Korea. Most of the decline in forest area has occurred since 1950 and has been concentrated in the tropical areas of developing countries to expand crop land, build cattle ranges, and extract timber. The temperate forests in industrial countries have essentially remained constant but consist largely of even-aged monoculture tree farms that do not support a high level of biodiversity as an ecologically complex natural forest does. Further, air pollution destroys large amounts of forests, particularly in Europe; specifically, 26% of this continent's trees have moderate to severe defoliation. The world production of various forest products in different geographic regions is listed in Table 18.3.

The potential for fabricating polymeric materials depends largely on the availability of **petroleum and coal**. To illustrate

Table 18.3 World wood production in 2000. Unit: 10^6 m^3, except wood pulp, which is given in 10^3 tons. Note: 1 ton of sawn wood coniferous= 1.82 m^3; 1 ton sawn wood broadleaved = 1.43 m^3; 1 ton veneer sheets = 1.33 m^3; 1 ton plywood = 1.54 m^3; 1 ton particle board = 1.54 m^3.

Country	Sawn wood coniferous	Sawn wood Broad-leaved	Veneer sheets	Plywood	Particle board	Wood pulp
Africa	2.4	5.3	0.7	0.7	0.5	0.3
N. America	156.0	37.1	0.6	19.6	31.6	17.5
S. America	14.1	15.5	3.6	3.1	2.8	1.2
Asia	29.7	24.6	2.7	27.5	8.7	2.4
Europe	109.6	17.2	7.2	5.6	37.0	15.0
Oceania	6.5	1.7	0.4	0.7	1.2	1.2
Total	**318.9**	**101.3**	**15.3**	**60.0**	**81.8**	**37.6**

Source: 2000 Industrial Commodities Statistics Yearbook, United Nations (2002).

[1] 100 ln 2.

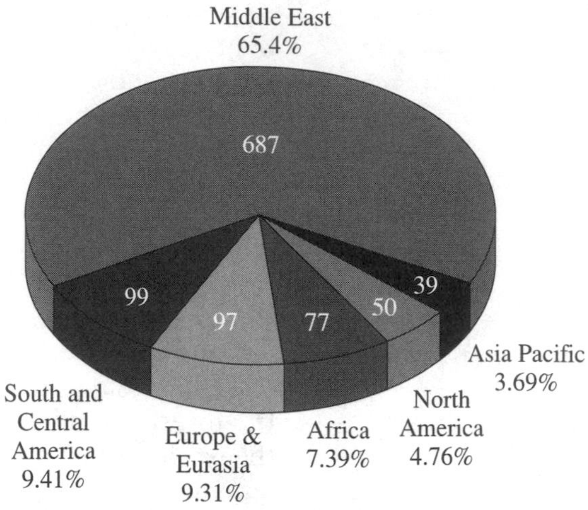

Middle East
65.4%
687
99
97
77
50
39
Asia Pacific
3.69%
South and
Central
America
9.41%
Europe &
Eurasia
9.31%
Africa
7.39%
North
America
4.76%

FIGURE 18.3. Proved oil reserves at the end of 2002 for various geographical regions. The reserves for the individual blocks (rounded) are given in 10^9 barrels. (They are printed in white numbers.) The total proved world oil reserves at the end of 2002 is estimated to be $1{,}047.5 \times 10^9$ bbl. *Source*: BP statistical review of world energy, 2003.

this, Table 18.2 also lists the production and consumption data for crude oil and coal. It needs to be emphasized that only 2% of the consumed oil goes into the manufacturing of plastics, and 1% is used for pharmaceutical products. The remainder is burned as fuel. Figure 18.3 displays the known petroleum resources for various geographic regions.

Table 18.4 lists the average daily oil production for the year 2002 for major oil-producing countries. It can be inferred from this table that the USA exploits its resources to a much larger degree in proportion to her known reserves, compared to most other countries (see Figure 18.3 and Table 18.4). Figure 18.4 depicts the world oil consumption from 1980 to 2002. The prices for plastics depend largely on the price of oil which fluctuates considerably over the years (mostly for political reasons). Figure 18.5 depicts crude oil prices from 1981 through 2003.

To complete the overall picture, Table 18.5 provides the production figures and reserves for **coal**.

It is alarming to note from Table 18.2 how fast some of our presently known **reserves** would deplete if the current consumption remains at the same level and if no new sources are discovered. This may be particularly true for oil, as shown in Figure 18.6. However, exploration efforts for the past 50 years have consistently yielded additional crude oil reserves that even exceed consumption at present, as depicted in Figure 18.7. **Reserves** are defined as deposits that can be profitably exploited using current technologies at current prices. In other words, the reserves are directly affected by the market price. Moreover, deposits that are not exploited within 20 years are considered to have little significant fi-

TABLE 18.4. Average daily oil production in 2002 for various countries, given in 10^6 barrels

Country	Daily average oil production
OECD[1]	
United States	9.00
Canada	2.93
Mexico	3.61
North Sea[2]	6.21
Other OECD	1.65
Total OECD	23.40
Non-OECD	
OPEC[3]	
Crude	
Algeria	1.31
Indonesia	1.27
Iran	3.44
Iraq	2.02
Kuwait	1.89
Libya	1.32
Nigeria	2.12
Qatar	0.68
Saudi Arabia	7.63
United Arab Emirates	2.08
Venezuela	2.60
Natural Gas Plant Liquids	2.10
Refinery Processing Gain	0.06
Total OPEC	28.71
Former USSR	9.38
China	3.39
Other Non-OECD	11.45
Total Non-OECD	52.93
Total	76.33

[1]OECD = Organization for Economic Co-Operation and Development.
[2]North Sea includes the United Kingdom Offshore, Norway, Denmark, Netherlands Offshore, and Germany Offshore.
[3]OPEC = Organization of Petroleum Exporting Countries.
Source: International Petroleum Monthly, June, 2003.

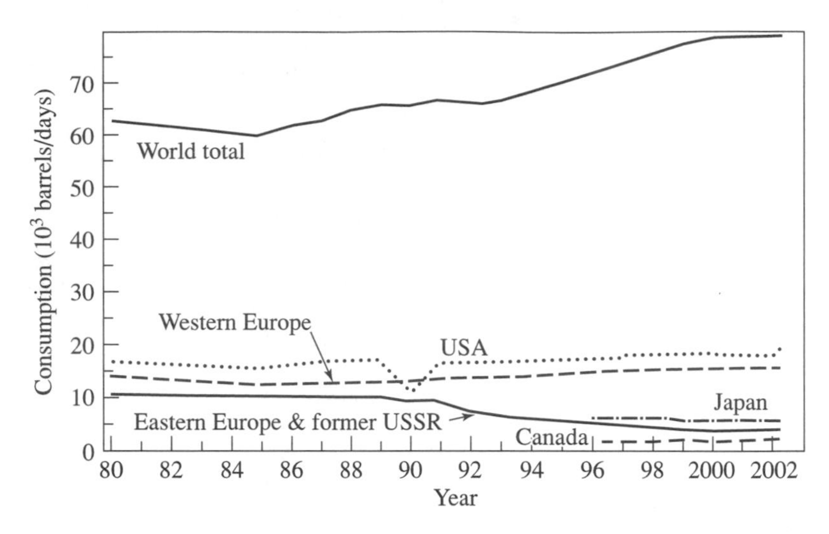

FIGURE 18.4. Daily world petroleum (crude oil) consumption in selected countries during 1980 through 2002. (Source: "International Energy Annual 2003," U.S. Department of Energy.)

nancial value. Thus, any exploration efforts to find reserves beyond 20 years are generally not performed for economic reasons. On the other hand, the present coal deposits seem to last for a much longer time period. However, mining will become increasingly expensive and dangerous once greater depths must be confronted.

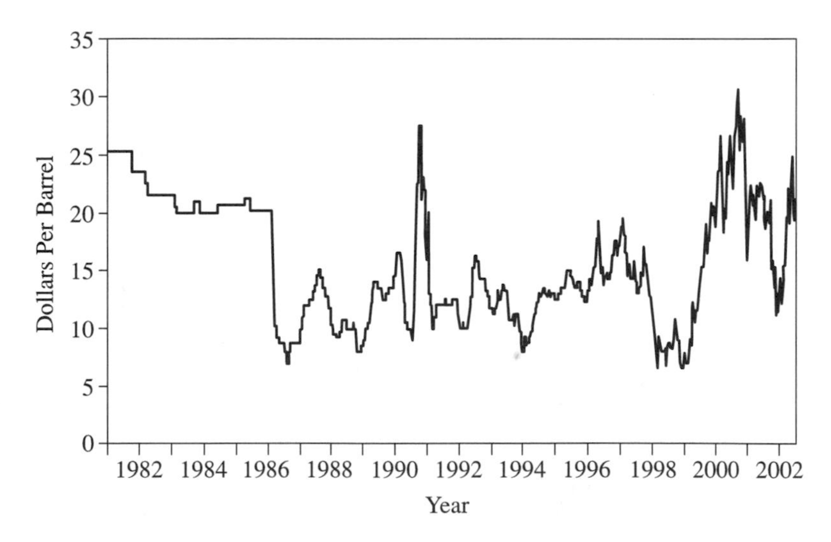

FIGURE 18.5. Crude oil posted prices from 1981–2003. (Kern River oil field, 13° API gravity). *Source*: Chevron USA Inc., crude oil price bulletins.

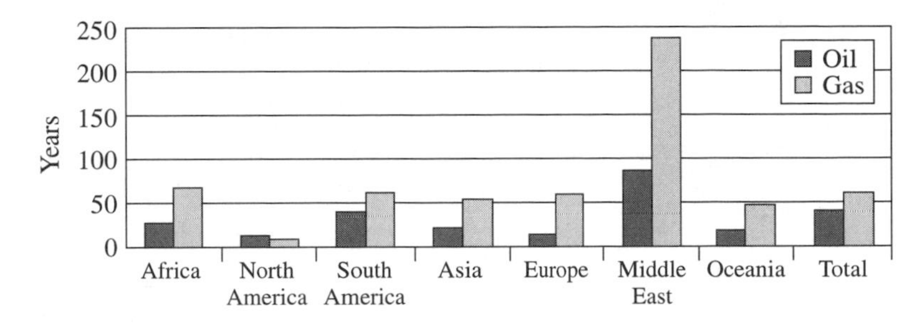

FIGURE 18.6. Schematic representation of the number of years the currently known oil (and gas) reserves will last assuming the present consumption rate. *Source*: World Energy Council, London, 2000.

Developments to promote the usage of alcohol, rapeseed oil (for diesel motors), and other renewable energy sources are useful for environmental and possible long-range economic benefits. The seed of the rape plant (*brassica napus*, also called *canola*, *swede* or *colza*) contains about 40% oil, which is mainly used for cooking, soap production, and technical applications. The rape cake that remains after pressing the seeds is rich in proteins and is used as animal fodder. The rape plant is native to Europe but is also cultivated in China and India. Rapeseed oil has been the top oil produced in the European Union and now accounts for more than one third of the total European vegetable oil production. It has passed the consumption of soy oil. European leaders contemplate obtaining a certain independence from foreign mineral oil by utilizing rapeseed oil for energy production and transportation. Since specialized engines need to be developed for burning rapeseed oil, a different avenue has been found,

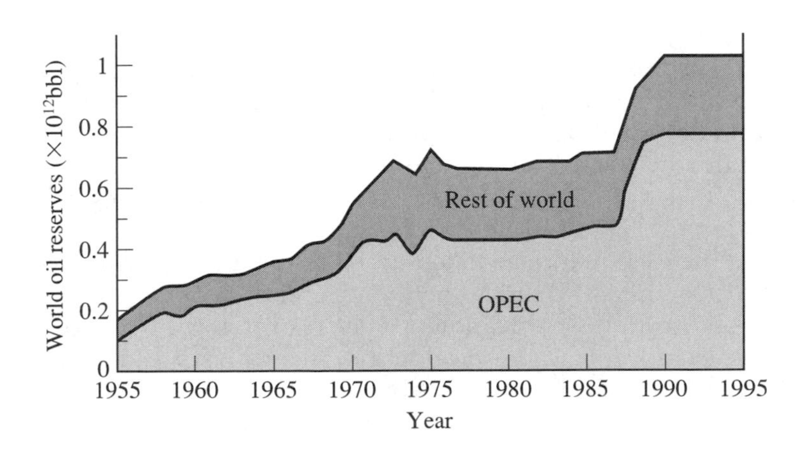

FIGURE 18.7. Worldwide crude oil reserves from 1955 to 1995. (Source: "Oil and Gas Journal," March 1996.) The world oil reserves have remained fairly stable since this figure was drawn. These reserves at the end of 2002 are listed as 1.05×10^{12} bbl; see Figure 18.3.

TABLE 18.5. Proven recoverable world reserves for coal (bituminous coal including anthracite, subbituminous coal, and lignite; data for peat are omitted). Also listed are world production figures. (The numbers are given in 10^6 metric tons and are from 1999.)

Region	Production/Year	Reserves	Years' supply
Asia	1,670	252,308	151
North America	1,080	257,906	239
Europe	1,007	312,686	310
Oceania	307	82,664	269
Africa	231	55,367	240
South America	46	21,752	473
Middle East	1	1,710	1710
Total World	**4,343**	**984,453**	**227**
Selected Countries			
USA	997	249,994	251
Russian Federation	249	157,010	630
China	1,030	114,500	111
India	314	84,396	269
Australia	304	82,090	270
Germany	202	66,000	327
South Africa	223	49,520	222
Ukraine	82	34,153	416
Poland	171	22,160	129
Canada	72	6,578	91
Mexico	10	1,211	121
Japan	4	773	193
North Korea	81	600	7

Source: "1999 Survey of Energy Resources," published by the World Energy Council, London.
Note: The estimates of coal reserves are rather vague. No standard for determining coal reserves exists.

namely, to manufacture *rapeseed methyl ester* (RME) which can be utilized in regular diesel engines, but requires, on the other hand, an additional production process that consumes some of the contained energy. The CO_2 emission of engines driven with RME is 60% less than when using diesel oil. Indeed, during the combustion of "bio-diesel" only as much carbon dioxide is generated as the rape plant has taken up from the atmosphere during the growth phase. On the other hand, emission of CO_2-equivalents result from the cultivation and the processing of rapeseed oil to RME and during the production of fertilizers and pesticides. (Note that natural gas combustion causes also less CO_2 emission compared to diesel, specifically, by 52%.) The main

TABLE 18.6. Average annual yield of certain materials produced on 1 hectare (10^4 m^2 or 2.47 acres) of land

30	kg of wool[a]
105	kg of jute[b]
500	kg of natural rubber
700	kg of linen (flax) fiber
1,000	kg of cotton
7,000	kg of wheat
10,000	kg of rice
70,000	kg of sugar beets

[a]Sheep may eventually end up as meat.
[b]Of which maximal 7.5% is fiber.

disadvantage of biodiesel is, however, that it is uneconomic. At present, the price of RME before taxes is about twice that of conventional diesel. Expressed differently, the energy price given in dollars per gigajoule is 10.2 for rapeseed oil, 13.8 for RME, and 5.2 for diesel oil.[2]

However, one should not forget in this context that crops for energy production may eventually compete for *prime land* on which food is grown for a steadily increasing world population. Mankind may eventually have to make a choice and decide how much of the available farm land is utilized for food and clothing, and how much is used for renewable energy, including solar energy. It is therefore instructive to know the quantity of various goods that can be harvested from a given piece of land. Table 18.6 provides this information. It is, of course, realized that not all land is equally suited for each of the listed raw materials, and that the availability of water is also a major factor for some crops, particularly for rice.

18.4 • Recycling and Domestic Waste

As already discussed, Table 18.2 contains the projected number of years at the end of which the currently known world reserves may be exhausted. These data strongly suggest the necessity for recycling if we want to preserve essential resources for future generations. This is particularly true for copper (pipes, wires), lead (batteries), zinc (galvanized steel, brass plumbing fixtures),

[2] *Source*: Bundesministerium für Umwelt, Naturschutz und Reaktorsicherheit, Berlin, Germany, (2000).

and tin (cans, alloys). Some technocrats see this, however, in a different light. They claim that there will always be enough resources available through price-driven explorations. Moreover, optimists believe that human ingenuity will eventually provide new avenues which have not yet been imagined that may satisfy our needs. However, recycling is also an obligation for protecting the environment, by reducing landfills (which emit methane, a powerful greenhouse gas) and removing toxic and other deleterious elements (mercury, lead, selenium, heavy metals) from soil and water. Additionally, recycling can have economic benefits by saving the energy and other costs for extracting valuable materials from minerals and agricultural sources.

Table 18.7 displays the amount of secondary materials which were recovered in the United States from scrap during 1990. It

TABLE 18.7. Recycling of metals in the United States in 1990

Metal	Secondary metal (metric tons)	Percent of total secondary metal	Value of secondary metal (million dollars)	Percent of value of total secondary metal
Iron (incl. steel)	55,500,000	91	25,000	68
Aluminum	2,400,000	4	3,900	11
Copper	1,300,000	2	3,600	10
Lead	920,000	2	930	2
Zinc	340,000	1	560	2
Manganese	60,000	—	3	—
Magnesium	54,000	—	170	—
Nickel	25,000	—	220	—
Antimony	20,400	—	35	—
Titanium	15,000	—	140	—
Tin	7,800	—	65	—
Molybdenum	3,000	—	17	—
Tungsten	2,200	—	9	—
Silver	1,700	—	260	1
Cobalt	1,600	—	30	—
Cadmium	700	—	5	—
Selenium	100	—	1	—
Vanadium	100	—	2	—
Chromium	90	—	580	1
Mercury	90	—	<1	—
Platinum Group	71	—	660	2
Tantalum	50	—	18	—
Gold	49	—	600	2
Totals (rounded)	60,600,000	100	36,800	100

Source: U.S. Department of the Interior, Bureau of Mines.

can be seen from this list that, while substantial amounts of aluminum, copper, lead, zinc, and tin were indeed recouped, by far the largest amount of recycling took place involving **iron and steel** (91% of total secondary metals). All taken, almost 37 billion dollars worth of metals have been returned into the industrial production this way (Table 18.7, Column 4). The recycling of steel scrap (automobiles, appliances, etc.) is a profitable business which is the major driving force for its application. The cost of recycled iron is about one-quarter of that for primary iron. (We discussed the mechanisms of reusing scrap for iron smelting at the end of Chapter 7.) The trend for iron and steel recycling is essentially rising. In 2001, it reached about 60% of the total industrial consumption, see Figure 18.8. For best economics, low alloy steel is usually separated from cast iron before remelting.

Aluminum recycling (mostly from beverage cans and construction scrap) is likewise a lucrative endeavor (netting $800 per ton of scrap) because it saves the energy that is expended for producing raw aluminum from aluminum ore (bauxite). This energy amounts to about 95% of the total cost of producing aluminum goods. In other words, it takes about 30 times more energy to produce aluminum from its ore than recycling used products taking the chemical reaction $Al_2O_3 \rightarrow 2Al + \frac{3}{2}O_2$ and remelting Al into consideration. For that reason, about 60% of all aluminum cans are presently recycled in the United States. About 45% of the aluminum manufactured in the United States (6.5×10^6 tons) stems, at present, from recycled products; see Figure 18.9.

For similar reasons as above, the recycling of **glass** is also quite profitablc in many parts of the world. In the United States, glass comprises nearly 6% of the household solid waste; see Figure 18.15(a). Approximately 2.8×10^6 tons of secondary glass (called *cullet*) is remelted annually in the United States. (Returnable bottles, as used in many European countries, would be even more cost-effective.) Since glass is heavy, transportation costs are high.

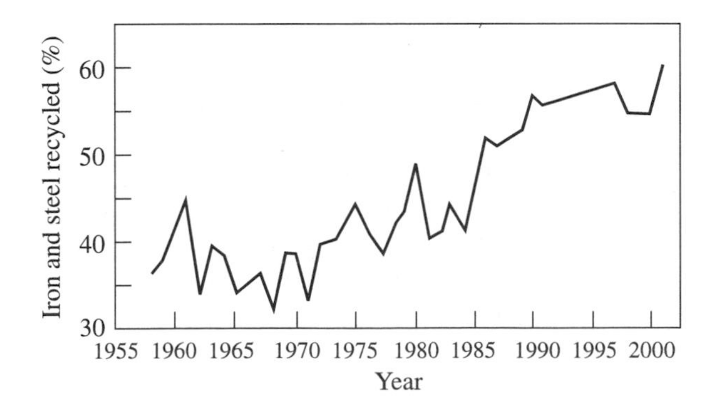

FIGURE 18.8. Iron and steel recycling in the United States in percent of total industrial consumption. (*Source*: United States Department of the Interior, Bureau of Mines.)

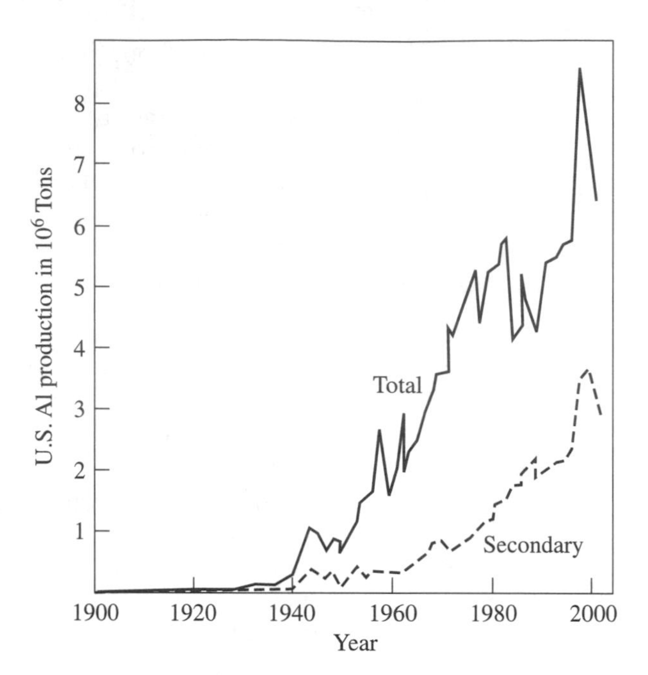

FIGURE 18.9. Aluminum production and recycling in the United States from 1920 until 2001. (*Source*: U.S. Department of the Interior, Bureau of Mines, August 2002; Statistical Compendium.)

Figure 18.10 depicts data for glass discarded in the municipal waste and recycling rates from 1960 to 2000. As can be seen there, glass recovery picked up considerably in the 1980s after which it held steady in the 1990s. The reason for this probably is that at present, not enough regional plants that remelt cullet exist. Glass is often (and not very cost-effectively) separated by color. However, for some products, such as fiberglass insulation, mixed colored glass is acceptable. In general, glass companies have learned to compensate for differently colored cullet. When 10% of cullet is added to new glass production, about 2.5% of the required energy is saved. Glass does not break down in landfills.

A further profitable and necessary recycled substance, namely **silver**, can be obtained from photographic fixing solutions into which a substantial amount of the silver from light-sensitive emulsions is dissolved. Moreover, electronic scrap (see below) yields substantial amounts of silver. In 2001, 1,060 tons of silver were recovered in the United States, valued at 150 million US dollars.

The **copper** recovered from refined or remelted scrap composed in 2001 about 34% of the total US copper supply and had an equivalent refined value of $1.9 billion. Twenty-eight percent originated from old scrap and 72% (833,000 tons) from new scrap. Lower copper prices in 2001 and secondary copper smelter closures led to a continued downward trend in recovery. The

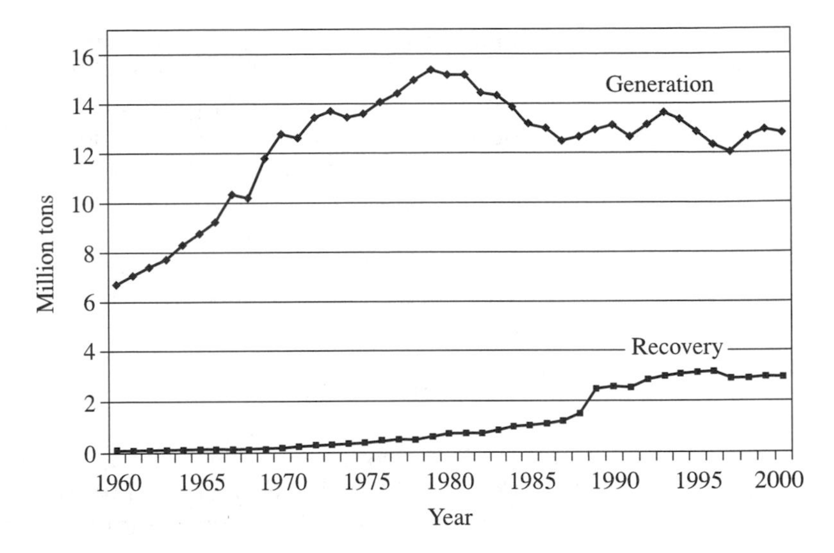

FIGURE 18.10. Glass "generation" in the municipal solid waste (MSW) (i.e., discarded into the garbage) and recovery from the MSW between 1960 and 2000 in 10^6 tons in the USA. Source: Environmental Protection Agency, "MSW in the US, 2000 Facts and Figures."

United States is, however, the largest international source for copper scrap, accounting for about 21% of all reported scrap exports, whereby China is the largest recipient with 3.4 million tons.

The recycling of **plastics** is, compared to glass and metal recycling, not yet at a desirable and cost effective level. As outlined in Chapter 16, *thermoplastic* materials, such as polyethylene teraphthalate (PET, e.g., Dacron[3], Mylar[3]), high-density polyethylene (HDPE), or polyvinyl chloride (PVC, e.g., house sidings, window frames, pipes) can be shredded and "reformed" into new goods. There is a secondary market for these "raw materials," as long as they are clean and not mixed with other plastics; see Table 18.1. Quite often, however, discarded items are transferred for convenience to "alternate use" applications, such as sweaters or insulation for sleeping bags and parkas. Other secondary uses are trash cans, flower pots, traffic A-frame barricade panels, or traffic cones.

On the other hand, *thermosets* are more difficult to recycle since the energy and cost to transfer them into alternate products would be substantially greater compared to using new material. They are more advantageously incinerated for heat pro-

[3]Trademarks of Dupont.

duction (see Table 18.14). A by-product of incineration is hydrochloric acid (recovered from filters in smoke stacks) which can be sold or re-used for the manufacturing of PVC. The amount of dioxin emission is claimed to be small.

In contrast to the just-mentioned *mechanical recycling*, an alternate technique has been introduced, called *chemical recycling*. It involves the decomposition of plastics, such as PET, to submolecular segments via processes called *glycolysis* or *alcoholosis*. This yields pure starting chemicals. In the year 2000, 35% of soft drink bottles made of PET and 30% of the manufactured milk and water bottles made of HDPE have been recovered in the United States from the municipal solid waste (MSW) stream and were channeled into recycling. (Switzerland recycled in 2001 82% of plastic containers; see Table 18.12). Moreover, the European Council of Vinyl Manufactures introduced in 2000 a new process in which PVC is dismantled into its basic components, namely hydrochlorics and synthesis gas, from which new PVC is manufactured (Schlackebad technique).[4]

Not all plastics are, however, recycled at such relatively high rates, as outlined above. Specifically, the total recycling of "packaging resins" in the United States amounted to only 7.6% in 2000.

Industry has developed a coding system that classifies plastics

TABLE 18.8. Plastic container codes and applications (compare to Chapter 16)

Code	Plastic	Applications
(1)—PET[a]	Polyethylene terephthalate	Beverage bottles, cosmetic jars, food containers
(2)—HDPE	High-density polyethylene	Milk, juice, and water bottles, toys, suitcases, telephones
(3)—V	Vinyl; polyvinylchloride (PVC)	Clear food packaging
(4)—LDPE	Low-density polyethylene	Grocery bags and wrappings
(5)—PP	Polypropylene	Yogurt containers, medicine bottles
(6)—PS	Polystyrene	Tableware, cases for compact disks, videocassettes, etc.
(7)—Other	All other resins and layered multi materials	

[a]A condensed milk company forced the name change from "PET" to "PETE," still "PET" is mostly used.

Source: Society of the Plastic Industry.

[4]*Source*: Arbeitsgemeinschaft PVC und Umwelt e.V.

TABLE 18.9. Plastics distribution (by weight) in municipal solid waste. The numbers in parentheses refer to the code listed in Table 18.8

47%	Polyethylene	(2,4)
16%	Polypropylene	(5)
16%	Polystyrene	(6)
6.5%	Polyvinylchloride	(3)
5%	Polyethylene terephthalate	(1)
5%	Polyurethane	(7)
4.5%	Other plastics	(7)

with respect to their chemical constituents for easier recycling. These codes are given in Table 18.8. Many municipalities collect only plastics having Codes 1 and 2 for the above-mentioned reasons. (Caps of jugs generally have a different code, that is, composition and should therefore not be included in the recycling stream of collectables because they could ruin an entire new product.)

In Table 18.9 the distribution of plastics in the municipal solid waste is listed. Figure 18.11 depicts the generation and recovery rate of plastics from 1960 to 2000 in the MSW.

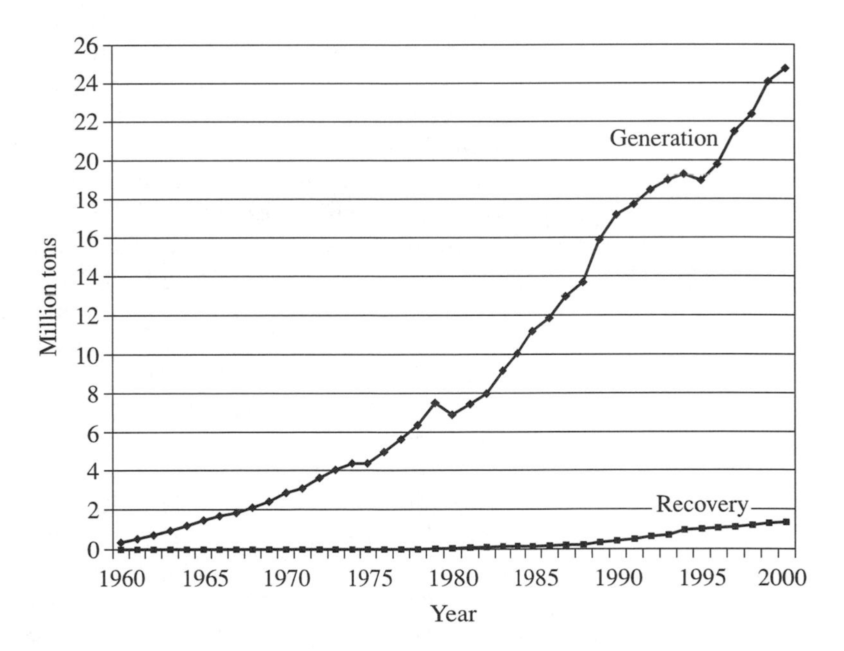

Figure 18.11. Plastics "generation" in MSW and recovery rate from MSW in the United States for the years 1960 until 2000. *Source*: Environmental Protection Agency.

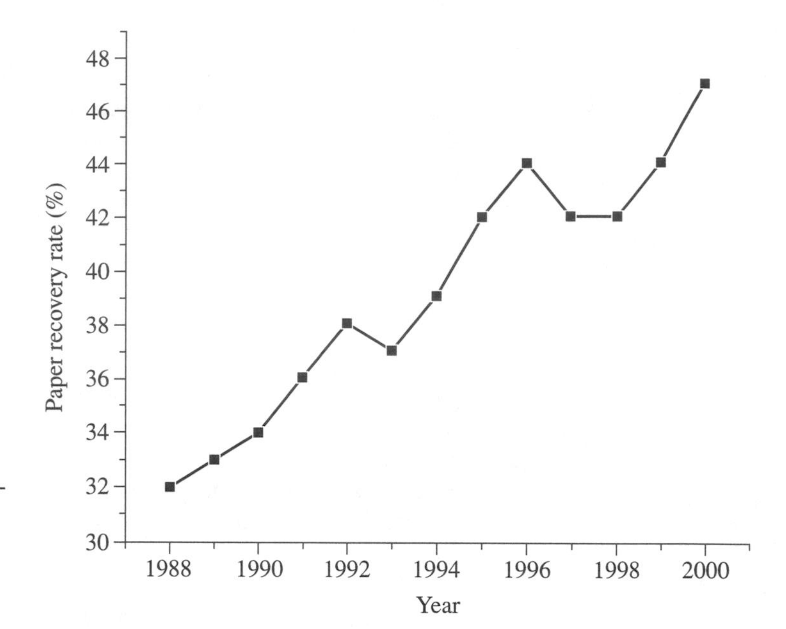

FIGURE 18.12. U.S. paper recovery rate between 1988 and 2000. *Source*: Am. Forest and Paper Association, Washington, DC. www.afandpa.org.

Paper recovery in the world is increasing. In 2002, Americans collected 48% of all paper used. Recovery rates for key grades, such as old newspapers (ONP) and old corrugated containers (OCC) were even higher in 2002, namely at 71% and 74%, respectively. The U.S. recovery rates between 1988 and 2000 are shown in Figure 18.12. The government goal for 2012 is 55%. Domestic paper mills consumed, in 2001, 34.9 million tons of scrap paper, which provided 37.7% of the industry's fiber. About 75% of all recovered paper in the United States is used for new paper and paperboard products at domestic mills. The rest is exported to foreign recyclers or channelled into alternate products such as animal bedding, hydromulch, compost, or insulation. Figure 18.13 depicts the distribution of major paper products which have been manufactured in 1995 in the United States from recycled paper. Paperboard and containerboard comprise the largest amount, as expected.

The scrap paper market is quite volatile. Some governments demand a larger secondary paper usage. Substantial amounts of waste paper are sold to Asia, particularly China and South Korea. Newspaper print brought in mid-2003 about $500 per metric ton (of the type "30 pound weight east coast"), whereas "22 lb directory paper" nets between $710 and 720 per short ton. One of the highest prices have to be paid for "coated publication paper, 70 lb, sheet" which brings $1,500 to $1,700 per short ton.

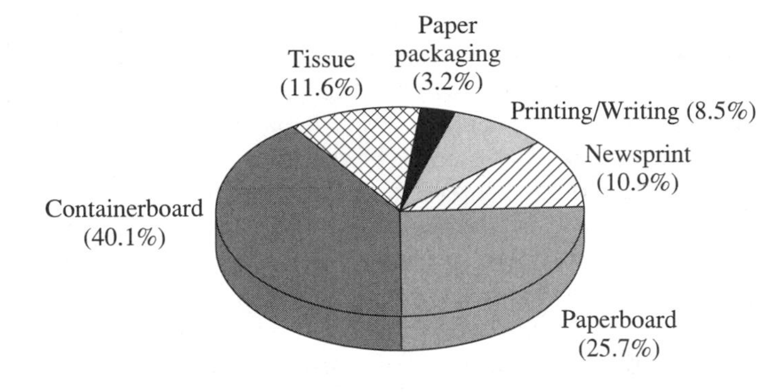

FIGURE 18.13. Distribution of products which have been manufactured in the United States in 1995 from recycled paper. (*Source*: American Forest and Paper Association, Inc., 1996.)

On the low side, "mixed paper and board sorted" can be sold for \$38 per short ton.[5]

Some paper mills specialize in utilizing only secondary fibers since different equipment needs to be employed for this case. Due to the present market situation, scrap paper can be profitably remanufactured into cardboard, wrapping paper, or grocery bags after the material has been mixed with water and beaten into pulp, as explained in Section 16.2. This process, however, shortens the size of the fibers, which causes some loss in strength. Thus, virgin fibers which are longer need to be added for binding, particularly if thin paper is desired. Another cost factor is the deinking process, which can be skipped if no contact with food is expected and the gray color of the paper or paper board is no problem. (Many cereal, etc., boxes are gray on the inside!) Interestingly enough, some recycled paper is priced at times somewhat higher than new paper, even though the production costs may be lower than from raw material. Apparently some consumers are willing to pay a little more when knowing it would benefit the environment.

Used **oil**, which is generated at an approximate amount of 1.2×10^9 gallons annually, is presently recycled at the rate of nearly 70% and is primarily used as fuel or is re-refined for lubrication purposes. Dumping oil in any form is environmentally unacceptable because of its possible contamination of ground water.

The recycling of **lead** is likewise a necessity for environmental and health reasons. At present (2000) 96.4% of automobile batteries are recycled in the United States. This number is probably quite similar in other industrialized countries.

The situation for recycling **household batteries** is, however, less positive. They contain mercury, lithium, and silver (hearing aids, pacemakers, cameras, calculators, watches etc.), or nickel and cad-

[5] Source: "Pulp and Paper Week", Paperloop Inc., San Francisco, August 2003.

mium (cameras, rechargeable power tools, and appliances). Some household batteries are sealed lead-acid devices. If disposed in landfills, heavy metals have the potential to leach slowly into the soil, groundwater, or surface water. Dry cell batteries contribute about 88% of the total mercury and 50% of the cadmium in the MSW stream. Because of this, mercury reduction in batteries began in 1984. During the last five years, the industry has reduced the mercury usage by about 86%. Moreover, some alkaline batteries contain no added mercury and several mercury-free, carbon zinc batteries are on the market. There are only a few recycling facilities in the United States that recover mercury from mercury oxide batteries. Nickel-cadmium batteries likewise can be reprocessed to claim the nickel. Still, most batteries collected through household collection programs are disposed of in "hazardous waste landfills." This appears to be a minor problem for the casual observer. However, statistically, the average person throws away eight household batteries per year, which amounts to about three billion batteries in the United States.

A new class of recyclable items is called **electronic scrap** mostly stemming from discarded, **obsolete computers**. They contain a variety of usable materials such as gold and other precious metals (silver, palladium, platinum), copper, and a variety of additional metals such as *antimony, arsenic, cadmium, chromium, cobalt, mercury, selenium,* zinc, iron, barium, gallium, manganese, and aluminum. (The hazardous materials are listed in italics.) Further, glass (from monitors), and plastics (casings) accumulate. Recycling of obsolete computers has created a completely new industry, mainly due to the gold and copper which is recovered this way. Specifically, 1 metric ton of electronic scrap from personal computers contains more gold than can be extracted from 17 t of gold ore. The value of 1 t of precious metals recovered from electronic scrap was valued in 1998 to be about 3.6 million dollars (60% silver, 40% gold, palladium, and platinum). Expressed differently, one ton of circuit boards may contain between 80 and 1,500 g of gold and between 160 and 210 kg of copper. The latter is 30 to 40 times the concentration of copper contained in copper ore mined in the United States. Another source for copper are monitors that contain between 4 and 7% of this metal. Recyclers in the United States, Canada, and Europe have extracted substantial amounts of useful materials from computer parts, as shown in Table 18.10.

The enormous amount of electronic scrap that hits the waste stream is caused by the rapid replacement rate of computers. Specifically, between 14 and 20 million PCs become obsolete every year in the United States. For every three computers purchased in the United States, two will be taken out of service. At present the

TABLE 18.10. Materials recovered in the United States from electronic scrap [X 10^3 metric tons]

Type of material	1997	1998
Glass	11.6	13.2
Plastic	3.7	6.5
Metals		
Aluminum	3.9	4.5
Steel	14.5	19.9
Copper	4.3	4.6
Combined precious metals (gold, palladium, platinum, and silver)	0.001	0.001
Other	3.1	3.6
Total	**41.1**	**52.3**

Source: National Safety Council (1999).

average lifetime of a computer is considered to be between 3 and 5 years when used in typical homes and about 2 years in businesses (at least as long the economy is booming). After that time span new, advanced computers are bought because they possess a substantially increased processing speed and storage capability, enjoying at the same time lower prices. Indeed, according to Moore's law, the number of transistors on a given chip (and thus its capability) doubles every 18 months. This may not persist so indefinitely, but the trend for replacing perfectly workable computers will probably continue for many years to come.

The glass from the cathode ray tubes (CRT's) of monitors is generally crushed and the lead that it contains (about 15–20%) is extracted. In other cases the lead-containing glass is utilized for the production of new CRTs. In any case, lead is considered a hazardous material and should not reach the environment, particularly the ground water. Alternatively, the silica may be reused as fluxing agent in iron production; see Chapter 7.

The plastic casings of computers rarely can be recycled for new plastic parts (see Chapter 16). However, plastics are often incinerated and the energy thus freed is used for heating purposes for example in cement kilns. Indeed, 1 t of plastics can replace nearly 2 t of coal, see Table 18.14. In 1998 about 6,500 t of plastics were recovered from computers.

Another class of recyclable products comprises materials that stem from **automobiles**. The recent trend is directed toward lighter cars since a 100 kg weight reduction results in a saving of about 10% of gasoline. As a result, cars that were built in 1965 from metals to the tune of 82%, now utilize much more plastics

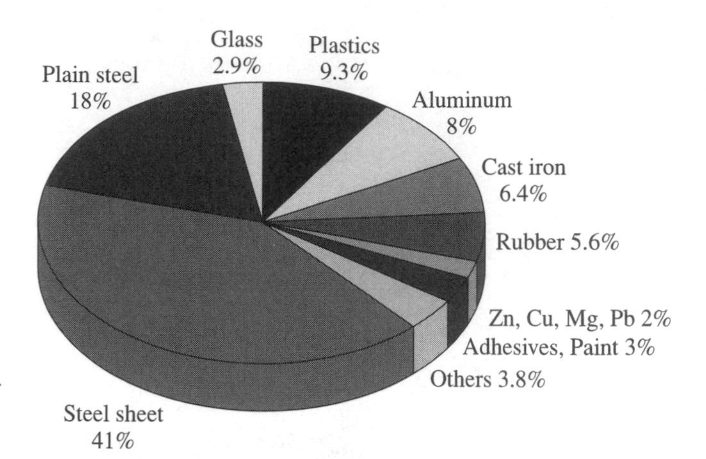

FIGURE 18.14. Materials used in European cars. *Source*: Journal of Metals, August 2003.

and aluminum parts. The typical composition of the materials found in European cars is depicted in Figure 18.14. It is shown there that plastics, such as polyvinyl chloride, polypropylene, and polyurethane rubber amount to nearly 10% and aluminum to about 8% of the total weight. The average car in Europe is used for 12 to 15 years, after which it is professionally dismantled, sorted, shredded, and recycled. Indeed, the recycling rate of automobiles in Europe is estimated to be 75–80%, which is higher than that for other consumer products ending up in the municipal waste. (The rest of the used cars are exported to East European countries, the former Soviet Union, or to North Africa.) Plastic parts are used as fuel, as already outlined before.

Of public interest is the **municipal solid waste (MSW)** and its distribution and recycling. Domestic waste amounts to about 55–65% of the total discards in the United States. The remaining part stems from schools, stores, and industrial sites, whereby regional factors, such as climate and levels of commercial activity contribute to some variations. Figure 18.15 depicts the composition of the average municipal solid waste in the United States in the year 2000, and for comparison, the MSW in a highly industrialized European country (Switzerland). The differences, particularly in the amount of paper and paperboard, are significant. It has been estimated that the average U.S. citizen discarded in 2000 about 2 kg of garbage each day which is up from 1.2 kg in 1960. This amounts to a total of 2.3×10^8 metric tons of solid waste per year for the entire United States. As an example, each year, Americans discard enough wood and paper to heat one billion homes for one year.

The question then arises, how much of this municipal solid waste is eventually recovered and recycled into secondary mate-

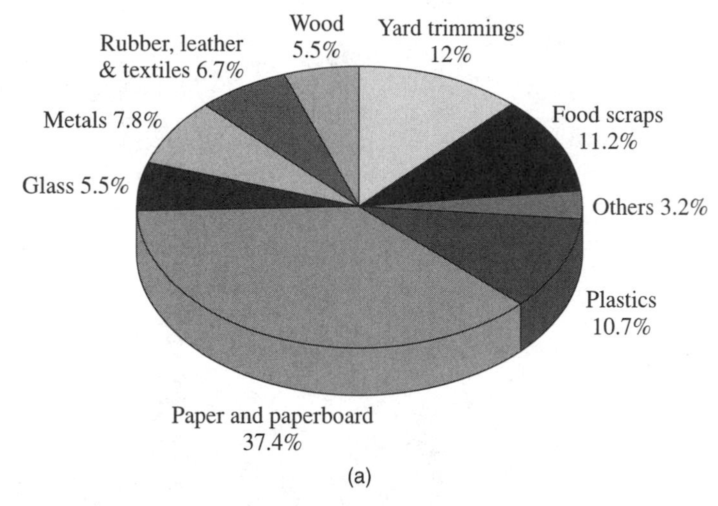

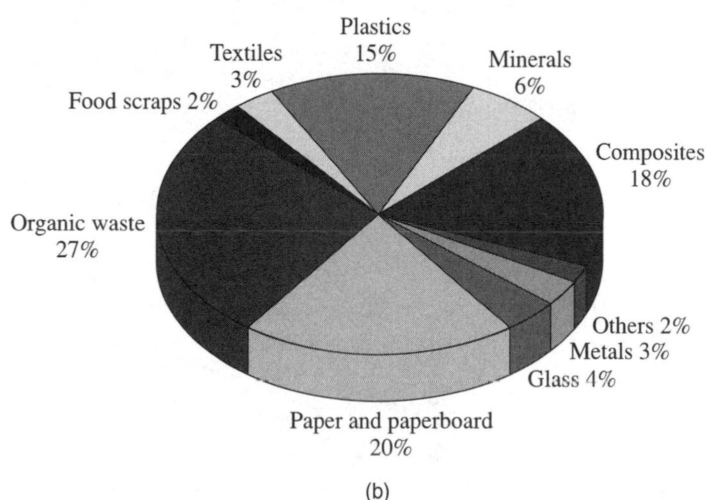

FIGURE 18.15. Distribution of the composition of municipal solid waste (a) in the United States (2000) and (b) in Switzerland (2001/02). (*Sources*: U.S. Environmental Protection Agency, March 2003, and Buwal, Bern, Switzerland 2003.)

rials? Table 18.11 presents a list of the recovery rates in the United States. As an average, the U.S. recycling rate of MSW in the year 2000 was about 30% (including composting). The federal government aims toward a goal of 35% by the year 2005 and of 40% by 2010. Yet many states are more ambitious. Fourteen have goals of 50% or more, including Rhode Island (70%) and New Jersey (65%). Some environmental groups, such as the Grass Roots Recycling Network based in Georgia, advocate "zero waste." These numbers seem to be not too unrealistic if one takes into account that San Francisco today has already a recycling rate of 46%. The recycling rate in Switzerland varied in 2001/02 between 30 and 70%, with an average of 50%.

TABLE 18.11. Recycling of MSW in the United States in the year 2000

Materials	Recovery in % of generation based on weight
Paper and Paperboard	45.4
Corrugated boxes	70.7
Newspapers	58.2
Glass, total	23.0
Container glass	25.9
Steel total	34.0
Steel, containers (mostly cans)	58.6
Steel from major appliances	73.5
Aluminum total	27.4
Aluminum cans	54.6
Wood	3.8
Plastics	5.4
Plastic containers	8.9
Rubber and leather	12.2
Textiles	13.5
Yard trimmings	56.9
Total recovery from MSW	**30.1**

Source: US Environmental Protection Agency: Municipal Solid Waste in the USA; 2000 Facts and Figures.

For comparison, Table 18.12 presents recycling data from Switzerland for the years 1995 and 2000. Again, remarkable differences can be detected in virtually all categories. It should be noted, however, that the Swiss (and some other European governments as well) allow the municipalities to levy upfront

TABLE 18.12. Domestic Recycling in Switzerland in 1985 and 2001

	Recycled amount in % of consumption (based on weight)	
Material	1995	2001
Paper	61	69
Container glass	85	92
PET plastics	74	82
Tin/iron cans	50	70
Aluminum cans	85	92
Total domestic aluminum	22	27
Batteries	54	67

Source: Buwal, Bern, Switzerland.

charges on new goods to benefit recycling and to render secondary raw materials more competitive. On the other hand, Germany in 2003 has instituted an exorbitantly large deposit fee (25–50 cents per container) concomitant with a ridiculously complicated and expensive refunding system of these deposit fees in order to discourage the sale of "one way" metal cans or plastic and glass containers, thus forcing customers into buying multiuse beer, milk, and soda bottles or paper containers. This practice stifles the lucrative part of recycling. (See, however, in this context the energy savings for refillable glass containers presented in Table 18.17.) Similar, but more hidden, subsidies (e.g., pickup charges to households) and small deposit fees are also practiced in some U.S. states or municipalities, with the consequence that the real costs of secondary raw materials is often obscured. An intense discussion is under way about whether the government should indeed mandate recycling or if the free market should decide what to economically recycle. Some waste managers estimate that the price tag of recycling large amounts of MSW would be between $20 and $135 a ton if all costs are included. Thus, new avenues for more economic recycling need to be found. One should take into account, however, that building, administering, and maintaining landfills have their price too in addition to the costs, which eventually may arise for cleanups a few generations later due to leaking of hazardous substances into the groundwater. At present, the landfill cost per ton is 50–60 dollars in the US and 60–170 dollars in Germany. Furthermore, the United States recycling industry consists presently of approximately 29,345 establishments that employ over 950,000 people, thus generating an annual payroll of nearly $34 billion and which grosses over $222 billion in annual sales. On a per ton basis, for every job created at a landfill, recycling sustains 10 more jobs just through sorting materials.[6] Moreover, recycling helps to reduce greenhouse gas emissions that affect the global climate (see in this context Table 18.18). In 1996, recycling of solid waste in the United States prevented the release of 33 million tons of carbon dioxide into the air, roughly the amount emitted annually by 25 million cars. In short, the recycling issue is not simply black or white but entails business as well as emotional and societal concerns in addition to the consideration of a cleaner environment, for which no price should be too high.

Heated public discussions arise over what to do with that part

---

[6] *Source*: USA Statistics, 2003.

TABLE 18.13. Incineration rates of remaining MSW in different countries

Switzerland	80%
Denmark	70%
Singapore	60%
Sweden	55%
Japan	50%
Netherlands	40%
France	35%
Germany	22%
United States	17%

Source: IWSA update Spring 1993.

of the MSW which cannot be economically recycled. Three major possibilities do exist, namely, composting, landfill, and incineration. Naturally, only a limited amount of the MSW can be composted, and the space for landfills is said to becoming scarce. **Incineration** seems to be a viable option that reduces the volume of the MSW to about 3% of its original amount and may provide 2% of the energy needs of a country. Indeed, Switzerland incinerated, in 1993, 80% of that MSW which was not recycled. A compilation of incineration rates in various countries is given in Table 18.13. A prerequisite for incineration is, of course, that the exhaust gases are filtered to the extent that little or no pollution is encountered. The ashes also need to be relieved of toxic substances. The same is, incidentally, also true for coal- or oil-fired electricity generation plants. It is said that the

TABLE 18.14. Energy content of some materials in (MJ/kg)

Plastics	44.2
Tires	33.8
Newspaper	18.6
Wood	15.8
Textiles	15.8
Food waste	6.0
Fuel oil	48.6
Coal	22.3
Average for MSW in the USA	10.5

TABLE 18.15. Remaining annual MSW in kg per person after recyclables have been removed

United States	492
Denmark	469
Netherlands	467
Switzerland	419
England and Wales	353
Germany	335
Spain	322
Greece	314
Belgium	313
Ireland	311
France	304
Italy	301
Portugal	231

Source: Unweltbericht 1993; BUWAL, Switzerland.

pollution problem can essentially be controlled if mandated by the government.

Table 18.14 lists the heat content of several materials. Incineration of plastics and tires by far provide the largest energy return. It is comparable to fuel oil and is even larger than that of coal.

Finally, Table 18.15 provides an overview of how much MSW remains for incineration, landfill, etc., after the recyclables have been sorted out or separately collected. The data suggest that the United States probably could do better.

The most ideal alternative to recycling and landfills would be if trash would never be generated and thus, reach the municipal waste stream in the first place. **Waste prevention** or **source reduction** can be achieved in a number of ways, such as by reducing the amount of packaging (particularly in the mail-order business or for simple consumer products), reducing the volume of newspapers (especially the Sunday editions), by using refillable bottles, pallets, and drums, by lengthening the lives of products (such as tires and household appliances), promoting onsite composting of yard trimmings, leaving grass clippings on the lawn, or by donating clothing to charities, etc. Between 1990 and 2000, inhabitants of the United States prevented a total of 55.1 million tons of MSW from entering the waste stream (that is, 2.4% per year), whereby almost half of the reduced waste in 2000 came from organics, particularly yard trimmings, and approximately one quarter from containers and packaging materials. However, some disposal rates have climbed between 1990 and

2000, such as for clothing, footwear, and plastic containers, probably due to the booming economy in the 1990s. As economic growth results in more products and materials being generated, there will be an increased need to invest in source reduction activities as outlined above.

18.5 • Substitution of Rare and Hazardous Materials

Recycling is only one means to save raw materials. Other avenues include substituting rare materials with somewhat more abundant ones, such as copper wires with aluminum wires (even though the conductivity is somewhat less favorable), or copper pipes with plastic pipes, or tin foils with aluminum foils, to mention a few. Further, plastic shopping bags are more economical in production, transportation (weight advantage), and disposal than paper bags.

A large number of materials should not be utilized for consumable products at all because of their toxicity and thus, their negative impact on the environment. In the previous section, we already gave pertinent examples, including toxic elements in batteries, and the efforts by industry to phase them out and replace them by nontoxic materials. The European Union has mandated in February 2003 the implementation of lead-free solders within 18 months, particularly for electronic equipment (see above). Moreover, mercury, cadmium, chromium, polybrominated biphenyls, and polibrominated diphenyl ethers are banned by January 2006. In the United States, *The Metallurgical Society* (TMS) has arranged several symposia to study lead-free solders, particularly their mechanical properties and their microstructure. The microelectronics industry is currently consuming 150 tons of lead-based solder per year (generally Pb with 2% Sn and 2.5% Ag, or Pb-Sn5), which is used to attach (bond) computer chips to their metal frames for packaging. Any replacement for these solders should meet or exceed the characteristic properties of the lead-based alloys. Specifically, they should melt easily; make a reliable electrical connection even under thermal cycling and high current densities (electromigration); should have an excellent wettability to copper, nickel, and silver substrates; a melting temperature somewhat above 260°C (to accommodate the reflow process); a good thermal conductivity; a wire manufacturability; and notably, they should be inexpensive. So far most lead-free solders, such as Sn-Ag-

Cu or the intermetallics Ag_3Sn, Cu_6Sn_5, Ni_3Sn_4, and $AuSn_4$ are tin-based. However, these solders have melting points below 240°C, they are brittle, and the thermal reliability of the bonds is low. A recently developed bismuth-based alloy containing 2.6% Ag fulfills most of the above-mentioned requirements, except thermal conductivity and sufficient wetting on Cu surfaces.

18.6 • Efficient Design and Energetic Considerations

Finally, a more efficient design may save materials. Quite often, equipment structures are overdesigned; that is, they contain more material than necessary. Moreover, only surfaces need to be covered with wear-resistant substances, while the interior can be made of a less costly and more abundant material.

Another factor that needs to be considered is the **energy expended** to extract raw materials from their ores. Table 18.16 lists the approximate energy (in gigajoules) used to obtain one ton of material. It is seen there that aluminum and plastics consume the most energy for their production. Still, manufacturing certain goods out of plastics may be quite cost-effective. For example, to fabricate 100 km of a 10-cm diameter cast iron drainage pipe requires 1,970 tons of oil but only 275 tons when PVC is used. Or, it takes 230 tons of oil to create 10^6 one-liter glass bottles out of sand, whereas the same amount of PVC bottles can be manufactured using only 66 tons

TABLE 18.16. Energy consumption in GJ to extract 1 metric ton of raw material from its ore or source

Aluminum	238
Plastics	100
Zinc	70
Steel	50
Glass	20
Cement	8
Timber	2
Oil	44
Coal	29

TABLE 18.17. Energy consumption per use for a 12-ounce beverage container given in (MJ) taking all factors such as mining, transportation, manufacturing, waste collection, etc., into consideration.

Aluminum can used once	7.4
Glass bottle used once	3.9
Recycled aluminum can	2.7
Recycled glass bottle	2.7
Refillable glass bottle used 10 times	0.6

of oil. Plastic bags require 40% less energy to produce than paper bags.

In this context, it may be of interest to know the energy that is used to produce or remanufacture a standard 12 ounce (0.35 liter) beverage container. Table 18.17 provides this information for aluminum and glass containers. The data reemphasize that recycling saves considerable amounts of energy.

Table 18.18 summarizes the **energy savings** relative to the energy required for virgin production, which can be achieved **by recycling** certain materials. It is noted that aluminum, steel, and recyclable plastics yield the largest energy saving. This has, of course, also considerable influence on reduced greenhouse gas emissions as listed in the same table.

TABLE 18.18. Energy Savings and reduced CO_2 emission through recycling of one ton of material relative to virgin production.

Materials	Grade	Reduction of energy in %	Equivalent in barrels of oil	Tons CO_2 Reduced
Aluminum	Recycle	95	37.2	13.8
Paper*	Newsprint	45	3.97	−0.03
	Print/writing	35	3.95	−0.03
	Linerboard	26	2.34	0.07
	Boxboard	26	2.43	0.04
Glass	Recycle	31	0.9	0.39
	Reuse	328	9.54	3.46
Steel	Recycle	61	2.71	1.52
Plastic	PET	57	11	0.985
	PE	75	10.8	0.346
	PP	74	10.2	1.32

*Energy calculations for paper recycling count unused wood as fuel.
Source: EPA (1996).

A somewhat different issue involves **energy savings through efficient design** of devices. As an example, the efficiency of standard incandescent light bulbs is only about 5%, or expressed differently, their *luminous efficacy* is 16 lumens per watt (lm/W). This is in contrast to fluorescent and high intensity discharge (HID) lamps whose luminous efficacies are 85 and 90 lm/W, respectively. It is anticipated that solid-state lighting (SSL) generated by light emitting diodes (LED, Section 13.4) eventually accomplishes 150 lm/W by 2012 and 200 lm/W by 2020. Naturally, the lifetime of the devices and the acquisition costs are also determining factors for using these new lighting products. Thus, an *ownership cost*, given in dollars per mega-lumen and hour ($/Mlm-h) is defined, which is, as of this writing (2003), clearly in favor of fluorescent and HID lamps, but is expected to be by 2020 as follows: SSL-LED: 0.48; HID: 0.83; fluorescent: 1.00; and incandescent light bulb: 5.63 $/Mlm-h. In the United States about one third of all primary energy is used to produce electricity, from which about one fifth is utilized for lighting. If only half of the electricity consumed for lighting can be saved by using more efficient devices, about 70 fewer power plants are needed. Moreover, 87 megatons less of C-equivalent greenhouse gases would be emitted every year, (3%) which brings the United States about 7.5% closer toward meeting the Kyoto protocol (which the United States has not yet ratified). In short, it is anticipated that with solid-state lighting devices eventually a 50% power conversion efficiency will be achieved. This would save roughly 35 billion dollars per year to the consumer. Worldwide, these figures are probably about a factor of three or four higher.

18.7 • Safety

Last but not least, a material has to be safe for the intended purpose. For example, casings for household appliances that generate heat during operation *must* be made of metal or flame-resistant thermoset plastics rather than out of slightly less expensive, but flammable thermoplastic materials. Serious injuries and even deaths have been reported when baby monitors overheated and caused the casing, made of a thermoplastic material, to catch fire and emit toxic fumes. The manufacturer saved 12 cents per device for using this inappropriate material. Another example: For safety reasons, the lenses for eyewear are required to be crafted out of tempered glass or plastics. Windshields for automobiles consist of laminated glass rather than ordinary window glass to prevent injuries in case of an accident. The list could be extended.

18.8 • Stability of Materials
(in light of the World Trade Center collapse)

Despite many speculations to the contrary, the steel skeletons of the twin towers in the World Trade Center did *not* melt on 9-11-2001 even at the enormous heat produced by burning jet fuel. Indeed, the temperature of burning hydrocarbons in air is estimated to be about 1,000°C, which is much less than the 1,500°C necessary to melt steel. Moreover, the fuel-rich, diffuse flame (as evidenced by the black smoke) suggests even lower temperatures in the 750–800°C range. However, as discussed in Chapter 8, structural steel undergoes stress relief at relatively low temperatures, which causes steel to loose half of its strength at 650°C. This loss in strength is still not more than one third of the allowable yield strength of steel for this design. Instead, the collapse is thought to have been triggered by the nonuniform thermal expansion of steel at different locations of the building. This caused some of the steel columns to buckle to the outside, which yielded in turn a loss of support of the floor joists. These joists are commonly held by "angle clips" that are mounted to the columns. Once one or two floors lost support and fell down, the underlying floors could not bear the extra weight and all the floors eventually cascaded down within ten seconds. In short, the WTC collapse on 9-11-2001 occurred because of thermally induced loss of about 50% of the strength of the steel and by the buckling of the steel columns as a consequence of thermal expansion that caused a lack of floor support.

18.9 • Closing Remarks

The above brief outlines may have demonstrated that engineers, and particularly materials engineers, shoulder a high responsibility for conserving the world's resources, to protect the environment, and to advise manufacturers on how to make strong, safe, energy-efficient, durable, and appropriately functional products at the lowest possible cost. Moreover, engineers should exercise the highest ethical standards because of their responsibility for humanity and nature.

Suggestions for Further Study

Abfallstatistik 2001, Bundesanstalt für Umwelt, Wald und Landschaft (Buwal), Bern, Switzerland.

Bottles, Cans, Energy, B.M. Hannon, Editor, *Environment*, **14** (2)(1972), p. 11.

CRC Handbook of Energy Efficiency, edited by F. Kreith and R.E. West, CRC Press, Boca Raton, FL (1996).

T.W. Eager and C. Musso, *Why did the World Trade Center Collapse*? Journal of Metals 53, 8(2001).

End-of-Life-Vechicle Recycling in the European Union, Journal of Metals (JOM), August (2003) pp. 15–19.

Handbook of Solid Waste Management, F. Kreith, Editor, McGraw-Hill, New York (1994).

Industrial Commodities Statistics Yearbook, United Nations, New York.

International Energy Annual, U.S. Department of Energy, Washington, DC.

International Petroleum Monthly (a monthly publication).

Lead-Free Solders and Processing Issues in Microelectronics, Journal of Metals (JOM) June (2003), pages 49–69.

Metallstatistik 1984–1994, Metallgesellschaft AG, Frankfurt, Germany (1995).

Metal Prices in the United States, U.S. Bureau of Mines, Washington, DC.

Mineral Commodity Summaries, U.S. Geological Survey, U.S. Bureau of Mines Washington, DC.

MSW in the US, 2000 Facts and Figures, Environmental Protection Agency, Washington, DC (yearly publication).

Net-Energy Analysis, D.T. Spreng, Editor, Praeger, New York (1988).

Plastics News, Crain Communications, Inc., (a weekly newspaper), Chicago, IL.

Platt's Metals Week Price Handbook, 21st Edition, McGraw-Hill, New York (1994).

Press releases, Umweltbundesamt.de Berlin/Germany (1999–2002).

Recovered Paper Statistical Highlights, American Forest and Paper Association, Washington, DC.

Statistical Compendium, United States Department of the Interior, Bureau of Mines Special Publication.

Survey of Energy Resources, World Energy Council, London.

The Aqueous Recovery of Gold from Electronic Scrap, Journal of Metals (JOM) August (2003) pp. 35–37.

S.Y. Tsao, "Readmap projects significant LED penetration of lighting market by 2010," *Laser Focus World* May (2003), 511.

C. Tschudin, O. Hutin, S. Arsalane, F. Bartels, P. Lambracht, and M. Rettenmayr. *Lead Free Soft Solder Die Attach Process for Power Semiconductor Packaging*, SEMICON China 2002 Techn. Symp., SEMI Shanghai (2002), AH1.

Vital Signs, 1996, Worldwatch Institute, W.W. Norton, New York, (1996).

World Oil, Gulf Publishing Co., Houston, TX (a monthly journal).

19

What Does the Future Hold?

Very few people, if any, probably would have predicted 50 years ago what kind of materials would dominate our technology today. Who could foresee the computer revolution and thus the preponderance of silicon in the electronics industry? How many scientists or engineers prognosticated the laser and its impact on communication, data-processing, data storage, and thus on optical materials? Was thcrc anybody who foretold, 50 years ago, the impact of superalloys, of composites, or of graphite fibers as important materials? Were ceramics not essentially perceived as clay and sand; that is, could anybody anticipate high-tech ceramics, including high T_c-superconductors, heat-resistant tiles (for space shuttles), silicon carbide engine parts, etc.? And finally, how many people could visualize 50 years ago the impact of plastics as one of the prime materials of the present day, or an accentuation on recycling and environmental protection?

On the other hand, nearly "everybody" predicted only 15 years ago that composites and ceramics for high-temperature engines would be the materials which would enjoy healthy growth rates in the years to come. This, however, has not happened. One of the major funding agencies in the United States (ARPA) has recently stated that the past 20 years have not brought the expected progress in structural ceramics, brittle matrix composites, or intermetallics. It is said that these "exotic" materials may probably never be used in critical parts such as turbine blades because they are too brittle, and a balance of properties may probably never be achieved. A 50% cut in support for the ensuing projects has therefore occurred. Instead, a resurgence of substantial interest in *metals* research may boost, for example, nickel-based superalloys (e.g., Ni–Al) which are often coated with μm-thick heat barriers. These coatings, consisting, for example, of zirconia or Ni–Cr–Al–Y are anticipated to prevent the respective alloys in turbine blades from melting or creeping. In other words, a change in research support from "exotic" materials to new alloys is presently thought to be more in the national interest. In essence, a shift from *structural materials* (such as ceramics, composites, etc.) to *functional materials* (such as smart materials, electromagnetic materials, and optical materials) will probably take place in the next couple of years. Specifically, future funding is anticipated for the development of compact lasers, solid-state lighting through inorganic and organic light-emitting diodes, holographic data storage, thermoelectric materials (used for cooling of high T_c superconductors, microprocessors, IR detectors, etc.) and for smart materials (which involve a series of sensors that control actuators). Further, a shift from empirical materials selection towards one based on model calculations and on a fundamental understanding of the physics and chemistry of materials science will probably take place. Finally, the exploration of nanostructures and nanotechnology will probably play a major role in future research funding.

Materials Science has expanded from the traditional metallurgy and ceramics into **new areas** such as electronic polymers, complex fluids, intelligent materials, organic composites, structural composites, biomedical materials (for implants and other medical applications), biomimetics, artificial tissues, biocompatible materials, "auxetic" materials (which grow fatter when stretched), elastomers, dielectric ceramics (which yield thinner dielectric layers for more compact electronics), ferroelectric films (for nonvolatile memories), more efficient photovoltaic converters, ceramic superconductors, improved battery technologies, self-assembling materials, fuel cell materials, optoelectronics, artificial diamonds, improved sensors (based on metal oxides, or conducting polymers), grated light valves, ceramic coatings in air (by plasma deposition), electrostrictive polymers, chemical-mechanical polishing, alkali metal thermoelectric converters, luminescent silicon, planar optical displays without phosphors, MEMS, and supermolecular materials. Some materials scientists are interested in *green* approaches, by entering the field of environmental-biological science, by developing environmentally friendly processing techniques and by inventing more recyclable materials.

Another emerging field is called **Nanomaterials by severe plastic deformation** (SPD) which involves the application of very high strains and flow stresses to work pieces. As the name implies, the respective new process yields microstructural features and properties in materials (notably metals and alloys) that differ from those known for conventional cold-worked materials. Specifically, pore-free grain refinements down to nanometer dimensions, and dislocation accumulations up to the limiting density of 10^{16} m^{-2} are observed. SPD yields an increase in tensile ductility without a substantial loss in strength and fatigue behavior. Furthermore, unusual phase transformations leading to highly metastable states have been reported and are associated with a formation of supersaturated solid solutions, disordering, amorphization, and a high thermal stability. Moreover, superplastic elongations in alloys that are generally not superplastic can be achieved. This affords a superplastic flow at strain rates significantly faster than in conventional alloys, enabling the rapid fabrication of complex parts. Finally, the magnetic properties of severely plastic deformed materials are different from their conventional counterparts. In particular, one observes an enhanced remanence in hard magnetic materials, a decrease of coercivity, (i.e. energy loss) in soft magnetic materials, and an induced magnetic anisotropy.

In short, the field of materials science is extending into new territory and this trend is expected to continue.

Still, "Predictions are quite difficult to make, particularly if they pertain to the future." As an example, a U.S. congressman suggested at the end of the 19th century that the patent office should be abolished, "since all major inventions have been made already." Moreover, it has been shown more than once that extrapolations of the present knowledge and accomplishments into the years which lay ahead were flatly wrong. Thus, utmost care needs to be exercised when projections are made. To demonstrate this, Figure 19.1 displays the development of essential materials properties during the 20th century.

One particular graph that demonstrates essentially correct predictions is worthy of some considerations. Figure 19.1 (d) depicts the number of transistors on a semiconductor chip in yearly intervals and reveals that initially the transistor density doubles about every 12 months. This plot, which was empirically deduced from earlier production figures (by an extrapolation of only three earlier data points), was dubbed *Moore's law* (after Gordon E. Moore at Fairchild Semiconductor) and depicts a log-linear relationship between device complexity and time. This type of prediction into the future is often referred to as a *controlling variable*, or a *self-fulfilling prophecy* since each computer chip manufacturer knows what the competitor will present in a given amount of time and acts accordingly. In other words, Moore's law involves human ingenuity for progress rather than physics. Higher transistor densities means higher processing speeds, lower power consumption, better reliability, and reduced cost. Beginning in the mid-seventies, the slope became less steep but still behaved in a log-linear fashion, and the rate of density doubling slowed down to every 18 months. (At the same time interval the *magnetic* storage density on hard disk drives doubled every 3 years.)

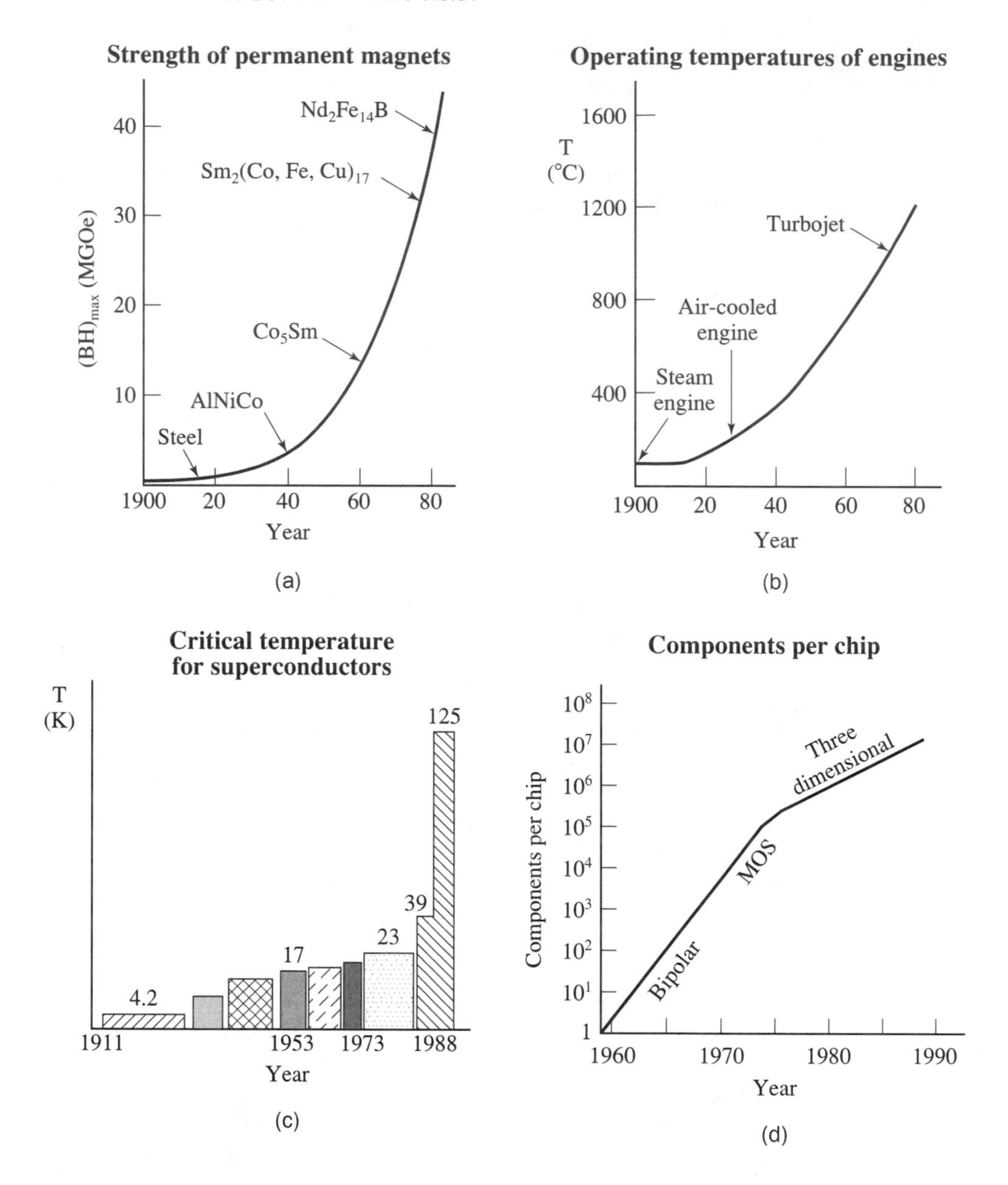

FIGURE 19.1. Improvements of materials properties during the 20th century.

On the other hand, for each doubling of performance, new and more sophisticated production facilities which may have price tags of about twice the previous factory, have to be built. Specifically, it is predicted that by 2005 a single chip fabrication facility will cost 10 billion dollars or 80 % of Intel's net worth. Thus, eventually there may be no economic incentive anymore to make transistors smaller unless computer chip companies team together (such as IBM and Toshiba, or Motorola and Siemens) and share a given facility.

While the substantial improvement of physical properties during the past 40 years is certainly impressive, it probably would be wrong to assume that a similar sharp rise would continue in the next decade or two. Indeed, certain limitations exist when, for example, atomistic dimensions are reached. This may stifle further progress as long as the same conventional methods are applied. In other words, new, innovative concepts are needed instead and probably will be found.

One may only speculate what kinds of discoveries might be made by the next generation of scientists and engineers if they would let their imagination roam freely into yet unexplored realms. Among these discoveries may be:

- A completely different family of materials which are not derived from already existing substances but are instead newly created by modifications of genes, that is, by gene technology. These biologically generated materials could possibly be custom-designed with respect to their physical properties, stability, and their recyclability. They may be created from renewable, inexhaustible resources or by bacteriologic transformations of already existing products.
- The energy of the future may not be generated by burning wood, coal, or oil, or by involving fissionable or fusionable elements, but by exploiting hitherto unknown "disturbances" that are neither electromagnetic nor of particle nature. This energy source may be tapped, should it exist, and it is hoped that mankind will have developed at that point a high degree of morality so that it may not be misused for destructive purposes.
- We are presently quite fixated on the concept that matter consists exclusively of atoms built from protons, neutrons, electrons, and a handful of other particles. Is it not possible that another type of matter does exist which is built of different particles beyond our present imagination? Maybe this alternate form of matter will be discovered in the next 50 years, should it exist.
- The abundance of radioactive waste surely will be a burden to future generations. New techniques may be found which are capable of manipulating the ratio of protons and neutrons in radioactive elements which will transform them into nonradiating isotopes.
- The transition temperature at which superconduction commences may again be substantially raised by employing new materials which have a striking similarity to fibers spun by animals or which are otherwise created in the body of animals or humans such as in nerve cells.
- Smart acoustic materials may be discovered which compensate incoming sound with a complementary sound, thus eliminating any noise. In particular, the acoustical properties of

materials will probably be studied more intensely in the future than they have been in the past.

- The storage of energy in batteries, etc., is at present most inefficient and requires bulky devices. New techniques will probably be discovered that raise the energy to mass ratio and increase the efficiency by involving a plasma technology, which is harnessed in containers consisting of high-temperature resistant materials.
- Materials may be found which, when weakened by fatigue or cracking, will activate a self-healing mechanism that returns the material, while in use, to its originally intended properties without external intervention.
- New types of trees or plant species will be genetically engineered which can be harvested in about 7–8 years rather than in the present 30-year time interval.
- The process of photosynthesis may be copied and used to create new materials and energy.

These few examples may serve as stimuli for future research and thus may lead materials science into new dimensions for the third millennium. It is hoped that future developments will be for the betterment of mankind (and not for its destrucion) since science will be increasingly linked to societal issues.

The public seems to be more and more dissatisfied by the fact that the substantial investments which are expended for science and education so far have not solved the problems of society. It almost seems that, with expanding technologies and the consequential rise in the amount of "desirable" consumer goods, the gap between the "haves" and the "have nots" steadily widens. This may be a contributing factor to the social unrest and the mindless crimes against property and human lives. Thus, a concerted effort by government, academia, and industry has to be initiated in the near future to find out how educational institutions in general and scholars in particular can contribute to the national welfare. It is imperative that our thinking and our deeds are less governed by money, rules, regulations, and mindless laws but instead by our forces of the heart, that is, our caring, compassion, and love for others, including the less fortunate individuals, in other parts of the world. It seems that the problems may be solved only when the next generation is educated in body, mind, *and* spirit, that is, beyond the factual knowledge of science. Specifically, the rising generation has to be taught to appreciate and especially *respect* the history of mankind, the cultures of other countries, the arts, in their diversity and their important place in life. Moreover, we need to appreciate the beauty but also the vulnerability of Planet Earth, which requires our caring responsibility for life in all its varied forms for generations to come. It is hoped that this book has made a contribution toward this goal.

APPENDICES

Appendix I

Summary of Quantum Number Characteristics

The energy states of electrons are characterized by four quantum numbers. The main quantum number, n, determines the overall energy of the electrons, i.e., essentially the radius of the electron distribution. It can have any integral value. For example, the electron of a hydrogen atom in its ground state has $n = 1$.

The quantum number, l, is a measure of the angular momentum L of the electrons and is determined by $|\mathbf{L}| = \sqrt{l(l+1)}\hbar$, where l can assume any integral value between 0 and $n - 1$.

It is common to specify a given energy state by a symbol which utilizes the n- and l-values. States with $l = 0$ are called s-states; with $l = 1$, p-states; and with $l = 2$, d-states, etc. A $4d$-state, for example, is one with $n = 4$ and $l = 2$.

The possible orientations of the angular momentum vector with respect to an external magnetic field are again quantized and are given by the magnetic quantum number m. Only m values between $+l$ and $-l$ are permitted.

The electrons of an atom fill the available states starting with the lowest state and obeying the Pauli principle which requires that each state can be filled only with two electrons having opposite spin ($|\mathbf{s}| = \pm\frac{1}{2}$). Because of the just-mentioned multiplicity, the maximal number of electrons in the s-states is 2, in the p-states 6, in the d-states 10, and in the f-states 14.

The electron bands in solids are named by using the same nomenclature as above, i.e., a $3d$-level in the atomic state widens to a $3d$-band in a solid. The electron configurations of some isolated atoms are listed on the next page.

		K	L	M	N	O
Z	Element	1s	2s 2p	3s 3p 3d	4s 4p 4d 4f	5s 5p 5d 5f
1	H	1				
2	He	2				
3	Li	2	1			
4	Be	2	2			
5	B	2	2 1			
6	C	2	2 2			
7	N	2	2 3			
8	O	2	2 4			
9	F	2	2 5			
10	Ne	2	2 6			
11	Na	2	2 6	1		
12	Mg	2	2 6	2		
13	Al	2	2 6	2 1		
14	Si	2	2 6	2 2		
15	P	2	2 6	2 3		
16	S	2	2 6	2 4		
17	Cl	2	2 6	2 5		
18	Ar	2	2 6	2 6		
19	K	2	2 6	2 6	1	
20	Ca	2	2 6	2 6	2	
21	Sc	2	2 6	2 6 1	2	
22	Ti	2	2 6	2 6 2	2	
23	V	2	2 6	2 6 3	2	
24	Cr	2	2 6	2 6 5	1	
25	Mn	2	2 6	2 6 5	2	
26	Fe	2	2 6	2 6 6	2	
27	Co	2	2 6	2 6 7	2	
28	Ni	2	2 6	2 6 8	2	
29	Cu	2	2 6	2 6 10	1	
30	Zn	2	2 6	2 6 10	2	
31	Ga	2	2 6	2 6 10	2 1	
32	Ge	2	2 6	2 6 10	2 2	
33	As	2	2 6	2 6 10	2 3	
34	Se	2	2 6	2 6 10	2 4	
35	Br	2	2 6	2 6 10	2 5	
36	Kr	2	2 6	2 6 10	2 6	
37	Rb	2	2 6	2 6 10	2 6	1
38	Sr	2	2 6	2 6 10	2 6	2
39	Y	2	2 6	2 6 10	2 6 1	2
40	Zr	2	2 6	2 6 10	2 6 2	2
41	Nb	2	2 6	2 6 10	2 6 4	1
42	Mo	2	2 6	2 6 10	2 6 5	1
43	Tc	2	2 6	2 6 10	2 6 5	2

Appendix II

Tables of Physical Constants

The International System of Units (SI or mksA System)

In the SI unit system, essentially four base units, the meter, the kilogram (for the mass), the second, and the ampere are *defined.* Further base units are the Kelvin, the mole (for the amount of substance), and the candela (for the luminous intensity). All other units are *derived units* as shown in the table below. Even though the use of the SI unit system is highly recommended, other unit systems are still widely used.

			Expression in terms of	
Quantity	Name	Symbol	Other SI units	SI base units
Force	Newton	N	—	$kg \cdot m/s^2$
Energy, work	Joule	J	$N \cdot m = V \cdot A \cdot s$	$kg \cdot m^2/s^2$
Pressure	Pascal	Pa	N/m^2	$kg/m \cdot s^2$
El. charge	Coulomb	C	J/V	$A \cdot s$
Power	Watt	W	J/s	$kg \cdot m^2/s^3$
El. potential	Volt	V	W/A	$kg \cdot m^2/A \cdot s^3$
El. resistance	Ohm	Ω	V/A	$kg \cdot m^2/A^2 \cdot s^3$
El. conductance	Siemens	S	A/V	$A^2 \cdot s^3/kg \cdot m^2$
Magn. flux	Weber	Wb	$V \cdot s$	$kg \cdot m^2/A \cdot s^2$
Magn. induction	Tesla	T	Wb/m^2	$kg/A \cdot s^2$
Inductance	Henry	H	Wb/A	$kg \cdot m^2/A^2 \cdot s^2$
Capacitance	Farad	F	C/V	$A^2 \cdot s^4/kg \cdot m^2$

Physical Constants (SI and cgs units)

Mass of electron (free electron mass; rest mass)	$m_0 = 9.109 \times 10^{-31}$ (kg) $= 9.109 \times 10^{-28}$ (g)
Charge of electron	$e = 1.602 \times 10^{-19}$ (C)
Velocity of light in vac.	$c = 2.998 \times 10^{8}$ (m/s) $= 2.998 \times 10^{10}$ (cm/s)
Planck constant	$h = 6.626 \times 10^{-34}$ (J · s)
	$= 6.626 \times 10^{-27}$ (g · cm²/s) $= 4.136 \times 10^{-15}$ (eV · s)
	$\hbar = 1.054 \times 10^{-34}$ (J · s)
	$= 1.054 \times 10^{-27}$ (g · cm²/s) $= 6.582 \times 10^{-16}$ (eV · s)
Avogadro constant	$N_0 = 6.022 \times 10^{23}$ (atoms/mol)
Boltzmann constant	$k_B = 1.381 \times 10^{-23}$ (J/K)
	$= 1.381 \times 10^{-16}$ (erg/K) $= 8.616 \times 10^{-5}$ (eV/K)
Bohr magneton	$\mu_B = 9.274 \times 10^{-24}$ (J/T) $\equiv$ A · m² $= 9.274 \times 10^{-21}$ (erg/G)
Gas constant	$R = 8.314$ (J/mol · K) $= 1.986$ (cal/mol · K)
Permittivity of empty space	$\epsilon_0 = 1/\mu_0 c^2 = 8.854 \times 10^{-12}$ (F/m) $\equiv$ (A · s/V · m) $\equiv$ (N/A²)
Permeability of empty space	$\mu_0 = 4\pi \times 10^{-7} = 1.257 \times 10^{-6}$ (H/m) $\equiv$ (V · s/A · m)
Faraday constant	$F = 9.648 \times 10^{4}$ (C/mol)

Useful Conversions

1 (eV) $= 1.602 \times 10^{-19}$ (J) $= 1.602 \times 10^{-12}$ (g · cm²/s²) $= 1.602 \times 10^{-19}$ (kg · m²/s²)
$= 3.829 \times 10^{-20}$ (cal) $= 23.04$ (Kcal/mol)
1 (J) = 1 (kg · m²/s²) $= 10^{7}$ (erg) $= 10^{7}$ (g · cm²/s²) $= 2.39 \times 10^{-1}$ (cal)
1 (1/Ωcm) $= 9 \times 10^{11}$ (1/s)
1 (1/Ωm) $= 9 \times 10^{9}$ (1/s)
1 (C) = 1 (A · s) = 1 (J/V)
1 (Å) $= 10^{-10}$ (m)
1 (torr) = 133.3 (N/m²) ≡ 133.3 (Pa) = 1 (mm Hg)
1 (bar) $= 10^{5}$ (N/m²) $\equiv 10^{5}$ (Pa)
1 (Pa) = 1 (N/m²) $= 1.45 \times 10^{-4}$ (psi)
1 (psi) $= 6.895 \times 10^{3}$ (Pa)
1 (cal) $= 2.6118 \times 10^{19}$ (eV)

1 (mm) (milli) $= 10^{-3}$ (m)	1 km (Kilo) $= 10^{3}$ m
1 (μm) (micro) $= 10^{-6}$ (m)	1 Mm (Mega) $= 10^{6}$ m
1 (nm) (nano) $= 10^{-9}$ (m)	1 Gm (Giga) $= 10^{9}$ m
1 (pm) (pico) $= 10^{-12}$ (m)	1 Tm (Tera) $= 10^{12}$ m
1 (fm) (femto) $= 10^{-15}$ (m)	1 Pm (Peta) $= 10^{15}$ m
1 (am) (atto) $= 10^{-18}$ (m)	1 Em (Exa) $= 10^{18}$ m

Electronic Properties of Some Metals

Material	Fermi energy E_F [eV]	Number of free electrons, N_{eff} $\left[\frac{\text{electrons}}{\text{m}^3}\right]$	Work function (photoelectric) ϕ [eV]	Resistivity ρ [$\mu\Omega$ cm] at 20° C
Ag	5.5	6.1×10^{28}	4.7	1.59
Al	11.8	16.7×10^{28}	4.1	2.65
Au	5.5	5.65×10^{28}	4.8	2.35
Be	12.0		3.9	4.0
Ca	3.0		2.7	3.91
Cs	1.6		1.9	20.0
Cu	7.0	6.3×10^{28}	4.5	1.67
Fe			4.7	9.71
K	1.9		2.2	6.15
Li	4.7		2.3	8.55
Na	3.2		2.3	4.20
Ni			5.0	6.84
Zn	11.0	3×10^{28}	4.3	5.91

Electronic Properties of Some Semiconductors

	Material	Gap energy E_g [eV] 0 K	300 K	Conductivity $\sigma\left[\frac{1}{\Omega \cdot \text{m}}\right]$	Mobility of electrons $\mu_e\left[\frac{\text{m}^2}{\text{V} \cdot \text{s}}\right]$	Mobility of holes $\mu_h\left[\frac{\text{m}^2}{\text{V} \cdot \text{s}}\right]$	Work function (photoelectric) ϕ [eV]
Element	C (diamond)	5.48	5.47	10^{-12}	0.18	0.12	4.8
	Ge	0.74	0.66	2.2	0.39	0.19	4.6
	Si	1.17	1.12	9×10^{-4}	0.15	0.045	3.6
	Sn (gray)	0.08		10^{6}	0.14	0.12	4.4
III–V	GaAs	1.52	1.42	10^{-6}	0.85	0.04	
	InAs	0.42	0.36	10^{4}	3.30	0.046	
	InSb	0.23	0.17		8.00	0.125	
	GaP	2.34	2.26		0.01	0.007	
IV–IV	α-SiC	3.03	2.99		0.04	0.005	
II–VI	ZnO	3.42	3.35		0.02	0.018	
	CdSe	1.85	1.70		0.08		

Magnetic Units

Name	Symbol	em-cgs units	mks (SI) units	Conversions
Magnetic field strength	**H**	$Oe \equiv \frac{g^{1/2}}{cm^{1/2} \cdot s}$	$\frac{A}{m}$	$1\ \frac{A}{m} = \frac{4\pi}{10^3}\ Oe$
Magnetic induction	**B**	$G \equiv \frac{g^{1/2}}{cm^{1/2} \cdot s}$	$\frac{Wb}{m^2} = \frac{kg}{s \cdot C} \equiv \mathbf{T}$	$1\ T = 10^4\ G$
Magnetization	**M**	$\frac{Maxwell}{cm^2} \equiv \frac{g^{1/2}}{cm^{1/2} \cdot s}$	$\frac{A}{m}$	$1\ \frac{A}{m} = \frac{4\pi}{10^3}\ \frac{Maxwells}{cm^2}$
Magnetic flux	ϕ	$Maxwell \equiv \frac{cm^{3/2} \cdot g^{1/2}}{s}$	$Wb = \frac{kg \cdot m^2}{s \cdot C} = V \cdot s$	$1\ Wb = 10^8\ Maxwells$
Susceptibility	χ	Unitless	Unitless	$\chi_{mks} = 4\pi\chi_{cgs}$
(Relative) permeability	μ	Unitless	Unitless	Same value
Energy product	**BH**	MGOe	$\frac{kJ}{m^3}$	$1\ \frac{kJ}{m^3} = \frac{4\pi}{10^2}\ MGOe$

Conversions from the SI Unit System into the Gaussian Unit System

Quantity	mks (SI)	cgs (Gaussian)
Magnetic induction	B	B/c
Magnetic flux	Φ_B	Φ_B/c
Magnetic field strength	H	$cH/4\pi$
Magnetization	M	cM
Magnetic dipole moment	μ_m	$c\mu_m$
Permittivity constant	ϵ_0	$1/4\pi$
Permeability constant	μ_0	$4\pi/c^2$
Electric displacement	D	$D/4\pi$

Note: The equations given in this book can be converted from the SI (mks) system into the cgs (Gaussian) unit system and vice versa by replacing the symbols in the respective equations with the symbols listed in the following table. Symbols which are not listed here remain unchanged. It is imperative that consistent sets of units are utilized.
$\mu_0 = 4\pi \times 10^{-7} = 1.257 \times 10^{-6}$ (V · s/A · m) ≡ (Kg · m/C^2) ≡ (H/m).
$\epsilon_0 = 8.854 \times 10^{-12}$ (A · s/V · m) ≡ (F/m).

Appendix III The Periodic Table of the Elements

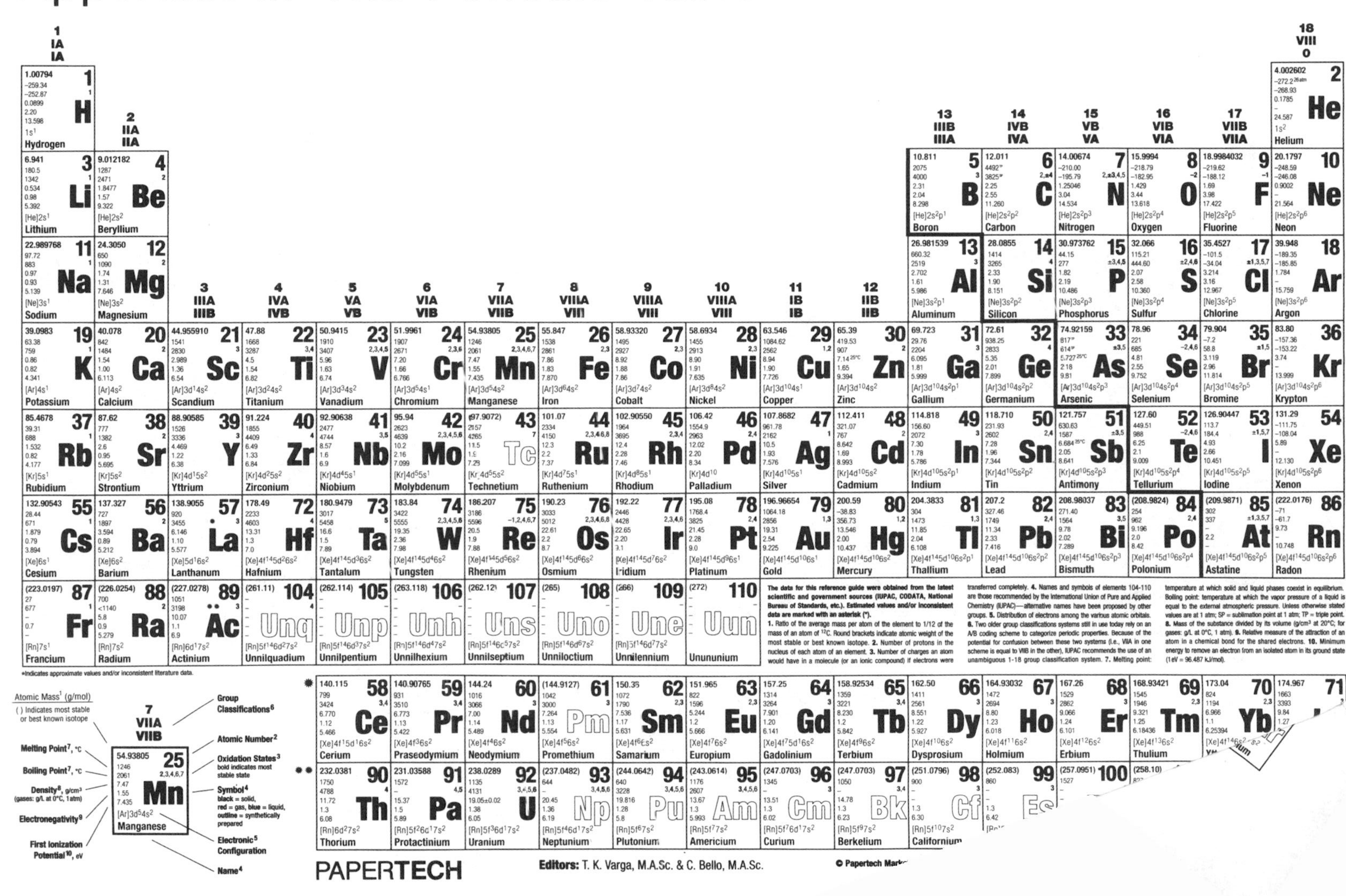

...ndix IV

Ap... ...ons to Selected ...ems

2

2.1. 0.77 (m)
2.2. 75%
2.3. 0.114 (cm) (Not 0.1195 cm!)
2.4. 1.26×10^4 (N)
2.5. 36%
2.6. 703.1 (MPa)
2.7. 0.233
2.8. $\Delta V = 0$
2.10. $\epsilon_t = 10.26\%$
$\epsilon = 10.8\%$
$\sigma_t = 105.8$ (MPa)
$\sigma = 95.49$ (MPa)
2.11. 3.998 (cm)
2.12. $l = 1.502$ (m)
$\sigma = 254.6$ MPa (<300 MPa)
2.13. 0.292
2.14. 0.5 (Disregard squares of small quantities).

Chapter 3

3.1. (220)
3.2. Counterclockwise sequence starting with the front plane:
$(10\bar{1}0)$; $(01\bar{1}0)$; $(\bar{1}100)$; $(\bar{1}010)$; $(0\bar{1}10)$; $(1\bar{1}00)$
3.3. (a) $\alpha = 60°$ (equilateral triangles!); $\beta = 120°$
(b) $\alpha = 109.47°$
3.4. 2 (not 6!)
3.5. 1/3, 2/3, 1/2
3.6. *Hint:* Draw two triangles, one vertically between the A and B planes and the other within the B plane.
3.7. $\{100\}\langle 100\rangle$

Appendix IV
Solutions to Selected Problems

Chapter 2

2.1. 0.77 (m)
2.2. 75%
2.3. 0.114 (cm) (Not 0.1195 cm!)
2.4. 1.26×10^4 (N)
2.5. 36%
2.6. 703.1 (MPa)
2.7. 0.233
2.8. $\Delta V = 0$
2.10. $\epsilon_t = 10.26\%$
$\epsilon = 10.8\%$
$\sigma_t = 105.8$ (MPa)
$\sigma = 95.49$ (MPa)
2.11. 3.998 (cm)
2.12. $l = 1.502$ (m)
$\sigma = 254.6$ MPa (<300 MPa)
2.13. 0.292
2.14. 0.5 (Disregard squares of small quantities).

Chapter 3

3.1. (220)
3.2. Counterclockwise sequence starting with the front plane: $(10\bar{1}0)$; $(01\bar{1}0)$; $(\bar{1}100)$; $(\bar{1}010)$; $(0\bar{1}10)$; $(1\bar{1}00)$
3.3. (a) $\alpha = 60°$ (equilateral triangles!); $\beta = 120°$
(b) $\alpha = 109.47°$
3.4. 2 (not 6!)
3.5. 1/3, 2/3, 1/2
3.6. *Hint:* Draw two triangles, one vertically between the A and B planes and the other within the B plane.
3.7. $\{100\}\langle 100\rangle$

Appendix III The Periodic Table of the Elements

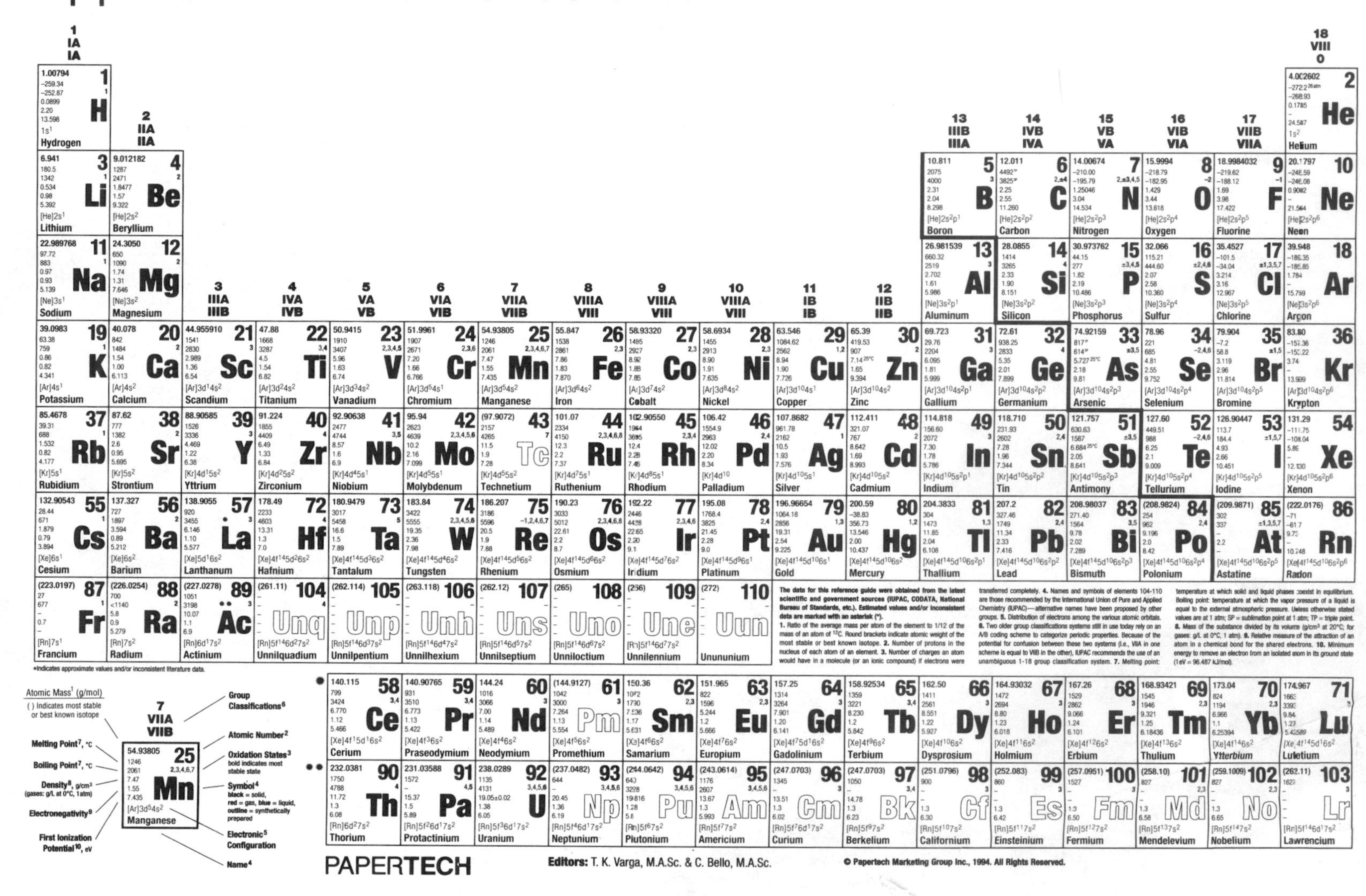

3.8. $[\bar{1}11]$ direction (close packed) lies in (110) plane (not close packed)

3.9. CsCl (a) = 8; (b) one Cs and one Cl ion (c) $a_0 = 2(r + R)/\sqrt{3}$
NaCl (a) = 6; (b) 4 Cl ions and 4 Na ions (c) $a_0 = 2(r + R)$

3.10. (a) Each Zn ion has four nearest S ions.
(b) Four Zn ions and four S ions, that is, eight ions.
(c) $a_0 = 4(r + R)/\sqrt{3}$

3.11. 0.34

3.12. (a) 0.707; (b) 0.866

3.13. (a) 0.340; (b) 0.907

3.14. 8.933 (g/cm^3)

3.15. $[1\bar{1}00]$

3.16. $a \dfrac{\sqrt{2}}{2}$

Chapter 5

5.1. Liquid: 54.5%
Solid: 45.5%

5.2.

Peritectic at 799°C:	$\alpha + L \rightarrow \beta$
Peritectic at 756°C:	$\beta + L \rightarrow \gamma$
Peritectoid at 640°C:	$\gamma + \epsilon \rightarrow \zeta$
Eutectic-type (not named) at 640°C:	$\gamma \rightarrow \epsilon + L$
Peritectoid at 590°C:	$\gamma + \zeta \rightarrow \delta$
Eutectoid at 586°C:	$\beta \rightarrow \alpha + \gamma$
Eutectoid at 582°C:	$\zeta \rightarrow \delta + \epsilon$
Eutectoid at 520°C:	$\gamma \rightarrow \alpha + \delta$
Peritectic at 415°C:	$\epsilon + L \rightarrow \eta$
Eutectoid at 350°C:	$\delta \rightarrow \alpha + \epsilon$
Eutectic at 227°C:	$L \rightarrow \eta + \beta$-Sn
Peritectoid at 189°C:	$\epsilon + \eta \rightarrow \eta'$
Eutectoid at 186°C:	$\eta \rightarrow \eta' + \beta$-Sn

Nonstoichiometric phase ϵ forms at 676°C
Allotropic transformation from β-Sn $\rightarrow$ α-Sn at 13.2°C
Note: Some transformations are hardly discernible in Figure 5.17.

5.3. (a) 8.8 mass % Cu in Ag at 780°C
(b) 8.0 mass % Ag in Cu at 780°C

5.4. negligible

5.5. $\alpha + \theta$
$C_\alpha = 1.5$ mass % Cu
$C_\theta = 52$ mass % Cu

5.6. About 850°C

5.7. No

5.8. (a) 780°C
(b) Ag–28.1 mass % Cu
(c) About 830°C
(d) About 6 mass % Cu

Chapter 6

6.1. $X = -\infty \rightsquigarrow C_x = 2C_i - C_0$
$X = 0 \rightsquigarrow C_x = C_i$
$X = +\infty \rightsquigarrow C_x = C_0$

6.2. $t \to \infty \rightsquigarrow C_x = C_i$
$t \to 0$ (x positive) $\rightsquigarrow C_x = C_0$
$t \to 0$ (x negative) $\rightsquigarrow C_x = 2C_i - C_0$

6.3. $\frac{f_g}{f_v} = 82.9$

6.4. (a) $D = 1.44 \times 10^{-21}$ cm²/s; No
(b) $n_v = 1.19 \times 10^{18}$ vacancies/cm³
$n_v/n_s = 1.97 \times 10^{-5}$ vacancies per lattice site
(c) $n_v = 3.57 \times 10^{17}$ vacancies/cm³
$n_v/n_s = 5.92 \times 10^{-6}$ vacancies per lattice site
(d) $D_{v0} = 8.3 \times 10^{-10}$ cm²/s; Yes
(e) D becomes smaller

6.5. $t = 3.47$ (hr)

6.6. (a) $x = 4\sqrt{Dt}$; (b) $x = \sqrt{Dt}$; (c) $C_x \approx 0.15\, C_i$

Chapter 8

8.1. γ (austenite) 52%; Fe_3C (cementite) 48%

8.2. 2% C

8.3. Pearlite: 84%
Primary α: 16%
Pearlite contains 0.77% C
Primary α contains 0.0218% C

8.4. $r_{int} = 0.534$ Å; latticc will cxpand

8.5. $r_{int} = 0.361$ Å (tetrahedral); $r_{int} = 0.192$ Å (octahedral)

8.6. Interstitial site for γ-iron is larger. Thus, higher solubility of C in γ-iron. See Figure 8.1.

8.7. Atomic radius is larger for higher coordination number.

Chapter 9

9.1. 0.05917 (note difference between ln and log!)

9.2. $E = 0.68$ (V)

9.3. $C_{ion} = 0.0445$ M
$m = 2.829$ (g)

9.4. P-B-ratio = 1.75 (between 1 and 2 $\rightsquigarrow$ continuous and adherent)

9.5. P-B-ratio = 1.58 (stable oxide)

9.6. $W = 2.96$ (g)

9.7. $t = 756.3$ (s) = 12.6 (min) (for $n = 1$)
$t = 2{,}269$ (s) = 37.8 (min) (for $n = 3$)

Chapter 11

11.1. 0.1 (V)

11.2. $2.73 \times 10^{22} \left(\frac{\text{electrons}}{\text{cm}^3}\right)$ or $1.07 \left(\frac{\text{electrons}}{\text{atom}}\right)$

11.3. $1.948 \times 10^{47} \left(\frac{\text{electrons}}{\text{J m}^3}\right) \equiv 3.12 \times 10^{28} \left(\frac{\text{electrons}}{\text{eV m}^3}\right)$
11.4. $5.9 \times 10^{22} \left(\frac{\text{electrons}}{\text{cm}^3}\right)$
11.5. $9.81 \times 10^{9} \left(\frac{\text{electrons}}{\text{cm}^3}\right)$
11.6. $T \approx 19{,}781$ (K)
11.7. 0.396 (eV)
11.8. $3.12 \times 10^{3} \left(\frac{1}{\Omega\text{m}}\right)$
11.9. 0.11 (A)
11.10. 0.44 ($m^2/V \cdot s$)

Chapter 12

12.1. 2×10^4 (A/m); 251.3 (Oe)
12.2. 694 (K)
12.3. $\mu = 1.001$ (in mks units)
12.4. $\mu_B = 9.272 \times 10^{-24}$ (J/T) or 9.27×10^{-21} (erg/Oe)
12.5. $FeO \cdot Fe_2O_3$; $\mu_m = 4\ \mu_B$
$CoO \cdot Fe_2O_3$; $\mu_m = 3\ \mu_B$
12.6. Yes, it can be easily done!
12.7. $B = 0.1$ (T)

Chapter 13

13.1. W = 27.8 (nm)
13.2. $R_{Ag} = 98.88\%$; $R_{glass} = 5.19\%$
13.3. $Z = 7.81$ (nm)
13.4. $T_2 = 94.3\%$
13.5. $(\nu_1)_K = 1.03 \times 10^{15}$ (s^{-1})
$(\nu_1)_{Li} = 1.92 \times 10^{15}$ (s^{-1})
13.6. $N_f = 5.49 \times 10^{22}$ (electrons/cm^3);
$N_a = 5.86 \times 10^{22}$ (atoms/cm^3);
$N_f/N_a = 0.94$ (electrons/atom)

Chapter 14

14.1. Average 5.26%
14.2. $T = 1{,}142$ (K) = 869 (°C)
14.3. $-J_Q$: Glass: 3.8×10^3 ($J/s \cdot m^2$)
Al: 4.74×10^5 ($J/s \cdot m^2$)
Wood: 3.2×10^2 ($J/s \cdot m^2$)
14.4. $l = 411$ (Å)
14.5. $\Delta L = 4.9$ (mm)

Index

Abaca, 328, 337
Abbé, E., 306
Aboriginal people, 10
ABS, 375
Absorbance, 250
Absorption, 249
  of light, 252–259
Absorption band, 254
Acceptor impurity, 201
Acceptor level, 201
Acoustic materials, 411
Acrylics, 358
Activation energy, 104, 105, 119
Active stainless steel, 162
Adamá, 287
Addition polymerization, 350
Additives to polymers, 360
Adobe, 289
Advanced ceramics, 317
Afterglow, 259
Age hardening, 89–95, 115
Aging, 89
Agricola, G., 131
Agriculture, 130
Akkadian tongue, 126
Al-Cu, 90
Al-Cu-Mg, 94
Alchemy, 132, 368
Alcohol, 381
Alkali halides, 27
Alkaline batteries, 163
Allophane, 298
Allotropic materials, 37
Allotropic transformation, 37, 141, 297
Alloyed steels, 149
Alloys, 66, 74–101
  conduction in, 193–194
Alluvial deposits, 68
Alnico alloys, 240
Alnico magnets, 241
Alpaca, 329
α-phase, 76
α-quartz, 297
Alternate-use applications, 387
Alumel, 112
Alumina, 296, 319
Aluminum, *see* Al *entries*
Aluminum recycling, 385
Amber, 4, 71, 332
Amber electricity, 174
Amorphous ferromagnets, 239
Amorphous polymers, 358
Amorphous state, 307
Amorphous structure, 25
Amp-turns per meter, 224
Ampère, A. 177
Anatolia, 5, 6, 8, 70, 126, 290
Anemia, 139
Angle of refraction, 247–248
Angora rabbit, 329
Angular frequency, 247
Anion, 114
Animal fiber, 327, 328, 333–335
Animal guts, 332
Animal skin, 3
Anisotropy, 43
Annealing, 51, 117, 144
Annealing point, 313
Annealing twins, 37
Anode, 158
Antiferromagnetic materials, 233
Antiferromagnetism, 223, 233, 234
Antimicrobal substances, 364
Antistatic agent, 360
Aqua regia, 371
Aramid, 362
Aristotle, 127
Armor, 321
Arrhenius diagram, 108
Arrhenius equation, 102, 107, 211
Arrow, 126
Arrow tips, 4, Plate 1.1
Arsenic, 64, 67
Arsenic-pentafluoride, 207
Artificial ivory, 346
Artificial silk, 328, 334, 346
Asbestos, 329
Asphalt, 289
Assyria, 70, 126, 291
Atom flow, 109
Atom flux, 109
Atomic bonding, 27
Atomic structure
  of condensed matter, 24–26
  of polymers, 43
Atomization, 116
Atoms, 24
  arrangement of, 31–46
  binding forces between, 26–31
  in motion, 102–123
Attapulsite clay, 302
Attenuation, 250
Austenite, 141, 142
Austenitizing treatment, 144
Auxetic materials, 408
Ax, 67, 71

Babylon, 126, 289
Backbone, 349
Baekeland, L. H., 347
Bainite, 145, 148

Baked clay, 7
Bakelite, 347
Ball clay, 293
Balsa wood, 341
Baltic Sea, 128
Bamboo structure, 122
Ban Chiang, 68
Band structure, 190, 191, 198, 200, 209
Band gap, 201
Band overlapping, 192
Barium titanate, 218, 220
Bark, 332
Barrier region, 203
Basal plane, 40
Basal slip, 52
Basalte, 292
Base, 203
Basic oxygen process (BOP), 133
Bast fiber, 328, 336–338
$BaTiO_3$, 218, 220
Battery, 207, 411
Bauxite, 376, 385
Bave, 335
Beads (glass), 291
Beating (paper), 343
Beer equation, 251
Bell Beaker Folk, 9
Bell metal, 87
Bellows, 6, 129
Bend test, 12, 13
Bentonite, 302
Benzene ring, 349
Bessemer converter, 134
Bessemer steel, 134
β-cristobalite, 46, 297
β-phase, 81
β-quartz, 297
β-tin, 38
β-tridymite, 297
Bias voltage, 204
Bifunctional system, 350
Billiard balls, 346
Binary phase diagrams, 76–87
Binding energy, 27
Binding forces between atoms, 26–31, 48
Bioceramics, 321
Bio-diesel, 382
Biodiversity, 377
Bioglass, 321
Biological fixation, 322
Biologically generated materials, 411
Biomimetics, 408
Biosensor, 207
Bipolar transistor, 203
Biringuccio, V., 132
Bismuth, 67
Bisqueware, 292
Bivalent metal, 192
Black ironwood, 341
Black Sea, 128
Blast furnace, 129, 133, 150
Bloch wall, 233
Bloom, 125, 129, 133
Blow tube, 6, 304
Body-centered cubic (BCC) crystal structure, 32
Body painting, 7, 288
Bohr magneton, 230
Boltzmann constant, 103, 275
Bombyx mori, 327
Bonding mechanism, 27
Bone, 3, 4
Bone china, 393
Boron carbide, 317
Boron nitride, 298
Böttger, J.F., 292, 293
Box bellows, 129
Bragg, W., 24
Branched polymers, 43
Brass, 70, 75
Bravais, A., 24
Bravais lattice, 32, 33
Breaking strength, 15
Bricks, 288
Bridging oxygens, 312
Brinell hardness number, 21
Brin, 325
Brittle materials, 12, 16
Brittleness, 10, 127
Bronze, 64, 67, 75, 94, 126
Bronze Age, 3, 64–72, 97
Brush fiber, 328
Bubbles (gas), 98
Buckyballs, 362
Building material, 287
Buna, 332
Burden, 133
Burgers vector, 50, 56
Burnishing, 295

c/a ratio, 32, 53
Calcination, 316
Caloric, 176
Cambium, 341
Camel hair, 329
Campfire, 7
Caoutchouc, 330
Capacitance, 215
Capacitor, 219
Capacity, 181
Carbohydrate, 328
Carbon, 127
Carbon fourteen dating, 295
Carbonizing, 129
Cardboard, 391
Case hardened iron, 127
Cashmere wool, 329
Cassiterite, 70
Cast iron, 129, 150–152
Casting, 97–99, 129
Catal Hüyük, 8, 290
Cathode, 157
Cathode rays, 182
Cathodic protection, 162, 163
Cathodoluminescence, 257
Cation, 114
Cave of the Treasure, 67
Cavity, 98, 261
CD player, 263
Cellophane, 353
Celluloid, 346, 347
Cellulose, 328, 336–337, 340, 345
Cellulose nitrate, 345–347
Celtic tribes, 128
Cement, 315–316
Cementation, 365
Cementite, 141, 142
Ceramic magnets, 241
Ceramics, 16, 45, 287–290, 317, 407
  creep in, 120–121
  crystal structures of, 44–46
  definition, 44
  diffusion in, 113–115
Ceremonial vessel, 68
Cermet, 320, 362
Cesium chloride crystal structure, 45–46
Chain armor, 128
Chain folding of polymers, 249
Chain reaction polymerization, 350
Chalcolithic Period, 3, 6–10, 66
Chalybes, 126
Chameleon fiber, 364
Characteristic penetration depth, 250
Characteristic X-rays, 266–267
Charcoal, 127

Charge density, 218
Chariot, 126
Chemical recycling, 388
Chill zone, 97
China, 66, 129
Chinese teapots, 291
Chlorophyll, 137
Chromel, 212
Chrysotile, 329
Clay, 3, 62, 287, 288, 296, 297
Clay crucible, 68
Cliff-dwelling, 289
Climbing of dislocations, 118
Clock (electronic), 220
Cloisonné technique, 295
Close-packed plane, 35
Coal, 378, 381
Coalescence, 120
Coating of metals, 165
Cobalt-samarium magnets, 236
Cobble creep, 119
Cocoon, 335
Coercive field, 219, 232
Coercive force, 236
Coercivity, 231, 239
Coherency strain field, 91
Coherent particles, 91
Coherent precipitates, 89
Coherent precipitation, 91, 92
Coherent scattering of electrons, 188
Coinage, 87, 367, 371
Coir fiber, 328
Coke, 133
Cold working, 19, 117, 127
Collector, 203
Collimation, 261
Colloidal gold, 369
Colonies, 143
Coloring of glazes, 295
Columnar grain, 98
Columnar zone, 98
Compact disk (CD), 268
Compass, 175
Component, 76
Composite material, 116, 288, 315, 340, 361, 407
Composition cell, 164
Composting, 398
Compound semiconductor, 201
Compounds, 74–101
Computer, 203
Computer revolution, 407
Concentration cell, 161
Concentration gradient, 109
Concrete, 27, 315–316, 375
Condensation polymerization, 352
Condensed matter, atomistic structure of, 24–26
Conducting polymers, 206–208
Conduction
  in alloys, 193–194
  in polymers, 206–208
  thermal, 278–282
Conduction band, 197
Conductivity, 185, 187, 189, 192
  of bivalent metals, 192
  classical, 189
  of monovalent metals, 192
  quantum mechanical, 192
Conductors, 172, 185
  ionic, 210–212
Congruent maximum, 87
Congruent phase transformations, 87
Conjugated organic polymer, 206
Constantan, 212
Contact potential, 214
Continuous casting, 99
Conventional unit cell, 32
Cooling curve, 83, 84
Coordination number, 34
Copolymer, 360
Copper, 5, 67, 128
See also Cu *entries*
Copper-arsenic, 67
Copper axe, 9
Copper ore, 6
Copper smelting, 125
Copper–Stone Age, 3, 6–10
Copper–tin, 67
Copper–tin phase diagram, 87
Copper–zinc, 70
Cordage, 336, 338
Core-forming technique, glass, 303
Core loss, 236
Core material, 238
Coring, 78, 99, 109
Cork, 332, 341
Corrosion, 130, 154–158, 268
  in glass, 167–168
  in low-alloy steel, 164
  in martensitic stainless steel, 164
  in polymers, 166–167
Corrosion pit, 161
Cortex, 341
Corundum, 319
Cosmic dust, 137
Cosmic iron, 137
Cotton, 327, 337
Covalent bond, 28–29
Cradle of civilization, 68
Creep, 75, 117–121, 408
  in ceramics, 120–121
  in glass, 120
  in polymers, 121
Creep constant, 119
Creep exponent, 119
Creep rate, 118
Creep strain, 118
Creep test, 22
Creep time, 118
Crevice corrosion, 162
Cristobalite, 300, 307
  crystal structure, 302
Critical resolved shear stress, 48, 60
  for single crystals, 52
Croesus, 372
Cross-linking, 352
Cross-slipping, 57
Crown, 67
Crown glass, 306
Crude cast iron, 129
Crude oil, 377–382
Crude steel, 135
  production, 136
Cryotron, 196
Crystal, 24
Crystal glass, 306
Crystal structure, 24, 32
  of ceramics, 44–46
  of cristobalite, 301
Crystal systems, 33
Crystalline polymer, 355, 358
Crystallography, 31–46
  of clay minerals, 298
Cu, *see also* Copper *entries*
Cu-Ag, 80
Cu-Ni, 76
Cubic crystal, 33
Cullet, 306, 385
Cuneiform, 126
Curie Constant, 229
Curie temperature, 220, 232, 234
Curie–Weiss law, 229
Currency, 366–367
Current, 185
Current density, 186
Cutting tools, 320

CW laser, 263
Cyanide leaching, 370
Cyprus, 6, 70
Czochralski crucible pulling, 198

Dagger, 131
Damascus steel, 131
Damping constant, 248
Damping strength, 189, 253
Davisson, C.J., 183
de Broglie, L., 183
De-inking, 391
De la pirotechnia, 132
De re metallica, 131
Dead Sea Scrolls, 67
Dealloying, 157
Debye–Scherrer technique, 24
Debye temperature, 274, 275
Decibel, 250
Decolorant, 304
Decomposition temperature (polymers), 359
Decorations, 295
Decortication, 336
Defect electron, 198
Defoliation, 377
Deformation twins, 37
Deglassing, 113
Degradation temperature, 359
Delft, 291
δ-ferrite, 141
Dendritic growth, 79
Denier, 337
Density, 35
  of (electron) states, 190, 191
Dental alloys, 369
Depleted zone, 103
Devitrification, 313, 315
Dezincification, 157
Diamagnetic materials, 225, 226
Diamagnetism, 223, 227–228
Diameter of universe, 184
Diamond, 18, 28, 29
Diamond cubic structure, 46
Diamonds, artificial, 408
Dicalcium silicate, 316
Dielectric constants of materials, 215, 216, 218
Dielectric displacement, 216
Dielectric loss, 219
Dielectric material, 214, 216
Dielectric polarization, 217
Dielectric properties, 214–218
Dielectrics, 181, 183, 220
Diffusion, 78, 103–107
  in amorphous solids, 114
  in ceramics, 114–115
  by interchange, 105
  through polymers, 115
Diffusion coefficient, 109, 211
Diffusion constant, 109
Diffusion creep, 119
Diffusion depth, 122
Diffusion distance, 112
Diffusion jump distance, 110
Diffusivity, 109
Digital video disk (DVD), 269
Dinitrate of cellulose, 345
Diode, 199
Dipole, 30, 216
Dipole moment, 216, 218
Direct reduction process, 133
Directional diffusion, 108–109
Directionality, 28
Directions in unit cells, 40
Disko (Greenland), 137
Dislocation, 25, 47–55, 103
Dislocation core, 50, 58
Dislocation-core diffusion, 107
Dislocation creep, 118
Dislocation density, 57, 96, 117
Dislocation line, 49
Dislocation loop, 93
Dislocation multiplication, 57
Dislocation pinning, 56, 118
Dispersion, 248, 251
Dispersion strengthening, 95, 99, 142
Displacive transformation, 297
Disproportionation, 351
Domain, 220, 232
Domain wall, 233
Domestic waste, 394–400
Donor atom, 199
Donor electron, 199
Doping, 196, 199, 206
Dragon's blood, 132
Drain, 204
Drift of atoms, 109
Driving force, 108
Drude, P., 189, 259
Drude theory, 254
Duality of electrons, 182
Ductile cast iron, 151
Ductile material, 12, 16
Ductile metals, 10
Ductility, 10, 16, 26, 116, 117, 127
  of clay, 298
Dulong–Petit value, 271, 274
Dumbbell, 103
Duralumin, 89
Dynamic recovery, 58
Dynamo machine, 179

Earphone, 220
Earthenware, 290
Earthenware clay, 290
Earth's magnetic field, 175
Easy direction, 238
Easy glide, 58
Economic considerations, 373–406
Eddy current, 236
Eddy current loss, 233, 236, 239
Edge dislocation, 48
Effective diffusion depth, 122
Efficient design, 401–402
Einstein, A., 261, 270, 276
Einstein relation, 211
Elastic materials, 10, 12, 16, 63
Elastic range, 14
Elastic strain energy, 103
Elasticity, 16, 63
Elastomeric polymers, 356
Elastomers, 347, 355, 357
Electric arc furnace, 134
Electric dipole, 30, 181, 254
Electric displacement, 216
Electric field, 188
Electric field lines, 217
Electric field strength, 216, 217, 249
Electric force, 189
Electric furnace smelting, 133
Electric motor, 178, 235
Electrical conduction in ionic crystals, 210
Electrical conductivity, 117, 184, 189
Electrical insulator, 205
Electrical phenomena in materials, 173–174
Electrical properties of materials, 185–222
Electrical steel, 231–232
Electricity, 173
Electrochemical cell, 157
Electrochemical corrosion, 158, 160

Electrochemical reactions, 157
Electrochromic display, 207
Electrode potential, 160
Electroluminescence, 260
Electrolysis, 180
Electromagnet, 224, 234
Electromagnetic cgs (emu) units, 227
Electromagnetic induction, 178
Electromagnetic materials, 408
Electromagnetic spectrum, 247
Electromigration, 109, 122
Electromotive force (emf), 159
Electron, 172
Electron band structure, 191, 198
Electron charge, 189
Electron collisions, 188
Electron energy band structure, 190
Electron hole, 198, 199
Electron microscopy, 24
Electron mobility, 198, 201
Electron-orbit paramagnetism, 229
Electron radius, 184
Electron scattering, 188
Electron sharing, 28
Electronegativity, 27–28
Electronic materials, 174, 184
Electronic Materials Age, 172
Electronic polarization, 218
Electronic properties, 173–184
Electronic switch, 204
Electroplating, 159
Electrostriction, 221
Electrum, 371
Embryo, 96
emf (electromotive force), 159
Emission of light, 259–275
Emitter, 203
Empirical materials, 408
End group, 348, 351
Energy band, 190
Energy barrier, 211
Energy consumption to extract raw materials, 401
Energy content of waste materials, 398
Energy level, 190
Energy product, 420
Energy state, 191
Engineering strain, 20
Engineering stress, 19
Environmental considerations, 373–406
Environmental interaction, 154
Epidermis, 341
Epoxy, 347, 354
Equiaxed zone, 98
Erosion–corrosion, 166
Ethylene, 350
Euclid, 176
Eutectic composition, 81, 129
Eutectic microconstituent, 83
Eutectic phase diagram, 80–83
Eutectic point, 82
Eutectic reaction, 141
Eutectic temperature, 81
Eutectoid point, 84
Eutectoid reaction, steel, 141
Eutectoid steel, 142
Eutectoid temperature, 149
Eutectoid transformation, 84
Excitation energy, 190
Exciton, 259
Exotic materials, 408
Expansion, thermal, 280–282
Expansion coefficients, 280
  of glass, 314
Extractives, 337
Extrinsic semiconductor, 199–201
Extrinsic stacking fault, 37

Fabrics (advanced), 363
Face-centered cubic (FCC) crystal structure, 32
Faïence, 292
Farad, 181, 215
Faraday, M., 178
Faraday cage, 181
Faraday constant, 164, 418
Faraday effect, 179
Faraday equation, 164
Fatigue test, 22
Fe, *see* Iron *entries*
Fe-Cr, 149
Fe-Fe_3C phase diagram, 142
Feathers, 3
Federsee, 71
Feldspar, 292, 297
Felt, 335
Felting, 326
Fermi energy, 189
  of intrinsic semiconductors, 199
Fermi velocity, 192
Ferric iron, 139
Ferrimagnetism, 173, 223, 234–235
Ferrite, 141, 142
Ferrite magnets, 241
Ferrite start temperature, 147
Ferroelectric materials, 219
Ferroelectricity, 219–220
Ferroelectrics, 219
Ferromagnetics, 226
Ferromagnetism, 173, 223, 230–233
Ferrous iron, 139
Fertility charm, 287
Fiber-reinforced composite, 361
Fibers, 3
  natural, 326
  properties of, 337
Fibrils, 336
Fibroin, 333
Fick's first law, 109
Fick's second law, 110
Fictive temperature, 313
Field effect transistor, 204, 207
Filler, 360
Filling, 344
Fine bainite, 146
Firing of clay, 300
Fishing hooks, 4
Flame-resistant plastics, 385
Flame retardant, 360
Flat panel glass, 305
Flax, 4, 327, 328, 337–338
Fleece, 327
Flint glass, 306
Flint stone, 4
Flow stress, 17, 96
Fluorescence, 259
Fluorescent lamp, 260
Flux, 133, 309
  atom, 109
Flux meter, 231
Fluxing agent, 6
Foam glass, 306
Folded chain model, 355
Folklore, 372
Fool's gold, 369
Forbidden band, 197
Fort Knox, 367
Forward bias, 203
Fourier Law, 279
Fracture toughness, 16
Frank–Read source, 57

Franklin, B., 173
Free electron, 29, 181, 187, 208, 253, 280
Free electron model, 189
Free electron theory, 254
Free energy, 115, 116
  of formation, 156
Free radical, 350
Frenkel defects, 103, 114
Frequency selective element, 221
Friction force, 253
Frictional electricity, 173
Fritted glaze, 291
Fruit-hair fiber, 328
Fulcrum, 77
Full annealing, 144
Functional ceramics, 317
Functional materials, 408
Fused quartz, 306
Fused silica, 46

Galalith, 347
Galileo, G., 176
Gallium arsenide, 201
Galvanic cell, 178
Galvanic corrosion, 161
Galvanic couple, 159, 160
Galvanic electricity, 174
Galvanic series, 161, 162
Galvanized steel, 167
Galvanometer, 178
γ-rays, 265
Garbage, 394
Gas sensor, 207
Gate, 204
Gauss (unit), 226, 420
Gaussian error function, 112
Generator, 234
Georgius Agricola, 131
Germanium, 29, 197
Germer, L.H., 183
Gibbs phase rule, 76
Gilding, 371
Glass, 16, 29, 46, 120
  creep in, 120
  recycling of, 307, 385
Glass beads, 71, 302
Glass-blowing, 304
Glass-ceramics, 315
Glass cutting, 314
Glass electricity, 174
Glass formers, 309
Glass-making
  history of, 302–307
  scientific aspects of, 307–315
Glass transformation range, 312
Glass transition range, 312
Glass transition temperature, 127
  polymers, 359
Glass vessel, 303
Glass wool, 306
Glassy amorphous solid, 359
Glassy stage, 359
Glaze, 7, 290, 294
Glazed bricks, 289
Glazing, 291, 302
Goethe, J.W. von, 245
Gold, 5, 71, 366–372
  properties of, 367
Gold panning, 370
Gold rush, 370
Gold standard, 368
Golden Age, 4
Gombroon, 292
Good iron, 126, 127
Goodyear, C., 331
GP zones (Guinier–Preston zones), 93–94, 115
Grafting, 360
Grain boundary, 43, 96, 103
Grain boundary diffusion, 106, 121
Grain boundary energy, 116
Grain boundary pinning, 116
Grain boundary sliding, 120
Grain growth, 116, 117
Grain orientation, 234–235
Grain-oriented silicon iron, 233, 235
Grain refiner, 96, 143
Grain size strengthening, 96–97, 99
Grains, 41
Graphite, 150, 207, 298
Graphite fibers, 407
Gravity/electricity relationship, 179
Gray cast iron, 151
Greenhouse gas, 384, 402–403
Greenstone, 4
Grocery bags, 391, 402
Groundwood pulp, 343
Guinier, A., 24
Guinier–Preston zones (GP zones), 93–94, 115
Gun cotton, 345
Gutenberg, Johannes, 89
Gypsum, 316

H-coefficient, 146
Hagen–Rubens equation, 252, 254
Hair, 327
Hall–Petch relation, 59, 96
Halloysite, 298
Hammering, 5, 64, 67, 125, 127, 129
Han dynasty, 130, 290
Harappan people, 72
Hard magnetic materials, 231, 238
Hard rubber, 331
Hard wood, 339
Hardenability, 149
Hardening, 142
  of glass, 314
  by hammering, 5
Hardness, 10, 66, 126, 127
Hardness test, 21
Hardwood, 340
Harmonic oscillator, 252, 269, 275
Harness buckle, 130
Haya people, 131
HDPE, 375, 388
Heartwood, 341
Heat, 176
Heat capacity, 272, 275
  at constant pressure, 273
  at constant volume, 273
  per mole, 274
Heat conduction, 271, 278
  in dielectric materials, 281
Heat conductivity, interpretation of
  classical, 280
  quantum-mechanical, 281
Heat flux, 279
Heat shield, 369
Helmholtz, H., 177
Hematite, 137
Hemoglobin, 137
Hemp, 4, 327, 328, 337
Henequen, 328, 337
Hertz, H., 246
Heterogeneous nucleation, 93, 96, 97
Hexagonal close-packed (HCP) structure, 32, 34

Hexagonal crystal structure, 33
Hieroglyphics, 126
High-Performance Concrete, 316
High-tech ceramics, 317–325
Hittites, 126
Hoe, 130
Hollander beater, 344
Homer, 72, 127
Homogeneous nucleation, 93
Homogenization, 79, 109
Homogenization heat treatment, 80
Hooke's Law, 15
Horns, 332
Horseshoe, 128
Hot dip zinc galvanizing, 167
Hot-rolling, 99
Hot shortness, 79
Hot working, 19, 80
Hume–Rothery rule, 80
Hund's rule, 229, 234
Hydration, 316
Hydraulic cement, 315
Hydrogen reference cell, 160
Hydrous silicate, 296
Hydroxyl (OH2) ions, 300
Hypereutectic alloy, 83
Hypereutectoid steel, 144
Hypoeutectic alloy, 82
Hypoeutectoid steel, 142
Hysteresis loop, 216, 231, 235
Hysteresis loss, 233, 237

Ice crystals, 31
Ideal resistivity, 189
Iliad, 72
Imagination, 411
Impact tester, 22
Inca, 72
Incineration, 398
Incoherent scattering of electrons, 188
Index of refraction, 248
India, 71–72
Indirect transition, 257
Indo-China, 68
Indus river, 71–72
Industry, 130
Inhibitor, 167
Initiator, 350
Innovative concepts, 411
Inoculant, 98
Inoculator, 96
Insulator, 173, 185, 191, 302
Integrated circuit, 201
Intensity of light, 249
Interatomic bonds, 26
Interatomic distance, 110
Interband transition, 256
Interdendritic segregation, 79
Interdendritic shrinkage, 98
Interdiffusion, 112
Intergranular corrosion, 165
Intermediate phase, 86–87
Intermediates (glass), 309
Intermetallic compound, 86–87
Intermetallic phase, 95
Internal energy, 276
International system of units, 417–418, 420
Interplanar spacing, 40–41
Interstitial, 25, 103
Interstitial diffusion, 104–105
Interstitialcy, 25, 103
Interstitialcy mechanism, 105
Intraband transition, 257
Intrinsic conduction, 199
Intrinsic semiconductor, 197–199
Intrinsic stacking fault, 37
Invariant point, 81
Inverse spinel structure, 234
Investment casting, 97
Ion, 26
Ion implantation, 103
Ionic bond, 26–28
Ionic conductors, 210–212
Ionic crystals, 209
Ionic dipole moment, 219
Ionic polarization, 217
Ionization energy, 201
Iron, 125, 141, 376
see also Fe *entries*
Iron Age, 3, 125–131
Iron belt, 136
Iron carbide, 141
Iron–carbon alloy, 127
Iron–carbon phase diagram, 142
Iron casting, 129
Iron deposits, 136
Iron–graphite phase diagram, 151
Iron Mountain, 137
Iron–neodymium–boron magnets, 239
Iron ore, 125, 126, 137
Iron oxide, 125, 159, 288
Iron pillar, 130
Iron recycling, 384, 385
Isomorphous binary phase diagram, 76–80
Isopleth plot, 88
Isothermal plot, 88
Isotropic materials, 43
Istle, 328, 337
Ivory, 332, 346

Jade, 4
Jewelry, 71, 371
Jomon ware, 295
Joule, J.P., 177
Jump frequency of atoms, 107, 109
Jute, 325, 328, 337

K-shell, 26
Kalevala, 139
Kaolin, 299
Kaolin clay, 293, 344
Kaolinite, 298, 302
Karat, 371
Keramos, 317
Keratin, 333
Kiln, 7, 290
Kirkendall shift, 113
Knitting needle, 71
Knoop microhardness technique, 21
Kraft process, 343

L-shell, 26
Lake Victoria, 131
Lamellar microstructure, 82
Laminar composite, 363
Laminated glass, 403
Landfill, 384, 398
Lapis lazuli, 302
Large-angle grain boundaries, 43
Laser, 201, 260, 261, 407
Laser materials, 262
Laser printer, 263
Latent heat of fusion, 81
Latex, 330, 341
Lattice, 32
Lattice constant, 32
Lattice defects, 102–103
Lattice distortions, 91
Lattice parameter, 32
Lattice vibration, 272

Lattice vibration quanta, 257
Laue, M. von, 24
LDPE, (low-density polyethylene) 388
Leaching, 157
Lead, 5
Lead glaze, 291
Leaf fiber, 328, 336, 337
Leathery stage, 359
Lens, 176
Lenz law, 227
Lever rule, 77
Levitation effect, 227
Light
  absorption of, 252–259
  bulb (invention), 328
  emission of, 259–267
Light-emitting diode (LED), 201, 263
Light quantum, 246
Lightning, 174
Lignin, 336, 338
Lime, 306
Limestone, 133, 302, 306
Limonite, 137
Line-defects, 48
Line fibers, 336
Linear hardening region, 58
Linear packing fraction, 35
Linear polymer, 348, 354
Lining, 134
Liquidus line, 76
Liquidus plot, 87
Llama, 329
Lodestone, 139, 183
Long line current, 163
Lorentz, H.A., 255
Lost wax method, 97
Lower bainite, 145
Lower yield point, 17
Lucite, 358
Lüders bands, 19
Lumen, 336, 341
Luminescence, 259
Luminous efficacy, 403
Luster decoration, 295

Mace head, 67
Machinability, 144
Mackintoshes, 331
Macromolecule, 43, 348
Magnesia, 174, 296
Magnesium, *see* Mg *entries*
Magnetic anisotropy, 237
Magnetic compass, 139
Magnetic dipole moment, 420
Magnetic disk, 240
Magnetic domain, 228, 238
Magnetic electricity, 174
Magnetic field parameters, 225
Magnetic field strength, 223, 226, 420
Magnetic flux, 226, 420
Magnetic flux density, 223–224, 226
Magnetic induction, 223, 420
Magnetic materials, 174, 223
Magnetic properties of materials, 223–243
Magnetic recording, 241–242
Magnetic Resonance Imaging system, 240
Magnetic sensor, 238
Magnetic tape, 222, 242
Magnetic units, 420
Magnetism, 174, 222
Magnetite, 137
Magnetization, 225, 420
Magnetorestrictive transducer, 238
Magnifying glass, 176
Maiolica, 290
Majority carrier, 202
Malachite, 6, 71, 125, 290
Malleable iron, 153
Martensite, 139, 146
Martensitic transformations in nonferrous alloys, 147
Marver, 304
Mass-average molar mass, 350
Mastication, 331
Materials
  electrical properties of, 185–220
  fundamental mechanical properties of, 12–22
  magnetic properties of, 223–244
  optical properties of, 245–269
  price of, 373–375
  production volumes of, 374
  strength of, 46–61
  thermal properties of, 271–284
  world reserves of, 376–383
Materials science, 4
Matrix, 74
Matthiessen rule, 189, 193
Maximum energy product, 238
Maxwell, J.C., 179
Maxwell equations, 232
Mayer, L.L., 177
Mechanical properties, 12–22
  of polymers, 352, 357
  of steel, 144
Mechanical recycling, 388
Mehrgarh (Pakistan), 72
Meißen Porcelain, 291, 292
Meissner effect, 227
Melting temperature (point),
  of glasses, 313
  of metals and alloys, 76, 128
  for polymers, 359
Memory, 205
Mer, 348
Mercury, 5
Merino sheep, 327
Mesolithic phase, 3
Mesopotamia, 70, 126, 291
Metallic bond, 29
Metallic glass, 307
Metals
  electronic properties of, 419
  prices of, 373–374
  resistivity of, 187–189
Metalworking, 9
Metastyrene, 347, 348
Meteoric iron, 125
Metglas, 307
Methyl rubber, 332
Mg-partially stabilized zirconia, 95
Mg_2Zn, 94
Mica, 297
Microcomputer, 205
Microconstituent, 83
Microminiaturization, 156
Microphone, 220
Microprobe, 113
Middle East, 67
Milk glass, 293
Millefiori glass, 303
Miller–Bravais indices, 40
Miller indices, 38–40
Mineral fiber, 329
Mining, 8
Minting, 372
Mirror, 176
Miscibility gap, 85
Mixed bonding, 31
Mixed dislocation, 55
mksA system of units, 417–418

Mobility, 198
of electrons, 198
of ions, 210
Modifiers, 121, 309
Modulus of elasticity, 16, 17
wood, 342
Mohair, 328–329
Mohenjo-daro, 72
Molar heat capacity, 272, 274
Molar mass, 348
Molecular currents, 226
Molecular polarization, 217
Molecular weight, *see* Molar mass
Monochromatic light, 261
Monochromatic X-rays, 266
Monoclinic crystal, 33
Monomer, 43, 348, 350
Monotectic reaction, 85–86
Monotectoid reaction, 85–86
Monovalent metal, 192
Moore's law, 393, 409
Mosaic glass, 303
MOSFET, 205
Mud, 287
Mud brick, 7
Mulberry silkworm, 327
Mullite, 87, 299, 300
Mumetal, 236, 238
Municipal solid waste, 394
Murano glass, 305
Mythology, 139, 372

n-p-n transistor, 204
n-type semiconductor, 199
Nabarro–Herring creep, 119
Nanomaterials by severe plastic deformation, 408
Nanostructure, 408
Nanotubes, 362
Native gold, 369
Native metals, 5
Natural aging, 90
Natural fibers, properties of, 333–345
Nature of electrons, 181
Near close-packed slip systems, 54
Necking, 14–16, 18, 20, 119
Néel temperature, 232
Neodymium-boron-iron, 240
Neolithic phase, 3
Neomagnets, 240
Neoprene, 332
Nernst equation, 160
Network formers, 309
Network modifiers, 309
Network polymers, 352, 354
New Zealand flax, 328
Newton, I., 245
Newtons (unit), 14
Ni-Al, 94
Nitrocellulose, 345
Noble gas configuration, 26
Nodular cast iron, 152
Nomads, 130
Nonbridging oxygen, 312
Noncoherent precipitates, 91, 92
Nonconductors, 185
Nonequilibrium solidus line, 78
Nonlinear elastic deformation, 357
Nonporous oxide layer, 157
Nonsteady-state diffusion, 110–111
Normalizing, 144
North pole, 175
Nubia, 370
Nuclear fission, 136
Nuclear fusion, 136
Nucleation and growth, 96
Nucleus, 96
Number
of atoms per unit cell, 34
of free electrons, 189, 254
of ions, 255
Number-average molar mass, 351
Nylon, 328, 337

Obsidian, 4, 302
ODS alloys, 95
Odyssey, 127
Oersted (unit), 226, 420
Oersted, H.C., 174, 177
Offset yield strength, 16
Ohm, G.S., 186
Ohm's law, 186
Oil, 376
recycling of, 391
Olivine, 297
One-molar solution, 159
Opacifier, 249
Opaque substances, 173, 252
Open-hearth process, 134
Operational amplifier, 205
Optical constants, 247–252
Optical disk, 268
Optical fiber, 258, 306
Optical glass, 306
Optical materials, 408
Optical phenomena, 175
Optical properties of materials, 245–269
Optical pumping, 261
Optical storage devices, 267–269
Optical telecommunication, 263
Optics industry, 305
Opto-electronics, 208
Organic polymers, 44
Orientation polarization, 217
Ornaments, 5
Orowan mechanism, 93, 99
Orthorhombic crystal structure, 33
Orthosilicate, 297
Oscillator, 188
Ötzi, 9
Ovenware, 315
Overaging, 90
Oxidation, 156, 158
Oxidation rate, 157
Oxidation site, 158
Oxides, 27

p-electrons, 26
p-*n*-p transistor, 204
p-type semiconductor, 200
p-B ratio, 156
Packing factor, 34
Paleolithic phase, 3
Palestine, 126
Palm-type fiber, 328
Palmyra, 328
Paper, 342–345, 377
recycling of, 390–391
Paper board, 377
Paper-making, 343
Papyrus, 342
Paramagnetic materials, 224
Paramagnetism, 222, 228–229
Parchment, 342–343
Parenchyma, 340
Particle board, 377
Particulate composite, 362
Pascal (unit), 14, 417
Passivation, 165
Passive stainless steel, 165
Pauli principle, 229, 415

Pearlite, 141, 142
Pearlite finish time, 145
Pearlite start time, 145
Pectic substances, 336
Peierls stress, 49, 50
Peltier effect, 212
Penetration curve, 111
Percussion flaking, 4
Periderm, 340, 341
Periodic Table, 26, 421
Peritectic transformation, 84, 141
Peritectoid transformation, 85
Permalloy, 236, 238
Permanent magnets, 238, 240
Permeability, 223, 224, 420
  of free space, 223
Permeability constant, 420
Permittivity constant, 420
Permittivity of empty space, 214
Perovskite crystal structure, 195
Perpetual motion, 178
Peru, 10, 72
PET (PETE) (polyethylene terephthalate), 375, 388
Petroleum, 377
Phase, 76
Phase-coherent light, 260
Phase diagrams, 75–88
Phase transformation- congruent/incongruent, 87
Phenolic resin, 347
Phloem, 340
Phonograph pickup, 230
Phonon, 257, 272, 276–277, 280
Phosphor, 260
Phosphorescence, 259
Photochromic glass, 306
Photoelectric effect, 246
Photoluminescence, 260
Photon, 243
Photosynthesis, 137, 328, 412
Photovoltaic device, 203
Physical constants, table of, 418
Piassava, 328
Piezoelectricity, 219–220
Pig iron, 129, 133–135, 150
Pigment, 7, 288
Pilling–Bedworth (P-B) ratio, 158
Pin, 71
Pipe diffusion, 107
Planar packing fraction, 35
Planar transistor, 201
Planck, M., 243
Planck constant, 243
Plant fiber, 327, 328
  composition of, 338
Plasma frequency, 254
Plaster, 315
Plastic container codes, 388
Plastic deformation, 10, 48, 103, 118
Plastic region, 14, 16
Plasticity, 47–62
Plasticizer, 360
Plastics, 345–354
  recycling of, 387–389
Plate glass, 305
Platinum/Pt-13%Rh, 211
Plowshare, 71, 128, 130
Plywood, 342, 377
Point contact transistor, 201
Poisson ratio, 13
Polarization, 215, 217, 218, 249
Pollution, 377, 398, 402
Poly-sulfur nitride, 206
Polyaniline, 207
Polychrome vessels, 303
Polycrystallinity, 41, 42
Polycrystals, 43, 59–61
  slip in, 57–58
Polyester, 337, 347
Polyethylene, 349
Polygonization, 117
Polymer chains, 31
Polymerization, 348, 350
Polymers, 43, 347
  additives to, 360
  atomic structure of, 43
  conduction in, 205–209
  creep in, 121
  degradation of, 167, 168
  diffusion through, 115
  history of, 353
  mechanical properties of, 17, 18, 354–359
  prices of, 375
  properties of, 354–359
Polymethyl-methacrylate, 358
Polymorphic materials, 37
Polypyrrole, 207
Polystyrene, 348, 359, 375
Polyvinyl chloride, 358
Pontil, 305
Population density, 192
Population inversion of electrons, 260
Porcelain, 292, 293, 294
Porosity, 64, 67, 98, 99, 113, 116, 120
Porous film, 157
Portland cement, 316
Potter's wheel, 290, 294, 295
Pottery, 3, 7, 29, 70, 287, 303
Pounds per square inch (psi), 14
Powder metallurgy, 95, 133
Power law creep, 119
PP, (polyproylene) 375
Pre-pottery era, 286
Precipitation hardening, 88, 89–95
Price of materials, 373–374
Printing (movable letters), 89
Primary α, 83
Primary creep, 118
Prism planes, 41
Probability integral, 112
Production volumes of materials, 374–375
Proeutectic constituent, 83
Propagation (in polymerization), 350
Prosenchyma, 340
Protective layer, 154–155
Protein, 333
PS (polystyrene), 375, 388
Ptolemaic theory, 176
Puddling, 130
Pulp, 343, 391
Pulping, 343
Pulsed laser, 262
Pulverization, 116
Pumping (laser), 261
Pumping efficiency, 262
PVC resin, 375
Pyrex, 315
Pyrite, 137
Pyrrhotite, 139
Pythagoras, 176

Q-switching, 264
Quantum mechanics, 190
  of absorption of light, 256
Quantum numbers, 415–416
Quantum theory, 192, 246
Quartz, 46, 219
Quartz crystal resonator, 220
Quartz sand, 302
Quaternary phase diagram, 88

Quench rate, 146
Quenching, 89, 103, 127, 129, 144, 307
Qutub Minar tower (India), 130

Radiation damage, 103
Radio frequency signal, 220
Radioactive tracer element, 113
Radioactive waste, 411
Raffia fiber, 328
Ramie fiber, 328, 337, 338–339
Random access memory, 267
Rapeseed oil, 381–383
Rapid solidification, 80, 97, 98, 306
Rare-earth magnets, 240
Rate equation, 107
Rate of oxidation, 157
Rayon, 328
Read-only memory (ROM), 268
Rechargeable battery, 207
Recording head, 241–242
Recovery, 51, 117, 194
Recovery rate of municipal solid waste, 383–397
Recrystallization, 51, 117
Recrystallization temperature, 19
Rectifier, 201
Recycling, 383–400
  of aluminum, 385
  of automobiles, 393
  of bronze, 128
  of copper, 386
  of electronic scrap, 392
  of glass, 306, 385
  of household batteries, 391
  of lead, 391
  of municipal waste, 394
  of oil, 391
  of paper, 390
  of plastics, 387
  of scrap steel, 134
  of silver, 386
Red stoneware, 292, 293
Reduction process, 158
Reed, 332
Refining, 343
Reflection, 176
Reflectivity, 248, 369
Refraction, 176
Refractive power, 248
  of glass, 306
Refractory materials, 302
Regenerated cellulose fibers, 328
Rekh-Mi-Re, 67
Relative abundance of iron, 136
Relative permeability, 224
Relative permittivity, 214
Relaxation time, 189
Relief ornaments, 295
Remalloy, 239
Remanence, 230, 238
Remanent magnetization, 230
Remanent polarization, 218
Repeat unit, 43, 44, 348
Reserves (oil), 378–382
Residual resistivity, 188, 189, 193
Resinous electricity, 174
Resistance, 185
Resistivity, 185
  of metals, 187–189
  of nickel–chromium alloys, 194
  of ordered alloys, 193
  in superconductors, 194
  of two-phase alloys, 193
Resolved shear force, 59
Resonance frequency, 255, 258
Restoring force, 255
Retained austenite, 148
Retting, 336, 338
Reverse bias, 202
Reverse osmosis, 364
Rhombohedral crystal structure, 33
Richardson-Ellingham diagram, 157
Rigveda, 131
Ring exchange, 105
Rinman, S., 127
Riser, 99
Roasting, 8
Rockwell hardness, 21
Röntgen, W.C., 265
Rotation axis symmetry, 38
Rubber, 330–332, 356
Rubber tree, 330
Rubbery stage, 121, 350
Ruby, 319
Rudna Glava, 8
Rupture lifetime, 119
Rust, 159
Rusting, 155
Rutherford, E., 183

s-electrons, 26
Sacrificial anode, 163, 164
Safety, 403
Safety pin, 71
Salt glaze, 292
Sand casting, 97
Sapphire, 319
Sapwood, 341
Sarnath (India), 131
Saturation current, 203
Saturation magnetization, 230
Saturation polarization, 218
Scale, 156
Scattering of light, 252
Schmid factor, 60
Schmid's law, 60
Schönbein, C.F., 345
Schott, O., 306
Schottky defect, 103, 114, 210
Schrödinger, E., 183
Schrödinger equation, 190
Scrap paper market, 390
Screw dislocation, 54
Scutching, 338
Sea of Tranquility (moon), 135
Sea Peoples, 128
Secondary recrystallization, 238
Seebeck coefficient, 211
Seed fiber, 328, 337, 338
Segregation, 78, 99
Selective leaching, 158
Selective steeling, 127
Self-assembling materials, 408
Self-diffusion, 106
Self-healing mechanism, 412
Self-interstitial, 25, 103
Semi-infinite solid, 111
Semi-synthetic polymers, 347
Semiconductor devices, 201–205
Semiconductor laser, 261
Semiconductors, 173, 185, 191, 196–205
  electronic properties of, 419
Semicrystalline polymer, 359
Semipermeable membrane, 159
Semiprecious stones, 302
Sequential storage, 268
Sericin, 333, 335
Sericulture, 327
Setting time, 316
Shang dynasty, 68, 291
Shape anisotropy, 240
Shape memory effect, 147
Shaping, 125
Shear modulus, 16, 56
Shear strain, 16
Shear stress, 13, 16, 57
Sheet metal, 99

Shell, 3, 332
Shives, 338
Short line current, 163
Short-range order, 307
Shrinkage, 98, 301
SI unit system, 226, 417–418, 420
Siderite, 137
Siemens, W., 187
Siemens–Martin process, 134
Silica, 46, 168, 296, 297, 302, 307
Silicate ceramics, 29
Silicates, 297
Silicon, 28, 29, 196
Silicon carbide, 317
Silicon dioxide, *see* SiO_2 *entries*
Silicon nitride, 317
Silicon nitride ceramics, 318
Silk, 327, 329, 333–336
Silk gum, 333
Silkworm, 335
Silver, 5
Silver recycling, 386
Sinai peninsula, 6, 8
Sinan compass, 175
Single crystal, 41
Sintering, 95, 116
SiO_2, 46, 156, 297
SiO_2-Al_2O_3 phase diagram, 300
Sisal fiber, 328, 337
Sizing, 344
Skin, 332, 360
Slag, 6, 125
Slime, 344
Slip
  of dislocations, 48, 99, 118
  in polycrystals, 59–61
  for pottery, 295
Slip direction, 50, 53
Slip plane, 48, 49, 50, 54
Slip system, 51
Small-angle grain boundaries, 43
Smart materials, 363, 364, 408
Smectide, 298
Snell, W., 177
Snell's law, 248
Societal issues, 412
Socket tool, 66
Soda, 302, 306
Soda–lime–silica glass, 46, 306
Sodium chloride structure, 45
Soft magnetic materials, 231, 234
Soft-paste porcelain, 294
Soft porcelain, 293
Softening point, 313
Softwood, 339, 340
Sol-gel technique, 258, 323
Solar cell, 203
Solder (lead free), 400
Solenoid, 223
Solid solution strengthening, 74–75, 99, 142
Solidification, 98
Solidus line, 76
Soliton, 257
Solute atoms, 74
Solution heat-treatment, 89
Solvent, 74
Solvus line, 80
Sonar detector, 220
Source (in transistor), 204
Space charge region, 202
Spear, 4, 68, 126
Specific gravity, 319
Specific heat capacity, 272, 273
Specific volume, 358
  of glass, 312
Spheroidizing, 144
Spherulites, 355
Spider drag line, 357, 358
Spin paramagnetism, 228
Spinning, 327
Sponge iron, 125, 131, 133
Sponges, 332
Spontaneous light emission, 260
Spring constant, 252
Stabilizer, 168, 360
Stack casting, 130
Stacking faults, 37
Stacking sequence, 35
Stainless steel, 149, 158, 165
Standard emf series, 161
Static electricity, 174
Statuary bronze, 87
Staudinger, H., 348
Steady-state creep, 118
Steady-state diffusion, 109–110
Steady-state flow, 110
Stealth aircraft, 362
Steatite, 291
Steel, 127, 141
Steel making, 133
Steel production data, 135–136
Step-growth reaction, 352
Stiffness, 16
Stimulated emission, 260, 261
Stoichiometric intermetallic compounds, 86–87
Stone, 29
Stone Age, 3–5
Stoneware, 292
Stovetop, 315
Strain, 13
Strain field, 57
Strain gage, 220
Strain hardening, 20, 51, 58
Strain hardening coefficient, 19
Strain hardening exponent, 21
Strain hardening rate, 21
Strain point, 313
Straw, 331
Strength, of materials, 47–62, 116, 117
Strength coefficient, 21
Strengthening, 19
Stress, 12, 119
Stress corrosion cracking, 165
Stress raiser, 319
Stress relief anneal, 117, 194
Stress–strain diagram, 14, 16, 19
Strontium titanate, 219
Structural ceramics, 317
Structural materials, 408
Styrene, 347, 349
Substituting rare materials, 400–401
Substitutional solid solution, 74
Sunn fiber, 328
Superalloy, 94, 407
Superconductivity, 194–196
Superconductors, 185, 225, 227
  1-2-3 compound, 195
  high-T_c, 195
  Type I, 195
  Type II, 195
Supercooled liquid, 313
Supermalloy, 236, 238
Supermendur, 236
Superplacticity, 121
Supersaturated solid solution, 90
Surface charge density, 216
Surface diffusion, 106–107
Surface emitter, 265
Susceptibility, 223, 420
Swelling, 168
Switching device, 203
Sword, 126, 131
Synthetic fiber, 337
Synthetic polymer, 347
Synthetic rubber, 332

Tall, 288
Tschirnhaus, E.W. von, 293
Technical saturation
magnetization, 231
Telescope, 176
Tell, 288
Temperature coefficient of
resistivity, 187
Temperature dependence of
ferromagnetism, 232
Tempered glass, 314, 403
Tempering, 127, 129, 144, 147
Tensile strength, 15
of glass, 314
Tensile stress, 13
Tensile tester, 12
Termination by combination,
351
Ternary phase diagrams, 87–89,
89
Terra cotta, 289
Tertiary creep region, 119
Tesla (unit), 224, 420
Tetracalcium aluminoferrite, 316
Tetragonal crystal structure, 33
Tetrahedron, 28
Texture, 235
Thailand, 68
Thales of Miletus, 174
Thermal arrest, 81
Thermal conduction, 270,
276–280
Thermal conductivity, 269, 279
Thermal electricity, 174
Thermal emission, 260
Thermal energy, 103
Thermal expansion, 269, 280–282
Thermal expansion coefficient,
156
Thermal insulation, 271
Thermal properties of materials,
177, 271–284
Thermal shock, 305
Thermocouple, 211–212
Thermoelectric phenomena,
211–213
Thermoelectric power, 211
Thermoelectric power generator,
212
Thermoelectric refrigerator, 213
Thermoluminescence dating, 296
Thermomigration, 109
Thermoplastic polymer, 345,
352, 354, 357
Thermoplastics, 357
Thermosets, 347, 352, 356, 357
Thermosetting polymer, 352, 354
Thin magnetic films, 241
Thomson, G. P., 183
Thomson, J. J., 182
Threshold energy for photon
absorption, 255
Tie line, 77
Timber, 375, 377
Time between collisions, 190
Timna Valley, 8, 10
Tin, 64, 67, 128
Tin glaze, 291
Tin plague, 38
Titanic, 154
Titanium diborate, 317
Toledo (Spain), 131
Tools, 287
Toughness, 22
Tow, 338
Transducer, 220
Transformation-toughened
zirconia, 318
Transformer, 234
Transient creep, 118
Transistor, 201, 204
Translucent dielectric, 249
Transmissivity, 251
Transoidal polyacetylene, 206
Transparent matter, 173, 252
Tricalcium aluminate, 316
Tricalcium silicate, 316
Triclinic crystal structure, 33
Trinitrate of cellulose, 345
True strain, 20
True stress, 20
TTT diagram, 144, 145
for hypoeutectoid plain-carbon-steel, 148
for noncutcctoid steels, 147
Tungstate, 260
Tungsten carbide, 317
Turkey, 126
Turquoise, 302
Tutankhamen, 371
Tuyère, 6, 133
Twinning, 37, 53, 146
Two-phase region, 76, 81
Two-tone bell, 69–70

Ultimate tensile strength, 15
Ultimate tensile stress, 15
Ultra hard ceramics, 321
Ultra-large-scale integration
(ULSI), 201
Uneticians, 71
Unpaired electron, 350
Unsaturated bond, 350
Unvitrified ware, 290
Upper bainite, 145
Upper yield point, 17

Vacancy, 25, 102, 103
Vacancy concentration, 103
Valence angle, 29
Valence band, 197
Valence electron, 29, 181, 190
Van der Waals bond, 30–31
Varna, 371
Vegetable fiber, 328, 336–339
Vellum, 343
Venus of Vestonice, 7, 288, 289
Vermiculite, 298
Vibrations of lattice atoms, 255
Vicalloy, 237
Vickers hardness, 21
Virgin iron, 231
Viscoelastic region, 19
Viscose, 328
Viscosity, 120, 348
Viscosity-temperature diagram
(glass), 313
Viscous stage, 359
Vision, 176
Vitreous electricity, 174
Vitreous silica, 46
Vitrification, 292, 300
Vitrified ware, 289
Voltage, 185
Voltaic electricity, 174
Volume diffusion, 106
Von Stahel und Eysen, 132
Vortex state, 196
Vulcanization, 331, 356

Waste, 394–396
Waste prevention, 399
Waterline corrosion, 163
Wave-particle duality
of electrons, 182
of light, 247
Wavelength of light, 247
Weapon, 71, 286
Weathering steel, 156
Weaving (linen), 327

Wedgwood, J., 290, 291
Wheels, 128
Whisker, 320, 362
White cast iron, 152
White graphite, 298
White X-radiation, 266
Widmanstätten structure, 93
Wiedemann-Franz law, 271
Wieland, 133
Wilm, A., 89
Wood, 3, 339–342
  mechanical properties of, 15, 17, 339
  tensile strengths of, 342
Wood production, 377
Wool, 4, 326, 333–334, 337
Wootz steel, 131
Work hardening, 19, 51, 62, 99, 122, 194
Working point, 313
Working range, 313
World reserves of materials, 376–383
World Trade Center, 404
Wrapping paper, 390
Wrought iron, 126, 129, 131

X-rays, 263
Xylem, 340

Yarn, 336, 338
Yield of land, 383
Yield strength, 14, 74, 75
Young's modulus, 15

Zeiss, C., 306
Zheng-Zhou, 129
Zinc blende structure, 46
Zirconia, 318
Zirconia-toughened alumina, 319
Zirconium dioxide, 318
ZnS, 261
Zone refining, 198
ZrO_2-MgO, 94

ATLANTIC OCEAN
British Isles
Cologne
Rhine R.
Danube R.
Vestonice
Hallstatt
Venice
Etruria
Rome
Varna
Black Sea
Caucasus Mts.
Caspian Sea
Aral
Constantinople
Anatolia
Byzantium
Greece
Magnesia
Catal Hüyük
Toledo
Rio Tinto
Carthage
Mediterranean Sea
Cyprus
Euphrates R.
Mesopotamia
Tigris R.
Assyria
Persia
Damascus
Babylon
Sinai
Timna
Nile R.
Luxor
EGYPT
Red Sea
AFRICA
Lake Victoria